WILLIAM F. MAAG LIBRARY
YOUNGSTOWN STATE UNIVERSITY

COMPREHENSIVE CHEMICAL KINETICS

COMPREHENSIVE

Section 1. THE PRACTICE AND THEORY OF KINETICS (3 volumes)

Section 2. HOMOGENEOUS DECOMPOSITION AND ISOMERISATION REACTIONS (2 volumes)

Section 3. INORGANIC REACTIONS (2 volumes)

Section 4. ORGANIC REACTIONS (5 volumes)

Section 5. POLYMERISATION REACTION (3 volumes)

Section 6. OXIDATION AND COMBUSTION REACTIONS (2 volumes)

Section 7. SELECTED ELEMENTARY REACTIONS (1 volume)

Section 8. HETEROGENEOUS REACTIONS (4 volumes)

Section 9. KINETICS AND CHEMICAL TECHNOLOGY (1 volume)

Section 10. MODERN METHODS, THEORY, AND DATA

CHEMICAL KINETICS

EDITED BY

R.G. COMPTON

M.A., D. Phil. (Oxon.)
Oxford University
The Physical Chemistry Laboratory
Oxford, England

G. HANCOCK
Oxford University
The Physical Chemistry Laboratory
Oxford, England

VOLUME 34

MODERN ASPECTS OF DIFFUSION-CONTROLLED REACTIONS
COOPERATIVE PHENOMENA IN BIMOLECULAR PROCESSES

E. KOTOMIN
V. KUZOVKOV
University of Latvia

1996
ELSEVIER
AMSTERDAM - LAUSANNE - NEW YORK - OXFORD - SHANNON - TOKYO

ELSEVIER SCIENCE B.V.
Sara Burgerhartstraat 25
P.O. Box 211, 1000 AE Amsterdam, The Netherlands

ISBN: 0 444 41631 5 (Series)
ISBN: 0 444 82472 3 (Vol. 34)

© 1996 Elsevier Science B.V. All rights reserved.

No part of this publication may be reproduced, stored in a retrieval system or transmitted in any form or by any means, electronic, mechanical, photocopying, recording or otherwise, without the prior written permission of the publisher, Elsevier Science B.V., Copyright & Permissions Department, P.O. Box 521, 1000 AM Amsterdam, The Netherlands.

Special regulations for readers in the U.S.A. – This publication has been registered with the Copyright Clearance Center Inc. (CCC), 222 Rosewood Drive, Danvers, MA 01923. Information can be obtained from the CCC about conditions under which photocopies of parts of this publication may be made in the U.S.A. All other copyright questions, including photocopying outside of the U.S.A., should be referred to the copyright owner, Elsevier Science B.V., unless otherwise specified.

No responsibility is assumed by the publisher for any injury and/or damage to persons or property as a matter of products liability, negligence or otherwise, or from any use or operation of any methods, products, instructions or ideas contained in the material herein.

This book is printed on acid-free paper.

Printed in The Netherlands.

COMPREHENSIVE CHEMICAL KINETICS

ADVISORY BOARD

Professor C.H. BAMFORD
Professor S.W. BENSON
Professor LORD DAINTON
Professor G. GEE
Professor G.S. HAMMOND
Professor K.J. LAIDLER
Professor SIR HARRY MELVILLE
Professor S. OKAMURA
Professor Z.G. SZABO
Professor O. WICHTERLE

Volumes in the Series

| | Section 1. | THE PRACTICE AND THEORY OF KINETICS (3 volumes) |

Volume 1 The Practice of Kinetics
Volume 2 The Theory of Kinetics
Volume 3 The Formation and Decay of Excited Species

Section 2. HOMOGENEOUS DECOMOSITION AND ISOMERSATION REACTIONS (2 volumes)

Volume 4 Decomosition of Inorganic and Organometallic Compounds
Volume 5 Decomposition and Isomerisation of Organic Compounds

Section 3. INORGANIC REACTIONS (2 volumes)

Volume 6 Reactions of Non-metallic Inorganic Compounds
Volume 7 Reactions of Metallic Salts and Complexes, and Organometallic Compounds

Section 4. ORGANIC REACTIONS (5 volumes)

Volume 8   Proton Transfer
Volume 9   Addition and Elimination Reactions of Aliphatic Compounds
Volume 10 Ester Formation and Hydrolysis and Related Reactions
Volume 12 Electrophilic Substitution at a Saturated Carbon Atom
Volume 13 Reactions of Aromatic Compounds

Section 5. POLYMERISATION REACTIONS (3 volumes)

Volume 14   Degradation of Polymers
Volume 14A Free-radical Polymerisation
Volume 15   Non-radical Polymerisation

Section 6. OXIDATION AND COMBUSTION REACTIONS (2 volumes)

Volume 16 Liquid-phase Oxidation
Volume 17 Gas-phase Combustion

Section 7. SELECTED ELEMENTARY REACTIONS (1 volume)

Volume 18 Selected Elementary Reactions

Section 8. HETEROGENEOUS REACTIONS (4 volumes)

Volume 19 Simple Processes at the Gas-Solid Interface
Volume 20 Complex Catalytic Processes
Volume 21 Reactions of Solids with Gases
Volume 22 Reactions in the Solid State

Section 9. KINETICS AND CHEMICAL TECHNOLOGY (1 volume)

Volume 23   Kinetics and Chemical Technology

Section 10. MODERN METHODS, THEORY, AND DATA

Volume 24   Modern Methods in Kinetics
Volume 25   Diffusion-limited Reactions
Volume 26   Electrode Kinetics: Principles and Methodology
Volume 27   Electrode Kinetics: Reactions
Volume 28   Reactions at the Liquid-Solid Interface
Volume 29   New Techniques for the Study of Electrodes and their Reactions
Volume 30   Electron Tunneling in Chemistry. Chemical Reactions over Large Distances
Volume 31   Mechanism and Kinetics of Addition Polymerizations
Volume 32   Kinetic Models of Catalytic Reactions
Volume 33   Catastrophe Theory
Volume 34   Modern Aspects of Diffusion-Controlled Reactions
             Coopareative Phenomena in Bimolecular Processes

Contributors to Volume 34

All chapters in this volume have been written by

E. KOTOMIN	University of Latvia Solid State Physics 8 Kengaraga Street Riga LV 1063 Latvia
V. KUZOVKOV	University of Latvia Solid State Physics 8 Kengaraga Street Riga LV 1063 Latvia

Preface

> He used to talk so much and so fast that – the following day – he was hardly able to remember what he had said – much less analyze it.
>
> V. Klyuchevsky, old Russian historian

A wide range of condensed matter properties including viscosity, ionic conductivity and mass transport belong to the class of thermally activated processes and are treated in terms of *diffusion*. Its theory seems to be quite well developed now [1–5] and was applied successfully to the study of radiation defects [6–8], dilute alloys and processes in highly defective solids [9–11]. Mobile particles or defects in solids inavoidably interact and thus participate in a series of *diffusion-controlled reactions* [12–18]. Three basic bimolecular reactions in solids and liquids are: dissimilar particle (defect) recombination (annihilation), $A + B \to 0$; energy transfer from donors A to unsaturable sinks B, $A + B \to B$ and exciton annihilation, $A + A \to 0$.

This theory, as originated from the early work of Smoluchowski [20], nowadays has numerous applications in several branches of chemistry, such as colloidal chemistry, aerosol dynamics, catalysis and the physical chemistry of solutions as well as in the physics and chemistry of the condensed state [21–24]. Until recently, its branch called *standard chemical kinetics* [12, 15, 16] based on the law of mass action seemed to be quite a complete and universal theory. However, because of their entirely phenomenological character, theories of this kind always operate with the *reaction rates* K which are postulated to be time-independent parameters.

Since the 1960s the first attempts were undertaken to develop a more rigorous theoretical formalism employing different techniques: the hierarchy of equations for many-particle distribution functions [25–28], field theory [29–32], and multiple scattering [33, 34]. Both theoretical studies and computer simulations [35–40] carried out in the 70s–80s have clearly demonstrated the principal shortcoming of the Smoluchowski-type theories (which are the basis of the standard chemical kinetics): they are two-particle approaches and thus neglect any effects related to *fluctuations in the reactant densities*. As an example of the limitations of these over-averaging theories we would mention the formation of such quite complex spatial structures as *fractal clusters* and reactions in restricted geometries [41, 42]. A very general review of a role of fluctuations in physical, chemical and biological processes has been presented recently in review articles [43, 44]. The aim of our book is to consider the kinetics and peculiarities of *fluctuation-controlled bimolecular reactions*.

The treatment, done for the first time in the 70s at the intermediate *mesoscopic* level (concentration distributions are continuous but fluctuating) by Balagurov and Vaks [45] and Ovchinnikov and Zeldovich [46] for the A + B → B and A + B → 0 reactions, respectively, has demonstrated their considerable deviations from the generally-accepted laws of the standard chemical kinetics resulting in an essentially non-Poisson spectrum of density fluctuations. Such a mesoscopic level of theory as well as macro- and microscopic approaches are discussed in detail in the book for the two above-mentioned kinds of fundamental bimolecular reactions in condensed matter, corresponding, e.g., to Frenkel defect recombination (annihilation) and energy transfer, respectively as well as for a third type of reactions of exciton annihilation. We will demonstrate that careful analysis of the kinetics of these simple bimolecular reactions, based on the second step in the cut-off of the infinite hierarchy of equations for many-particle densities, reveal their capability for *self-organization* (*cooperative phenomena*) which could be described in terms of the correlation length and critical exponents in the very same way as is done in statistical physics.

Among the important conclusions arising from this new formalism, we mention that, in contrast to what one might intuitively assume, similar particles (reactants), initially distributed at random and non-interacting with one other, after some reaction time become aggregated into domains containing particles of one kind only, A or B. Pattern formation results from the lateral interaction of similar particles (A–A and B–B) via their reaction with particles of another kind (A–B). This leads to a time-dependence of the reaction rate, not at short reaction times during the transient process (which is well known in chemical kinetics) but at *long times*. It occurs because similar particle aggregation leads to greater mean distances between dissimilar particles and thus hinders their reaction rate and concentration decay in time. This new theory reveals such previously *hidden parameters* as the ratio of diffusion coefficients, $D_A/(D_A + D_B)$, the spatial dimension $\bar{d}$, and the initial particle concentrations, $n_i(t = 0)$.

In the 70s, a new class of static long-range reaction in irradiated liquids, glasses and solids, called *tunnelling recombination* was discovered [47, 48]. Tunnelling recombination has a purely quantum-mechanical nature and results in the reaction between dissimilar particles separated by distances as large as 20–30 Å, despite the fact that at sufficiently low temperature these particles are immobile. We will discuss how tunnelling reactions manifest themselves in self-organisation phenomena.

A number of quite different techniques have been presented in the last few years for studying self-organisation phenomena in the bimolecular reactions in condensed matter. At present those are covered in the review article [49] and Proceedings of the conference [50] only; we discuss their advantages and shortcomings, and the principal approximations involved (in particular, that by Kirkwood). Where possible, analytical results are compared with computer simulations, since very limited experimental data are known at present in this field. Those that do exist are also considered and the conditions for the experimental observation of cooperative effects under study are predicted theoretically. We hope that this book may stimulate new experimental studies in this very important field.

Until recently, only complex reactions with dozens of intermediate products were known to produce self-ordering effects characterized by a formation of *spatio-temporal structures* [51]; they were studied mainly in terms of the universal theory developed in the 1970s and known as *synergetics* [52–54]. A formation of these structures in active extended media is now of great interest for multidisciplinar studies in physics, chemistry and biology. Synergetics studies quite general laws determining the processes of creation, migration and recombination (decay) of particles (excitations) of arbitrary nature leading to the pattern formation [55]. Considerable success in this field has been achieved by making use of a *stochastic description* of irreversible processes far from equilibrium [18, 52–54, 56]. Among the main subjects of these studies are condensed media and diffusion-controlled reactions therein. The studies by means of stochastic methods are based on the standard approach of Markov chains with transition probabilities which are non-linear functions of stochastic variables [25, 26, 28, 52, 53]. As is known from statistical physics, the corresponding infinite set of coupled equations cannot be solved exactly. An approximate description of many-particle problems in terms of synergetics usually leads to a finite set of several non-linear partial differential equations. Their non-linearities arise from the same causes as in the Boltzmann equation, and often these equations have more than a single solution.

Self-organization manifests itself only in systems far from equilibrium and consisting of a large number of objects, whose cooperative behaviour is sometimes considered in terms of the *non-equilibrium critical phenomena* (phase transitions) [53, 55, 57–62]. At present there are several complementary approaches to the physics of open systems far from equilibrium. A simple phenomenological description in terms of Langevin or Fokker–Planck equations was presented in [63]. Unfortunately, the structure and stationary distribution functions and the associated fluctuation-dissipative theorem are in general unknown, so these often represent an unwarranted extrapolation from some underlying deterministic approximation. An alternative scheme is provided by the stochastic master equations. Their additional linearization shows that the mentioned above phenomenological approach could be seriously in error. However, it is practically impossible to obtain systematic approximations in the vicinity of the critical points (see, however, [64, 65] where the *exact* Fokker–Planck or Langevin descriptions were derived from the master equation by means of the developed there "Poisson transformation"). Up to now the scaling and RG approaches [66–70] have not been used systematically because of the absence of a simple Landau–Ginsburg-type description. Their use requires modification of the statement of the problem, since only a rather limited class of stochastic problems defined by a set of the probability transitions could be solved by these methods alone. The formalism presented in [64, 65] seems to be in keeping with generalization of the scaling and RG approaches.

Recent stochastic studies of the role of reactant density fluctuations in the bimolecular kinetics revealed an unexpected appearance of *microscopic* patterns characterized by a short-range and intermediate-range order, thus not violating the homogeneity of a system, rather than the long-range ordered structures which are the usual object of studies in synergetics. The effects in which we are interested are to be accounted for by the violation of the large number law [52] when the description of a system in terms of average quantities is no longer sufficient. It is so because the self-organizing systems are characterized by anomalous fluctuations which govern the behaviour of the average quantities and *qualitatively* change their time development. For the reactions controlled by reactant motion, coherent behaviour of the fluctuations is typical, i.e., the existence of spatial correlation of closely spaced reactants in a system. An analogy between the fluctuations in a system far from the equilibrium and such critical phenomena as the phase transitions at equilibrium has been pointed out more than once [54, 55]. It is well known that the fluctuation terms cannot be neglected in the equations for average quantities in the vicinity of instability points, since after a long enough time the non-Poisson fluctuation spectrum determines the behaviour of the average quantities [69, 71]. Here one cannot neglect a perturbation which equals the deviation of the fluctuation spectrum from the Poisson one and thus increases in time. For systems with asymptotically unstable solutions (e.g., the well-known Lotka–Volterra model in its stochastic statement with interacting populations [52–54]) the *fluctuation dispersion cannot be neglected*. Thus the cut-off of the infinite hierarchy of equations for the distribution functions, which is of our interest, permits to obtain the approximate solution which is valid only at *short reaction times*. Higher-degree approximations are of key importance and often allow us to "discover" the latent degrees of freedom (e.g., dispersions) characterizing the kinetic behaviour of a system. Recent pioneering studies [45, 46] in fact discovered a *new class of self-organization phenomena*.

The key questions arose in this book are – are three above mentioned simplest bimolecular reactions complex enough to expect the appearance of self-organization or not? What is a marginal complexity for step reactions with several products?

For example, the standard synergetic approach [52–54] denies the possibility of any self-organization in a system with with two intermediate products if only the mono- and bimolecular reaction stages occur [49]; it is known as *the Hanusse, Tyson and Light theorem*. We will question this conclusion, which in fact comes from the qualitative theory of non-linear differential equations where coefficients (reaction rates) are considered as constant values and show that these simplest reactions turn out to be complex enough to serve as a basic models for future studies of non-equilibrium processes, similar to the famous Ising model in statistical physics. Different kinds of auto-wave processes in the Lotka and Lotka–Volterra models which serve as the two simplest examples of chemical reactions will be analyzed in detail. We demonstrate the universal character of cooperative phenomena in the bimolecular reactions under study and show that it is reaction itself which produces all these effects.

The considerable progress made in the studies of simple bimolecular reactions (which has led to such fundamental conclusions) was achieved by a more rigorous mathematical treatment of the problem, avoiding the use of the simplest approximations which linearize the kinetic equations. We focus main attention on the *many-point density formalism* developed in [26, 28, 49] since in our opinion it seems at present to be the only general approach permitting treatment of *all* the above-mentioned problems, whereas other theoretical methods so far developed, e.g., those of secondary quantization [19, 29–32], and of multiple scattering [72, 73], as well based on the electrostatic analogy [74] could be applied to particular problems only. For example, the diagrammatic perturbation technique, being based on the analogy between the master equation and the quantum field theory with non-Hermitian Hamiltonian describing Bose particles, could be applied only to the steady-state of a system with particle source [32, 75, 76]. Moreover, this approach allows us to clarify the analogy between the kinetic equations derived for description of bimolecular diffusion-controlled reactions [25, 77] and those commonly used in the self-organization theory and its applications. We will use widely the analogy between our problems and physics of critical phenomena and treat kinetics under study in terms of *correlation lengths and critical exponents*.

We restrict ourselves to the case of classical particles and we thus disregard all quantum effects. The particle motion (if any) occurs by thermally-activated hops in continuum medium.

References

[1] A.S. Nowick and J.J.Burton (eds), Diffusion in Solids: Recent Developments (Associated Press, New York, 1975).
[2] G.E. Murch and A.S. Nowick (eds), Diffusion in Crystalline Solids (Associated Press, New York, 1984).
[3] G.E. Murch, H.K. Birnbaum and J. Cost (eds), Nontraditional Methods in Diffusion (Metallurgical Soc. of AIME, Warrendale, 1984).
[4] R.J. Borg and G.J. Dienes, An Introduction to Solid State Diffusion (Associated Press, Boston, 1988).
[5] R. Ghez, A Primer of Diffusion Problems (Wiley, New York, 1988).
[6] C.P. Flynn, Point Defects and Diffusion (Clarendon Press, Oxford, 1972).
[7] S. Mrowec, Defects and Diffusion in Solids (Elsevier, Amsterdam, 1980).
[8] R.A. Jonson and A.N. Orlov (eds), Physics of Radiation Effects in Crystals (North-Holland, Amsterdam, 1986).
[9] G.E. Murch, Atomic Diffusion Theory in Highly Defective Solids (North-Holland, Amsterdam, 1980).
[10] A.S. Laskar, G. Brebec and C. Monty (eds), Diffusion in Materials, NATO ASI Ser. in Appl. Sciences, Vol. 179 (1990).
[11] J. Chem. Soc. Faraday Trans. (special issue) 86(8) (1990).
[12] S.W. Benson, The Foundations of Chemical Kinetics (McGraw, New York, 1960).
[13] K.J. Laidler, Chemical Kinetics (McGraw, New York, 1965).
[14] C.W. Gardiner, Rates and Mechanisms of Chemical Reactions (Benjamin, New York, 1969).
[15] H. Schmalzried, Solid State Reactions (Associated Press, New York, 1974).
[16] H. Eyring, S.H. Lin and S.M. Lin, Basic Chemical Kinetics (Wiley, New York, 1980).

References

[17] S.A. Rice, Diffusion-Controlled Reactions (Elsevier, Amsterdam, 1985).
[18] C.W. Gardiner, Handbook of Stochastic Methods in Natural Sciences (Springer, Berlin, 1985).
[19] A.A. Ovchinnikov, S.F. Timashev and A.A. Belyi, Kinetics of Diffusion-Controlled Chemical Processes (Nuova Science, New York, 1989).
[20] M.V. Smoluchowski, Z. Phys. Chem. B 92 (1917) 129.
[21] R.M. Noyes, Progr. React. Kinet. 1 (1961) 129.
[22] S.A. Rice and M. Pilling, Progr. React. Kinet. 9 (1978) 93.
[23] L. Kantorovich, E. Kotomin, V. Kuzovkov, I. Tale, A. Shluger and Yu. Zakis, Models of Processes in Wide-Gap Solids with Point Defects (Zinatne, Riga, 1991).
[24] G. Yablonskii, V. Bykov, A. Gorban' and V. Elokhin, Kinetic Models of Catalytic Reactions (Elsevier, Amsterdam, 1991).
[25] T.R. Waite, Phys. Rev. 107 (1957) 463; J. Chem. Phys. 28 (1958) 103.
[26] G. Leibfried, in: Bestrahlanseffekte in Festkorpern (Teubner, Stuttgart, 1965) p. 266.
[27] A. Suna, Phys. Rev. B: 1 (1970) 1716.
[28] E. Kotomin and V. Kuzovkov, Chem. Phys. 76 (1983) 479.
[29] M. Doi, J. Phys. A: 9 (1976) 1465; 9 (1976) 1479.
[30] Ya.B. Zeldovich and A.A. Ovchinnikov, Sov. Phys. JETP 74 (1978) 1588.
[31] A.S. Mikhailov, Phys. Lett. A 85 (1981) 214; A 85 (1981) 427.
[32] A.S. Mikhailov and V.V. Yashin, J. Stat. Phys. 38 (1985) 347.
[33] M. Muthukumar, J. Chem. Phys. 76 (1982) 2667.
[34] D.F. Calef and J. Deutch, J. Chem. Phys. 79 (1983) 203.
[35] I.A. Tale, E.A. Kotomin and D.K. Millers, J. Phys. C: 8 (1975) 2366.
[36] D. Toussaint and F. Wilczek, J. Chem. Phys. 78 (1983) 2642.
[37] Yu. Kalnin and E. Kotomin, Probl. At. Nauki Tekh. 1(29) (1984) 18 (Probl. Atom Sci. Technol., in Russian).
[38] R. Kopelman, J. Stat. Phys. 42 (1986) 185.
[39] K. Lindenberg, B.J. West and R. Kopelman, Phys. Rev. Lett. 60 (1988) 1777.
[40] V.L. Vinetsky, E.A. Kotomin, Yu.H. Kalnın and A.A. Ovchınnıkov, Sov. Phys. Usp. 33(10) (1990) 1.
[41] A. Blumen, J. Klafter and G. Zumofen, in: Optical Spectoscopy of Glasses (Reidel, Dordrecht, 1986) p. 199.
[42] L. Pietronero and E. Tosatti (eds), Fractals in Physics (North-Holland, Amsterdam, 1986).
[43] Ya.B. Zeldovich and A.S. Mikhailov, Sov. Phys. Usp. 153 (1987) 409.
[44] A.S. Mikhailov, Phys. Rep. 184 (1989) 307.
[45] B. Balagurov and V. Vax, Sov. Phys. JETP 38 (1974) 968.
[46] A.A. Ovchinnikov and Ya.B. Zeldovich, Chem. Phys. 28 (1978) 215.
[47] E. Kotomin and A. Doktorov, Phys. Status Solidi B: 114 (1982) 287.
[48] R.F. Khairutdinov, K.I. Zamaraev and V.P. Zhdanov, Electron Tunnelling in Chemistry: Chemical Reactions over Large Distances (Elsevier, Amsterdam, 1989).
[49] V. Kuzovkov and E. Kotomin, Rep. Prog. Phys. 51 (1988) 1479.
[50] J. Stat. Phys. (special issue) 65(12) (1991).
[51] A.M. Zhabotinsky, Concentrational Auto-Oscillations (Nauka, Moscow, 1984).

[52] G. Nicolis and I. Prigogine, Self-Organization in Non-Equilibrium Systems (Wiley, New York, 1977).
[53] H. Haken, Synergetics (Springer, Berlin, 1978).
[54] H. Haken (ed.), Chaos and Order in Nature (Springer, Berlin, 1984).
[55] P. Schuster (ed.), Stochastic Phenomena and Chaotic Behaviour in Complex Systems (Springer, Berlin, 1984).
[56] G. Careri, Ordine e Disordine Nella Materia (Latenza, Roma, 1982).
[57] T. Riste (ed.), Nonlinear Phenomena at Phase Transitions and Instabilities (Plenum, New York, 1982).
[58] K. Tomita, Phys. Rep. 86 (1982) 113.
[59] A.S. Mikhailov and I.V. Uporov, Sov. Phys. Usp. 114 (1984) 79.
[60] W. Horsthemke and D. Kondepudi (eds), Fluctuations and Sensitivity in Non-Equilibrium Systems (Springer, Berlin, 1984).
[61] D. Jasnow, Rep. Prog. Phys. 47 (1984) 9.
[62] A. Hasegawa, Adv. Phys. 34 (1985) 11.
[63] A. Nitzan, P. Ortoleva, J. Deutch and J. Ross, J. Chem. Phys. 61 (1974) 1056.
[64] C.W. Gardiner and S.J. Chaturvedi, J. Stat. Phys. 17 (1977) 429.
[65] D. Elderfield, J. Phys. A: 18 (1985) 2049.
[66] H. Mori and K. McNeil, Prog. Theor. Phys. 57 (1977) 770.
[67] I. Goldhirsch and I. Procaccia, Phys. Rev. A: 24 (1981) 572.
[68] K. Kang and S. Redner, Phys. Rev. A: 32 (1985) 435.
[69] D. Elderfield and D.D. Vvedensky, J. Phys. A: 18 (1985) 2591.
[70] G. Dewel, D. Walgraef and P. Borckmans, Z. Phys. B: 288 (1987) 235.
[71] G. Nicolis and M. Malek-Mansour, Phys. Rev. A: 29 (1984) 2845.
[72] Yu.B. Gaididei and A.I. Onipko, Mol. Cryst. Liq. Cryst. 62 (1980) 213.
[73] M. Bixon and R. Zwanzig, J. Chem. Phys. 75 (1981) 2354.
[74] B.F. Felderhof and I.M. Deutch, J. Chem. Phys. 64 (1976) 4551.
[75] S.F. Burlatsky, A.A. Ovchinnikov and K.A. Pronin, Sov. Phys. JETP 92 (1987) 625.
[76] A.M. Gutin, A.S. Mikhailov and V.V. Yashin, Sov. Phys. JETP 92 (1987) 941.
[77] V.V. Antonov-Romanovkii, Kinetics of Photoluminescence of Crystal-Phosphors (Nauka, Moscow, 1966).

Contents

Preface			IX
1	**Guide to the book**		1
	1.1	What is this book written for?	1
		1.1.1 What is absent in the book	1
		1.1.2 What you will find in this book	2
	1.2	Order and disorder	4
		1.2.1 The order parameter	4
		1.2.2 The critical point	5
		1.2.3 The molecular field theory and the equation of state	7
		1.2.4 The order parameter in chemical kinetics and the kinetic law of mass action	9
		1.2.5 The standard chemical kinetics as mean-field theory	10
	1.3	How many particles are necessary to create a many-particle problem?	12
		1.3.1 What defines a heap?	12
		1.3.2 A pair problem	13
		1.3.3 Random walks and encounters	14
		1.3.4 The model $A + B \to B$ reaction	15
		1.3.5 The Smoluchowski equation	16
		1.3.6 Symmetry of particle mobilities and space dimension	18
		1.3.7 Survival probability	19
		1.3.8 Diffusion asymmetry	20
		1.3.9 Ensembles of particles	21
		1.3.10 Model of the $A + B \to 0$ reaction	22
		1.3.11 The Smoluchowski theory of the diffusion-controlled reactions	23
	1.4	Intermediate order	25
		1.4.1 Classical ideal gas and the Poisson distribution	25
		1.4.2 Chemical kinetics and non-Poisson fluctuations	26
		1.4.3 The intermediate order parameter	27
		1.4.4 Transformation and propagation of the ordering	28
		1.4.5 Spatial particle correlations and correlation functions	30
	1.5	Critical phenomena	31
		1.5.1 Spatial correlations and the order parameter	31
		1.5.2 Fluctuations of the order parameter in chemical reactions	33
		1.5.3 Dynamical interaction between particles	38
		1.5.4 Correlation analysis and fluctuation-controlled kinetics	41
		1.5.5 The Waite–Leibfried theory	43
		1.5.6 Kinetic phase transitions	45
		1.5.7 Conclusion	50
References			51

Contents

2 Basic methods for describing chemical kinetics in condensed media: continuum models ... 53

2.1 Macroscopic approach ... 53
2.1.1 Standard chemical kinetics: systems with complete reactant mixing ... 53
- 2.1.1.1 Basic equations of formal kinetics ... 53
- 2.1.1.2 Simplest bimolecular reactions in condensed media ... 54
- 2.1.1.3 Qualitative study of ordinary differential equations ... 57
- 2.1.1.4 The Lotka model ... 59
- 2.1.1.5 The Lotka–Volterra model ... 61
- 2.1.1.6 Synergetic aspects of the chain of bimolecular reactions ... 63

2.1.2 Smoothing out of local density fluctuations ... 67
- 2.1.2.1 Balance equations ... 67
- 2.1.2.2 Diffusion equations and diffusion length ... 69
- 2.1.2.3 Local and full equilibrium ... 70
- 2.1.2.4 Irreversible $A + B \to C$ reaction. Fluctuations of the concentration difference ... 73
- 2.1.2.5 Moment method ... 79
- 2.1.2.6 Irreversible $A + B \to B$ reaction. Survival in cavities ... 81

2.2 The treatment of stochasticity on a mesoscopic level ... 84
2.2.1 The stochastic differential equations and the Fokker–Planck equation ... 84
- 2.2.1.1 Stochastic differential equations ... 84
- 2.2.1.2 Explosive instability of a linear system ... 86
- 2.2.1.3 Spatially extended systems ... 88
- 2.2.1.4 The $A + B \to C$ reaction. Stochastic particle generation ... 90

2.2.2 The birth–death formalism and master equations ... 93
- 2.2.2.1 The Chapmen–Kolmogorov master equation ... 93
- 2.2.2.2 Generating function ... 95
- 2.2.2.3 The Lotka–Volterra model ... 99
- 2.2.2.4 The Lotka model ... 104
- 2.2.2.5 Non-equilibrium critical phenomena ... 106
- 2.2.2.6 Spatially-extended systems ... 107

2.3 Microscopic treatment of stochasticity ... 108
2.3.1 Many-point densities ... 108
- 2.3.1.1 Statistics of many-particle systems ... 108
- 2.3.1.2 Cell formalism ... 116
- 2.3.1.3 Master equation ... 118
- 2.3.1.4 Superposition approximation ... 123
- 2.3.1.5 Many-point densities and probability densities ... 128

2.3.2 The field-theoretical formalism ... 129
- 2.3.2.1 Probability densities and quantum-mechanical analogy ... 129
- 2.3.2.2 Secondary quantization ... 132
- 2.3.2.3 Diagrammatic formalism ... 134

References ... 134

3	**From pair kinetics toward the many-reactant problem**	139
	3.1 Basic defects and processes in solids	139
	3.1.1 Mechanisms of defect creation in solids	139
	3.1.2 Mechanisms of defect recombination	140
	3.1.3 Defect dynamic interaction	142
	3.1.4 Mechanisms of defect migration	143
	3.1.5 Classification of processes in solids	145
	3.2 Annealing of geminate pairs of defects	149
	3.2.1 Smoluchowski equation	149
	3.2.2 Noninteracting particles	154
	3.2.3 Drift in the potential field	156
	3.2.4 Recombination on a discrete crystalline lattice	164
	References	167
4	**The linear approximation in bimolecular reaction kinetics**	171
	4.1 The shortened superposition approximation ($A + B \to 0$ reaction)	171
	4.1.1 Initial conditions	171
	4.1.2 The shortened superposition approximation	173
	4.1.3 The infinitely diluted system	177
	4.1.4 Waite's interpolation equations	179
	4.1.5 Recombination of immobile reactants	181
	4.1.6 Diffusion-controlled reactions	184
	4.2 Tunnelling recombination	188
	4.2.1 Static reactions	188
	4.2.2 Diffusion-controlled reactions	190
	4.2.3 Non-stationary process	194
	4.2.4 Effects of defect interaction	198
	4.2.5 The anisotropic effects	205
	4.3 Non-diffusion defect recombination and non-stationary kinetics	207
	4.3.1 The hopping recombination	207
	4.3.2 Estimates of the effective reaction radius and reaction rate	209
	4.3.3 General solution for the arbitrary hopping length	211
	4.3.4 Non-stationary hopping kinetics	214
	4.3.5 Illustrative examples: doped alkali halides and silica glasses	218
	4.3.6 Effects of defect anisotropy	225
	References	230
5	**The fluctuation-controlled kinetics: the basic formalism of many-point particle densities**	235
	5.1 The $A + B \to 0$ reaction	235
	5.1.1 The non-linear correlation dynamics	235
	5.1.1.1 Kinetic equations	235
	5.1.1.2 Infinitely diluted system and a pair problem	238
	5.1.1.3 Small parameter concept	240
	5.1.1.4 Reaction of immobile particles	243

			5.1.1.5	Diffusion-controlled reactions. Black sphere model	244
			5.1.1.6	Reactant dynamical interaction	249
		5.1.2	The superposition approximation		253
			5.1.2.1	General comments	253
			5.1.2.2	Computer simulations of immobile particles	256
			5.1.2.3	Tunnelling recombination	258
			5.1.2.4	Multipole interaction	265
			5.1.2.5	Diffusion-controlled reactions	267
	5.2	The A + B → B reaction			269
		5.2.1	Correlation dynamics		269
			5.2.1.1	Prehistory	269
			5.2.1.2	The basic equations	271
			5.2.1.3	The molecular field approximation	273
			5.2.1.4	The black sphere model	274
			5.2.1.5	The Kirkwood superposition approximation	275
			5.2.1.6	Discussion	276
		5.2.2	Other approaches		277
			5.2.2.1	Density expansion	277
			5.2.2.2	The $\bar{d} = 1$ case	280
			5.2.2.3	The Wiener trajectories approach	283
			5.2.2.4	The probability density function	287
			5.2.2.5	Reversible diffusion-controlled reactions	288
	5.3	The A + A → 0 and A + A → A reactions			291
References					296

6 The many-particle effects in irreversible A + B → 0 reaction ... 301

6.1	Long-range recombination of immobile particles		301
	6.1.1	Tunnelling recombination	301
	6.1.2	Correlation length ξ and critical exponents	303
	6.1.3	Reactions on fractals	309
	6.1.4	Multipole interaction	315
	6.1.5	The nearest-available-neighbour approximation	320
6.2	Kinetics of diffusion-controlled reactions of mobile non-interacting particles		330
	6.2.1	Spatial structure of a system of mobile non-interacting particles	330
	6.2.2	Reaction asymptotic behaviour at long times	334
	6.2.3	Correlation length and critical exponent	340
	6.2.4	Asymmetric particle mobilities	347
	6.2.5	Reaction rate $K(t)$	350
	6.2.6	Other formalisms and approaches	352
6.3	Dynamic particle aggregation induced by elastic interactions		356
	6.3.1	Asymmetric particle mobilities	360
	6.3.2	Equally mobile A and B particles	366
	6.3.3	An elastic attraction of dissimilar particles	367
6.4	Effect of non-equilibrium charge screening (Coulomb interaction)		371
References			383

Contents

7	**The many-particle effects in A + B → 0 reaction with particle generation**		387
7.1	The kinetics of defect accumulation under irradiation		387
	7.1.1	Introduction	387
	7.1.2	Virtual particle configurations	391
	7.1.3	The asymptotically exact equations	393
	7.1.4	The superposition approximation	397
	7.1.5	The basic kinetic equations	403
	7.1.6	Results for immobile particles	405
	7.1.7	Correlated particle creation	408
	7.1.8	Incorporation of diffusion	409
7.2	Diffusion-controlled particle aggregation under permanent source		415
	7.2.1	The basic kinetic equations	416
	7.2.2	Aggregate characterization	419
	7.2.3	Results	421
	7.2.4	Discussion	428
7.3	Other approaches		429
	7.3.1	Particle accumulation on fractals	430
	7.3.2	Long-wavelength approximation	434
	7.3.3	Single-species accumulation kinetics	436
7.4	Probabilistic models and computer simulations		438
	7.4.1	Simple models	438
	7.4.2	Computer simulations	446
		7.4.2.1 Quasi-continuum model in particle accumulation	446
		7.4.2.2 The discrete model	447
		7.4.2.3 Taking account of tunnelling recombination of defects	453
	7.4.3	Analysis of the accumulation equations	456
References			461
8	**Systems under birth and death conditions: Lotka and Lotka–Volterra models**		467
8.1	A novel criterion of the marginal complexity insuring self-organization in the course of chemical reactions		467
	8.1.1	Autowaves in non-equilibrium extended systems	467
	8.1.2	Basic models	468
	8.1.3	Systems with a complete stirring	469
	8.1.4	Role of diffusion	470
	8.1.5	The criterion of the marginal complexity of self-organized systems	472
8.2	The Lotka–Volterra model		473
	8.2.1	The stochastic Lotka–Volterra model	473
	8.2.2	Kinetic equations	474
	8.2.3	Superposition approximation	478
	8.2.4	Boundary conditions	480
	8.2.5	Standing waves	482
	8.2.6	Concentration dynamics	484
	8.2.7	Correlation dynamics	489
8.3	The Lotka model		493

	8.3.1	The stochastic Lotka model	493

- 8.3.1 The stochastic Lotka model ... 493
- 8.3.2 Kinetic equations ... 494
- 8.3.3 Superposition approximation ... 497
- 8.3.4 Basic equations ... 500
- 8.3.5 Standing waves ... 501
- 8.3.6 Concentration dynamics and oscillations of $K(t)$ ... 502
- 8.3.7 Correlation dynamics ... 509
- 8.3.8 Conclusions ... 512

References ... 512

9 Catalytic reactions on solid surfaces ... 515

- 9.1 A stochastic model for surface reactions without energetic interactions ... 515
  - 9.1.1 A general stochastic model of surface reactions ... 515
    - 9.1.1.1 Introduction ... 515
    - 9.1.1.2 Master equations ... 518
    - 9.1.1.3 The superposition approximation ... 521
    - 9.1.1.4 Method of solution ... 526
  - 9.1.2 The $A + \frac{1}{2}B_2 \to 0$ surface reaction: island formation and complete particle segregation ... 526
    - 9.1.2.1 Particular bimolecular reaction ... 526
    - 9.1.2.2 Basic equations ... 529
    - 9.1.2.3 The diffusion and desorption processes ... 535
    - 9.1.2.4 Results ... 535
    - 9.1.2.5 The extended ZGB-model incorporating diffusion and desorption processes ... 539
    - 9.1.2.6 Conclusion ... 542
  - 9.1.3 The influence of surface disorder on the $A + \frac{1}{2}B_2 \to 0$ reaction ... 544
    - 9.1.3.1 Problem statement ... 544
    - 9.1.3.2 Effects of disorder ... 545
    - 9.1.3.3 Summary ... 549
  - 9.1.4 A stochastic model for complex surface reaction systems: Application to the NH_3 catalytic formation ... 550
    - 9.1.4.1 Introduction ... 550
    - 9.1.4.2 The stochastic model ... 554
    - 9.1.4.3 The system behaviour for $S = 1$ ... 557
    - 9.1.4.4 The system behaviour for $S = 1/8$ ... 559
    - 9.1.4.5 The correlation functions ... 561
    - 9.1.4.6 Discussion ... 561
    - 9.1.4.7 Conclusion ... 563
- 9.2 A stochastic model for surface reactions including energetic particle interactions ... 564
  - 9.2.1 A general stochastic model ... 564
    - 9.2.1.1 Introduction ... 564
    - 9.2.1.2 The model ... 565
    - 9.2.1.3 Master equations ... 566

		9.2.1.4	Equation of motion for the two-point probabilities	569
		9.2.1.5	The definition of a standard model	573
		9.2.1.6	Examples of the standard model	575
	9.2.2	The $A + \frac{1}{2} B_2 \to 0$ reaction with energetic interactions		577
		9.2.2.1	Introduction	577
		9.2.2.2	The model	578
		9.2.2.3	Results for repulsive interactions	579
		9.2.2.4	Results for attractive interaction	585
		9.2.2.5	The effect of B-desorption	587
		9.2.2.6	Conclusion	589

References ... 591

10 General conclusion ... 593

References ... 594

Author Index ... 597

Subject Index ... 607

Chapter 1

Guide to the Book

1.1 WHAT IS THIS BOOK WRITTEN FOR?

1.1.1 What is absent in the book

> "What is that?" – "It is a compass to travel the world."
> ("Was haben Sie hier?" – "Ein Kompaß, um durch die Welt zu reisen")
> Lichtenberg

In this book we summarize the state of the art in the study of *peculiarities of chemical processes in dense condensed media*; its aim is to present the unique formalism for a description of *self-organization phenomena* in spatially extended systems whose structure elements are coupled via both matter diffusion and nonlocal interactions (chemical reactions and/or Coulomb and elastic forces). It will be shown that these systems could be described in terms of nonlinear partial differential equations and therefore are complex enough for the manifestation of wave processes. Their spatial and temporal characteristics could either depend on the initial conditions or be independent on the initial as well as the boundary conditions (the so-called *autowave processes*).

The thorough readers familiar with books on synergetics would expect to see after such an introduction several standard and bright illustrations of the spatio-temporal pattern formations in irreversible processes like a test tube with separated red and blue reactants, spiral waves on the surface of a solution etc. To their disappointment, we do not consider here the spatial structures violating system's homogeneity (because of the reasons explained below); the illustrations like Fig. 1.1 [1] and further figures also should not be interpreted in the traditional synergetic manner. All kinds of autowave motions considered in the book – both single running waves and periodic spherical waves – are not directly observable macroscopic material motions but these are more internal synchronous motion processes called below *correlation dynamics*. The experimentally observable quantity is the *concentration dynamics*, i.e., the change of reactant macroscopic concentrations which

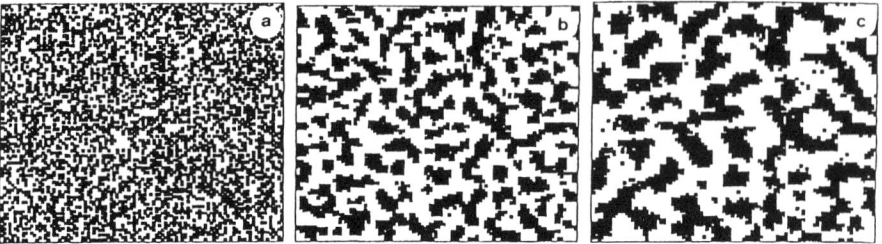

Fig. 1.1. Three stages in the island formation of A atoms with attractive NN interactions and coverage $n_A = 0.4$: (a) $t = 0$; (b) $t = 1000$ MCS; (c) $t = 5000$ MCS (MCS = 1 Monte Carlo step). Shown are 100×100 Sections of the 500×500 lattice used in the calculations.

could be monotonic or oscillatory functions of time. To stress this peculiarity of the problem under study, we have called earlier this process *microscopic self-organization* [2], in order to focus reader's attention on the fact that our systems reveal the short- and intermediate-range rather than long-range order in their structure elements. (It should be mentioned here that even in these systems the spatial scale defining the distinctive *correlation length* can infinitely increase thus making a classification of systems into microscopically and macroscopically ordered to be sometimes quite uncertain.)

1.1.2 What you will find in this book

> It is not an art to express something briefly if one has something to say as Tacitus. But if one is writing a book while having nothing to say thus transforming the truth with its *ex nihilo nihil* to a lie – this I call a merit.
>
> (Es ist keine Kunst, etwas kurz zu sagen, wenn man etwas *zu* sagen hat wie Tacitus. Allein wenn man nichts zu sagen hat und schreibt dennoch ein Buch und macht die Wahrheit mit ihrem *ex nihilo nihil* fit zu Lügnerin, das heißich Verdienst)
>
> Lichtenberg

We understand very well that any book inavoidably reflects authors' interests and scientific taste; this fact is, first of all, usually seen in the selection of material which in our case is very plentiful and diverse. For instance, Chapter 2 gives examples of different general approaches used in chemical kinetics (macroscopic, mesoscopic and microscopic) and numerous methods for solving particular problems. We focus here on the *microscopic* approach based on the concept of active particles (structure elements, reactants, defects) whose spatial redistribution arises due to their diffusion affected by the dynamic interaction of particles through the long-range potentials. In due course, macroscopic and mesoscopic approaches operating with average quantities – local or macroscopic concentrations defining the *concentration field* – are sketchen omitting details.

The authors of this book started working on chemical kinetics more than 10 years ago focusing on investigations of particular radiation – induced processes in solids and liquids. Condensed matter physics, however, treats *point (radiation) defects* as active particles whose individual characteristics define kinetics of possible processes and radiation properties of materials. A study of an ensemble of such particles (defects), especially if they are created in large concentrations under irradiation for a long time, has lead us to *many-particle problems*, common in statistical physics. However, the standard theory of diffusion-controlled reactions as developed by Smoluchowski [3] turned out to be not suited for the case of complex spatial distributions of defects created by high-energy radiation. Attempts to generalize this theory undertaken by a number of authors were not quite successful.

It is clear that a complete treatment of the recombination kinetics requires careful incorporation of mutual particle distribution, including *fluctuations* in their local concentrations due to diffusion and reaction. At present the role of fluctuations in reactant concentrations in chemical kinetics is well known [4]; it is intensively studied in the theory of self-organized nonequilibrium systems. However, these studies are based on macroscopic and mesoscopic classes of models. In other words, these models focus on fluctuations of quantities as a total number of particles in a whole system or of concentrations within cells; these cells are assumed to be large enough to consider numbers of particles therein as macrovariables, but on the other hand small enough to consider spatial particle distribution as inhomogeneous. The distinctive feature of these models is a treatment of the reactions inside cells exactly as in the *standard chemical kinetics*, i.e., in terms of the *law of mass action* with the constant reaction rate multiplied by a product of concentrations. This class of models neglects the real mechanism of defect recombination events, using only the average recombination characteristics, i.e., the reaction rate.

Therefore, we tried to develop the *adequate mathematical formalism of the fluctuation-controlled chemical kinetics* based on a concept of active particles. Simultaneously, the *mesoscopic* theory of concentration field fluctuations was developed by a number of investigators (see Chapter 2) having more qualitative character. Undoubtedly, these two approaches – microscopic and mesoscopic – overlap, since a lot of fundamental results like asymptotic law of concentration decay at long times are not sensitive to details of particle interaction. Nevertheless, we prefer to use here microscopic (ever simplest) models allowing us to get numerical estimates for actual systems (defects) rather than to make very general and uncertain conclusions. When presenting results of calculations for reactant concentration dynamics and formation of the relevant spatio-temporal particle structures, we discuss not only the general formalism but also details of the derivation and methods of the solution of the basic microscopic kinetic equations for various model problems.

A careful study of the fluctuation-controlled kinetics performed in recent years has led us to numerous deviations from the results of generally-accepted standard chemical kinetics. To prevent readers from getting "lost" in details of different formalisms and the ocean of equations presented in this book, we present in this introductory Chapter a brief summary, explain the necessity of developing the fluctuation kinetics and demonstrate its peculiarities compared with techniques presented earlier. We will use here the simplest mathematical formalism and focus on basic ideas which will be discussed later on in full detail.

1.2 ORDER AND DISORDER

1.2.1 The order parameter

Order leads to all virtues! But what leads to order?
(Ordnung führet zu allen Tugenden! Aber was führet zur Ordnung?)

Lichtenberg

The history of the study of macroscopic systems containing huge numbers of elements demonstrates clearly the necessity of developing and specifying the fundamental concept of an *order parameter* [5, 6]. For clarity, we consider here dense gases and liquids in thermodynamical equilibrium as standard systems whose properties are well known. Despite the fact that sometimes the concepts of equilibrium order and disorder for these systems are not very convincing and require the use of additional analogs with other (e.g., magnetic) systems, it suits well to the scope of our book.

It is convenient to begin with a simple one-component system having three phases – gas, liquid and solid. At the primitive level these phases differ by a density n, i.e., the number of particles per unit volume. If we fix now for simplicity the pressure in this system, in the thermodynamical equilibrium,

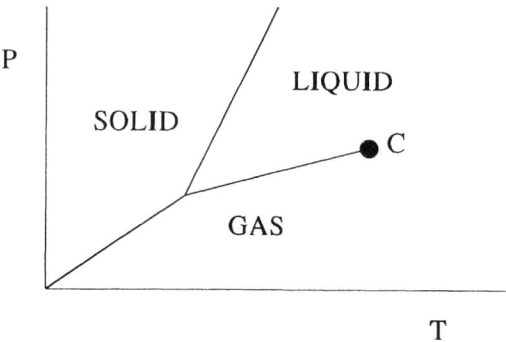

Fig. 1.2. Phase diagram of a simple solid.

the density is a function of the temperature T only, $n = n(T)$, and could be treated as an *order parameter*. For a multicomponent system one has several densities and the density becomes the multicomponent order parameter, $\boldsymbol{n} = \boldsymbol{n}(T)$. When the temperature decreases, n, as a rule, increases and changes monotonously. (The well-known exceptions are water near 0°C and phase transitions, respectively). We consider here neither the mechanism of the equilibrium structure formation which is based on the Boltzmann ordering principle (this problem was discussed more than once [4, 7]), nor the subtle concept of entropy and its connection with the ordering. For a certain combination of the pressure p and the temperature T two phases (e.g., solid and liquid) can coexist in equilibrium with each other. When passing the temperature point $T = T_c = T_c(p)$, the (hidden) heat is released and the density changes abruptly. This process is called the *first-order phase transition*. The gas–liquid equilibrium curve shown in Fig. 1.2 has very interesting property: it ends at some point called *critical*.

1.2.2 The critical point

> ...Catholics and other people.
> (Die Katholiken und die andern Menschen)
> Lichtenberg

The critical point is one of many examples of higher-order phase transitions including the second-order transitions in ferromagnetics and ferroelectrics and λ-transition in liquid He. Unlike the first-order transition, the heat of the

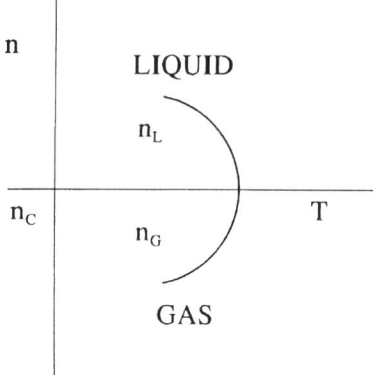

Fig. 1.3. Density near the critical point in the gas–liquid system.

phase transition is not released here (despite the fact that the specific heat reveals a certain singularity: its diagram looks like the Greek letter λ); the density also changes smoothly. Such higher-order phase transitions are also called *critical phenomena* (or order–disorder transitions) [6, 8, 9]. Below we use these transitions as a test system for illustrations of the problems of chemical kinetics.

Let us consider now the gas-liquid system near the critical point (Fig. 1.3). At $T \leqslant T_c$ both phases coexist, their densities n_L (liquid) and n_G (gas) could be formally written as $n_L = n_c + \delta n/2$, $n_G = n_c - \delta n/2$, where n_c is density at the critical point. Note that in the physics of critical phenomena the order parameter is often defined subtracting the background value of n_c, i.e., as the order parameter the *difference* of densities, $\delta n = n_L - n_G$, could be used rather than these individual densities themselves. Such an order parameter is zero at $T > T_c$ and becomes nonzero at $T < T_c$. Another distinctive feature of the order parameter is that for all simple systems the algebraic law $\delta n \propto (T_c - T)^\beta$ holds, where β is constant.

It is useful to check whether this kind of relations is valid for other systems like ferromagnetics and ferroelectrics too. Here the order parameters are the magnetization M and the polarization P, respectively. At high temperatures ($T > T_c$), and zero external field these values are $M = 0$ (paramagnetic phase) and $P = 0$ (paraelectric phase) respectively. At lower temperatures close to the phase transition point, however, spontaneous magnetization and polarization arise following both the algebraic law: $M, P \propto (T_c - T)^\beta$.

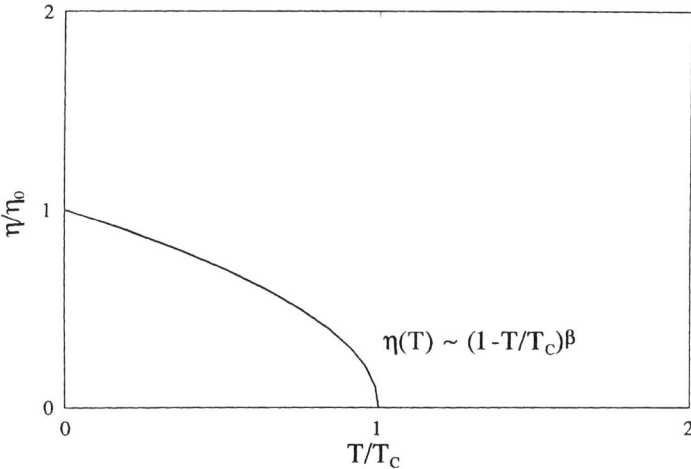

Fig. 1.4. Schematic behaviour of the order parameter in the vicinity of the critical point.

A more complicated but solvable problem is a definition of the order parameter for antiferromagnetics, binary alloys, superconductors etc. The dimensionless units T/T_c and η/η_0 (Fig. 1.4) allow us to present the behaviour of the order parameter $\eta = \eta(T)$ in a form *universal* for many quite different systems. Moreover, in some cases even quantitative similarities hold which concerns in particular the value of the exponent β. (The value of $\eta = 0$ characterizes always disordered phase.)

1.2.3 The molecular field theory and the equation of state

> The former borders of science are now its middle.
> (Wo damals die Grenzen der Wissenschaft waren, da ist jetzt die Mitte)
> Lichtenberg

First-principle calculations of the thermodynamic properties are more or less hopeless enterprise. One of the most famous *phenomenological* approaches was suggested by van der Waals [6, 8, 9]. Using the dimensionless pressure $\pi = p/p_c$, the density $\nu = n/n_c$ and the temperature $\tau = T/T_c$, the equation of state for the ideal gas reads $\pi = 8\nu\tau/(3-\nu) - 3\nu^2$. Its r.h.s. as a function of the parameter ν has no singularities near $\nu = 1$ ($\nu = \pi = \tau = 1$ is the critical point) and could be expanded into a series in the small parameter $\eta = (n - n_c)/n_c$ with temperature-dependent coefficients. Solving this equation after expansion at $p = p_c$ with respect to the order parameter one gets $\eta \propto (T_c - T)^{1/2}$, i.e., one obtains a value of the exponent $\beta = 1/2$. For a given pressure the equation of state for a multicomponent system (defining again the multicomponent order parameter $\boldsymbol{n}$ as a function of the temperature T) has the general form $\boldsymbol{F}(\boldsymbol{n}, T) = 0$, where the non-linear (vector) function $\boldsymbol{F}$ could be expanded into a series in the order parameter $\boldsymbol{n}$.

The van der Waals equation discussed above was derived in the same spirit as the so-called *molecular field* introduced by Weiss [8] in the theory of ferromagnetism. In this theory, magnetic interaction of dipoles can be formally expressed as resulting from some internal magnetic field depending on the nearest dipole orientation. In the Weiss approach the fluctuating internal field is replaced by the *mean (molecular) field* thus making this many-particle problem solvable. It is important to note here that the exponent β obtained from the van der Waals equation remains the same not only for all other model equations of state of real gases capable to describe the critical point, but also for other physical systems (e.g., ferromagnetics) provided that the equation defining the temperature dependence of the order parameter η was derived in an approximation similar to the molecular field.

An important step in developing the mean-field concept was done by Landau [8, 10]. Without discussing the relation between such fundamental quantities as disorder–order transitions and symmetry lowering, we just want to note here that his theory is based on thermodynamics and the derivation of the temperature dependence of the order parameter via the thermodynamic potential minimization (e.g., the free energy $A(\eta, T)$) which is a function of the order parameter. It is assumed that the function $A(\eta, T)$ is analytical in the parameter η and thus near the phase transition point could be expanded into the series in η; usually it is a polynomial expansion with temperature-dependent coefficients. Despite the fact that such a thermodynamical approach differs from the original molecular field theory, they are quite similar conceptually. In particular, the r.h.s. of the equation of state for the pressure of gases or liquids and the external field in ferromagnetics, respectively, have the same polynomial form.

It is known that the "classical" molecular field theory discussed above is not suited for describing a close vicinity of the critical point. Experimentally obtained values of the parameter β (called the *critical exponent*) are *essentially less* than $\beta_0 = 1/2$ predicted by the mean-field theory. On the other hand, the experimental values of $\beta = 0.33$–0.34 turn out to be universal for many different systems (except for quantum liquid–helium where β

is larger than the just quoted value). Obviously, this fact reflects some universal properties of different systems undergoing phase transitions whereas deviation from the value of β predicted by mean-field theory demonstrates its limitation.

1.2.4 The order parameter in chemical kinetics and the kinetic law of mass action

> Doubt everything at least once, even if it is the sentence "two by two is four".
> (Zweifle an allem wenigstens einmal, und wäre es auch der Satz: zwei mal zwei ist vier)
>
> Lichtenberg

Before continuing studies of the gas–liquid systems, let us first analyze the foundations of chemical kinetics in terms of the order parameter [11]. The *standard* chemical kinetics is based on the description of the homogeneous system of reacting particles by a set of first-order differential equations with respect to reactant concentrations or densities of particle numbers. In a general case of the multicomponent system characterized by the multicomponent order parameter n the rate of its change is assumed to be a function of n only: $\dot{n} = F(n)$. In other words, the state of the chemical system at any time t is assumed to be characterized *uniquelly* by its order parameter. One of the main goals of chemical kinetics is to determine the time development of the order parameter, $n = n(t)$. Another assumption of chemical kinetics is *the kinetic law of mass action* [11, 12], which specifies the form of the function $F(n)$ and is based on the intuitive and at first glance self-evident microscopic model of the chemical reactions. It assumes that thermal motion of particles leads to their collisions; non-elastic collisions result in particle transformations (reactions). It seems to be obvious that the number of collisions of particles of two kinds, A and B, is proportional to the *product* of their densities n_A and n_B, i.e., it is governed by the quantity $kn_A n_B$, where k is a proportionality coefficient. (It was one of the basic ideas of chemical kinetics from its early days [13].) Restricting ourselves to the so-called *bimolecular processes* (pair collisions only), the law of mass action defines uniquelly the function $F(n)$ as a polynomial of second order; the reaction constant k could be obtained by experiment.

Let us discuss critically the assumptions made above. For the gas–liquid system the statement that the order parameter is a function of the temperature, $n = n(T)$, is quite correct. (Both the experimental and the theoretical definition of this dependence are problematic.) Despite the fact that the order parameter is time-dependent, $n = n(t)$, this dependence is postulated here in the form of the non-self-evident differential equation $\dot{n} = F(n)$. Thus, the theory of irreversible processes [9], which attempts to describe the time development, seems to be far from its completion. A development of a macroscopic theory – that is thermodynamics of irreversible processes – is also problematic: the conditions under which parameters determining the time-development of irreversible processes are functions of macroscopic quantities only (like the density n) are not clear. However, especially on the statement just quoted the equation $\dot{n} = F(n)$ rests! Moreover, there exist exactly solvable models in chemical kinetics showing the insufficiency of macroscopic parameters thus arguing for the necessity of an improvement of the law of mass action. In the chemical physics of gas-phase reactions this postulate was questioned the last years only. The existence of this problem was realized also in condensed phases where delivery of reactants each to other is diffusive and too slow for smoothing of nonuniform particle distribution in the reaction volume. This results in the formation of local inhomogeneities in the spatial distribution of the reactants, thus making a use of macroscopic reactant densities as a single order parameter obviously incomplete.

1.2.5 The standard chemical kinetics as mean-field theory

> Weaknesses do not harm us anymore, as soon as we know them.
> (Schwachkeiten schaden uns nicht mehr, sobald wir sie kennen)
>
> Lichtenberg

The above mentioned assumptions of standard chemical kinetics have a certain analog with the mean field theory, i.e., both use the order parameters only and assume validity of the polynomial expansion in this parameter as it is shown in Fig. 1.5.

In particular, it is useful to define the critical point through $F(n_c) = 0$ (the stationary state). Since multicomponent chemical systems often reveal quite complicated types of motion, we restrict ourselves in this preliminary treatment to the *stable stationary states*, which are approached by the system without oscillations in time. To illustrate this point, we mention the simplest reversible and irreversible bimolecular reactions like A+A → B, A+B → B, A + B → C. The difference of densities $\eta(t) = n(t) - n_c$ can be used as the redefined order parameter η (Fig. 1.6). For the bimolecular processes the

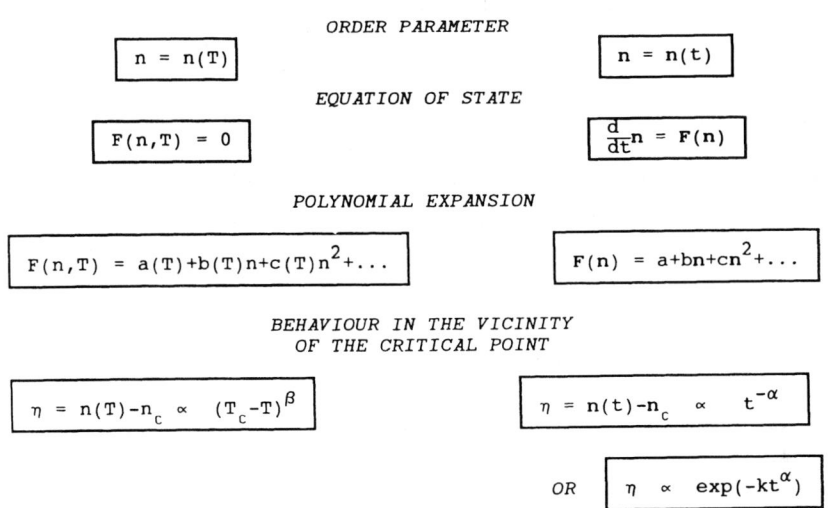

Fig. 1.5. A schematic analogy between gas–liquid system and chemical kinetics.

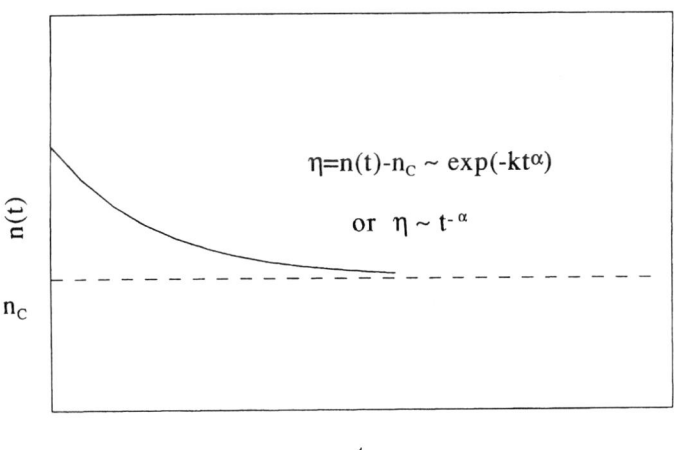

Fig. 1.6. The order parameter in the chemical system nearby the steady-state point (density relaxation).

approach to the steady-state is characterized by one of the two basic laws: it is either exponential, $\eta \propto \exp(-kt^\alpha)$, or algebraic, $\eta \propto t^{-\alpha}$; in both cases the standard kinetics predicts the same "classical" value of $\alpha_0 = 1$.

Pioneering papers [14, 15] (see also reviews [2, 16]) as well as theoretical studies done in the last years argue convincingly that the exponent α is less than α_0; this conclusion is confirmed by numerous computer simulations [17]. This shows that basic principles of chemical kinetics have to be critically analyzed – in analogy to careful studies done in physics of equilibrium systems in the vicinity of the critical points. The deviation from standard chemical kinetics could be expected not only in the *asymptotic laws* of the steady-state approach, but also in types of a possible system's motion. As it is known, complex chemical systems reveal often auto-oscillating processes with both retaining and spontaneous breaking of the system's homogeneity as well as more complicated chaotic motions. Appearance of these complicated motions is closely related to the *complexity of the system*; its measure is a type of non-linearity of the relevant kinetic equations and the number of components in the order parameter giving number of degrees of freedom. When going beyond the limitations of the mean-field theory, we in fact take into account additional degrees of freedom neglected in its simplified analysis.

1.3 HOW MANY PARTICLES ARE NECESSARY TO CREATE A MANY-PARTICLE PROBLEM?

1.3.1 What defines a heap?

"Who is there?" – "It's just me." – "Oh, that's too much!"
("Wer ist da?" – "Nur ich." – "Oh, das ist überflüssig genug!")

Lichtenberg

The transition from a macroscopic description to the microscopic level is always a complicated mathematical problem (the so-called *many-particle* problem) having no universal solution. To illustrate this point, we recommend to consider first the motion of a single particle and then the interaction of two particles, etc. The problem is well summarized in the following remark from a book by Mattuck [18] given here in a shortened form. "For the Newtonian mechanics of the 18th century the three-body problem was unsolvable. The general theory of relativity and quantum electrodynamics created unsolvable two-body and single-body problems. Finally, for the modern quantum field theory even vacuum (no bodies) turns out to be unsolvable problem. That is, if we are interested in exact solutions, even a single-body is already too many..." Despite the fact that methods of quantum field theory are widely spread in statistical physics and chemical kinetics, the situation is not yet dramatic. However, a really hard problem is to clarify the number of particles whose interaction has to be taken into account for an adequate description of many-particle (cooperation) phenomena, in which *each* particle affects the behaviour of all the rest particles. It is natural to try to restrict ourselves to considering two particles only as a minimal ensemble since in many cases a pair problem can be solved analytically.

Let us consider again a gas-phase system as the simplest illustration. Experimental data for not very dense gas are usually treated in terms of the equation of state $p/(nk_BT) = 1 + nB_2(T) + n^2B_3(T) + \cdots$ (the so-called *virial expansion* or expansion in a series of density powers n). Here the $B_n(T)$ are called virial coefficients: they depend on the temperature only. Consider now a pair of particles and ascribe to it the Boltzmann factor $\exp(-U(r)/(k_BT))$, where $U(r)$ is their potential energy. It is known from statistical physics [9] that upon neglecting three- and higher-order particle collisions the second virial coefficient can be expressed via the Boltzmann factor given above. Moreover, further terms of the virial expansion can also be formally written down, thus considering collisions of any number of particles, using systematically an expansion in the functions $f = \exp(-U(r)/(k_BT)) - 1$ (the so-called *Mayer's group expansion*).

For some time it was believed that the virial expansion allows to describe the critical point, but this hope failed. The point is that at high densities typical for the liquid phase, the most important terms of the expansion are those which describe the formation of a large cluster of interacting particles (the many-particle effect) whereas an approach based on the system's treatment as an ensemble of interacting *pairs* fails. From the point view of such a pair approach, formation of the *ordered* crystalline structure is a complete puzzle.

1.3.2 A pair problem

> The shorter the sergeant the bigger the proud.
> (Die kleinsten Unteroffiziere sind die stolzesten)
> Lichtenberg

In the chemical kinetics a pair problem is the first preliminary step in studying the many-particle problem; its results cannot be overestimated and extrapolated automatically for dense systems. On the other hand, this pair problem is of independent interest, e.g., for radiation-induced chemical processes. As it is known, high-energy irradiation (e.g., ions or neutrons) creates tracks in solids, whose structure depends on incident-particle energy, particular material etc. In the limiting case (or under electron beam- or X-irradiation) we have just an ensemble of *pairs* of complementary defects called also Frenkel pairs (vacancy–interstitial) [16] which are well-correlated in space. The kinetics of their recombination is of great interest for radiation physics of real materials and serves as the first step in understanding relaxation processes in tracks. It is also important for studying the mechanisms of radiation damage and effects of different kinds of irradiation.

1.3.3 Random walks and encounters

> And with the wine no longer in bottles but in their heads, they went out to the street.
> (Und mit dem Wein, der nun nicht mehr in den Bouteillen, sondern im Kopf war, gingen sie auf die Straße)
>
> Lichtenberg

Let us consider now just a pair of immobile point particles A and B which are chemically interacting with each other (that is, a given pair AB transforms into another pair A′B′). This pair reaction could be described by the simplest decay law: $\exp(-t/\tau)$, their lifetime $\tau = \tau(r)$ is defined by their

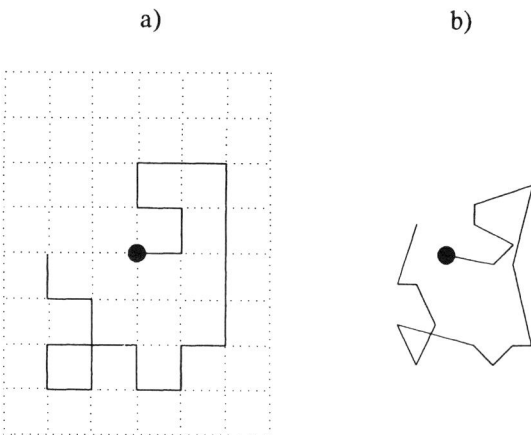

Fig. 1.7. A random trajectory on a square lattice (a) and the Brownian motion in continuum (b).

How many particles are necessary to create a many-particle problem?

Fig. 1.8. A drunken gentleman undergoing random walks.

relative distance r. It is convenient to introduce the reaction rate per unit time $\sigma(r) = \tau^{-1}(r)$. The first complication of this trivial problem comes from an incorporation of particle motion. Assuming that the medium is viscous, particle could be treated as undergoing the Brownian motion characterized by the *diffusion coefficients* D_A and D_B. In crystals this motion transforms into walks on the lattice (Fig. 1.7), where particle hops with equal probability to one of the nearest lattice sites. As it is known, particle motion after many such random walks could be treated in terms of the diffusion equation. (Note that some modern experiments have demonstrated necessity to go beyond the diffusion approximation.) The distinctive feature of diffusion is the linear dependence of the mean square displacement of a particle as a function of time, the diffusion coefficient is just a co-factor of time. One of the Chapters of Haken's book "Synergetics" [7] has subtitle: "How far a drunken man can walk", since random chaotic walks of a toper shown in Fig. 1.8 remind us the Brownian motion of a particle.

1.3.4 The model $A + B \to B$ reaction

> "How are you getting on?" – the blind man asked the lame.
> – "As you can see" – was his answer.
> ("Wie geht's?" sagte ein Blinder zu einem Lahmen.
> "Wie Sie sehen", antwortete der Lahme)
>
> Lichtenberg

Let us consider the analogy between an unpredictable toper and the Brownian motion of a particle A, adding now a policeman B to accompany him

Fig. 1.9. Reaction A + B → B and its visualization.

(Fig. 1.9). Assume first that the policeman is standing immobile on his post but he grasps a toper when they collide (lattice walks) or it catches policeman's eye restricted by some distance r_0 (the trapping radius in the continuous medium). We are interested in the following problems: (i) whether a toper will be *always* grasped; (ii) what is the probability not to be caught if a toper start walking at a given distance l from the policeman; (iii) how the results will be changed if the policeman also starts to move (his Brownian walks could be caused only by a dense London fog). The situation described above could be described by the A+B → B reaction (particles A are trapped by unsaturable sinks B); another example is the energy transfer from mobile excited molecules A to energy acceptors B which rapidly emit light and are again ready to get next excitation.

1.3.5 The Smoluchowski equation

> The American who was the first to discover Columbus made a bad discovery.
> (Der Amerikaner, der den Kolumbus zuerst entdeckte, machte eine böse Entdeckung)
>
> Lichtenberg

The above-described pair problem is treated by the Smoluchowski equation [3, 19] – see Fig. 1.10. It operates with the *probability densities* (Fig. 1.11) and contains the recombination rate $\sigma(r)$ which is a function of coordinates and the parameter $D = D_A$ characterizing particle motion. Knowledge of the probability density to find a particle at a given point at time moment t gives us (by means of a trivial integration over reaction volume) the quantity of our primary interest – *survival probability* of a particle in the system with

SCHRÖDINGER EQUATION

$$i\hbar \frac{\partial}{\partial t}\Psi(\vec{r},t) = -\frac{\hbar^2}{2m}\nabla^2 \Psi(\vec{r},t) + U(r)\Psi(\vec{r},t)$$

$\Psi(\vec{r},t)$ - PROBABILITY AMPLITUDE

SMOLUCHOWSKI EQUATION

$$\frac{\partial}{\partial t}w(r,t) = D\nabla^2 w(r,t) - \sigma(r)w(r,t)$$

$w(r,t)$ - PROBABILITY DENSITY

DRIFT IN THE POTENTIAL

$$\frac{\partial}{\partial t}w(r,t) = D\nabla\left[\nabla w(r,t) + \frac{w(r,t)}{k_B T}\nabla U(r)\right] - \sigma(r)w(r,t)$$

Fig. 1.10. A comparative analysis of the Schrödinger and Smoluchowski equations.

a trap. In terms of mathematics, the Smoluchowski equation is quite similar to the Schrödinger equation describing motion of a *quantum* particle having mass m in the potential $U(r)$. Analogously to the Schrödinger equation, one can study numerous kinetic problems for different choices of $\sigma(r)$. Another complication comes if one replaces the isotropic diffusion coefficient for anisotropic. In this case the diffusion coefficient D is no longer a numerical parameter but a tensor $\mathbf{D}$.

New difficulties arise when we try to take into account *the dynamical interaction* of particles caused by pair potentials $U(r)$; mutual attraction (repulsion) leads to the preferential drift of particles towards (outwards) sinks. This kind of motion is described by the generalization of the Smoluchowski equation shown in Fig. 1.10. In terms of our "illustrative" model of the chemical reaction $A + B \rightarrow B$ the drift in the potential could be associated with a search of a toper by his smell (Fig. 1.12). An analogy between Schrödinger and Smoluchowski equations is more than appropriate; indeed, it was used as a basis for a new branch of the chemical kinetics operating with the mathematical formalism of quantum field theory (see Chapter 2).

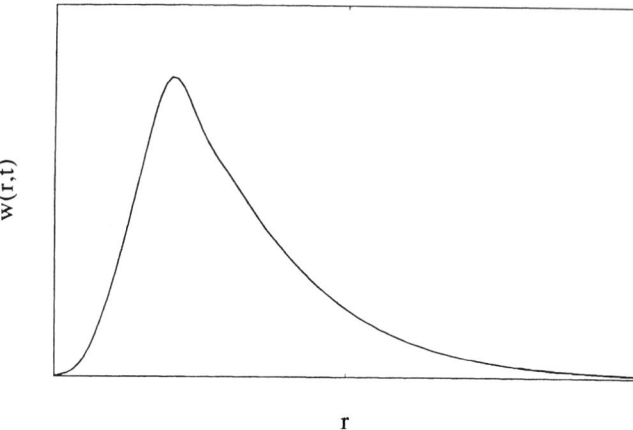

Fig. 1.11. The probability density obtained as a solution of the Smoluchowski equation.

Fig. 1.12. Reaction A + B → B with the particle interaction and its visualization.

1.3.6 Symmetry of particle mobilities and space dimension

> A fool imagining that he is a king does not differ from a real king, except that the former is a *negative* king and the latter is a *negative* fool. Neglecting minus signs, the two are equal.
> (Ein Narr, der sich einbildet, ein Fürst zu sein, ist von dem Fürsten, der in der Tat ist, durch nichts unterschieden, als daß jener ein negativer Fürst und dieser ein negativer Narr ist. Ohne Zeichen betrachtet, sind sie gleich)
>
> Lichtenberg

The Smoluchowski equation demonstrates the principal feature of the standard chemical kinetics: the latter is defined by a coefficient of the *relative* diffusion $D = D_A + D_B$; for a fixed value of D no matter, which of particles undergoes the Brownian motion! This result seems to be self-evident: irrespective who of two is performing random walks – a policeman or a toper – their approach (reaction) arises due to their *relative motion*. Another principal consequence of the Smoluchowski equation is related to the *space dimension* $\bar{d}$, and the question is: what is the difference between reaction kinetics in the bulk ($\bar{d} = 3$), on the surface ($\bar{d} = 2$) or in thin capillaries ($\bar{d} = 1$)? Since we consider here diffusion-controlled kinetics, the question is addressed mainly to the peculiarity of diffusion in different dimensions.

Many years ago Pólya [20] formulated the key problem of random walks on lattices: does a particle always return to the starting point after long enough time? If not, how its probability to leave for infinity depends on a particular kind of lattice? His answer was: a particle returns for sure, if it walks in one or two dimensions; non-zero *survival probability* arises only for the *three*-dimensional case. Similar result is coming from the Smoluchowski theory: particle A will be definitely trapped by B, irrespectively on their mutual distance, if A walks on lattices with $\bar{d} = 1$ or $\bar{d} = 2$ but it survives for $\bar{d} = 3$ (that is, in three dimensions there exist some regions which are never visited by Brownian particles). This illustrates importance in chemical kinetics of a new parameter $\bar{d}$ which role will be discussed below in detail.

1.3.7 Survival probability

> A. – "You became very old". B. – "It happens, when one lives long".
> (A. – "Sie sind sehr alt geworden". B. – "Ja, das ist gewönlich der Fall, wenn man lange lebt")
>
> Lichtenberg

Now let us try to extrapolate a solution of the Smoluchowski equation for the case of more than two particles. The simplest complication is well seen from the following example: consider a particle B participating in the A + B → B reaction and surrounded by many particles A, say, randomly distributed with the density n_A. Since particles A are point-like and are assumed not to interact, this many-particle problem could be reduced to a pair problem of a fate (survival) of some particular particle A*, irrespectively of other A particles. That is, we can consider survival probability of A placed at the beginning at some distance l from a trap; the result obtained should be averaged over the initial distribution of particles. In other words, if one knows the "law" of catching a single toper by policeman, a general law for a group of topers could also be easily found.

Fig. 1.13. A drunken man surrounded by policemen.

1.3.8 Diffusion asymmetry

> When I am standing, I have one opinion, but when I am lying I have another one.
> (Ich habe oft die Meinung, wenn ich liege, und eine andere, wenn ich stehe)
>
> Lichtenberg

At first sight, the problem just discussed seems to be symmetric, i.e., should be true also for a problem of the catching a single toper by a *group* of policemen – Fig. 1.13; that is all we need is to solve a pair problem. However, it is not the case except that the toper is immobile, $D_A = 0$ (the toper is dead-drunk). In the latter particular case the problem indeed could be analytically solved: at long time the probability that the toper did not meet the policemen is proportional to $\exp(-kt^\alpha)$ with the exponent α equal to the "classical" value of unity for $\bar{d} \geqslant 2$ but $\alpha = 1/2$ if $\bar{d} = 1$.

In the opposite case of $D_B = 0$ the problem turns out to be *principally* many-particle and cannot be reduced to a pair situation, as before. This problem was solved for the first time by Balagurov and Vaks quite recently, in 1973 [14]. The probability to avoid policemen again turns out to be exponential, $\exp(-kt^\alpha)$, but now with the value of the α "non-classical" for *any* dimension: $\alpha = \bar{d}/(\bar{d}+2)$!

It should be stressed that the probability for randomly walking toper not to be caught is *greater* than for immobile one (surely for a fixed $D = D_A + D_B$). The impression is created that a walking toper evades policemen and somehow watches closely after them! This example demonstrates perfectly an unexpected *asymmetry in the diffusion coefficients* (*mobilities*) which is entirely ignored in standard chemical kinetics. A reasonable question arises: how many particles are necessary indeed to take into account for what we call many-particle effect? For the one-dimensional case and particular $A+B \to B$ reaction the problem could be solved exactly and the answer is very simple: it is enough to investigate *three-particle* evolution in time – configurations of the BAB-type (particle A placed between two B's). Complete solution of a problem could be obtained then considering an ensemble of such three-particle configurations which differ by the distance between two B particles and position of A between them. For higher-dimensions the problem, strictly speaking, is not longer reduced to the three-particle one.

1.3.9 Ensembles of particles

> The history of a century is made of the histories of single years. In order to describe the spirit of a century, one cannot plug together the histories of hundred single years.
> (Die Geschichte eines Jahrhunderts ist aus den Geschichten der einzelnen Jahre zusammengesetzt. Den Geist eines Jahrhunderts zu schildern, kann man nicht die Geister der hundert einzelnen Jahre zusammenflicken)
>
> Lichtenberg

The next reasonable step in studying our "chemical games" is to consider ensembles of A's and B's (e.g., topers and policemen), when they are randomly and homogeneously distributed in the reaction volume and are characterized by macroscopic densities of a number of particles. The peculiarity of the $A + B \to B$ reaction is that the solution of a problem with a single A could be extrapolated for an *ensemble* of A's (in other words, a problem is linear in particles A). As it was said above, it is analytically solvable for $D_A = 0$ but turns out to be essentially many-particle for $D_B = 0$. It is useful to analyze a form of the solution obtained for the particle concentration $n_A(t)$ in terms of the basic postulates of standard chemical kinetics (i.e., the mean-field theory).

For immobile particles A the density of traps n_B enters into solution in such a way that the kinetic law of mass action holds formally (the concentration decay is proportional to the product of two concentrations) but replaces the constant reaction rate (the coefficient of this product) for the time-dependent function. Therefore, an exactly solvable problem of the bimolecular $A + B \to B$ reaction gives us an idea of the generalization of the mean-field theory – a use of the equation $\dot{\boldsymbol{n}} = \boldsymbol{F}(\boldsymbol{n}, t)$ instead of $\dot{\boldsymbol{n}} = \boldsymbol{F}(\boldsymbol{n})$ where dependence of the function $\boldsymbol{F}$ on the particle densities obeys the law of mass action but with time-dependent coefficients of the polynomial expansion. Such a perspective of a formal introduction of some function of time does not sound inspiring if it is not found from the independent microscopic analysis. The more so, since some particular results obtained for the same A + B → B reaction put under question the realization of this idea even in principle. As it is known, for immobile B particles [14] and long times the decay kinetics contains the specific combination $n_B^2 t$, which has nothing to do with the law of mass action! If we even assume that it results from the dependence of the expansion coefficients on the particle density, it would mean that our universal approach to the kinetics description fails.

Up to now we neglected dynamical interaction of particles. In a pair problem it requires the use of the potential $U(r) = U_{AB}(r)$ specifying the A–B interaction; in an ensemble of different particles interaction of *similar* particles described by additional potentials $U_{AA}(r), U_{BB}(r)$, could be essential. However, incorporation of such dynamic interactions makes a problem unsolvable analytically for *any* diffusion coefficients, analogously to the situation known in statistical physics of condensed matter.

1.3.10 Model of the A + B → 0 reaction

> Such people should be forced to wear buttons with a figure "zero" to highlight them.
> (Solche Leute sollte man Knöpfe mit dem Buchstaben Null tragen lassen, damit man sie kennte)
>
> Lichtenberg

In the examples considered above traps B were assumed to serve as sinks of the *infinite* capacity. Its simple modification for the case of saturable traps (the policeman who caught some toper is no longer on his post but is bringing him to the police station) leads us immediately to the *new class of reactions* known as A + B → C (C is a neutral reaction product – the policeman bound to a toper is out of his duty) or A+B → 0 (reaction product leaves the system). This latter reaction is typical for the so-called Frenkel defects in solids when complementary defects A and B (interstitial atoms and vacancies) annihilate each other, thus giving no products and restoring the perfect crystalline lattice.

Another example gives us the reactions on the catalyst surface, considered in the last Chapters of this book, where products depart to the gas phase. In all cases symbol '0' used for products reflects new situation and immediately makes a problem unsolvable for any particle mobilities.

1.3.11 The Smoluchowski theory of the diffusion-controlled reactions

> We are familiar with the art to make a new book from several old ones.
> (Wir verstehen die Kunst, aus ein paar alten Büchern ein neues zu machen)
> Lichtenberg

Let us return now to the attractive idea of using the solution of a pair problem for the evidently non-solvable exactly kinetics in many-particle systems. Such an approach was realized for the first time by Smoluchowski in his theory of diffussion-controlled reactions. He has considered an ensemble of particles A with density n_A surrounding a *single* sink – particle B at the origin. Irrespective of the particle mobilities, the kinetics is governed by the coefficient of relative diffusion $D = D_A + D_B$. Due to the presence of a trap, the spatial distribution of A's could be characterized by their time-dependent *local concentration* $C(r,t)$; $C(r,0) = n_A$ and r is a distance from a trap – Fig. 1.14. The time evolution of concentration is assumed to obey the same Smoluchowski equation as in Fig. 1.15 but with the modified boundary condition – $C(\infty, t) = n_A = n_A^0$. It means that at long distances concentration remains constant, n_A^0, i.e., the reservoir of A's is unexhaustable which, strictly speaking, is not adequate to the problem under study.

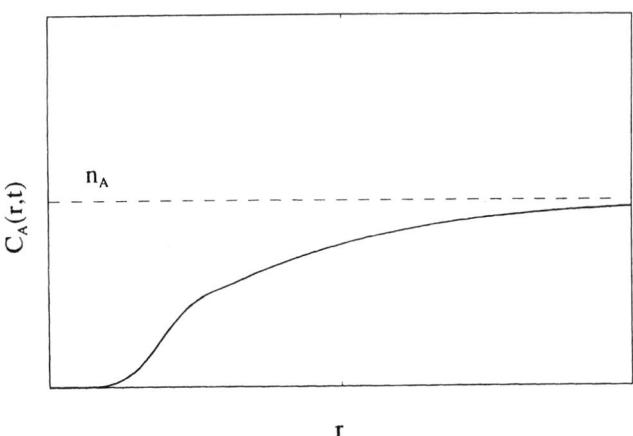

Fig. 1.14. Concentration front in the Smoluchowski theory.

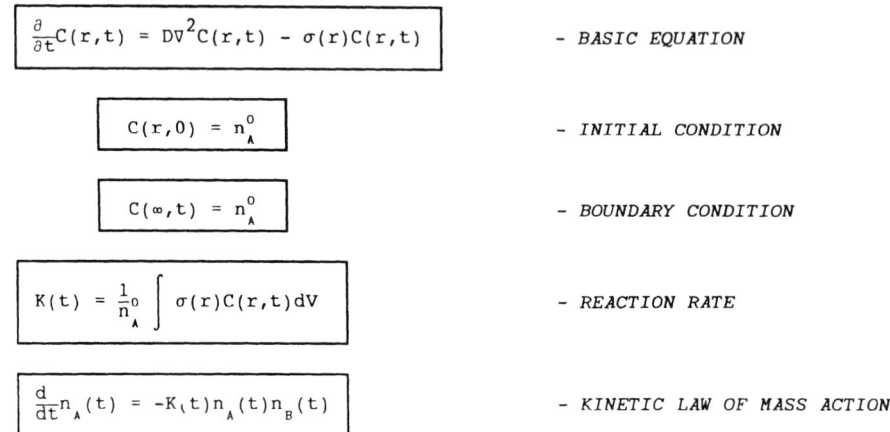

Fig. 1.15. The Smoluchowski approach to the bimolecular kinetics.

Solution of this problem could be easily obtained. Of the main interest is the expression for the flux of particles A, $j_A(t)$, falling to a sink B. This flux is proportional to the particle density n_A^0, $j(t) = K(t)n_A^0$, where the time-dependent co-factor $K(t)$ is the so-called *effective reaction rate*. It characterizes the trapping rate by a single B per concentration unit of A's. Note that at long reaction time, $t \to \infty$, $K(t)$ strives for the steady-state, $K_0 = K(\infty)$, only in the *three*-dimensional case. Thus, for the above-discussed "policeman–toper" game the steady-state value of the reaction rate, found by Smoluchowski, is $K_0 = 4\pi D r_0$ where r_0 is the trapping (reaction) radius. This relation for K_0 is extremely widely used in literature, even in the cases when it is for sure inadequate; say, in low dimensions ($\bar{d} = 1, 2$)! Here $K(t)$ is non-stationary and always decays in time *never* reaching the steady-state value. According to Smoluchowski, the dependence $K = K(t)$ is defined uniquelly by the reaction probability within a pair, $\sigma(r)$, and the coefficient of the relative diffusion.

This value of $K(t)$ is silently transferred in the Smoluchowski approach to *any* bimolecular reactions, including three basic processes discussed above: A+B → B, A+B → 0 and A+A → B (for the first time it was applied to the latter reaction describing the colloid formation in liquids). Such an approach retains the law of mass action and defines therein the time-dependent reaction rate as a solution of the diffusion equation. In other words, this method is nothing but an attempt to generalize the basic equation of the chemical kinetics in the form $\dot{n} = F(n,t)$, but it fails if one is interested in non-elementary bimolecular reactions, e.g., in chains of several reactions. This situation reminds Bohr's theory of atom which gave a good description of hydrogen atom but still could not be generalized for many-electron atoms which has stimulated development of the completely new universal approach called *quantum mechanics*. It is clear that in the Smoluchowski approach both mentioned asymmetry in reactant mobilities and dynamic interaction of similar particles cannot be treated just in principle. However, in many cases this method, despite its intuitive derivation, gives physically quite reasonable results. It is interesting that this approach was used without any theoretical justification for a half of a century until studies by Waite [21] and Leibfried [22].

They have demonstrated that the Smoluchowski approach could indeed be derived rigorously as a cut-off of the set of rigorous kinetic equations discussed in Chapter 4. However, the accuracy of this approximation and the range of applicability of the Smoluchowski approach remained unclear until recently [16].

1.4 INTERMEDIATE ORDER

1.4.1 Classical ideal gas and the Poisson distribution

> A strange noise, as if a whole regiment sneezed simultaneously.
> (Ein sonderbares Geräusch, als wenn ein ganzes Regiment auf einmal niesete)
>
> Lichtenberg

Let us try to understand deeper the nature of the order parameter. As usually, we start with a gas as a simplest one-component system. An important role in theoretical physics belongs to a model of *classical ideal gas* in which molecules (particles) obey the laws of Newtonian mechanics and do not interact with each other.

Consider physically small volume v. Due to discreteness of the matter distribution in space a number of particles N_v in a given volume is a random variable; $N_v = 0, 1, 2 \ldots$. However, on the average each volume contains $\langle N_v \rangle = nv$ particles. Define now microscopic, *local* density of the particle number within the small volume v centered at $\vec{r}$ as $\widehat{n}(\vec{r}) = N_v/v$. It describes natural inhomogeneity in the particle distribution arising due to their thermal motion (the so-called *density fluctuations*). The macroscopic particle density is just its mean value, $n = \langle \widehat{n}(\vec{r}) \rangle$.

It is useful to find a quantity that could serve us as a measure of these density fluctuations. Its simplest characteristic is the *dispersion* of a number of particles N in some volume V; i.e., $\langle N^2 \rangle - \langle N \rangle^2$. The distinctive feature of the classical ideal gas is a simple relation between the dispersion and macroscopic density: $\langle N^2 \rangle - \langle N \rangle^2 = \langle N \rangle = nV$. Moreover, all other fluctuation characteristics of the ideal gas, related to the quantity $\langle N^m \rangle$, could also be expressed through $\langle N \rangle$ or density n. Therefore, in the model of ideal gas the density n is the *only parameter* characterizing the fluctuation spectrum. Such the particle distribution is called the *Poisson distribution*. It could be easily generalized for the many-component system, e.g., a mixture of two ideal gases. Each component is characterized here by its density, n_A and n_B; density fluctuations of different components are statistically independent, $\langle N_A N_B \rangle = \langle N_A \rangle \langle N_B \rangle$.

1.4.2 Chemical kinetics and non-Poisson fluctuations

> The hypotheses of some innovators do not contradict our experience but I suspect that one day our experience will contradict them.
> (Die Hypothesen einiger Neuern laufen noch nicht gegen die Erfahrung, aber ich fürchte, die Erfahrungen werden einmal gegen sie laufen)
>
> Lichtenberg

The spatial distribution of a gas mixture with such properties can serve us as the simplest microscopic model of the standard kinetics; however, real chemical systems can differ considerably from the properties of ideal gases. It could be easily illustrated by the same bimolecular $A + B \to B$ reaction (e.g., in terms of the policeman–toper game). If the trapping radius is r_0, the fluctuations in a number of particles, N_A and N_B, in small volumes separated inside sphere of such a radius r_0 are no longer statistically independent; if particle B is placed in the origin, the probability of finding dissimilar particle A not far from it will be much less than for a *free* A particle. In other words, such closely spaced dissimilar particles are mutually *correlated*, which is characterized by the inequality $\langle N_A N_B \rangle \neq \langle N_A \rangle \langle N_B \rangle$. Considering as a simplest model system another bimolecular reaction, $A + A \to 0$ (one-component system) which is characterized similarly by the trapping radius r_0, we come to the same conclusion that similar particles A are also spatially correlated; if some particle is in the origin, no other particles can stationary exist in its vicinity, within the sphere of the radius r_0. It contradicts obviously the ideal gas properties. Therefore, distribution of similar particles in the chemical system does not obey the Poisson distribution, $\langle N^2 \rangle - \langle N \rangle^2 \neq \langle N \rangle$. This situation reminds more the *real* gas model, where particles (molecules) are considered as rigid spheres; a large amount of such particles cannot be placed into small volume purely due to geometrical restrictions.

1.4.3 The intermediate order parameter

> Order is a daughter of thoughts.
> (Ordnung ist eine Tochter der Überlegung)
> Lichtenberg

Before analyzing in detail particle distribution in real systems, let us introduce more informative and flexible system's characteristics rather than dispersion, $\langle N^2 \rangle - \langle N \rangle^2$, which is only *an integral property* of the fluctuation in a number of particles in a given small volume. Another, differential and more transparent fluctuation characteristic could be defined in the following way. Consider again one-component system, put some particle at the origin and count a number of particles in a small volume v at the distance r from it. Their numbers $N_v = 0, 1, 2, \ldots$ are random variables, but presence of a particle at the origin (even due to its finite size) puts some restrictions on particle distribution within volume v (the so-called *effect of the excluded volume*). That is, defining now the microscopic density $\widehat{n}(r) = N_v/v$, we no longer have the equality $\langle \widehat{n}(r) \rangle = n$, but some function of r, $\langle \widehat{n}(r) \rangle = C(r)$.

When drawing the dependence $C = C(r)$, we see how at long r the influence of the central particle is diminishing, i.e., asymptotically $C(\infty) = n$. The function $C(r)$ characterizes somehow *ordering* in the system. At long relative distances it is naturally called *long-range order*, described by the order parameter $n = C(\infty)$. At shorter distances $C(r) \neq n$ characterizes the *intermediate* order. Such an approach could be easily applied not only to continuous, but also to the discrete (lattice) systems, e.g., magnetic dipoles centered on the lattice sites. For these systems the vector $\vec{r}$ entering the function $C(\vec{r})$ has a discrete spectrum of values; its minimum value corresponds to the nearest neighbours where $C(r)$ characterizes the *short-order parameter* [8].

1.4.4 Transformation and propagation of the ordering

> I knew people who were drinking in privacy but were drunk in public.
> (Ich habe Leute gekannt, die haben heimlich getrunken und sind öffentlich besoffen gewesen)
>
> Lichtenberg

In the case of classical ideal gas $C(r) = n$, i.e., the intermediate order coincides with the long-range one! Before discussing real systems of interacting particles, remind ourselves that an ideal gas does not necessarily consists of classical particles only, the ideal gas model could be extended also for *quantum* noninteracting particles. In quantum statistics particles differ by the symmetry properties of their *wave functions*; they are either asymmetric with respect to the permutation of particle coordinates (*fermions* obeying the Fermi–Dirac statistics) or symmetric (bosons and Bose–Einstein statistics) respectively. The first case takes place for particles with non-integer spins whereas the second – for those with integer spins.

Symmetry or asymmetry of wave functions introduce also certain ordering into the system of ideal particles. An interesting question is how it is reflected in their spatial distribution (Fig. 1.16). This problem could be solved

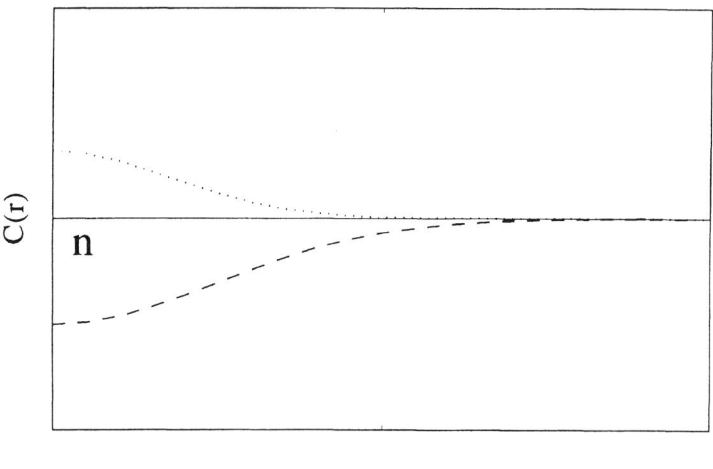

Fig. 1.16. The intermediate order in ideal gas: (i) ideal classical gas (full line); (ii) ideal Fermi gas (broken line); (iii) ideal Bose gas (dotted line).

exactly and the intermediate ordering indeed is observed at short relative distances. For fermions $C(r) < n$, i.e., the presence of some particle at the origin decreases probability of finding other particles nearby due to their effective "repulsion" (a kind of the excluded volume effect). In contrast, for bosons $C(r) > n$, i.e., particles reveal a kind of "attraction". These examples demonstrate different kinds of ordering in the system of quantum particles and show how abstract order in wave-function symmetry transforms into the transparent *spatial* ordering of particles.

Consider now the system of interacting particles. In low-density systems (diluted gas) a pairs approximation works well and the Boltzmann distribution is valid giving $C(r) = n \exp(-U(r)/(k_B T))$. If molecules are treated as rigid spheres with the diameter R, we have for the pair potentials $U(r) = \infty$, $C(r) = 0$, as $r \leqslant R$ and $U(r) \equiv 0$, $C(r) = n$, if $r > R$. Such kind of ordering could be called *elementary*, it arises due to excluded volume effect caused by finite sizes of particles. As we increase density of particles in the system, the pair approach becomes gradually incorrect. The intermediate order parameter $C(r)$ changes and the transformation of the elementary order into intermediate one occurs; the latter is characterized by the distinctive scale larger than R, including appearance of maxima and minima in the $C(r)$ curve as shown in Fig. 1.17. This example shows how the ordering is propagated in space.

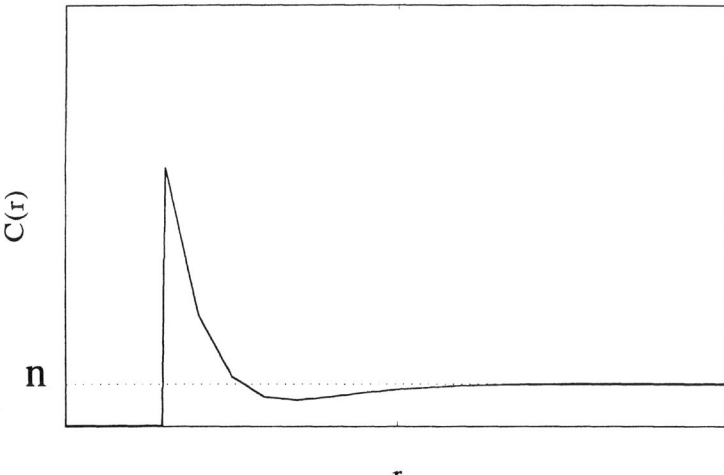

Fig. 1.17. The intermediate order in a dense system of rigid balls.

The use of more realistic potentials, including both repulsive and attractive contributions, results in the more smooth intermediate order curves $C(r)$. As previously, they reveal oscillations thus reflecting the specific law of mutual particle distribution in dense systems.

1.4.5 Spatial particle correlations and correlation functions

> To turn into a bull is not yet suicide.
> (Sich in einen Ochsen verwandeln ist noch kein Selbstmord)
>
> Lichtenberg

The definition of the intermediate order parameter through the density $C(r)$ is not quite convenient, especially for many-component systems. In the latter case one has to specify the central-particle type, as well as the nature of analyzed particles. A reasonable question raises – what is the difference between mean density of particles A at the distance r from the central particle B and vice versa. It shows the necessity to introduce more symmetric definition of the intermediate order where two spatial points under analysis separated by the distance r were treated equivalently. Such an approach treats the intermediate order in the molecular system as correlation of its parts. Consider the two points $\vec{r}_1$ and $\vec{r}_2$ (with the relative separation r) which are surrounded by small non-overlapping volumes and calculate the mean value of the product of microscopic densities $\langle \hat{n}(\vec{r}_1)\hat{n}(\vec{r}_2) \rangle$. Since the mean value of the microscopic density is the macroscopic density, let us separate the dimensional co-factor through the relation $\langle \hat{n}(\vec{r}_1)\hat{n}(\vec{r}_2) \rangle = n^2 \chi(r)$, where information on the long-range order is given by n^2 whereas a function $\chi(r)$ is the *joint correlation function* depending on the relative distance r only and characterizing the intermediate order. It could be easily shown that for one-component system new and old definitions of the intermediate order are very simply related: $C(r) = n\chi(r)$; the long-range order corresponds to the asymptotics of the correlation function $\chi(\infty) = 1$. Many-component system is characterized in quite a similar way by a set of the correlation functions: $\chi_{AA}(r), \chi_{BB}(r)$ and $\chi_{AB}(r) = \chi_{BA}(r)$. It is generally assumed that mean-fied theories operating only with the long-range order parameters neglect the fluctuations of the order parameter. However, an approach using some fluctuation characteristics (in particular, the joint correlation functions) takes also into account the order parameter fluctuations.

1.5 CRITICAL PHENOMENA

1.5.1 Spatial correlations and the order parameter

> He used to give blessings on Sundays and often punishments already on Mondays.
> (Er teilte des Sonntags Segen und oft schon des Montags Prügel aus)
>
> Lichtenberg

Let us consider now behaviour of the gas–liquid system near the critical point. It reveals rather interesting effect called the *critical opalescence*, that is strong increase of the light scattering. Its analogs are known also in other physical systems in the vicinity of phase transitions. In the beginning of our century Einstein and Smoluchowski expressed an idea, that the opalescence phenomenon is related to the density (order parameter) fluctuations in the system. More consistent theory was presented later by Ornstein and Zernike [23], who for the first time introduced a concept of the *intermediate order as the spatial correlation in the density fluctuations*. Later Zernike [24] has applied this idea to the lattice systems.

As it is known [5], the intensity of the scattered light gives us an information about the system's disorder, e.g., presence therein of pores, impurities etc. Since macroscopically liquid is homogeneous, critical opalescence arises due to local *microscopic* inhomogeneities – an appearance of small domains with different local densities. In other words, liquid is ordered inside these domains but still disorded on the whole since domains are randomly distributed in size and space, they appear and disappear by chance. Fluctuations of the order parameter have large amplitude and involve a wide spectrum of the wavelengths (which results in the milk colour of the scattered light).

Of the greatest interest in this critical scattering phenomenon are studies at the long wavelengths; according to the Ornstein–Zernike theory [23], a linear size of ordered domains increases *infinitely*, when one approaches to the critical point. This prediction is drawn from the analysis of the asymptotic behaviour of the joint correlation function $\chi(r)$. Despite the fact that its asymptotic value always equals to unity, $\chi(\infty) = 1$, the intensity of the scattered light is defined by an integral containing the factor $\chi(r) - 1$, i.e., the effect is governed by an *approach of $\chi(r)$ to the asymptotic value*. It has been shown that near the critical point this approach obeys the algebraic law.

It should be reminded, that in gases under low pressure the correlation function is defined practically by the Boltzmann factor $\chi(r) = \exp(-U(r)/(k_B T))$. At high densities the potential $U(r)$ should be replaced with the effective potential $W(r)$, characterizing the cooperative way of the interaction propagation. The algebraic law for the correlation function just mentioned means that the same is true for the effective potential too. Potentials of this kind are characterized in mechanics by the *infinite action radius*. Note that nevertheless the original two-particle potential $U(r)$ could be short-range one, i.e., change of the asymptotic behaviour of the *effective* potential turns out to be typical many-particle effect. When leaving the vicinity of the critical point, the effective potential $W(r)$ could be characterized by a finite action radius – the so-called *correlation radius of fluctuations*, $\xi = \xi(T)$. It should be stressed once more that the *correlation length* ξ defines the distinctive spatial scale of the correlation in the local density fluctuations. In other words, it gives a typical size of the domain inside which local density is nearly constant. Surely, such domains differ in their sizes, including those domains which are compared to the correlation length, but large domains are more stable and important.

In particular, in magnetics the correlation length characterizes a size of the domain where all dipoles are oriented in the same direction; it results in the *local* magnetisation (which is compensated by other domains with different dipole orientation). The divergence of the correlation length allows us to give a new definition of the critical point: $\xi(T_c) = \infty$. Despite the fact that the Ornstein–Zernike theory takes into account fluctuations in the order parameter, in some respects it shows distinctive features of the mean-field approach: an idea of averaging the fluctuating characteristics is just applied here to the higher level of approximation, corresponding to the joint correlation functions. As a result, this theory is capable to describe qualitatively the effect of critical opalescence, but fails to analyze correctly the vicinity of the critical point. Analogously the long-range order parameter, the correlation length reveals the algebraic law as a function of the temperature increment but with an incorrect exponent.

New aspects of the problem arise when studying the *temporal* behaviour near the critical points. The general statement is as follows: the relaxation time with respect of any external perturbation of the system becomes anomalously long. It is clear that such anomalies result from long-range fluctuations of the order parameter thus reflecting existence of the ordered domains with the linear size ξ. Due to their large sizes, these domains slowly decay and are created by thermal motion of particles; the long wavelength exitations are the most stable.

The existence of the correlation length ξ gives a proof to the *hypothesis of the scale invariance* [8, 9]; in the vicinity of the critical point physical effects are governed namely by *large-scale (but not small-scale) fluctuations*. According to this hypothesis, the critical phenomena are characterized by the correlation length ξ only; all other length scales, including that of the microscopic interaction of the structure elements in the system, become unimportant. The divergence of the correlation length at the critical point, $\xi(T_c) = \infty$, leads also to the divergence of all other quantities whose dimension could be expressed through the length dimension.

The concept of scale invariance makes also clear such important feature of the critical phenomena, as dependence of the critical exponents on the space dimension $\bar{d}$ (as well as on certain symmetry properties of the system). Indeed, if due to scale invariance details of particle interactions become unimportant, the leading role should belong to the *fundamental* properties of a system – space dimension $\bar{d}$ and dimension of the order parameter (dependent on a number of components and their relation). As it was mentioned above, the behaviour of the correlation functions $\chi(r)$ is determined by the effective propagation of interactions in the many-particle system. It is easy to imagine that in one-dimensional case due to a topology of the system the propagation of the correlation, as some signal, could be trivially interrupted by any noise event in the chain but it hardly can happen in three dimensions; here a signal is distributed simultaneously in several directions which guarantees its propagation.

1.5.2 Fluctuations of the order parameter in chemical reactions

The rain was so heavy that it made all pigs clean and people – dirty.
(Es regnete so stark, daß alle Schweine rein und alle Menschen dreckig wurden)

Lichtenberg

Consider now the fluctuations of the order parameter in the system possessing the chemical reaction; this problem could be perfectly illustrated by computer simulations on lattices. We start with the bimolecular $A + B \to 0$ reaction discussed above, and first of all froze particle diffusion. Let the recombination event happen *instantly* when a pair AB of dissimilar particles occupies the nearest lattice sites (assume lattice to be squared). Immobile particles enter into reaction as a result of their creation with the equal probabilities in empty lattice sites; from time to time a newly created particle A(B) finds itself nearby pre-created B(A) and they recombine. (Since this recombination event is *instant*, the creation rate is of no importance.) This model describes, in particular, Frenkel defect accumulation in solids under irradiation. We are interested both in the values of steady-state particle concentrations under saturation (known within accuracy of their natural fluctuations) and spatial distribution of particles. Without discussing here the first problem (to be considered in Chapter 7 on a full scale), let us analyze just structure evolution of the system of particles [1].

An instant recombination (disappearance) of a pair AB of the nearest neighbours produces the simplest element of the ordering: such nearest pairs are absent at any time moment. (This statement of the problem has no other length (geometrical) parameters.) We shall clarify now, how this element of ordering is propagated and transformed into the *intermediate* ordering. The two possible cases are illustrated in Fig. 1.18 [1]. In the former case a whole lattice is completely covered by particles A and B and *ordered*, i.e., it is divided into two square domains containing only particles A and B, respectively. In the latter case the lattice is initially empty. Irrespective the initial conditions, at long times results of computer simulations are essentially the same: the lattice is divided into two big domains of the complicated shape (their positions and shape can vary); each domain consists of a single-kind particles. In other words, one observes the *reaction-induced segregation* of dissimilar particles. Sizes of these domains could be characterized by the correlation length $\xi = \xi(t)$; as it is seen from Fig. 1.18, in the steady-state it is close to the linear size of the simulated system.

The presented stationary distributions of particles could be statistically developed; the corresponding correlation functions $\chi(r)$ are shown in Fig. 1.19. It shows that the ordering within AB pairs not only propagated to the longer distances (at $r < \xi$ a number of AB pairs is less than that for a random distribution of particles) but also changed – non-interacting *similar* particles are now *spatially correlated* as if they efficiently attract each other, which has clear analogy with quantum ideal boson gas discussed above.

The existence of the (quasi) steady-state in the model of particle accumulation (particle creation corresponds to the reaction reversibility) makes its analogy with dense gases or liquids quite convincing. However, it is also useful to treat the possibility of the *pattern formation* in the $A + B \to 0$ reaction without particle source. Indeed, the formation of the domain structure here in the diffusion-controlled regime was also clearly demonstrated [17]. Similar patterns of the spatial distributions were observed for the irreversible reactions between *immobile* particles – Fig. 1.20 [25] and Fig. 1.21 [26] when the *long range (tunnelling) recombination* takes place (recombination rate $\sigma(r)$ exponentially depends on the relative distance r and could

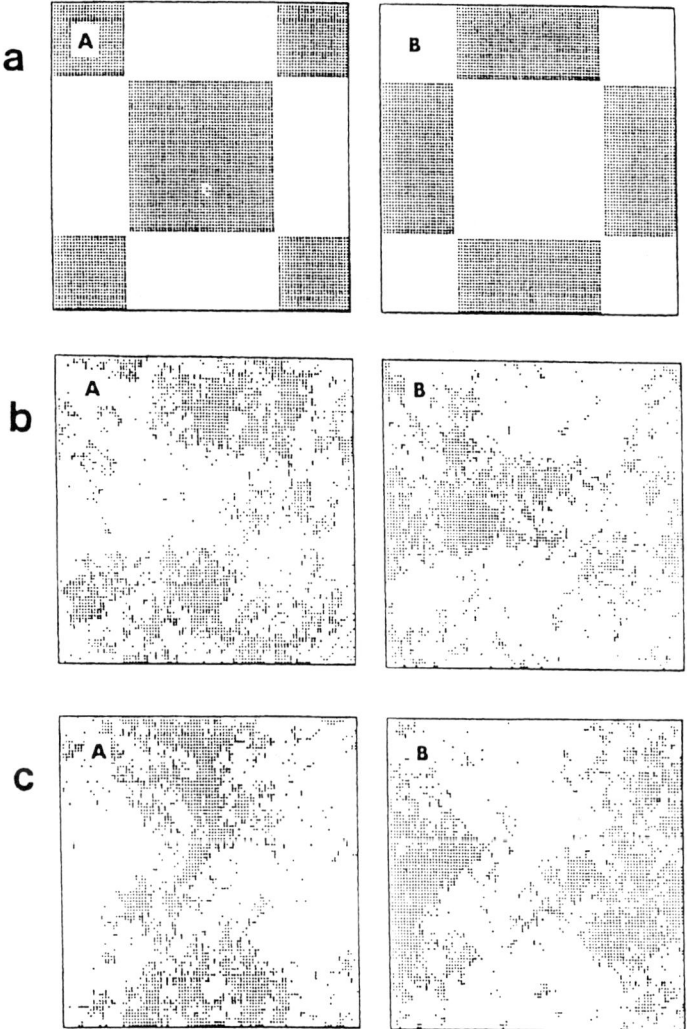

Fig. 1.18. Distribution of A and B particles on the surface in the annihilation reaction $A + B \to 0$. For clarity, the distributions of A's and B's have been separated and are shown in the left-hand column and in the right-hand column of the figure, respectively. The results shown correspond to constant and equal fluxes of A and B. The simulation were carried out on a 100×100 square lattice. (a) The A and B distribution are complementary. A narrow lane of empty sites separates between them. (b) The long-time (near steady-state) structure of the overlayer developing from the initial condition in (a). (c) The long-time overlayer pattern developing from an initially empty lattice.

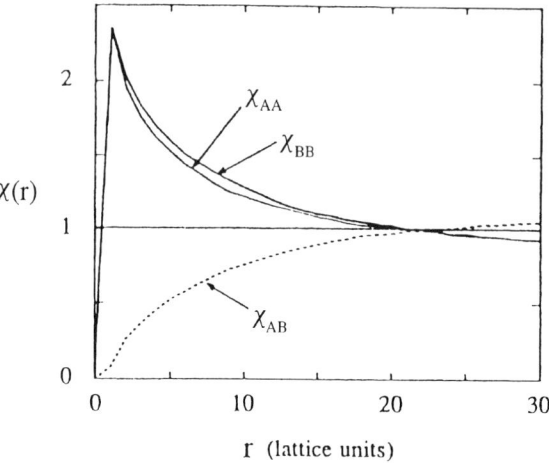

Fig. 1.19. The radial pair correlation function of the steady-state overlayer generated by the $A + B \to 0$ annihilation reaction, with no particle diffusion. Averaged over five simulations.

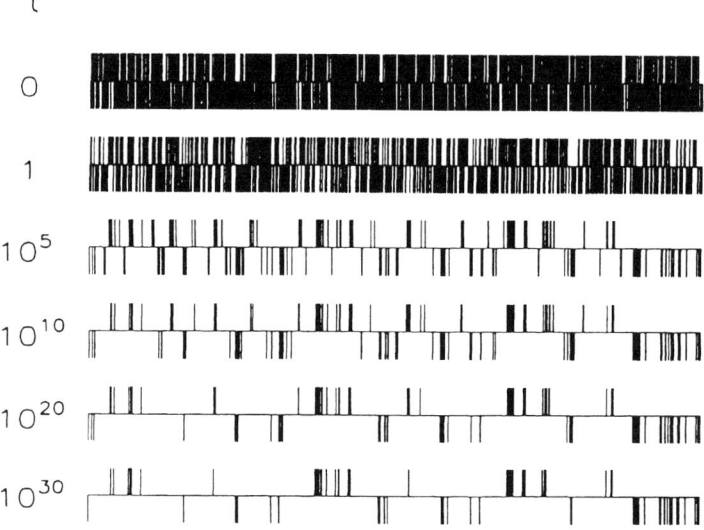

Fig. 1.20. Distribution of A and B particles (initially 1000 of each kind on chain with $L = 10\,000$ sites) during a numerical simulation. Each vertical line represents one A (up) or B (down) particle.

Critical phenomena

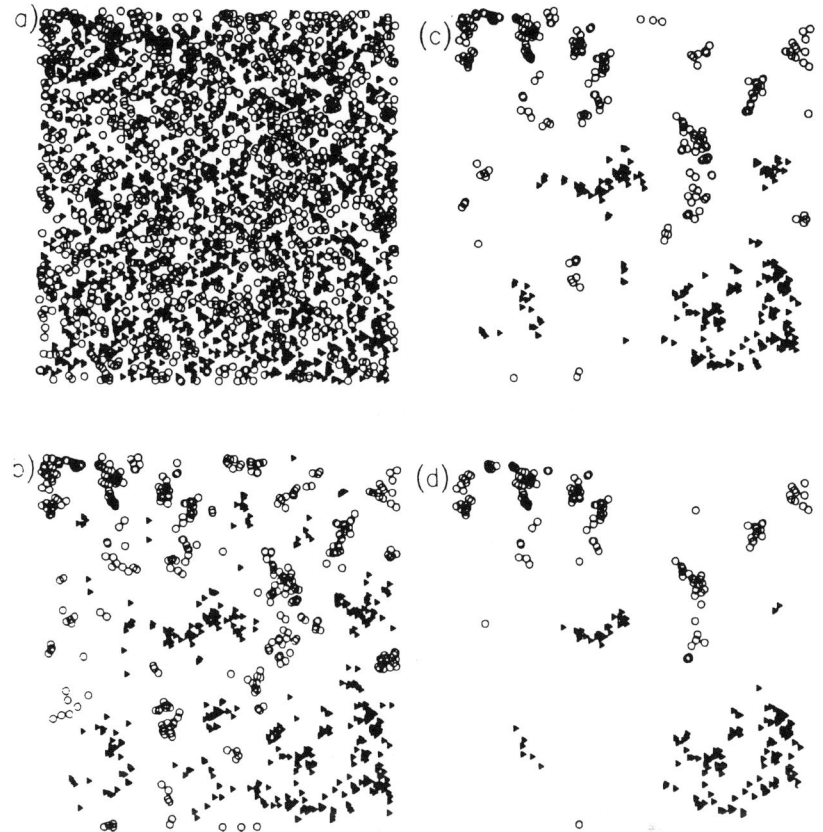

Fig. 1.21. Spatial distribution of A and B particles during a reaction process with $r_0 = 5$ on a 1000×1000 square lattice (with periodic boundary conditions) after starting with 10^4 particles of each kind. The distributions are shown for (a) $t = 10$, (b) $t = 10^5$, (c) $t = 10^{10}$, and (d) $t = 10^{15}$.

be characterized by some action radius r_0 – see Chapter 3 for more details). In this case we observe only certain quantitative difference with Fig. 1.18 – in the irreversible reaction the steady-state means *absence* of any particles, i.e., their density decreases monotonously and domains are rather loose. It should be stressed, however, that small densities of particles do not prevent their segregation; therefore the correlation in density fluctuations in a system of interacting particles is governed by mechanism which differs from that known for the equilibrium gas–liquid system. In the irreversible reaction its non-equilibrium nature is a source of the ordering, which has been demonstrated more than once in the *self-organization* phenomena [4]. The analysis of the time development of the decay of particle density has indicated clearly that their critical exponents are dependent on the space dimension $\bar{d}$; therefore, chemical reactions are similar to the stationary systems near the critical points.

Undoubtedly, it means that the chemical kinetics being a kind of the meanfield theory, should be improved through an introduction of the fluctuations in the order parameter.

1.5.3 Dynamical interaction between particles

> Father: – My daughter, do you known that Solomon says "If bad boys attract you, don't follow them." Daughter: – But what should I do, if good boys attract me?
> (Der Vater: – mein Töchterchen, du weiß, Solomon sagt: "Wenn dich bösen Buben locken, so folge ihnen nicht." Die Tochter: – Aber Papa, was muß ich dann tun, wenn mich die guten Buben locken?)
>
> Lichtenberg

New aspects of the problem arise when we consider *dynamical interaction* of particles (along with their chemical interaction, i.e., the reaction). This dynamical interaction can both stimulate bringing together dissimilar reactants (e.g., for oppositely charged particles with the Coulomb attraction) as well as to prevent their approach (for similarly charged particles). Note that namely dynamical interaction of particles introduces directly the *temperature* into the kinetics; before it affected kinetics only indirectly through diffusion coefficients. Of special interest here are correlations of *similar* particles – the effect neglected in the Smoluchowski approach for the *two*-component systems. It should be reminded that this diffusion-controlled theory was developed originally for the study of particle coagulation in one-component system. Examples of such structures are plotted in Fig. 1.22 and Fig. 1.23 [27], where the chemical reaction does not occur. In the lattice statement of a problem the dynamical interaction of particles is described in terms of the transition probabilities between nearest sites; they are defined through energy of pair interactions of particles with nearest neighbours.

Figure 1.24 shows a more complicated process of the $A + B \to 0$ reaction simulation, being divided into several intermediate stages. At the first stage, similarly to Figs 1.22 and 1.23, particles A are randomly created with certain density n_A and then, in a course of their diffusion and attraction, produce

Critical phenomena

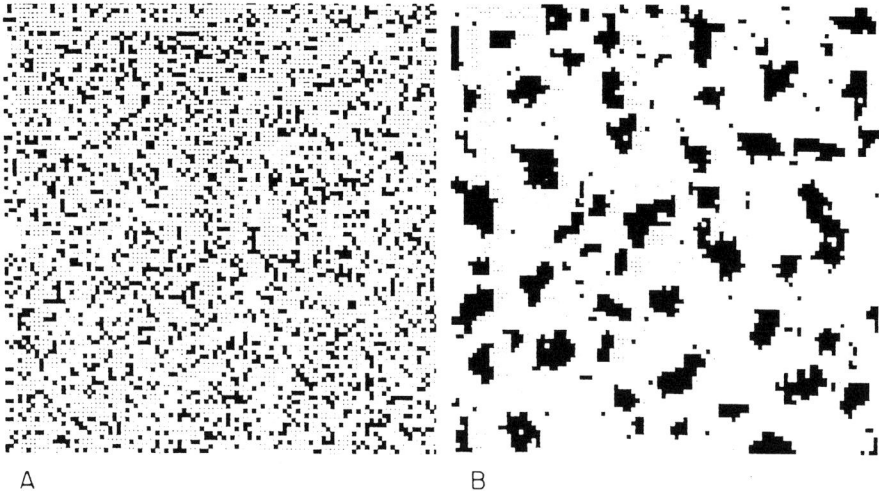

Fig. 1.22. Typical distribution of A atoms on the lattice at: (A) $t = 0$ (random distribution), (B) $t = 5000$ MCS of A diffusion; A atoms are aggregated in long lived islands. In both cases $n_A = 0.2$. Shown are 100×100 Sections of 500×500 square lattice used in the calculations.

Fig. 1.23. The same as Fig. 1.22 but with $n_A = 0.4$.

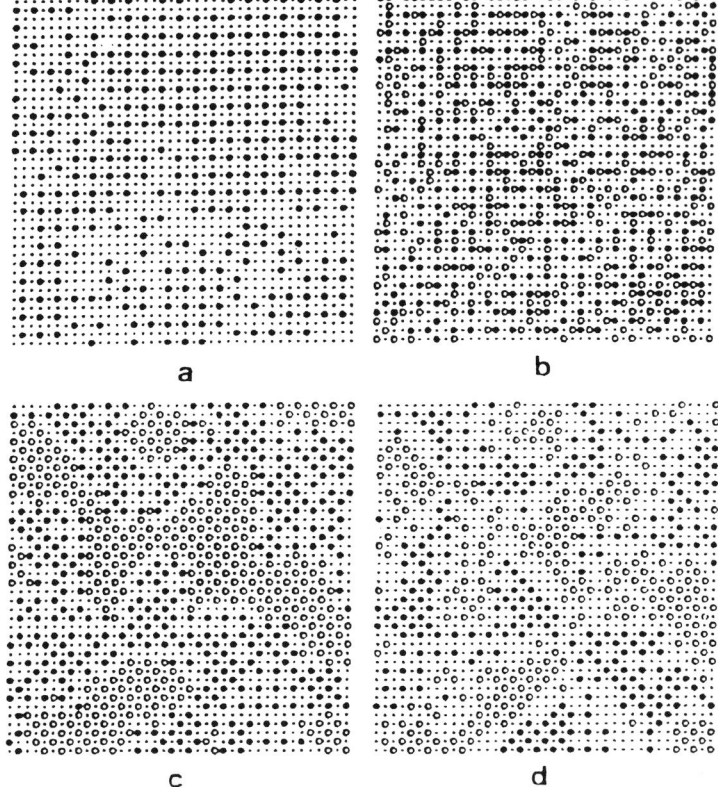

Fig. 1.24. Several stages of the aggregation and reaction process for the $A + B \to 0$ system. The coverages of the A and B atoms are $n_A = n_B = 0.2$. Shown are 40×40 Sections of the 100×100 lattice used in the calculation. (a) The configuration after A adsorbtion and 10 000 MCS of diffusion. (b) Shortly after adsorbtion of B into available sites. (c) Following 10 000 MCS of mutual diffusion and segregation. (d) The configuration while the reaction is in progress (after a temperature ramp). "The contact regime" has almost ended (boundary atoms have reacted) and further reaction will be diffusion-initiated.

some typical pattern with A-rich domains. At the second stage, in their turn, particles B are randomly created with the same density $n_B = n_A$. (Reaction between particles is forbidden.) Introduction into the system of particles B produces new kinds of particle interactions (within BB and AB pairs). The third stage (diffusion of B's without reaction) leads to the pattern (domain) formation of particles A and B. Finally, when reaction of nearest particles A

1.5.4 Correlation analysis and fluctuation-controlled kinetics

> One man could be tried in Britain for bigamy but was saved by his advocate proving that his client had three wives.
> (In England wird ein Mann der Bigamie wegen angeklagt und von seinem Advokaten dadurch gerettet, daß er bewies, sein Klient habe drei Weiber)
>
> Lichtenberg

How could we take into account the fluctuations of the order parameter? Let us return to the well-studied example of the gas–liquid system. A general equation of the state of gases and liquids proved in statistical physics [9] has a form $p = nk_{\mathrm{B}}T - n^2 G(\chi)$ where $G(\chi)$ is some integral containing the interaction potential of particles and the joint correlation function $\chi(r)$. Therefore, the equation for *the long-range order parameter n contains in itself the functional of the intermediate-order parameter $\chi(r)$*.

The "only" problem necessary for developing the condensation theory is to add to the above-mentioned equation of the state the equation defining the function $\chi(r)$. Unfortunately, it turns out that the exact equation for the joint correlation function, derived by means of basic equations of statistical physics, contains *three*-particle correlation function $\chi^{(3)}$, which relates the correlations of the density fluctuations in three points of the reaction volume. The equation for this three-particle correlations contains four-particle correlation functions and so on, and so on [9]. This situation is quite understandable, since the use of the joint correlation functions only for description of the fluctuation spectrum of a system is obviously not complete. At the same time, it is quite natural to take into account the density fluctuations in some approximate way, e.g., treating correlation functions in a spirit of the mean-field theory (i.e., assuming, in particular, that three-particle correlations could be expanded in two-particle ones).

On this way we arrive at Born–Green–Ivon, Percus–Yevick and hyperchain equations [5, 9], all having a general form $\Phi(\chi, \nabla\chi, n, T) = 0$. These non-linear integro-differential equations are close with respect to the joint correlation function, and Percus–Yevick equation gives the best approximation amongst known at present. An important point is that the *accuracy* of the method used for deriving the equation for $\chi(r)$ cannot be estimated in the framework of the same method, but only using the so-called *computer experiments* (*simulations*). Two basic approaches for studying properties of large clusters of particles are called the *molecular dynamics* method and *Monte-Carlo* method.

Due to high accuracy of numerical methods, statistical physics of gas–liquid systems is in a good shape, but theory in general still has unsolved problems. In particular, attempts to describe phase transition to the condensed phase face serious problems demonstrating inadequate theoretical treatment used at present for the spectrum density fluctuations. Indeed, the joint correlation function being nothing but the radial distribution function, is unable to predict arise of *a new type of order* corresponding to the crystalline lattice. Its geometric characteristics (lattice parameter and crystalline symmetry) are very likely hidden in the neglected correlation functions of the *higher order*. Using concept of the correlation functions, statistical physics permits systematic improvement of results, which however is restricted in practice by computational facilities available.

However, a question arises – could similar approach be applied to chemical reactions? At the first stage the general principles of the system's description in terms of the fundamental kinetic equation should be formulated, which incorporates not only macroscopic variables – particle densities, but also their fluctuational characteristics – the correlation functions. A simplified treatment of the fluctuation spectrum, done at the second stage and restricted to the joint correlation functions, leads to the *closed set* of non-linear integro-differential equations for the order parameter n and the set of joint functions $\chi(r,t)$. To a full extent such an approach has been realized for the first time by the authors of this book starting from [28]. Following an analogy with the gas–liquid systems, we would like to stress that treatment of chemical reactions *do not copy* that for the condensed state in statistics. The basic equations of these two theories differ considerably in their form and particular techniques used for simplified treatment of the fluctuation spectrum as a rule could not be transferred from one theory to another.

It is convenient to divide a set of fluctuation-controlled kinetic equations into two basic components: equations for time development of the order parameter n (*concentration dynamics*) and the complementary set of the partial differential equations for the joint correlation functions $\chi(r,t)$ (*correlation dynamics*). Many-particle effects under study arise due to *interplay of these two kinds of dynamics*. It is important to note that equations for the concentration dynamics coincide formally with those known in the standard kinetics

GAS-LIQUID SYSTEM	CHEMICAL SYSTEM
EQUATION OF STATE	DYNAMICS OF CONCENTRATIONS
$F(n,\chi,T) = 0$	$\frac{d}{dt}n = F(n,\chi)$
JOINT CORRELATION	DYNAMICS OF CORRELATIONS
$\Phi(\chi,\nabla\chi,n,T) = 0$	$\frac{\partial}{\partial t}\chi = D\nabla^2\chi + \Phi(\chi,\nabla\chi,n)$

Fig. 1.25. Correlation analysis in the gas–liquid system and in chemical kinetics.

(the kinetic law of mass action). The "only" difference is that in the latter theory the coefficient before the concentration product (the reaction rate) is assumed to be a constant value whereas in a new theory it is a functional of the intermediate order parameters $\chi(r,t)$ which values change in time due to evolution of the spatial distribution of particles. Fig. 1.25 illustrates an analogy in the description of gas–liquid and chemical systems.

1.5.5 The Waite–Leibfried theory

> To attach a beard to the holy father – do you call this to reform?
> (Dem Papst einen Bart machen, heißt das reformieren?)
>
> Lichtenberg

The said allows us to understand the importance of the kinetic approach developed for the first time by Waite and Leibfried [21, 22]. In essence, as is seen from Fig. 1.15 and Fig. 1.26, their approach to the simplest $A + B \to 0$ reaction does not differ from the Smoluchowski one! However, coincidence of the two mathematical formalisms in this particular case does not mean that theories are basically identical. Indeed, the Waite–Leibfried equations are derived as some approximation of the *exact* kinetic equations; due to a simplified treatment of the fluctuational spectrum a complete set of the joint correlation functions $\chi(r,t)$ for *all* kinds of particles is replaced by the only function $\chi_{AB}(r,t)$ describing the correlation of chemically reacting dissimilar particles. Second, the equation defining the correlation function $\chi = \chi_{AB}(r,t)$ is linearized in the function $\chi(r,t)$. This is analogous to the

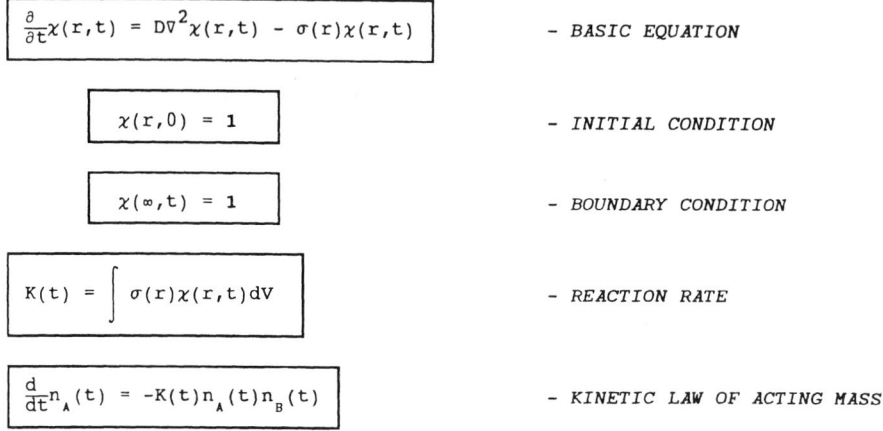

Fig. 1.26. An idea of the Waite–Leibfried approach.

Ornstein–Zernike theory of the critical opalescence which operates also with a linear equation for the joint correlation function.

A new *principal element* of the Waite–Leibfried theory compared to the Smoluchowski approach is the relation between the effective reaction rate $K(t)$ and the intermediate order parameter $\chi = \chi_{AB}(r,t)$. In its turn, the Smoluchowski approach is just an heuristic attempt to describe the simplest irreversible bimolecular reactions $A + B \to B$, $A + B \to B$ and $A + B \to 0$ and cannot be extended for more complicated reactions. The Waite–Leibfried approach is not limited by these simple reactions only; it could be applied to the reversible reactions and reaction chains. However, in the latter case the particular linearity in the joint correlation function $\chi = \chi_{AB}(r,t)$ does not always mean linearity of equations since additional non-linearity caused by particle densities can arise.

For a long time in chemical kinetics the preference was given exclusively to analytical methods of problems solution; these methods were used mainly for searching the steady-states of linear equations describing the correlation functions. However, nowadays when we know very well that the many-particle effects – the main problem studied in this book – are closely related to the non-linearity of the kinetic equations, much more attention is paid to the *numerical* solution of these equations and thus the linearization procedure used by Waite and Leibfried is now more of historical rather than practical interest.

1.5.6 Kinetic phase transitions

> To convert a fool into a genius or beech firewood into oak is probably the same trouble as to convert lead into gold.
> (Dummköpfe in Genies zu verwandeln oder Büchenholz in Eichen ist wohl so schwer als Blei in Gold)
>
> Lichtenberg

It is useful to discuss the incorporation of the *fluctuational degrees of freedom* into the kinetics of chemical reactions from the point of view of the self-organization theory. As it is known, the standard chemical kinetics could be also applied to the *spatially nonuniform systems* where the mean value of the local density $\langle \hat{n}(\vec{r}) \rangle = C(\vec{r}, t)$ does not coincide with the macroscopic density $n(t)$. For the diffusion-controlled processes the necessary generalization of the kinetic equations is adding of the relevant diffusion terms, provided the structure of reaction terms remains the same. That is, the polynomial expansion of the non-linear terms in equations is retained; individual terms of the expansion obeys the kinetic law of mass action. The typical form of the kinetic equations is given in Fig. 1.27 (the terms ∇C can arise if the external field is applied). Compare these equations with those for the *correlation dynamics* in the fluctuation-controlled kinetics. Restricting ourselves exclusively to the bimolecular processes, the pattern formation here is described by a set of nonlinear partial differential equations of the second order. In a chemical system the complex and interesting for us spatio-temporal particle structures can arise only for a large number of degrees of freedom (or number of components of the order parameter n). For the same systems equations describing the fluctuation-controlled processes have much greater degree of non-linearity which argues for their ability for describing the spatio-temporal structures *under much weaker restrictions* imposed on the order parameter n (e.g., number of its components). Note that the joint correlation functions are called in physics of gas–liquid systems the *radial distribution functions* since they depend on a modulus of the vector $\vec{r}$ only. This is why the auto-wave processes, which are expected to be described in

SELF-ORGANIZATION IN SPATIALLY INHOMOGENEOUS SYSTEMS	THE SAME IN HOMOGENEOUS SYSTEMS
STANDARD KINETICS	FLUCTUATION-CONTROLLED KINETICS
$\frac{\partial}{\partial t} c = D\nabla^2 c + \Phi(c, \nabla c)$	$\begin{cases} \frac{d}{dt} n = F(n, \chi) \\ \frac{\partial}{\partial t} \chi = D\nabla^2 \chi + \Phi(\chi, \nabla\chi, n) \end{cases}$

Fig. 1.27. Self-organization in spatially homogeneous and inhomogeneous media.

terms of a set equations for the correlation dynamics, will be quite specific radial waves of correlations.

Let us consider for the illustration the following process. Sites of a square lattice are occupied (completely or partially) by particles B. These particles decay, B → 0, with the lifetime q leaving their sites empty. In their turn, particles A are created with the constant rate p in empty sites, 0 → A. If by a chance any A turns out to be the nearest neighbour of any B, it is transformed instantly into another particle B, i.e., *the catalytic process* of particle transformation, A + B → B + B takes place. Using the time scale corresponding to $q = 1 - p$, we find that the process is governed by a single parameter p. It is also clear that in the presented statement of the problem contact of any A with some B particle can result in the transformation of not a single A, but a whole *cluster* consisting of several nearest A particles. In other words, the *wave of chemical transformation* arises (Fig. 1.28) and the problem is reduced to the well-known *site percolation* [5]. From the point of view of standard chemical kinetics this model is nothing but extension of the so-called *Lotka model* (the step reaction 0 → A, A + B → 2B, B → 0) for the discrete lattice system and has a simple solution. It is shown that small number of components of the order parameter and weak non-linearity lead (at long times) to a stable steady-state solution despite the fact that the *approach* to this steady-state can reveal character of the damped oscillations (under certain parameters of a problem).

Let us try now to predict qualitatively possible behaviour types of this chemical system as a function of the parameter p. For very large values of $p \sim 1$ most of lattice sites are covered by particles B and empty sites for A's arise only due to decay of some B's. Newly created A's are densely surrounded by particles B and are transformed immediately into them, A → B.

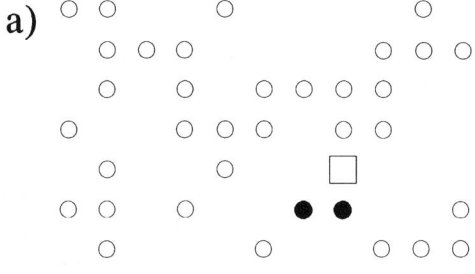

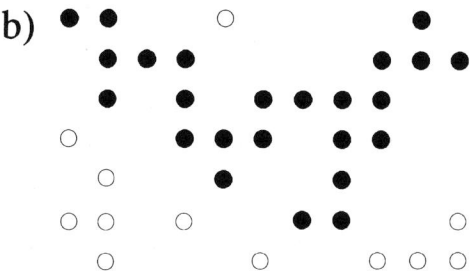

Fig. 1.28. Wave of chemical transformations taking place when newly created particle A falls into site marked by square in (a). Empty circles are A particles, whereas full circles are B particles.

Therefore, we can expect existence of the steady-state with a high density of B's but very small density of A's.

Consider now *the opposite case* of $p \ll 1$. Since the creation rate of new particles A is very small, it does not compensate the decay of B's whose density decreases down to a very small value. (However, the existence itself of B particles, even in small concentrations, is of vital importance for the future waves of A particle transformations.) If particles B were absent, the concentration of A's after time t would be $n_A \approx pt$. It is also clear that even at small densities n_A large cluster of these particles could be formed as a result of the statistical fluctuation. Touch of this cluster by some particle B will create a local wave of the A → B transformations which is *propagated* for a finite distance and thus retains finite number of particles B in the system. It is important to note that for small densities n_A a number of such clusters is also small.

One of the well-established principal results of the percolation theory [5] says that there exist some *marginal value of density* n_c (dependent on the lattice type; in particular, $n_c = 0.59$ for the square lattice) such that as $n_A \geqslant n_c$, the probability that a given atom A belongs to the *infinite* cluster is non-zero. It means that a touch of any B to this A-rich cluster creates a transformation wave propagating throughout a whole crystal. Since the critical density n_c is rather high, the catalytic transformation inside some infinite cluster will affect with time not only particles of a given cluster but also other particles A entering other finite-size clusters. As a result, density of particles B will increase abruptly to the high value whereas that of particles A will decrease. When the system consists mainly of particles B, their native decay will be predominant process reducing B concentration. At the next stage the next oscillation in particle concentrations, as described above, takes place etc.

What is described above is an idea of the so-called *chemical clock*, that is a reaction with periodic (oscillating) change of reactant concentrations; its period could be estimated as $\delta t > n_c/p$. In the condensed matter theory a leap in densities is interpreted as phase transitions of the first order. From this point of view, the oscillations correspond to a sequence in time of phase transitions where the two phases (i.e., big clusters of A's containing inside rare and small clusters of B's and vice versa) differ greatly in their structures.

The transition from a stable steady-state solution observed at large p to the oscillatory regime assumes the existence of the *critical value* of the parameter p_c, which defines the point of *the kinetic phase transition*: as $p > p_c$, the fluctuations of the order parameter are suppressed and the standard chemical kinetics (the mean-field theory) could be safely used. However, if $p < p_c$, these fluctuations are very large and begin to dominate the process. Strictly speaking, the region $p \sim p_c$ at $p > p_c$ is also fluctuation-controlled one since here the fluctuations of the order parameter are abnormally high.

What was said above is illustrated by Fig. 1.29 and Fig. 1.30 corresponding to the cases $p > p_c$ and $p < p_c$ respectively. To make the presented kinetic curves smooth, in these calculations the transformation rate A $\to$ B was taken to be *finite*. To make results physically more transparent, the effective reaction rate $K(t)$ of the A $\to$ B transformation is also drawn. The standard chemical kinetics would be valid, if the value of $K(t)$ tends to some constant. However, as it is shown in Fig. 1.30, $K(t)$ reveals its own and quite complicated time development; namely its oscillations cause the fluctuations in particle densities. The problems of kinetic phase transitions are discussed in detail in the last Chapter of the book.

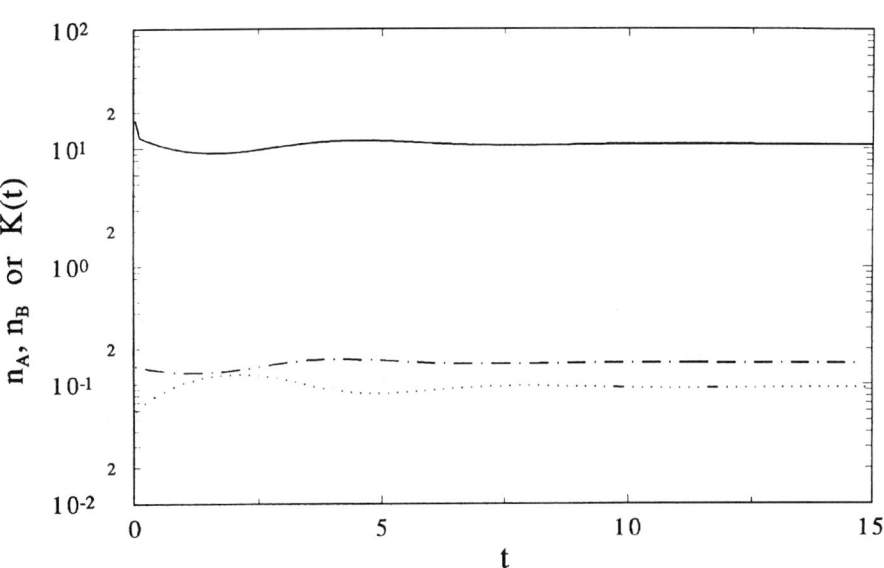

Fig. 1.29. The steady state in the Lotka model. Control parameter $p = 0.2$.

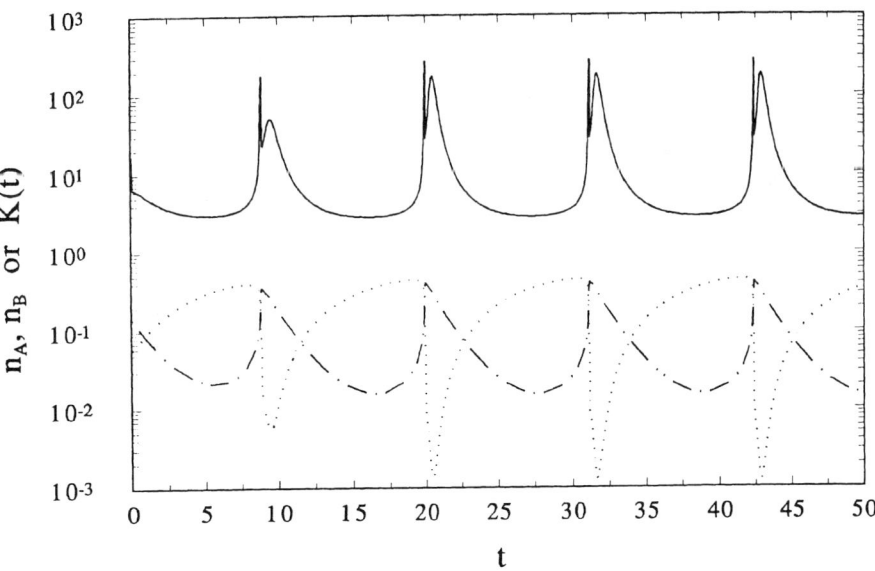

Fig. 1.30. Oscillations observed in the Lotka model. Control parameter $p = 0.1$.

1.5.7 Conclusion

> If you have a couple of trousers, sell one and buy this book.
> (Wer zwei Paar Hosen hat, mache eins zu Geld und schaffe sich dieses Buch an)
>
> Lichtenberg

The scope of this book is as follows. Chapter 2 gives a general review of different theoretical techniques and methods used for description the chemical reactions in condensed media. We focus attention on *three principally different levels of the theory*: macroscopic, mesoscopic and microscopic; the corresponding ways of the transition from deterministic description of the many-particle system to the stochastic one which is necessary for the treatment of density fluctuations are analyzed. In particular, Section 2.3 presents the method of *many-point densities* of a number of particles which serves us as the basic formalism for the study numerous fluctuation-controlled processes analyzed in this book.

Chapter 3 deals with *the pair problem* in terms of the Smoluchowski equation.

The consistent derivation and analysis of the *Waite–Leibfried equations* is presented in Chapter 4. We show that this theory is the *linear* approximation of the exact many-particle formalism. Its relation to the Smoluchowski theory is also established.

Chapter 5 deals with derivation of the basic equations of the *fluctuation-controlled kinetics*, applied mainly to the particular bimolecular $A + B \to 0$ reaction. The transition to the simplified treatment of the density fluctuation spectrum is achieved by means of the *Kirkwood superposition approximation*. Its accuracy is estimated by means of a comparison of analytical results for some "test" problems of the chemical kinetics with the relevant computer simulations. Their good agreement permits us to establish in the next Chapters the range of the applicability of the traditional Waite–Leibfried approach.

The theory of *irreversible diffusion-controlled reactions* is discussed in Chapter 6; the effects of particle Coulomb and elastic interactions are analyzed in detail. The many-particle effects (which in principle cannot be explained in terms of the linear theory) are demonstrated. Special attention is paid to the *pattern formation* and similar particle aggregation in systems of interacting and noninteracting particles.

In its turn, Chapter 7 deals with *reversible reactions* between both immobile and mobile particles in *the systems with particle sources*. This theory is of great importance for describing the process of the radiation (Frenkel) defect accumulation in solids. The self-organization phenomena and reaction-induced *dissimilar particle segregation* are analyzed in detail.

Lastly, *non-elementary several-stage reactions* are considered in Chapters 8 and 9. We start with the Lotka and Lotka–Volterra reactions as simple model systems. An existence of the undamped density oscillations is established here. The complementary reactions treated in Chapter 9 are *catalytic surface* oxidation of CO and NH_3 formation. These reactions also reveal undamped concentration oscillations and kinetic phase transitions. Their adequate treatment need a generalization of the fluctuation-controlled theory for the *discrete (lattice) systems* in order to take correctly into account the geometry of both lattice and absorbed molecules. As another illustration of the formalism developed by the authors, the kinetics of reactions upon *disorded* surfaces is considered.

We would like to conclude this introductory Chapter by the following general comment. Most of the papers dealing with the fluctuation-controlled reactions, focus their attention on the simplest bimolecular $A + B \to B$ and $A + B \to 0$ reactions. To our mind, main results in this field are already obtained and the situation is quite clear. In the nearest future the most prospective direction of kinetic theory seems to be *many-stage catalytic processes;* the first results are discussed in Chapters 8 and 9. Their study (stimulated also by the technological importance) should be continued using in parallel both refined mathematical formalisms of the fluctuation-controlled kinetics and full-scale computer simulations.

The authors are greatly indebted to Georg Christoph Lichtenberg (1742–1799) for his epigraphs used in this Chapter; we share a hope expressed in the last of his quoted statements.

References

[1] O.M. Becker, M. Silverberg and A. Ben-Shaul, Israel J. Chem. 30 (1990) 179.
[2] V.N. Kuzovkov and E.A. Kotomin, Rep. Prog. Phys. 51 (1988) 1479.
[3] M. Smoluchowski, Zs. Phys. Chem. Abt. B: 92 (1917) 129.
[4] G. Nicolis and I. Prigogine, Self-Organization in Non-Equilibrium Systems (Wiley, New York, London, 1977).
[5] J.M. Ziman, Models of Disorder (Cambridge Univ. Press, London, 1979).
[6] G. Careri, Ordine e Disordine Nella Materia (Laterza, Roma, 1982).
[7] H. Haken, Synergetics (Springer, Berlin, 1978).
[8] H.E. Stanley, Introduction to Phase Transition and Critical Phenomena (Oxford Univ. Press, New York, 1971).

[9] R. Balescu, Equilibrium and Non-Equilibrium Statistical Mechanics (Wiley, New York, 1975).
[10] E.M. Lifshitz and L.P. Pitaevskii, Statistical Physics, Vol. 2: Condensed Matter Theory (Nauka, Moscow, 1987).
[11] S.W. Benson, The Foundations of Chemical Kinetics (Wiley, New York, 1960).
[12] A.M. Zhabotinsky, Concentration Auto-Oscillations (Nauka, Moscow, 1974).
[13] I. Prigogine, From Being to Becoming: Time and Complexity in the Physical Sciences (Freeman, San Francisco, 1980).
[14] B. Balagurov and V. Vax, Sov. Phys. JETP 38 (1974) 968.
[15] A.A. Ovchinnikov and Ya.B. Zeldovich, Chem. Phys. 28 (1978) 215.
[16] E.A. Kotomin and V.N. Kuzovkov, Rep. Prog. Phys. 55 (1992) 2079.
[17] D. Toussaint and F. Wilczek, J. Chem. Phys. 78 (1983) 2642.
[18] R.D. Mattuck, A Guide to Feynmam Diagrams in the Many-Body Problem (McGraw-Hill, London, 1967).
[19] C.W. Gardiner, Handbook of Stochastic Methods for Physics, Chemistry and the Natural Sciences (Springer, Berlin, 1983).
[20] G. Pólya, Math. Ann. 84 (1921) 149.
[21] T.R. Waite, Phys. Rev. 107 (1957) 463.
[22] G. Leibfried, in: Bestrahlungseffekte in Festkörpern (Teubner, Stuttgart, 1965).
[23] L.S. Ornstein and F. Zernike, Proc. Acad. Sci. (Amsterdam) 17 (1914) 793.
[24] F. Zernike, Physica 7 (1940) 565.
[25] H. Schnörer, V. Kuzovkov and A. Blumen, Phys. Rev. Lett. 63 (1989) 805.
[26] H. Schnörer, V. Kuzovkov and A. Blumen, J. Chem. Phys. 92 (1990) 2310.
[27] M. Silverberg and A. Ben-Shaul, J. Chem. Phys. 87 (1987) 3178.
[28] V.N. Kuzovkov and E.A. Kotomin, J. Phys. C: 13 (1980) L499.

Chapter 2

Basic Methods for Describing Chemical Kinetics in Condensed Media: Continuum Models

2.1 MACROSCOPIC APPROACH

> The cognition process consists in the replacement of rough mistakes by more refined mistakes.
>
> Wittgenstein

2.1.1 Standard chemical kinetics: systems with complete reactant mixing

2.1.1.1 Basic equations of formal kinetics

There exist three qualitatively different levels for describing the kinetics of particle (reactant) migration and interaction in solids and liquids: *microscopic*, *mesoscopic* and *macroscopic* [1]. In the first, the microscopic level, the individual particles are considered as basic structural elements. The state of a whole system at any moment t is defined by all particle coordinates and this state changes due to particle migration or dynamic interaction (via Coulomb, elastic (for defects in solids) or van der Waals forces), as well as due to the birth and death (recombination) processes. To describe the spatio-temporal structure of a system consisting of interacting particles, we naturally can use well-developed formalism of statistical physics. For the sake of simplicity, in this Chapter we restrict ourselves to a *continuous* treatment of diffusion; the effects of finite hop lengths are discussed in Section 3.1. In this respect the homogeneous system consisting of interacting particles is similar to the condensed media without long-range order (dense gases, liquids or glasses). The only principal *peculiarity* is that the number of particles is varied due to the birth–death processes.

The simplest way widely used for describing the recombination process is called *formal chemical kinetics* [2–6]. Here a system of particles is considered as structureless, the only parameters used are *macroscopic concentrations* $n_i(t)$. The relevant kinetics is described by a set of ordinary differential equations

$$\frac{dn_i(t)}{dt} = F_i(n_1, \ldots, n_s), \quad i = 1, \ldots, s, \tag{2.1.1}$$

where s is a number of reactant kinds, whereas the actual expression for F_i, which in general are nonlinear functions of macroscopic concentrations, is defined by actual process under study and comes from the *law of mass action* [7, 8]. The most realistic models of birth–death and migration process are based on the *binary* (two-particle) approximation. The equations of formal kinetics in the case of mono- and bimolecular reactions read

$$\frac{dn_i(t)}{dt} = a_i^j n_j(t) + b_i^{jk} n_j(t) n_k(t), \tag{2.1.2}$$

where a_i^j are the reaction rates for the monomolecular contribution, whereas b_i^{jk} those for the bimolecular part. In equation (2.1.2) summation runs over repeated indices.

The more complicated phenomenological approach also uses *tri*molecular and higher-order reaction stages, which strictly speaking is nothing but a shortened form to describe a set of mono- and bimolecular reactions with a large number of intermediate products [7]. Unlike the linear terms in equation (2.1.2) whose meaning is self-evident (the elementary decay processes), the bilinear terms need more attention. In fact they arise due to a qualitative pattern, the *law of mass action*, [7] in which particles undergoing thermally activated hops collide from time to time with each other inelastically which leads to their transformation into other particles (reaction products). It is clear that a number of both elastic and inelastic collisions of particles of the jth and kth kind are proportional to the product of their concentrations n_j and n_k. These reaction rates b_i^{jk} are nothing but phenomenological parameters which obviously cannot be obtained in the framework of the structureless pattern considered here.

2.1.1.2 Simplest bimolecular reactions in condensed media

The mathematical technique of formal chemical kinetics is very useful for qualitative estimates and general analysis of processes in condensed matter. The treatment of a problem begins usually with the analysis of the reaction

Macroscopic approach

scheme and the corresponding set of kinetics equations (2.1.2). In the simplest case $b_i^{jk} = 0$ and equations (2.1.2) describe *monomolecular reaction*, A → B (e.g., radiative decay of excited molecules),

$$\frac{dn_A(t)}{dt} = -kn_A(t), \qquad (2.1.3)$$

whose solution is *exponential*

$$n_A(t) = n_A(0)e^{-kt} = n_A(0)e^{-t/\tau_0}. \qquad (2.1.4)$$

It contains the phenomenological parameter k called *reaction rate*. (The particle *lifetime* $\tau_0 = 1/k$.)

An example of monomolecular reaction is the radiative decay of triplet self-trapped excitons in alkali halide crystals which are produced by UV-irradiation [9]. On the other hand, high-energy irradiation of solids of arbitrary nature – both metals and insulators – results in creation of complementary radiation defects, are called also *Frenkel* defects – vacancies and interstitial atoms. At sufficiently high temperatures, the latter start to migrate through short-range hopping between the nearest interstitial positions and recombine vacancies approaching to within certain critical radius r_0 called *annihilation* radius – restoring perfect crystalline lattice, A + B → 0.

Such simplest *bimolecular* reaction, A + B → C, obeys the equation

$$\frac{dn_A(t)}{dt} = \frac{dn_B(t)}{dt} = -Kn_A(t)n_B(t), \qquad (2.1.5)$$

with the solution, setting $\delta n = n_B(0) - n_A(0)$,

$$\frac{1 + \delta n/n_A(t)}{1 + \delta n/n_A(0)} = e^{\delta n K t}. \qquad (2.1.6)$$

From equation (2.1.5) we obtain in the case of equal concentrations, $n_A(t) = n_B(t) = n(t)$

$$n(t) = \frac{n(0)}{1 + n(0)Kt}, \qquad (2.1.7)$$

which gives the *algebraic time* dependence at long time $t \gg 1/(n(0)K)$

$$n(t) \propto t^{-1}. \qquad (2.1.8)$$

If one of the reactants is in excess, $n_B(0) \gg n_A(0)$, we get from equation (2.1.6) at long time exponential decay law

$$n_A(t) = n_A(0)e^{-n_B(0)Kt}. \tag{2.1.9}$$

This case is called degenerate bimolecular or *pseudo-first order* reaction.

Another example of simple bimolecular reaction is mobile *exciton annihilation* which is well studied in molecular crystals; $A + A \to 0$ (zero means that usually we are not interested what is happening with reaction products) [10]. In this case the kinetic equation is obvious

$$\frac{dn_A(t)}{dt} = -Kn_A^2(t), \tag{2.1.10}$$

with the solution very similar to that of equation (2.1.7)

$$n_A(t) = \frac{n_A(0)}{1 + n_A(0)Kt}. \tag{2.1.11}$$

A general phenomenon is observed in many solids that due to interaction of the electronically excited particles (donors) D* with acceptors A the *energy transfer* occurs: $D^* + A \to D + A^*$ [11, 12]. Its probability depends on the actual kind of interaction (dipole–dipole or triplet–triplet transfer). When molecules are mobile, their reaction rate K is defined by their mobility, i.e., energy transfer becomes diffusion-controlled [13].

The latter reaction is widely used as a proving ground for new theories. Particles B (called also *scavengers*) are unsaturable energy sinks and thus after rapid light emission can absorb it anew. In this case the reaction scheme is $A + B \to B$ (concentration of B's remains constant n_B) which could be described by the following kinetic equation

$$\frac{dn_A(t)}{dt} = -Kn_A(t)n_B, \tag{2.1.12}$$

whose solution is exponential

$$n_A(t) = n_A(0)e^{-n_B Kt}. \tag{2.1.13}$$

The conclusion itself suggests that from the point of view of formal chemical kinetics both energy transfer and the Frenkel defect recombination with unequal concentrations obey the same exponential time decay law, see equations (2.1.9) and (2.1.13).

The basic feature of these kinetic equations is the assumption that reaction rate K is a *constant*, i.e., time-independent parameter due to which all spatial dependencies resulting from real particle motion or interaction are completely ignored. This would be the case if relative spatial distribution of dissimilar reactants remained the same in a course of reaction. We demonstrate below that it is not always the case.

2.1.1.3 Qualitative study of ordinary differential equations

Chemical processes in condensed media often cannot be reduced to simple mono- and bimolecular reactions simply because chains of reaction take place. Therefore their kinetics is described by a set of ordinary differential equations (2.1.1) which are generally nonlinear due to bimolecular stages. Independent variables $n_i(t)$, $i = 1, \ldots, s$ (intermediate reactions products) define a number of equations under study.

Solutions $n_i(t)$ of the set (2.1.1) depend on the initial conditions only because reaction rates are constant.

For a qualitative description of kinetics, of great interest are both *asymptotical* ($t \to \infty$) solutions independent on initial conditions and how particular solutions approach them. Complicated systems can reveal several such asymptotic solutions. Initial conditions define a choice of one of several possible asymptotic solutions. A general scheme for investigating a set of ordinary differential equations was very well described in a number of monographs [4, 7, 14–16], it includes.

(i) Search for the *stationary* points, i.e., solution of a set of equations

$$F_i(n_1, \ldots, n_s) = 0, \quad i = 1, \ldots, s. \tag{2.1.14}$$

Obviously, if as $t \to \infty$ the stationary solution $dn_i(t)/dt = 0$ exists, indeed the asymptotic solution $n_i(\infty)$ of (2.1.1) is one of the solutions n_i^0 of the set (2.1.14). Here we have an example of a simple but very important case of a stable stationary solution. Other stationary points cannot be ascribed to the asymptotic solutions, i.e., $n_i^0 \neq n_i(\infty)$, but they are also important for the qualitative treatment of the set of equations. Note that striving of the solutions for stationary values is not the only way of chemical system behaviour as $t \to \infty$: another example is *concentration oscillations* [4, 7, 16]. Their appearance in a set (2.1.2) depends essentially on a *nature of* stationary points, mentioned above. An asymptotic solution of the oscillating type is connected with the concept of the *limit cycle*. Complicated chemical systems reveal also irregular or chaotic concentration oscillations [8].

(ii) Determination of the *singular point type*. For this purpose new variables – deviations from stationary values are determined by

$$x_i(t) = n_i(t) - n_i^0. \tag{2.1.15}$$

In the vicinity of the stationary point $\{n_i^0\}$ non-linear functions F_i in (2.1.1) could be expanded into the Taylor series

$$F_i \approx \left.\frac{\partial F_i}{\partial n_j}\right|_{n_j^0} x_j + \left.\frac{\partial F_i}{\partial n_j \partial n_k}\right|_{n_m^0} x_j x_k + \cdots . \tag{2.1.16}$$

Since in the vicinity $\{n_i^0\}$ $x_i \ll 1$, we restrict ourselves to the case of linear system: thus the substitution of (2.1.15) and (2.1.16) into (2.1.1) and neglect of non-linear terms yields

$$\frac{\mathrm{d}x_i(t)}{\mathrm{d}t} = f_i^j x_j(t), \tag{2.1.17}$$

where

$$f_i^j = \left.\frac{\partial F_i}{\partial n_j}\right|_{n_j^0}. \tag{2.1.18}$$

When restricting ourselves to mono- and bimolecular stages, (2.1.2) we arrive at

$$f_i^j = a_i^j + b_i^{jk} n_k^0 (1 + \delta_{jk}), \tag{2.1.19}$$

where δ_{jk} is the Kronecker delta. We search the solution in the vicinity of the stationary point in a form $x_i(t) \propto \mathrm{e}^{\varepsilon t}$. In a qualitative theory of differential equations the key problem is to determine the eigenvalues of the coefficient matrix (2.1.18), since substitution of $x_i(t)$ into (2.1.17) yields

$$|f_i^j - \delta_{ij}\varepsilon| = 0. \tag{2.1.20}$$

If the real parts of all eigenvalues ε_i, $\mathrm{Re}\,\varepsilon_i < 0$ are negative, according to the Lyapunov theorem [14, 15] the stationary point n_i^0 is asymptotically stable and $n_i^0 = n_i(\infty)$. If at least one $\operatorname{Re} \varepsilon_i > 0$, the stationary point is unstable, $n_i^0 \neq n_i(\infty)$.

The best studied is the case of a second order system ($s = 2$). Since in this book we consider systems with a maximal number of intermediate products $s \leqslant 2$, let us illustrate what was said with two examples. (For $s > 2$ the qualitative analysis is much less obvious.)

2.1.1.4 The Lotka model

The study of mechanisms of auto-oscillating chemical reactions based on (2.1.1) has a long history. In his pioneering paper published in 1910, Lotka [17] suggested a mathematical model for damped concentration oscillations. Furthermore, it has been shown [18, 19] that a simple modification of this model permits the description of undamped oscillations too. In fact, the Lotka model (now half-forgotten) is a fragment of a wide class of mathematical models [7, 16], used for describing auto-oscillating chemical reactions [7] (for more details see Chapter 8).

The Lotka model is based on the following reaction chain

$$\begin{aligned} \mathrm{E} &\xrightarrow{k_0} \mathrm{A}, \\ \mathrm{A} + \mathrm{B} &\xrightarrow{K} 2\mathrm{B}, \\ \mathrm{B} &\xrightarrow{\beta} \mathrm{P}. \end{aligned} \quad (2.1.21)$$

Here E is the infinite reservoir of matter. It is assumed that in an open system concentration n_E is constant and E is linearly transformed into A followed by an autocatalytic transformation of A into B and its decay. The product P does not affect the reaction rate. The model (2.1.21) is described by a set of equations

$$\frac{dn_\mathrm{A}(t)}{dt} = p - K n_\mathrm{A}(t) n_\mathrm{B}(t), \quad (2.1.22)$$

$$\frac{dn_\mathrm{B}(t)}{dt} = K n_\mathrm{A}(t) n_\mathrm{B}(t) - \beta n_\mathrm{B}(t), \quad (2.1.23)$$

with $p = k_0 n_\mathrm{E}$.

A biological interpretation of the model could be easily formulated in terms of *prey animals* A and *predators* B living on them. Let $n_\mathrm{A}(t)$ be a population density of prey animals who stimulate reproduction of predators

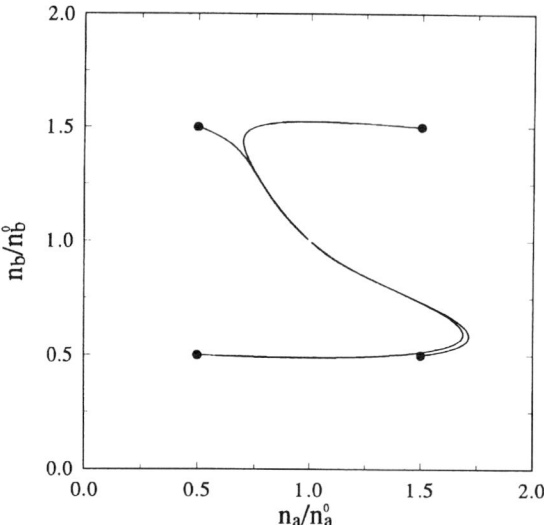

Fig. 2.1. A stable node. Four solutions of the Lotka model, equations (2.1.22)–(2.1.23) with the distinctive parameter $Kp/4\beta^2 = 2$ are presented. The starting point of each trajectory is marked with a black circle.

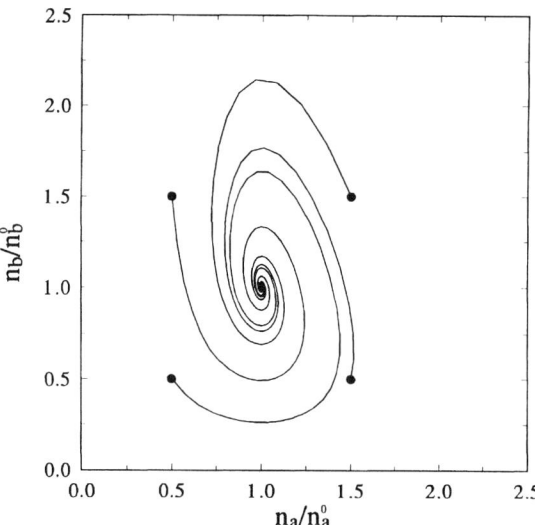

Fig. 2.2. A stable focus. The solution of the same equation as presented in Fig. 2.1, but with the parameter $Kp/\beta^2 = 1/16$.

by dividing a medium with a spontaneous production of food E for them. The decay is B $\xrightarrow{\beta}$ P modelling the natural decay of the predator population.

The use of (2.1.14) gives the Lotka model a single solution for the stationary point

$$n_A^0 = \frac{\beta}{K}, \qquad n_B^0 = \frac{p}{\beta}. \tag{2.1.24}$$

The characteristic equation (2.1.20) reads

$$\varepsilon^2 + \frac{pK}{\beta}\varepsilon + pK = 0, \tag{2.1.25}$$

and its roots are

$$\varepsilon_{1,2} = -\frac{pK}{2\beta} \pm \left[\left(\frac{pK}{2\beta}\right)^2 - pK\right]^{1/2}. \tag{2.1.26}$$

If $pK > 4\beta^2$ holds, the roots are real and $\varepsilon_{1,2} < 0$. That is, the stationary point (2.1.14) is asymptotically stable.

The way in which the solution $n_i(t)$ approaches its stationary value $n_i^0 = n_i(\infty)$ for a system with two degrees of freedom can be easily illustrated in a *phase space* (n_A, n_B) after eliminating time t – Fig. 2.1. This type of singular point is called a *stable node*.

When $pK > 4\beta^2$ holds, the singular point remains stable, $\mathrm{Re}\,\varepsilon_{1,2} < 0$, but the roots (2.1.16) have imaginary parts: $\mathrm{Im}\,\varepsilon_1 = -\mathrm{Im}\,\varepsilon_2$. In this case the phase portrait reveals a stable focus – Fig. 2.2. This regime results in damped oscillations around the equilibrium point (2.1.24). The damping parameter pK/β is small, for large β, in which case the concentration oscillation frequency is just $\omega = \sqrt{pK}$.

The Lotka model is an example of a rough system: deviations of concentrations from their asymptotic values (2.1.24) occur independently on chosen parameters p, K, β, i.e., small variations of these parameters cannot affect the way a system strives for the equilibrium state.

2.1.1.5 The Lotka–Volterra model

The better known Lotka–Volterra model [18, 19] unlike (2.1.21) is based on two autocatalytic stages

$$E + A \xrightarrow{k_0} 2A,$$
$$A + B \xrightarrow{K} 2B, \quad (2.1.27)$$
$$B \xrightarrow{\beta} P.$$

Here the infinite food E supply is assumed. Its biological interpretation is similar to the Lotka model: predators B live on prey animals A, both are reproduced by division.

The scheme (2.1.27) is described by the following set of equations

$$\frac{dn_A(t)}{dt} = \alpha n_A(t) - K n_A(t) n_B(t), \quad (2.1.28)$$

$$\frac{dn_B(t)}{dt} = K n_A(t) n_B(t) - \beta n_B(t), \quad (2.1.29)$$

with $\alpha = k_0 n_E$.

The singular point seen in the first quadrant of the phase plane (n_A, n_B)

$$n_A^0 = \frac{\beta}{K}, \qquad n_B^0 = \frac{\alpha}{k} \quad (2.1.30)$$

corresponds to the characteristic equation

$$\varepsilon^2 + \alpha\beta = 0 \quad (2.1.31)$$

having the roots

$$\varepsilon_{1,2} = \pm i\omega_0, \qquad \omega_0 = (\alpha\beta)^{1/2}. \quad (2.1.32)$$

This type of a pattern of singular points is called a *centre* – Fig. 2.3. A centre arises in a *conservative system*; indeed, eliminating time from (2.1.28), (2.1.29), one arrives at an equation on the phase plane with separable variables which can be easily integrated. The relevant phase trajectories are *closed*: the model describes the undamped concentration oscillations. Every trajectory has its own period $T > 2\pi/\omega_0$ defined by the initial conditions. It means that the Lotka–Volterra model is able to describe the continuous frequency spectrum $\omega < \omega_0$, corresponding to the infinite number of periodical trajectories. Unlike the Lotka model (2.1.21), this model is not rough since

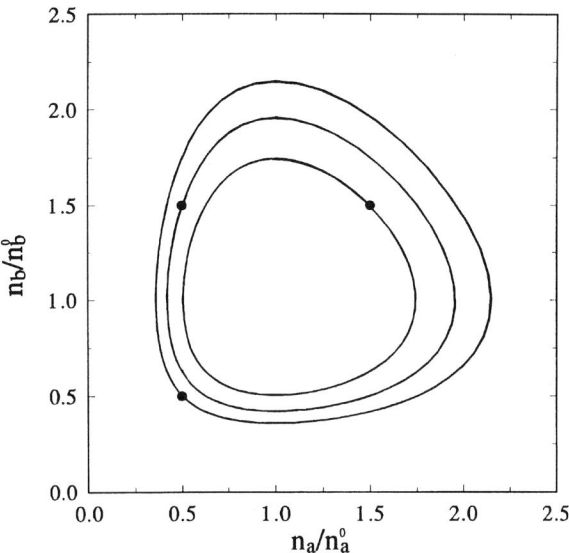

Fig. 2.3. A centre. Three solutions of the Lotka–Volterra equations (2.1.28)–(2.1.29) are presented; the distinctive parameter $\alpha/\beta = 1$. The starting point of each trajectory is shown by a black circle.

there is no fluctuation damping here: any perturbations of parameters α, β, K result in a system's selected orbit and change of frequencies become irregular. This is why at present oscillations in the Lotka–Volterra model (2.1.27) are interpreted as a *noise* [16] – unlike the so-called *chemical clock* with well defined frequencies and amplitudes of the concentration oscillations [20].

2.1.1.6 Synergetic aspects of the chain of bimolecular reactions

In the case of a chemical clock, the asymptotic $(t \to \infty)$ solution depends on time, there are not only singular points but also singular trajectories. An example is the *stable limit cycle* – Fig. 2.4, i.e., a closed trajectory to which all phase trajectories existing in its vicinity strive.

Of interest is the study of conditions under which such a limit cycle emerges in a system. The chemical clock serves as an example of the so-called *temporary structures* study which was stimulated by a fundamental problem of order emerging from chaos. In the last decade it became a central part of a new discipline called *synergetics* [1, 21, 22].

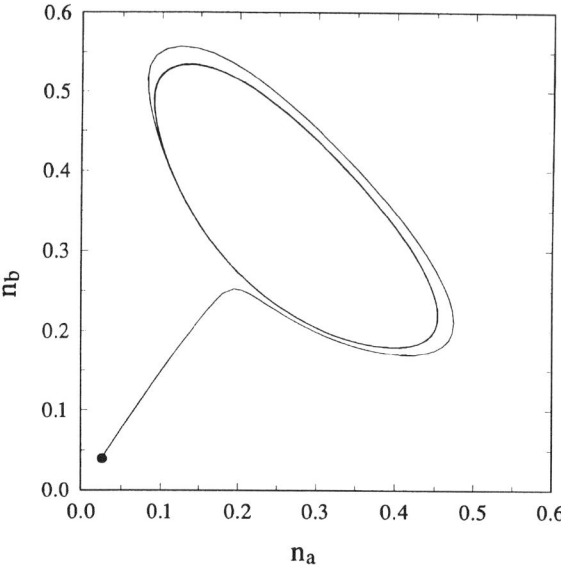

Fig. 2.4. A stable limit cycle.

Note here, without proof, one of the synergetic theorems about limit cycles [14, 15]: a *stable limit cycle contains at least one singular point or the unstable node of focus-type exists.*

Such singular points shown in Figs 2.5 and 2.6 did not emerge in the above discussed examples. It has a universal reason since according to the quite general Hanusse theorem [23] (see also [16]) dealing with unstable singular points, *the limit cycle surrounding unstable node or focus cannot arise in a system with two intermediate products if only the mono- and bimolecular reaction stages occur.*

Its proof is trivial and in fact is based on a restriction imposed on coefficients (2.1.19) of the relevant characteristic equation.

Therefore, systems with two intermediate products treated in terms of kinetic equations (2.1.2) reveal at $t \to \infty$ only the stationary solutions $n_i(\infty)$. To observe non-trivial time behaviour of concentrations (meaning auto-oscillations or *temporary structure*), either more freedom degrees (accompanied with lost transparency of the qualitative analysis) or greater non-linearity are required according to this theorem.

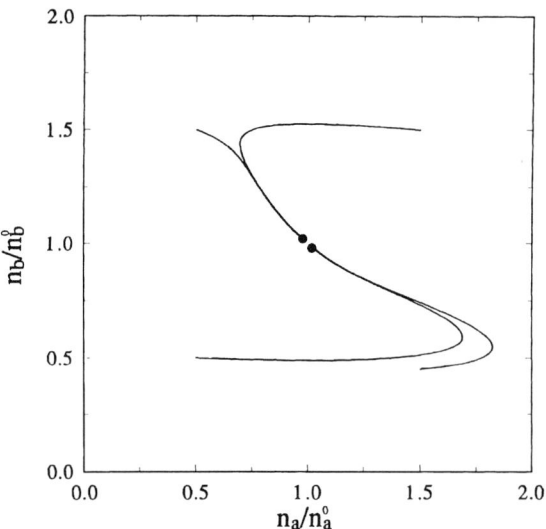

Fig. 2.5. An unstable node is obtained as a formal solution of the Lotka equations (2.1.22)–(2.1.23) with time inversion, $t \to -t$, and parameter $pK/\beta^2 = 2$. Note that these equations cannot be associated with a set of mono- and bimolecular reactions.

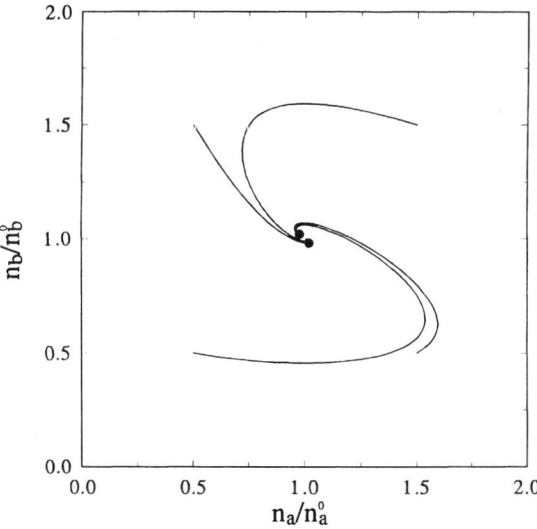

Fig. 2.6. Unstable focus is obtained as in Fig. 2.5 but for the parameter $pK/\beta^2 = 1/2$.

This is namely the main reason why *tri*molecular models (e.g., Brusselator [16]) being often physically non-transparent but having only two intermediate stages have attracted such great attention in synergetic studies.

As an illustration, consider the following trimolecular model [24]:

$$\begin{aligned} E &\xrightarrow{k_1} A, \\ C + A &\xrightarrow{k_2} B + D, \\ 2A + B &\xrightarrow{k_3} 3A, \\ A &\xrightarrow{k_4} P, \end{aligned} \quad (2.1.33)$$

which corresponds to a set of kinetic equations

$$\frac{dn_A(t)}{dt} = k_1 n_E - (k_2 n_C + k_4) n_A(t) + k_3 n_A^2(t) n_B(t), \quad (2.1.34)$$

$$\frac{dn_B(t)}{dt} = k_2 n_C n_A(t) - k_3 n_A^2(t) n_B(t). \quad (2.1.35)$$

The trimolecular stage $2A+B \xrightarrow{k_3} 3A$ gives cubic non-linearity $k_3 n_A^2(t) n_B(t)$ in (2.1.34), (2.1.35). From the point of view of a variety of different type of solutions, the model (2.1.33) turns out to be ideal for the investigation of temporary structures in chemical systems. At the same time, physical reasonability of the trimolecular stage is not evident. Of key importance here are the shortcomings of mono- and bimolecular stages in describing the chemical clock in a system with a rigidly fixed number of intermediate products rather than the wide spread of these trimolecular stages in real situations. This limitation is lifted, however, for a greater number of intermediate products. It could be well demonstrated by the *Lorenz model* [25] (see also [26]) that three degrees of freedom in a system with quadratic non-linearity is enough for emerging complex chaotic behaviour.

From the point of view of the Hanusse theorem just discussed, a system with two intermediate stages and mono- and bimolecular reactions are not capable to reveal any temporary, spatial and spatio-temporal structures. However, results obtained in the past few years permit reconsideration of such an absolute point of view.

Spatial homogeneity of a system (needed for making use of the formal chemical kinetics) is secured, first of all, by complete particle mixing. On

Macroscopic approach

the other hand, despite the fact that homogeneous distribution of interacting particles could be produced for the uniform external source, the perfect reactant mixing cannot be reached *in principle*: the stochastic interaction and recombination of particles by itself can create spatial correlations of reactants thus leading to emergence of microscopic structures which, nevertheless, do not violate the global macroscopic homogeneity of solids and liquids. However, until recently, these effects of non-uniform reactant distributions were completely neglected in the chemical kinetics [20]. Let us consider this point in more detail.

2.1.2 Smoothing out of local density fluctuations

2.1.2.1 Balance equations

Described in Section 2.1.1 the formal kinetic approach neglects the spatial fluctuations in reactant densities. However, in recent years, it was shown that even formal kinetic equations derived for the spatially extended systems could still be employed for the qualitative treatment of reactant density *fluctuation effects* under study in homogeneous media. The corresponding equations for *fluctuational* diffusion-controlled chemical reactions could be derived in the following way. As any macroscopic theory, the formal kinetics theory operates with physical quantities which are averaged over some physically infinitesimal volumes $v_0 = \lambda_0^3$, neglecting their dispersion due to the atomistic structure of solids. Let us define the *local* particle concentrations

$$C_i(\vec{r}, t) = \frac{N_i(\vec{r})}{v_0}, \qquad (2.1.36)$$

where $N_i(\vec{r})$ is a number of i-kind particles in volume v_0 centered at $\vec{r}$. Next we can use the *balance equations* [16, 27]

$$\frac{\partial C_i(\vec{r}, t)}{\partial t} = -\nabla \vec{j}_i(\vec{r}, t) + F_i, \qquad (2.1.37)$$

where $\vec{j}(\vec{r}, t)$ is the diffusive flux, F_i is the particle production rate. Following the *first Fick law* [2, 3, 27] this flux is

$$\vec{j}(\vec{r}, t) = -D_i \nabla C_i(\vec{r}, t). \qquad (2.1.38)$$

Due to macroscopic homogeneity of the systems under study, the diffusion coefficients D_i are assumed to be coordinate-independent. (Incorporation of the anisotropic diffusion coefficient was performed by [28–31]. In this case the scalar D_i in equation (2.1.38) should be replaced by a tensor.) In the case of particle production

$$F_i = F_i(C_1, \ldots, C_s) = a_i^j C_j(\vec{r}, t) + b_i^{jk} C_j(\vec{r}, t) C_k(\vec{r}, t) \quad (2.1.39)$$

equation (2.1.37) could be written in a form

$$\frac{\partial C_i(\vec{r}, t)}{\partial t} = D_i \nabla^2 C_i(\vec{r}, t) + F_i(C_1, \ldots, C_s), \quad i = 1, \ldots, s. \quad (2.1.40)$$

All local concentrations C_i of particles entering the non-linear functions F_i in equation (2.1.40) are taken at the *same* space points, in other words, the chemical reaction is treated as a *local* one. Taking into account that for extended systems we shouldn't consider distances greater than the distinctive microscopic scale λ_0, the choice of equation (2.1.40) means that inside infinitesimal volumes v_0 particles are well mixed and their reaction could be described by the phenomenological reaction rates earlier used for systems with complete reactant mixing. This means that λ_0 value must *exceed* such distinctive scales of the reaction as contact recombination radius, effective radius of a dynamical interaction and the particle hop length, which imposes quite natural limits on the choice of volumes v_o used for averaging.

The Hanusse theorem [23] discussed in Section 2.1.1 was later generalized for the case of diffusion by Tyson and Light [32]. Therefore, the mono- and bimolecular reactions with one or two intermediate products are expected to strive asymptotically, as $t \to \infty$, for the stationary spatially-homogeneous solution $C_i(\vec{r}, \infty) = n_i(\infty)$ corresponding to equations (2.1.2) for a system with the complete particle mixing.

Another problem arises: what is the kinetics of the approach to this stationary state. There are two cases:

(i) If the diffusing particles act to smooth concentration inhomogeneities *quicker* than chemical reaction occurs, the asymptotic laws under question are defined completely by the kinetic equations derived for the case of complete particle mixing. Many similar semiqualitative methods have been used more than once [33–44]. They have demonstrated the principal importance of the diffusive approach of reactants in the kinetics of bimolecular processes in condensed media.

The definition (2.1.36) of local particle concentrations as quantities averaged over small volumes v_0 *does not* mean, however, their equivalence to the macroscopic concentrations $n_i(t)$, since the latter are averaged over the volume of a *whole* system:

$$\overline{C_i(\vec{r},t)} = n_i(t). \tag{2.1.41}$$

The point here is that the procedure of averaging over small volume v_0 does not exclude the long-range fluctuations of a number of reactants in a system.

(ii) Even when reactants are distributed in a whole reaction volume at random (as it occurs in classical perfect gases), their Poisson distribution does not contradict *spatial inhomogeneities* in particle densities governed by the relation well known in the statistical physics:

$$\overline{(N_i - \overline{N_i})^2} = \overline{N_i} \tag{2.1.42}$$

(bars denote averaging over arbitrary volume). This unavoidable statistical inhomogeneity is enough to affect the reaction rate.

2.1.2.2 Diffusion equations and diffusion length

Chemical reactions are classified usually as diffusion-controlled, whose rate is limited by a reactant spatial approach to each other, and reaction-controlled (kinetic stage), whose rate is limited by a reaction elementary event. For systems with ideal reactant mixing considered in Section 2.1.1, there is no mechanism of reactant mutual approach. On the other hand, the kinetic equations (2.1.40) distinguish between reaction in physically infinitesimal volumes and the distant reactant motion in a whole reaction volume. In the absence of reaction particle diffusion is described by equation

$$\frac{\partial C(\vec{r},t)}{\partial t} = D\nabla^2 C(\vec{r},t). \tag{2.1.43}$$

For the initial condition $C(\vec{r},0) = \delta(\vec{r} - \vec{r}_0)$, (particle is placed at $t=0$ at point $\vec{r}_0$, $\delta(\vec{r})$ is the Dirac function) the solution of (2.1.43) is $C(\vec{r},t) = G(\vec{r},\vec{r}_0;t)$, where

$$G(\vec{r},\vec{r}_0;t) = (4\pi Dt)^{-3/2} \exp\left(-\frac{(\vec{r}-\vec{r}_0)^2}{4Dt}\right) \tag{2.1.44}$$

is Green's function of the diffusion equations (2.1.43). Equation (2.1.44) gives the probability density to find particle at moment t at coordinate $\vec{r}$

provided the particle started motion from point $\vec{r}_0$. The solution of (2.1.43) with an arbitrary initial condition $C(\vec{r}, 0)$ is also expressed through (2.1.44):

$$C(\vec{r}, t) = \int C(\vec{r}_0, 0) G(\vec{r}, \vec{r}_0; t) \, d\vec{r}_0. \qquad (2.1.45)$$

The exponent entering (2.1.44) could be rewritten as

$$\exp\left(-\frac{(\vec{r} - \vec{r}_0)^2}{4 l_D^2}\right), \qquad (2.1.46)$$

where

$$l_D = \sqrt{Dt} \qquad (2.1.47)$$

is the so-called diffusion *length*. When integrating (2.1.45), the exponent (2.1.46) differs essentially from zero on a spatial scale $l = |\vec{r} - \vec{r}_0| \leqslant l_D$ only; i.e., the length l_D defining the linear size of the region (sphere), to which boundaries at moment t perturbation is spread. The smoothing out of inhomogeneities in concentration distribution occurs rather slowly. This is especially the case for large-scale (*long-wavelength*) fluctuations in concentration distribution: thus, for a region with a distinctive size l this process requires smoothing time l^2/D.

These peculiarities of the diffusion task are valid for *any* spatial dimension $\bar{d}$, in which case (2.1.44) should be generalized as

$$G(\vec{r}, \vec{r}_0; t) = (4\pi D t)^{-\bar{d}/2} \exp\left(-\frac{(\vec{r} - \vec{r}_0)^2}{4Dt}\right). \qquad (2.1.48)$$

The expression for the diffusion length remains the same.

2.1.2.3 Local and full equilibrium

Following Zeldovich and Ovchinnikov [35], let us consider the role of reactant diffusion in establishing equilibrium in a reversible A $\rightleftarrows$ B + B reaction. In terms of formal kinetics, it is described by the equations

$$\frac{dn_B(t)}{dt} = k_0 n_A(t) - k_1 n_B^2(t), \qquad 2n_A(t) + n_B(t) = \text{const.} \qquad (2.1.49)$$

Macroscopic approach

Its steady-state solution (2.1.49) obeys the law of mass action

$$\frac{n_B^2(\infty)}{n_A(\infty)} = K_p(T),$$

where $K_p(T) = k_0/k_1$ is the temperature (T) dependent equilibrium constant. The law of mass action could also be presented in the following form

$$n_B(\infty) = (K_p(T)n_A(\infty))^{1/2}. \qquad (2.1.50)$$

Let us consider a hypothetical experiment: assume that particles A are kept at such a low temperature that there is no dissociation and the fluctuational distribution in a system obeys (2.1.42). Then the reaction volume is monotonously heated up to the temperature when dissociation begins to occur and the dependence $n_B = n_B(t)$ is monitored. According to (2.1.49), the approach to equilibrium is exponential, i.e.,

$$n_B(t) - n_B(\infty) \propto e^{-kt} \qquad (2.1.51)$$

with $k = k_0/2 + 2k_1 n_B(\infty)$.

However, this passage in fact involves several stages. At the first stage a system strives to a local equilibrium

$$C_B(\vec{r}, t) = (K_p(T)C_A(\vec{r}, t))^{1/2} \qquad (2.1.52)$$

according to the law of mass action and since diffusion had no time to change the concentration distribution. Such kinetics of local chemical equilibrium formation seems to be quite rapid. The second stage is associated with the formation of an equilibrium fluctuation spectrum (2.1.42). After time t, diffusion tends to produce equilibrium fluctuations in domains with linear sizes l_D which equals the diffusion length (2.1.47). That is why (2.1.42) is valid only inside these domains rather than in whole system's volume.

Let us divide the whole volume into blocks with a distinctive size l_D as it is shown in Fig. 2.7. Denote an average over block volume by a bar:

$$\overline{C_A} = \overline{C_A(\vec{r}, t)}, \qquad \overline{C_B} = \overline{C_B(\vec{r}, t)}, \qquad (2.1.53)$$

whereas an average over an ensemble of blocks – by $\langle \cdots \rangle$. Thus macroscopic concentrations are

$$n_A(t) = \langle \overline{C_A} \rangle, \qquad n_B(t) = \langle \overline{C_B} \rangle. \qquad (2.1.54)$$

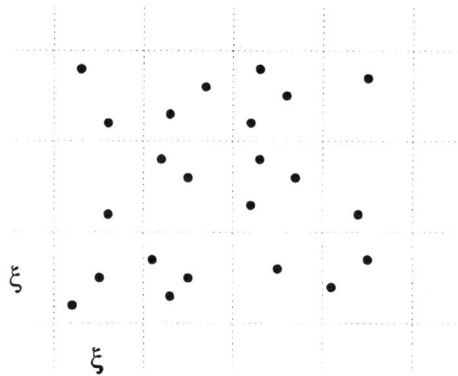

Fig. 2.7. Domain structure of the solution of the diffusion problem; the distinctive size of the domain equals $\xi = l_D$ (the so-called diffusive length).

Let us make double averaging of (2.1.52): over block volume and then over an ensemble of blocks. We find

$$n_B(t) = \left\langle \overline{(K_p(T)C_A(\vec{r},t))^{1/2}} \right\rangle. \tag{2.1.55}$$

Writing down

$$C_A(\vec{r},t) = n_A(t) + \delta C_A(\vec{r},t), \qquad \delta C_A(\vec{r},t) \ll n_A(t), \tag{2.1.56}$$

we arrive at

$$C_A^{1/2}(t) \approx n_A^{1/2}(t) + \frac{\delta C_A(\vec{r},t)}{2n_A^{1/2}(t)} - \frac{[\delta C_A(\vec{r},t)]^2}{8n_A^{3/2}(t)}. \tag{2.1.57}$$

Making average, we take into account that $\langle \overline{\delta C_A(\vec{r},t)} \rangle = 0$. In each block (2.1.42) is valid, i.e.,

$$\overline{[C_A(\vec{r},t) - \overline{C_A}]^2} = \frac{1}{V^2}\overline{(N_A - \overline{N_A})^2} = \frac{\overline{N_A}}{V^2} = \frac{\overline{C_A}}{V}. \tag{2.1.58}$$

On the other hand, an average over an ensemble of blocks is equivalent to that over a whole system which is suggested to be so large that one can neglect fluctuations of macroscopic quantities. In particular,

$$\langle(\overline{C_A})^2\rangle \approx \left(\langle\overline{C_A}\rangle\right)^2 = n_A^2(t). \tag{2.1.59}$$

Calculation of (2.1.55) gives

$$n_B(t) = \left(K_p(T)n_A(t)\right)^{1/2}\left(1 - \frac{1}{8n_A(t)V}\right) \tag{2.1.60}$$

provided $2n_A(t) + n_B(t) = \text{const}$. Since $V = l_D^3 = (Dt)^{3/2}$, with the help of (2.1.60) we find that the asymptotic law for reactant density approach to the equilibrium (2.1.50) is

$$n_B(t) - n_B(\infty) \propto (Dt)^{-3/2}, \tag{2.1.61}$$

which differs essentially from the exponential law (2.1.51) obtained earlier for the kinetic control. A similar conclusion was drawn by Zeldovich and Ovchinnikov [34] for other reversible reaction, $A + B \rightleftarrows C$. These observations demonstrate once more that late stages of any bimolecular reactions are diffusion-controlled, i.e., reaction rate is defined by complementary reactant migration each to the other.

2.1.2.4 Irreversible $A + B \to C$ reaction. Fluctuations of the concentration difference

Let us consider now the irreversible $A + B \to C$ reaction in the case of equal reactant concentrations, $n_A = n_B$, following Ovchinnikov and Zeldovich [35]. The reaction kinetics obeys the equations

$$\frac{\partial C_A(\vec{r},t)}{\partial t} = D_A\nabla^2 C_A(\vec{r},t) - kC_A(\vec{r},t)C_B(\vec{r},t), \tag{2.1.62}$$

$$\frac{\partial C_B(\vec{r},t)}{\partial t} = D_B\nabla^2 C_B(\vec{r},t) - kC_A(\vec{r},t)C_B(\vec{r},t). \tag{2.1.63}$$

Due to inevitable fluctuations – both thermodynamical or related to the system's prehistory, quantities $C_A(\vec{r},t)$ and $C_B(\vec{r},t)$ are not identical but their volume averages coincide:

$$\overline{C_A(\vec{r},t)} = \overline{C_B(\vec{r},t)} = n_A(t) = n_B(t) = n(t).$$

The distinctive feature of the $A + B \to C$ reaction is existence of the constant quantity $z = n_A(t) - n_B(t) = n_A(0) - n_B(0)$. Assuming that reactants have equal diffusion concentration difference $D_A = D_B = D$, introduce the quantity

$$Z(\vec{r},t) = C_A(\vec{r},t) - C_B(\vec{r},t), \qquad (2.1.64)$$

then subtract (2.1.63) from (2.1.62) thus arriving at the equation

$$\frac{\partial Z(\vec{r},t)}{\partial t} = D\nabla^2 Z(\vec{r},t). \qquad (2.1.65)$$

In other words, the kinetics of smoothing out of $Z(\vec{r},t)$ fluctuations is governed by a simple linear diffusion equation which solution and, in particular, the Green function, are well known. Similarly to (2.1.43), (2.1.45) we can now write down

$$Z(\vec{r},t) = \int Z_0(\vec{r}_0) G(\vec{r},\vec{r}_0;t)\, d\vec{r}_0, \qquad (2.1.66)$$

where $Z_0(\vec{r}) = Z(\vec{r},0)$.

The equation for the function *complementary* to $Z(\vec{r},t)$, i.e.,

$$U(\vec{r},t) = C_A(\vec{r},t) + C_B(\vec{r},t) \qquad (2.1.67)$$

reads as

$$\frac{\partial U(\vec{r},t)}{\partial t} = D\nabla^2 U(\vec{r},t) + \frac{k}{2}\left[Z^2(\vec{r},t) - U^2(\vec{r},t)\right]. \qquad (2.1.68)$$

Let us consider now the extreme case of an *instant* reaction, $k \to \infty$. In this case particles A and B do not coexist in physically infinitesimal volumes. The solution of (2.1.68) comes from putting there the cofactor of k to zero, i.e.,

$$U(\vec{r},t) = |Z(\vec{r},t)|. \qquad (2.1.69)$$

In other words, the reaction kinetics at $k \to \infty$ is entirely determined by *diffusive* encounter of reactants since in the degenerate problem, described by (2.1.65) and (2.1.69), a role of diffusion is self-evident. Equation (2.1.69) means that whole volume is indeed divided into blocks characterized by a *sign* of $Z(\vec{r},t)$: for $Z(\vec{r},t) > 0$, $C_A(\vec{r},t) = Z(\vec{r},t)$, $C_B(\vec{r},t) = 0$ and vice versa.

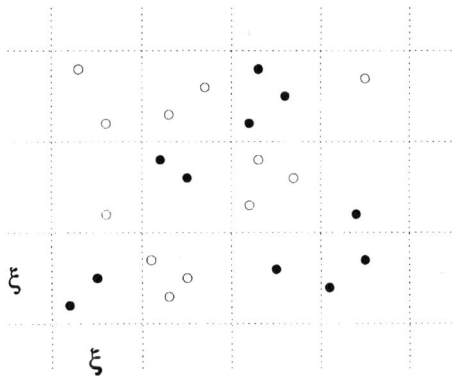

Fig. 2.8. Same as in Fig. 2.7 for the bimolecular reaction $A + B \to C$ ($n_A = n_B$).

Averaging the $C_A(\vec{r},t) = (Z(\vec{r},t) + |Z(\vec{r},t)|)/2$ over volume and using the condition $\overline{Z(\vec{r},t)} = 0$, one gets

$$n(t) = n_A(t) = \frac{1}{2}\overline{|Z(\vec{r},t)|}. \qquad (2.1.70)$$

For the long reaction times, $l_D \gg \lambda_0$, this new scale dominates over all other dimensions of the process and thus we can consider a model where the whole reaction volume is divided into such alternating blocks with linear sizes l_D, and each of them contains either A or B particles predominantly (Fig. 2.8).

It means physically that striving to survive as long as they could, similar particles (A's or B's) are forming *dynamical aggregates* which are more stable against recombination with particles of another kind than random mixture of isolated single particles A and B. Due to this aggregation (we stress that similar particles *do not* interact as it occurs e.g., in colloid formation) average distance between dissimilar particles becomes essentially greater than for random distribution and the reaction rate thus is *reduced* as compared to the case of complete reactant mixing (well stirred system).

Diffusion stimulates the dissimilar particle approach to each other and reaction between them, in this process, as it was said above, diffusion acts to smooth out density inhomogeneities, but on the other hand, the reaction creates them and often the latter trend is predominant.

Let us focus now our attention on (2.1.70). Assume that initial distribution of reactant concentration satisfies the thermal equilibrium (2.1.42). Then averaging $Z_0^2(\vec{r})$ over volume could be done similarly to (2.1.58)

$$\overline{Z^2(\vec{r})} = \overline{\left((C_A(\vec{r},0) - \overline{C_A}) - (C_B(\vec{r},0) - \overline{C_B})\right)^2}$$
$$= 2n(0)/V. \qquad (2.1.71)$$

When deriving (2.1.71), statistical independence of fluctuations of both reactant concentrations is used. An average quantity entering (2.1.71) could be expressed as

$$\overline{\left(C_A(\vec{r},0) - \overline{C_A}\right)^2} + \overline{\left(C_B(\vec{r},0) - \overline{C_B}\right)^2} -$$
$$- 2\overline{\left(C_A(\vec{r},0) - \overline{C}\right)\left(C_B(\vec{r},0) - \overline{C_B}\right)}.$$

An initial distribution $Z_0(\vec{r})$ is assumed to be random variable with the Gauss distribution

$$W = c \exp\left[-\frac{1}{2n(0)} \int Z_0^2(\vec{r}) \, dV\right], \qquad (2.1.72)$$

where c is normalization constant. Equation (2.1.72) agree with (2.1.71). Using the integral representation

$$|y| = \frac{2}{\pi} \int_0^\infty \zeta^{-2}(1 - \cos \zeta y) \, d\zeta \qquad (2.1.73)$$

we can write down $|Z(\vec{r}, t)|$ in a from

$$|Z(\vec{r}, t)| = \frac{2}{\pi} \int_0^\infty \zeta^{-2} \left(1 - \cos[\zeta Z(r,t)]\right) \, d\zeta. \qquad (2.1.74)$$

Thus averaging in (2.1.70) is reduced to the average of the exponential with the distribution (2.1.72)

$$\langle \exp\left[i\zeta Z(\vec{r}, t)\right]\rangle$$
$$= \int \exp\left[i\zeta \int G(\vec{r}, \vec{r}_1; t) Z_0(\vec{r}_1) \, dV_1\right] W \, d\{Z_0\}, \qquad (2.1.75)$$

where $d\{Z_0\}$ denotes functional integration. In fact, it is performed over independent real and imaginary parts of the Fourier expansion components of the random variable $Z_0(\vec{r})$, as it has been demonstrated by Toussaint and Wilczek [45] employing quite similar technique. The functional integral entering (2.1.75) could be exactly calculated:

$$\langle \exp[i\zeta Z(\vec{r},t)] \rangle = \exp\left[(-\zeta^2 n(0)/2) \int G^2(\vec{r},\vec{r}_1;t)\,dV_1\right]. \quad (2.1.76)$$

Substituting (2.1.76) into (2.1.74) and integrating over ζ, we obtain the asymptotics of $n(t)$ as $t \to \infty$ to be [33]

$$n(t) \propto (Dt)^{-3/4}. \quad (2.1.77)$$

The conclusion suggests itself that the decay law $n(t) \propto t^{-1}$ obtained earlier in terms of standard chemical kinetics (2.1.8) is replaced by a *slower* decay.

As it was first noted by Zeldovich [33] it is not easy to distinguish experimentally between exponents 1 and 3/4 (equations (2.1.8) and (2.1.77)). The approach just presented cannot be applied to *charged* reactants since their electrostatic attraction cuts off spatial fluctuation spectrum at the Debye radius.

From a formal point of view (2.1.77) could be generalized for an arbitrary space dimension $\bar{d}$. To do it, it is enough to replace in (2.1.76) the Green function (2.1.44) into (2.1.48), which results in

$$n(t) \propto (Dt)^{-\bar{d}/4}. \quad (2.1.78)$$

At last, note logical inconsistency of the method presented. Non-uniform concentration distribution, corresponding to the Poisson fluctuation spectrum (2.1.42), is introduced through initial condition imposed on $Z(\vec{r},t)$ – see (2.1.71), (2.1.72). However, equation (2.1.42) disagrees with the starting kinetic equation (2.1.40): the solution of the latter in the absence of reaction, $F_i = 0$, is $C_i(\vec{r}, t \to \infty) = n_i(0)$. Consequently, we can find dispersion of a number of particles within an arbitrary volume:

$$\overline{(N_i - \overline{N_i})^2} = 0. \quad (2.1.79)$$

Therefore, the standard chemical kinetics *overestimates* the diffusive smoothing out of initial density inhomogeneities as compared to the thermal fluctuation level (2.1.42).

Equation (2.1.78) holds for large diffusion lengths l_D, $l_D \gg l$, where $l = n(0)^{-1/\bar{d}}$ is the mean initial distance between reactants. This condition is met if $n(t) \ll n(0)$. On the other hand, (2.1.78) is derived providing very large reaction rate ($k \to \infty$) at the kinetic stage which permitted to detect more slow diffusion terms acting to smooth fluctuations. However, for $\bar{d} > \bar{d}_0 = 4$, (2.1.78) predicts more *rapid* decay law than (2.1.8)! This is why (2.1.78) holds for space dimensions $\bar{d} \leqslant \bar{d}_0 = 4$, i.e., the marginal dimension $\bar{d}_0$ is observed in this problem. The same conclusion was drawn by Kang and Redner [46, 47] in terms of a scaling formalism.

The approach discussed could be generalized for the case of unequal concentrations. In this case $\overline{Z(\vec{r}, t)} = -\delta n$, $\delta n = n_B(0) - n_A(0) \geqslant 0$ and instead of (2.1.70) we have

$$n_A(t) = \frac{1}{2}\left(\overline{|Z(\vec{r},t)|} - \delta n\right), \qquad (2.1.80)$$

whereas (2.1.71) is replaced by the dispersion of the fluctuating quantity $Z_0(\vec{r})$

$$\overline{\left[Z_0(\vec{r}) - \overline{Z_0(\vec{r})}\right]^2} = \frac{2n(0)}{V}, \qquad (2.1.81)$$

with $\overline{Z_0(\vec{r})} = -\delta n$, $n(0) = (n_A(0) + n_B(0))/2$.

The Gaussian distribution (2.1.72) could also be generalized for the case when the mean value of the fluctuating value is not zero:

$$W = c\exp\left[-\frac{1}{2n(0)}\int \left(Z_0(\vec{r}) - \overline{Z_0(\vec{r})}\right)^2 dV\right]. \qquad (2.1.82)$$

Equation (2.1.82) agrees with (2.1.81).

Averaging over distribution (2.1.82) can be performed analytically using the representation (2.1.74); the functional integral could be calculated exactly. After transformations one gets

$$n_A(t) = \frac{1}{\pi}\int_0^\omega \zeta^{-2} \cos[\zeta \delta n]\left(1 - \exp(-f\zeta^2)\right) d\zeta, \qquad (2.1.83)$$

where $f = \frac{n(0)}{2}(8\pi Dt)^{-\bar{d}/2}$.

As $\delta n = 0$, (2.1.83) transforms into the relevant equation derived above for equal concentrations revealing the asymptotics (2.1.78). If $\delta n \neq 0$, (2.1.83) demonstrates exponential decay at $t \to \infty$ (pre-exponential cofactor is missed)

$$n_A(t) \propto \exp\left[-\frac{2\delta n^2}{n(0)}(8\pi Dt)^{\overline{d}/2}\right]. \tag{2.1.84}$$

Equation (2.1.84) predicts more slow decay as compared with the chemical kinetics (kinetic stage – (2.1.9)) unless $\overline{d} \leqslant 2$. That is, marginal dimension $\overline{d}_0 = 2$ occurs. The distinctive feature of (2.1.84) defining the range of its applicability is cofactor δn^2. Taking into account that (2.1.83) is valid as $l_D \gg l$, where $l = n(0)^{-1/\overline{d}}$, the exponential kinetics (2.1.84) becomes essential as $l_D \gg l_{\delta n}^2/l$, $l_{\delta n} = \delta n^{-1/\overline{d}}$. In another extreme case the pre-exponential factor predominates resulting in the asymptotics $n_A(t) \propto t^{-\overline{d}/4}$, as is observed for equal concentrations. That is, the *crossover* takes place (transition from power asymptotics to the exponential). Note that Schnörer, Sokolov and Blumen [48] have obtained these results employing the combination theory.

2.1.2.5 Moment method

Let us consider now two small volumes v_0 centered at $\vec{r}_1$ and $\vec{r}_2$, respectively. The number of i-kind particles therein, $N_i(\vec{r}_1)$ and $N_i(\vec{r}_2)$, are stochastic quantities which averaged over v_0 are $\overline{N_i(\vec{r}_1)} = \overline{N_i(\vec{r}_2)} = n_i(t)v_0 = \overline{N_i}$. Define now the *correlation functions* of similar particles [36, 37]

$$S_i(r,t) = \overline{\bigl(C_i(\vec{r}_1,t) - n_i(t)\bigr)\bigl(C_i(\vec{r}_2,t) - n_i(t)\bigr)} \tag{2.1.85}$$

as well as that for the dissimilar particles $(i \neq j)$

$$S_{ij}(r,t) = \overline{\bigl(C_i(\vec{r}_1,t) - n_i(t)\bigr)\bigl(C_j(\vec{r}_2,t) - n_j(t)\bigr)}. \tag{2.1.86}$$

Here the average is taken over a whole volume with a fixed $\vec{r} = \vec{r}_1 - \vec{r}_2$ distance. Due to the system's homogeneity and isotropy the S_i and S_{ij} are functions of $r = |\vec{r}|$ only. We illustrate below how the equations for the correlation functions could be derived from the set of equations (2.1.40) and the averaging procedure. First, let us define the initial conditions for $S_i(r,0)$ and $S_{ij}(r,0)$.

For non-overlapping volumes, $r > \lambda_0$, the fluctuations in particle numbers are statistically independent and thus

$$S_i(r,0) = \frac{1}{v_0^2}\overline{\left(N_i(\vec{r}_1) - \overline{N_i}\right)\left(N_i(\vec{r}_2) - \overline{N_i}\right)} = 0. \qquad (2.1.87)$$

For completely overlapped volumes, $r = 0$, we use (2.1.42) which yields

$$S_i(0,0) = \frac{\overline{N_i}}{v_0^2} = \frac{n_i(0)}{v_0}. \qquad (2.1.88)$$

Taking into account the statistical independence in coordinates of dissimilar particles, we arrive at

$$S_{ij}(r,0) = 0. \qquad (2.1.89)$$

In the limiting case $v_0 \to 0$ equations (2.1.87), (2.1.88) could be formally expressed via the Dirac δ-function

$$S_i(r,0) = n_i(0)\delta(\vec{r}). \qquad (2.1.90)$$

Let us illustrate the moment method with the $A + B \to C$ reaction. In the stoichiometric case, $n_A(t) = n_B(t) = n(t)$ averaging of equations (2.1.62), (2.1.63) over volume gives

$$\frac{dn(t)}{dt} = -k\left(n^2(t) + S(0,t)\right), \qquad (2.1.91)$$

where $S(r,t) = S_{AB}(r,t)$. Equation (2.1.91) can be also rewritten in a form

$$\frac{dn(t)}{dt} = -K(t)n^2(t), \qquad (2.1.92)$$

quite similar to (2.1.5) of the formal chemical kinetics but now having the time-dependent effective reaction rate

$$K(t) = k\left(1 + \frac{S(0,t)}{n^2(t)}\right). \qquad (2.1.93)$$

Neglecting third-order momenta, the joint correlation functions (2.1.85), (2.1.86) become [36, 37]

$$\frac{\partial S(r,t)}{\partial t} = (D_A + D_B)\nabla^2 S(r,t) -$$
$$- kn(t)\left(2S(r,t) + S_A(r,t) + S_B(r,t)\right), \qquad (2.1.94)$$

$$\frac{\partial S_\nu(r,t)}{\partial t} = 2D_\nu \nabla^2 S_\nu(r,t) - 2kn(t)\left(S(r,t) + S_\nu(r,t)\right),$$
$$\nu = A, B. \qquad (2.1.95)$$

The limiting case $v_0 \to 0$ used in deriving the initial conditions (2.1.90) for $S_i(r,t)$ is not quite correct since the short-wavelength fluctuations with $\lambda < \lambda_0$ become no longer valid. We can avoid them, considering instead δ-function in (2.1.90), which defines $S_i(r,0)$, something less singular. It is important, however, that equation (2.1.90) well-suited for description of the long-time reaction asymptotics are governed namely by the long-wavelength fluctuations.

The solution of equations (2.1.92), (2.1.94) and (2.1.95) for equal reactant concentrations, $D_A = D_B$ was obtained in [36, 37]. They demonstrated that as $t \to \infty$, the reaction rate tends to zero, $K(t) \to 0$, whereas the concentration

$$n(t) \approx \left(-S(0,t)\right)^{1/2}. \qquad (2.1.96)$$

They have also estimated that $S(0,t) \propto t^{-3/2}$, i.e., $n(t) \propto t^{-3/4}$.

2.1.2.6 Irreversible $A + B \to B$ reaction. Survival in cavities

In the case of $A + B \to B$ reaction fluctuations in the concentration difference, $Z(\vec{r},t)$, play no essential role and the asymptotic behaviour of $n_A(t)$ as $t \to \infty$ is defined by the space fluctuations in B reactant concentration [49–51].

The point is that a system of B's reactants always has a small number of large volume cavities plotted schematicaly in Fig. 2.9. Reactants A, if found therein, have large lifetimes limited by their migration to the cavity boundaries. Note that namely these A particles define long-time asymptotics of the reaction.

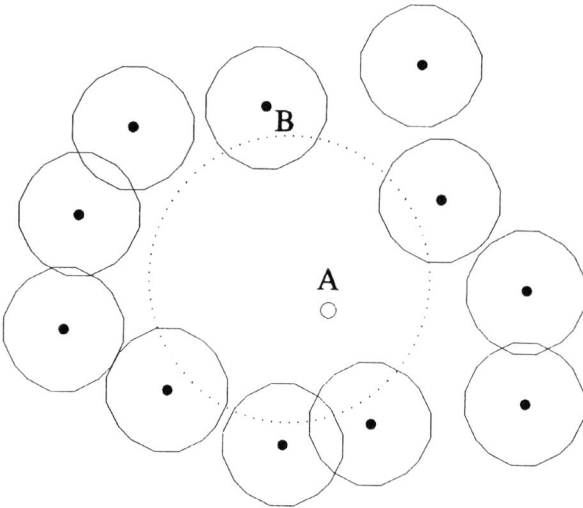

Fig. 2.9. Particle A is a cavity formed by the recombination spheres around particles B.

To illustrate this idea, let us consider a simple model. Let there be a cavity V of an arbitrary shape. The concentration distribution $C_A(\vec{r}, t)$ obeys the diffusion equation

$$\frac{\partial C_A(\vec{r}, t)}{\partial t} = D \nabla^2 C_A(\vec{r}, t). \qquad (2.1.97)$$

Reaction with B particles on the cavity boundary could be described in terms of a completely absorbing boundary

$$C_A(\vec{r}, t)\Big|_{\vec{r} \in S} = 0, \qquad (2.1.98)$$

where S is the internal surface of the cavity. It should be stressed that in this formalism only A particles are assumed to be mobile, i.e., $D = D_A$, $D_B = 0$. In the opposite case, $D_A = 0$, the problem can be solved exactly and there are *no* fluctuation effects under study. An intermediate case is estimated by [52].

The solution of (2.1.97) with the boundary condition (2.1.98) could be expressed by a series:

Macroscopic approach

$$C_A(\vec{r}, t) = \sum_j U_j(\vec{r}) e^{-\varepsilon_j t}, \qquad (2.1.99)$$

where ε_j are eigenvalues of the equation

$$\varepsilon U(\vec{r}) + D\nabla^2 U(\vec{r}) = 0 \qquad (2.1.100)$$

with the boundary condition

$$U(\vec{r})\Big|_{\vec{r} \in S} = 0. \qquad (2.1.101)$$

As it follows from (2.1.99), the decay of A's concentration inside the cavity is defined by the exponential $e^{-\varepsilon_0 t}$, where ε_0 is the lowest eigenvalue of (2.1.100). For a given cavity volume V minimal ε_0 value corresponds to the spherical cavity. In this case $\varepsilon_0 = D(\pi^2/l^2)$, where l is the sphere radius. The average of $C_A(\vec{r}, t)$ over V as $t \to \infty$ could be estimated to be

$$\overline{C_A(t)} \approx n_A(0) e^{-\varepsilon_0 t}. \qquad (2.1.102)$$

To estimate $n_A(t)$, (2.1.102) has to be averaged over all kinds of cavities, i.e., over l or V. The probability of finding a cavity with volume V in a system of B's particles with concentration n_B is known as the Perren formula

$$W(V) \, dV \approx n_B e^{-V n_B} \, dV. \qquad (2.1.103)$$

The averaging (2.1.102) over l gives

$$n_A(t) = \langle \overline{C_A(t)} \rangle$$
$$= 4\pi n_A(0) n_B \int_0^\infty l^2 \, dl \exp\left(-\frac{4}{3}\pi l^3 n_B - \frac{\pi^2 D t}{l^2}\right). \qquad (2.1.104)$$

The integral could be easily estimated by means of the steepest descent method which gives

$$n_A(t) \propto \exp\left(-\frac{5}{3}\pi^{8/5}(2n_B)^{2/5}(Dt)^{3/5}\right). \qquad (2.1.105)$$

In (2.1.105) the power cofactor being dependent on the approximations used (e.g., spherical cavities) is omitted. Note that according to (2.1.105), $n_A(t)$

decay is slower than it is expected from formal chemical kinetics, (2.1.13). Ovchinnikov and Zeldovich [50] extended (2.1.105) for the A+B → C reaction, provided $n_A(0) \ll n_B(0)$. This condition means that the concentration of B particles remains practically constant.

For the arbitrary space dimension $\bar{d}$ the relation $\varepsilon_0 \propto D/l^2$ is still valid. Taking into account that $V \propto l^{\bar{d}}$, (2.1.105) could be generalized:

$$n_A(t) \propto \exp\left(-c(Dt)^{-\bar{d}/(\bar{d}+2)}\right), \qquad (2.1.106)$$

where c is constant.

2.2 THE TREATMENT OF STOCHASTICITY ON A MESOSCOPIC LEVEL

> A truth is so delicate that any small deviation from it comes you to a mistake, but this mistake is also so delicate that after small retreat you find yourselves in a truth again.
>
> B. Pascal

2.2.1 The stochastic differential equations and the Fokker–Planck equation

2.2.1.1 Stochastic differential equations

As was shown in Section 2.1, in some cases thermal fluctuations of reactant densities affect the reaction kinetics. However, the equations of the formal chemical kinetics are not suited well enough to describe these fluctuations: in fact they are introduced *ad hoc* through the initial conditions to equations. The role of fluctuations and different methods for incorporating them into formal kinetics equations were discussed more than once.

One of the simplest methods to generalize formal kinetics is to treat reactant concentrations as *continuous stochastic* functions of time, which results in a transformation of deterministic equations (2.1.1), (2.1.40) into stochastic differential equations. In a system with completely mixed particles the macroscopic concentration $n_i(t)$ turns out to be the average of the stochastic function $c_i(t)$

$$n_i(t) = \langle c_i(t) \rangle. \qquad (2.2.1)$$

Here $c_i(t)$ are solutions of the stochastic differential Ito–Stratonovich equation [26, 34, 99]

$$dc_i(t) = F_i(c_1, \ldots, c_s)\, dt + G_{ij}\, dW_j(t), \qquad (2.2.2)$$

where the coefficients $G_{ij} = G_{ij}(c_1, \ldots, c_s)$ are functions of concentrations, whereas $dW_j(t)$ is the *Winer process* which depends on s variables. The second term in (2.2.2) describe fluctuations in particle production, which generally speaking, depends on concentrations. However, when this is not the case, equation (2.2.2) could be written in the *Langevin*-like form

$$\frac{dc_i(t)}{dt} = F_i(c_1, \ldots, c_s) + \varphi_i(t), \qquad (2.2.3)$$

here $\varphi_i(t)$ stochastic forces are usually expected to be normally distributed and δ-correlated in time.

Use of the stochastic differential equation (2.2.2) as the equation of motion instead of equation (2.1.1) results in the treatment of the reaction kinetics as a *continuous Markov process*. Calculations of stochastic differentials, perfectly presented by Gardiner [26], allow us to solve equation (2.2.2). On the other hand, an averaged concentration given by this equation could be obtained making use of the distribution function $f = f(c_1, \ldots, c_s; t)$. The latter is nothing but solution of the *Fokker–Planck* equation [26, 34]

$$\frac{\partial}{\partial t} f(c_1, \ldots, c_s; t) = -\frac{\partial}{\partial c_i}(F_i f) + \frac{1}{2}\frac{\partial^2}{\partial c_i \partial c_j}\left[(GG^\mathrm{T})_{ij} f\right], \qquad (2.2.4)$$

where G^T is a transposed matrix of coefficients. (In interpreting stochastic differential equation (2.2.2), we followed Ito.) Despite the stochastic differential equations like (2.2.2) and the Fokker–Planck equation (2.2.4) which are equivalent, in practical applications they turn out to be complementary since due to different approximations used for their solution they differ also in the applicability range.

Shortcomings of the above described approach are self-evident: the fluctuations entering equation (2.2.2) are independent of the deterministic motion, the passage from the deterministic description given by equation (2.1.1) to the stochastic one needs a large number of additional phenomenological parameters determining G_{ij}. To define them, the *fluctuation-dissipative theorem* should be used.

On the other hand, the stochastic treatment in terms of equation (2.2.2) is applied to the *macroscopic* values $c_i(t)$ characterizing a system. As is generally-accepted, thermal fluctuations of macroscopic quantities being very small are not of great interest. The only exception when fluctuations appear to be important (which leads to an untrivial result) is the case when the solution of a set of equations $F_i(c_1,\ldots,c_s) = 0$ finds itself near the bifurcation point [26, 34, 90]. This is why we are going to consider now the *spatially-extended* systems [26, 67, 68].

2.2.1.2 Explosive instability of a linear system

To illustrate new effects which could be observed employing this technique, consider a simple model [67, 118]. Start from the chemical kinetics equation

$$\frac{dn(t)}{dt} = -k_0 n(t) + k n(t). \tag{2.2.5}$$

The first term here is proportional to the kinetic coefficient k_0 and describes particle decay, whereas the second – their reproduction. Solution of (2.2.5) is

$$n(t) = n(0) e^{-\gamma t}, \tag{2.2.6}$$

where $\gamma = k_0 - k$, and is defined by the sign of γ: for $\gamma < \gamma_0 = 0$, we face infinite reproduction (an explosion), but for $\gamma > \gamma_0$ a system is stable (particles disappear).

Assume now that the kinetic coefficients k_0, k are exposed to random variations. Let us discuss how can it affect the margin of instability γ_0. Putting

$$k - k_0 = -\gamma + g(t), \tag{2.2.7}$$

where $g(t)$ is the fluctuating contribution, we arrive at stochastic equation instead of ordinary differential equation (2.2.5):

$$\frac{dc(t)}{dt} = -\gamma c(t) + g(t) c(t). \tag{2.2.8}$$

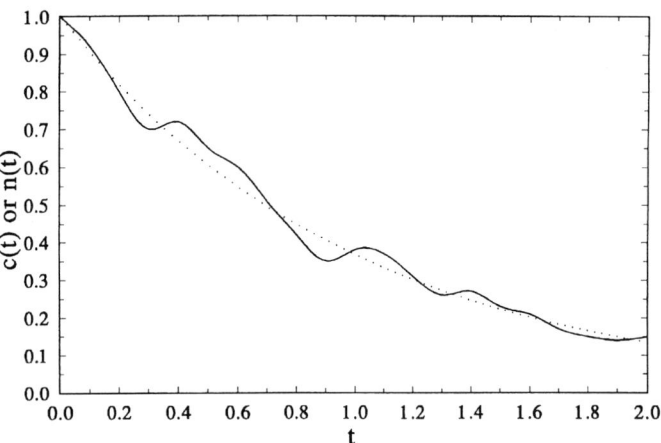

Fig. 2.10. The random function $c(t)$ (full line) is a solution of (2.2.8); their mean solution $n(t)$ is shown by the dotted line.

Figure 2.10 qualitatively illustrates the random solution $c(t)$ of (2.2.8). Taking into account that for the random value $\langle g(t) \rangle = 0$ and assuming that the Gauss distribution holds:

$$\langle g(t)g(t_1) \rangle = 2h\delta(t - t_1), \quad (2.2.9)$$

we can solve differential equation (2.2.8), interpreting it after Stratonovich. The general solution (2.2.8) could be given in a form

$$c(t) = n(0) \exp\big(-\gamma t + W(t)\big), \quad (2.2.10)$$

where $W(t)$ is a stochastic Winer process, $g(t) = dW(t)/dt$, with statistical characteristics

$$\langle W(t) \rangle = 0, \quad \langle W^2(t) \rangle = ht. \quad (2.2.11)$$

Taking into account the identity

$$\langle \exp(W(t)) \rangle = \exp(ht) \quad (2.2.12)$$

we find the first moment of the stochastic process

$$n(t) = \langle c(t) \rangle = n(0)e^{(h-\gamma)t}. \quad (2.2.13)$$

Therefore, the explosive instability boundary is shifted as compared with (2.2.6); now $\gamma_0 = h$.

2.2.1.3 Spatially extended systems

Let us start with the generalized balance equations (2.1.37). The stochastic differential equations arise due to a formal adding of the fluctuating particle sources. In a general case both the fluctuations of the diffusion flux

$$\vec{j}_i(\vec{r}, t) = -D_i \nabla C_i(\vec{r}, t) - \vec{\varphi}_i(\vec{r}, t), \qquad (2.2.14)$$

where the second term is the vector Langevin source, and the fluctuations of the particle production $\varphi^0(\vec{r}, t)$ could be taken into account leading to the following kinetic equations

$$\frac{\partial C_i(\vec{r}, t)}{\partial t} = D_i \nabla^2 C_i(\vec{r}, t) + F_i(C_i, \ldots, C_s) + \varphi_i(\vec{r}, t), \qquad (2.2.15)$$

with

$$\varphi_i(\vec{r}, t) = \varphi_i^0(\vec{r}, t) + \nabla \vec{\varphi}_i(\vec{r}, t). \qquad (2.2.16)$$

These fluctuations are assumed to be *local*, i.e., uncorrelated in space and δ-correlated in time. It is generally-accepted for a stochastic variables that

$$\langle \varphi_i^0(\vec{r}, t) \varphi_j^0(\vec{r}_1, t_1) \rangle = h_i \delta_{ij} \delta(\vec{r} - \vec{r}_1) \delta(t - t_1). \qquad (2.2.17)$$

Equation (2.2.6) could also be generalized for the situation when the sources are concentration-dependent [67, 68]. As an illustration, let us generalize solution (2.2.8) of the explosive instability. Chemical reaction is described by the following equation

$$\frac{\partial C(\vec{r}, t)}{\partial t} = D \nabla^2 C(\vec{r}, t) - \gamma C(\vec{r}, t) + g(\vec{r}, t) C(\vec{r}, t), \qquad (2.2.18)$$

where random field $g(\vec{r}, t)$ has the Gauss distribution with the mean value $\langle g(\vec{r}, t) \rangle = 0$ and the correlation function

$$\langle g(\vec{r}, t) g(\vec{r}_1, t_1) \rangle = 2H(\vec{r} - \vec{r}_1) \delta(t - t_1). \qquad (2.2.19)$$

Equation (2.2.18) is interpreted after Stratonovich. The correlation length $H(r)$ in (2.2.19) was taken in the exponential form [67], e.g.,

$$H(r) = h \exp\left[-\frac{1}{2}\left(\frac{r}{r_0}\right)^2 \right],$$

and is characterized by the correlation radius r_0. The marginal instability γ_0 in (2.2.18) depends on h, ratio D/r_0^2 as well as space dimension $\overline{d}$.

Therefore, the simplest procedure to get the stochastic description of the reaction leads to the rather complicated set of equations containing phenomenological parameters h_i (equation (2.2.17)) with non-transparent physical meaning. Fluctuations are still considered as a result of the external perturbation. An advantage of this approach is a useful analogy of reaction kinetics and the physics of equilibrium critical phenomena. As is well known, because of their nonlinearity, equations (2.1.40) reveal non-equilibrium bifurcations [78, 113]. A description of diffusion-controlled reactions in terms of continuous Markov process – equation (2.2.15) – makes our problem very similar to the static and dynamic *theory of critical phenomena* [63, 87]. When approaching the bifurcation points, the systems with reactions become very sensitive to the environment fluctuations, which can even produce new nonequilibrium transitions [18, 67, 68, 90, 108]. The language developed in the physics of critical phenomena can be directly applied to the processes in spatially extended systems.

The stochastic differential equation (2.2.15) could be formally compared with the Fokker–Planck equation. Unlike the complete mixing of particles when a system is characterized by s stochastic variables (concentrations $c_i(t)$), the local concentrations in the spatially-extended systems, $C(\vec{r}, t)$, depend also on the continuous coordinate $\vec{r}$, thus the distribution function $f(C_1, \ldots, C_s; t)$ turns to be a functional, that is real application of these equations is rather complicated. (See [26, 34] for more details about presentation of the Fokker–Planck equation in terms of the functional derivatives and problems of normalization.)

In recent years the diagrammatic technique of the perturbation theory found wide application in solving the stochastic differential equations, e.g., see a review article by Mikhailov and Uporov [68].

2.2.1.4 The $A + B \to C$ reaction. Stochastic particle generation

Let us apply general stochastic equations (2.2.15) to the simple $A+B \to C$ reaction with particle creation – the model problem discussed more than once ([84] to [93]). A relevant set of kinetic equations reads

$$\frac{\partial C_A(\vec{r}, t)}{\partial t} = D_A \nabla^2 C_A(\vec{r}, t) - k C_A(\vec{r}, t) C_B(\vec{r}, t) + \varphi_A(\vec{r}, t), \quad (2.2.20)$$

$$\frac{\partial C_B(\vec{r}, t)}{\partial t} = D_B \nabla^2 C_B(\vec{r}, t) - k C_A(\vec{r}, t) C_B(\vec{r}, t) + \varphi_B(\vec{r}, t). \quad (2.2.21)$$

The most general method to solve the set (2.2.20), (2.2.21) seems to be their transformation in terms of different-order correlations. Let us present reactant concentrations as a sum of the mean value and fluctuating term

$$C_i(\vec{r}, t) = n_i(t) + \delta C_i(\vec{r}, t). \quad (2.2.22)$$

Let us define the *joint concentration correlators* through

$$S_{ij}(r, t) = \langle \delta C_i(\vec{r}_1, t) \delta C_j(\vec{r}_2, t) \rangle, \quad (2.2.23)$$

where $r = |\vec{r}_1 - \vec{r}_2|$. Higher-order-correlators contain a corresponding number of the fluctuating values $\delta C_i(\vec{r}, t)$ in the averaged product (2.2.23). Averaging of a set (2.2.20), (2.2.21) results in equations for the mean values $n_i(t)$ where joint correlations (2.2.23) are also involved. Equations for them arise multiplying the basic equations (2.2.20), (2.2.21) by fluctuating values $\delta C_j(\vec{r}_1, t)$ with further averaging. In doing so, we get equations for joint correlators $S_{ij}(r, t)$ containing third-order correlators, etc. Therefore, we face the standard problem of decoupling correlators in order to reduce formally exact but infinite set of equations to the approximate but finite set. In practice only joint correlators are used [83].

To avoid bulky calculations, we restrict ourselves by the following problem statement: particles A and B have equal diffusion coefficients, $D_A = D_B = D$, fluctuating particle sources in $\varphi_i(\vec{r}, t)$ in (2.2.20), (2.2.21) are characterized by Poisson statistical properties:

$$\langle \varphi_A(\vec{r}, t) \rangle = \langle \varphi_B(\vec{r}, t) \rangle = p, \quad (2.2.24)$$

$$\langle \delta\varphi_A(\vec{r}, t) \delta\varphi_B(\vec{r}_1, t_1) \rangle = \langle \delta\varphi_A(\vec{r}, t) \rangle \langle \delta\varphi_B(\vec{r}_1, t_1) \rangle = 0, \quad (2.2.25)$$

$$\langle \delta\varphi_A(\vec{r},t)\delta\varphi_A(\vec{r}_1,t_1)\rangle = \langle \delta\varphi_B(\vec{r},t)\delta\varphi_B(\vec{r}_1,t_1)\rangle$$
$$= p\delta(\vec{r}-\vec{r}_1)\delta(t-t_1), \tag{2.2.26}$$

where $\delta\varphi_i(\vec{r},t) = \varphi_i(\vec{r},t) - p$.

Equation (2.2.24) means homogeneous generation of particles A and B with the rate p (per unit time and volume), whereas (2.2.25) comes from the statistical independence of sources of a different-kind particles. Physical analog of this model is accumulation of the complementary Frenkel radiation defects in solids. Note that depending on the irradiation type and chemical nature of solids (metal or insulator), dissimilar Frenkel defects could be either spatially correlated in the so-called *geminate pairs* (see Chapter 3) or distributed at random. We will focus our attention on the latter case.

As it has been done in Section 2.1.2, let us introduce the concentration difference $Z(\vec{r},t) = C_A(\vec{r},t) - C_B(\vec{r},t)$, obeying the equation

$$\frac{\partial Z(\vec{r},t)}{\partial t} = D\nabla^2 Z(\vec{r},t) + \delta\varphi(\vec{r},t), \tag{2.2.27}$$

with $\delta\varphi(\vec{r},t) = \varphi_A(\vec{r},t) - \varphi_B(\vec{r},t)$. Statistical properties of $\delta\varphi(\vec{r},t)$ are defined by (2.2.24) to (2.2.26).

This equation is linear and could be solved exactly, whereas the complementary equation for $U(\vec{r},t) = C_A(\vec{r},t) + C_B(\vec{r},t)$ is non-linear. We restrict ourselves to those problems whose solutions could be expressed through $Z(\vec{r},t)$.

Fourier components $Z(\vec{k},t)$ of (2.2.27) have correlators

$$\langle Z(\vec{k},t)Z(-\vec{k}_1,t_1)\rangle = \delta_{\vec{k},\vec{k}_1}\frac{2p}{Dk^2}\exp\left(-Dk^2|t-t_1|\right). \tag{2.2.28}$$

As $k \to 0$, (2.2.28) is large and can be interpreted as large-scale segregation of A and B particles, i.e., formation of loose aggregates containing predominantly similar particles, A or B only. Note that it is a purely statistical effect since similar particles are assumed not to interact! More details for calculating joint correlators and dissimilar particles confirming aggregation effect are available from the papers [57] to [83].

Calculating the correlator at the same time moment

$$\langle Z(\vec{k},t)Z(-\vec{k},t)\rangle = \frac{2p}{Dk^2} \tag{2.2.29}$$

shows that the total fluctuation spectrum

$$\int \langle Z(\vec{k},t)Z(-\vec{k},t)\rangle d\vec{k} = \int \langle Z^2(\vec{r},t)\rangle d\vec{r} \qquad (2.2.30)$$

becomes divergent in the small k region in one- and two dimensions $\bar{d} = 1, 2$. This means that aggregate size is macroscopic and comparable with the size of a system itself. From the formal point of view the integral (2.2.30) diverges also for large k. This kind of divergence takes place also for the fluctuation spectrum (2.1.72) considered by us in Section 2.1.2 and describing so called *white noise*. Divergence at large k is due to the extrapolation of diffusion equation on the space scale smaller than the distinctive scale λ_0. In fact one has to restrict himself by the components in the Fourier transformation with $k = k_{\max} < 1/\lambda_0$ and to introduce in (2.2.30) the upper integration limit $k_{\max}$. This large-k peculiarity does not affect results of Section 2.1.2, since an asymptotic reaction law is defined entirely by a *long-wavelength* part of spectral region (i.e., by small k).

The fluctuation spectrum just discussed differs considerably from (2.1.72). Let us estimate a role of similar particle aggregation in their decay kinetics after the source is switched off at time $t = 0$. Decay is described by (2.1.62), (2.1.63) ($D_A = D_B$). For concentration difference (2.1.64) we have the initial condition $Z(\vec{r}, 0) = Z_0(\vec{r})$. In this case the fluctuation spectrum corresponds to (2.2.29):

$$\langle Z_0(\vec{k})Z_0(-\vec{k})\rangle = \frac{2p}{Dk^2}. \qquad (2.2.31)$$

As is noted in Section 2.1.2, one has to distinguish the kinetic stage, at which $n(t) \propto t^{-1}$ ($n_A(t) = n_B(t)$) and the diffusive stage replacing it at longer times. For the latter we can use an estimate similar to (2.1.70):

$$n(t) = \frac{1}{2}\langle |Z(\vec{r},t)|\rangle. \qquad (2.2.32)$$

Its calculation with the fluctuation spectrum (2.2.31) could be performed employing the scheme used in Section 2.1.2, which results in [84]

$$n(t) = \frac{1}{2\pi}\left[\frac{p}{(2\pi)^{\bar{d}}}\int \exp(-2Dk^2 t)\frac{d\vec{k}}{Dk^2}\right]^{1/2}. \qquad (2.2.33)$$

In the $\overline{d} = 3$ case we arrive at

$$n(t) \propto p^{1/2} D^{-3/4} t^{-1/4}. \qquad (2.2.34)$$

Note that its asymptotics (2.2.34) gives essentially slower decay than (2.1.77) observed for the Poisson initial distribution. In the $\overline{d} = 2$ case the integral in (2.2.33) should be cut off at $k_{\min} \propto S^{-1/2}$ where S is surface square; practically complete disappearance of particles takes place after $t \simeq S/D$. Consider briefly the applicability of (2.2.34). Use of (2.2.34) for the initial fluctuation spectrum continues infinitely: similar particle aggregation takes a very prolonged time. Under finite excitation time, the peculiarity (2.2.31) at small k is not pronounced, (2.2.34) is not universal and it plays the role of the *intermediate* asymptotics and holds at $t < t_{\max}$, which depends on the irradiation time.

Stochastic aggregation does not emerge for *oppositely charged* particles, when electroneutrality holds due to conditions $n_A(t) = n_B(t) = n(t)$, particle charge $e_A = -e_B = e$. Let us introduce, following the Debye–Hückel method, the self-consistent potential ϕ through Poisson equation

$$\Delta \phi = \frac{4\pi e}{\varepsilon} \left[C_A(\vec{r}, t) - C_B(\vec{r}, t) \right], \qquad (2.2.35)$$

where Δ is the Laplace operator, ε is a static dielectric constant. The equations (2.2.20) and (2.2.21) have to be modified by adding force terms due to diffusion drift in the potential ϕ which is defined according (2.2.35) through the concentration difference $Z(\vec{r},t)$ (provided $D_A = D_B$).

The relevant calculations performed by Ovchinnikov and Burlatsky [84] showed that – in line with general physical ideas – the peculiarity of the fluctuation spectrum at small k disappears due to Coulomb repulsion. Automatically it transforms the long-time asymptotics into that known in formal kinetics (2.1.1).

2.2.2 The birth–death formalism and master equations

2.2.2.1 The Chapmen–Kolmogorov master equation

The approach developed in Section 2.2.1 is based on the independent treatment of the deterministic motion of a system and density fluctuations therein. The reaction description in terms of *random process* seems to be more consistent and logical. Equations (2.1.2) which were used above for a system with complete mixing of particles in a volume V could be rewritten through numbers of particles

$$\frac{\mathrm{d}N_i(t)}{\mathrm{d}t} = a_i^j N_j(t) + \frac{b_i^{jk}}{V} N_j(t) N_k(t). \tag{2.2.36}$$

Since the formal chemical kinetics operates with large numbers of particles participating in reaction, they could be considered as *continuous* variables. However, taking into account the atomistic nature of defects, consider hereafter these numbers N_i as random integer variables. The chemical reaction can be treated now as the birth–death process with individual reaction events accompanied by creation and disappearance of several particles, in a line with the actual reaction scheme [16, 21, 27, 64, 65]. Describing the state of a system by a vector $\boldsymbol{N} = N_1, \ldots, N_s$, we can use the *Chapmen–Kolmogorov master equation* [27] for the distribution function $P(\boldsymbol{N}, t)$

$$\frac{\partial P(\boldsymbol{N}, t)}{\partial t}$$
$$= \sum_{\boldsymbol{N}_1} \left[W(\boldsymbol{N} \mid \boldsymbol{N}') P(\boldsymbol{N}_1, t) - W(\boldsymbol{N}' \mid \boldsymbol{N}) P(\boldsymbol{N}, t) \right]. \tag{2.2.37}$$

Entering this equation transition probability $W(\boldsymbol{N} \mid \boldsymbol{N}')$ of the Markov process depends on the states $\boldsymbol{N}, \boldsymbol{N}'$ only. For mono- and bimolecular reactions these transition probabilities are not zero if vectors $\boldsymbol{N}$ and $\boldsymbol{N}'$ differ by several projections only. To specify $W(\boldsymbol{N} \mid \boldsymbol{N}')$ in equation (2.2.37), one has to start from the equations (2.2.36) for the formal kinetics accompanied with some probabilistic arguments.

To illustrate this approach, let us consider the $A + A \to B$ reaction. In this case the equation analogous to (2.1.10) reads (provided $\boldsymbol{N} = N_A = N$)

$$\frac{\mathrm{d}N(t)}{\mathrm{d}t} = -kN^2(t), \tag{2.2.38}$$

where $k = K/V$.

A number of collisions in the N-particle system is proportional to $N(N-1)$ rather than N^2 entering equation (2.2.38). The only transition has non-zero probability – which takes away particles from a system

$$W(N-2 \mid N) = \frac{k}{2} N(N-1). \tag{2.2.39}$$

The master equation reads

$$\frac{\partial P(N,t)}{\partial t} = W(N \mid N+2)P(N+2,t) - $$
$$- W(N-2 \mid N)P(N,t). \qquad (2.2.40)$$

Defining the average $\langle N \rangle = N(t)$ with the help of distribution function $P(N,t)$, $\langle N \rangle = \sum N P(N,t)$, one gets

$$\frac{dN(t)}{dt} = -k\langle N(N-1)\rangle. \qquad (2.2.41)$$

Returning now to the concentration $n(t) = N(t)/V$, and finding the dispersion of a number of particles in volume V,

$$\sigma_N^2 = \langle (N - \langle N \rangle)^2 \rangle, \qquad (2.2.42)$$

we finally get the kinetic equation

$$\frac{dn(t)}{dt} = -Kn^2(t) + \frac{K}{V^2}\left[\langle N \rangle - \sigma_N^2\right]. \qquad (2.2.43)$$

Its obvious peculiarity as compared with the standard chemical kinetics, equation (2.1.10), is the emergence of the fluctuational second term in r.h.s. The stochastic reaction description by means of equation (2.2.37) permits us to obtain the equation for dispersions σ_N^2 which, however, contains higher-order momenta. It leads to the distinctive infinite set of deterministic equations describing various average quantities, characterizing the fluctuational spectrum.

In terms of the master equation for the Markov process the formal kinetics is nothing but the mean-field theory where the fluctuation terms like that on the r.h.s. of equation (2.2.43) are neglected. Strictly speaking, the macroscopic description, equation (2.1.2), were correct if the fluctuation terms vanished as $V \to \infty$. In a general case the function $P(N,t)$ *does not* satisfy the Poisson distribution [16, 27]; in particular, $\sigma_N^2 \neq \langle N \rangle$.

2.2.2.2 Generating function

An efficient method permitting us to avoid the standard procedure of deriving, with the help of (2.2.37) the infinite set of equations for random value, dispersions and their higher momenta is the presentation of (2.2.37) in a form of the *generating function*. For a single random variable with the distribution function $P(N,t)$ the generating function $F(\zeta,t)$ is defined by the relation

$$F(\zeta,t) = \sum_{N=0}^{\infty} \zeta^N P(N,t). \qquad (2.2.44)$$

In terms of the method of the complex variable functions, $P(N,t)$ could be expressed through $F(\zeta,t)$:

$$P(N,t) = \frac{1}{2\pi i} \oint F(\zeta,t) \frac{d\zeta}{\zeta^{N+1}}, \qquad (2.2.45)$$

where the point $\zeta = 0$ falls into integration region. However, it is important to stress here that there is no need to use (2.2.45) since all statistical calculations could be carried out with the help of the generating function.

The normalization condition imposed on the distribution function gives the following condition for the generating function:

$$\sum_{N=0}^{\infty} P(N,t) = F(\zeta = 1, t) = 1. \qquad (2.2.46)$$

The mean value could be calculated as

$$\langle N \rangle = \left. \frac{\partial F(\zeta,t)}{\partial \zeta} \right|_{\zeta=1}, \qquad (2.2.47)$$

whereas dispersion

$$\sigma_N^2 = \left[\frac{\partial}{\partial \zeta} \zeta \frac{\partial}{\partial \zeta} F(\zeta,t) \right]_{\zeta=1} - \left[\frac{\partial F(\zeta,t)}{\partial \zeta} \right]_{\zeta=1}^2. \qquad (2.2.48)$$

To illustrate what was above said, consider the simple bimolecular process

$$\text{E} \to \text{A}, \qquad \text{A} + \text{A} = \text{B}, \qquad (2.2.49)$$

described in terms of the formal kinetics by the equations ($n_A(t) = n(t)$)

$$\frac{dn(t)}{dt} = p - K n^2(t). \qquad (2.2.50)$$

The treatment of stochasticity on a mesoscopic level

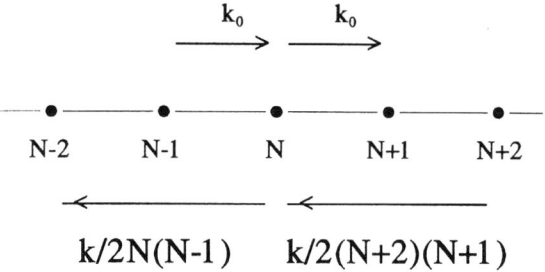

Fig. 2.11. Possible transitions on a site lattice and the corresponding transition rates.

Passing now to a new variable $N(t) = n(t)V$, we can rewrite it in a form

$$\frac{dN(t)}{dt} = k_0 - kN^2(t), \qquad (2.2.51)$$

where $k_0 = pV$, $k = K/V$. It has a single stationary state

$$N_0 = \sqrt{k_0/k}, \qquad (2.2.52)$$

which is asymptotically stable. The basic kinetic equation in a given case reads

$$\frac{\partial P(N,t)}{\partial t} = k_0 P(N-1,t) - k_0 P(N,t) +$$
$$+ \frac{k}{2}(N+1)(N+2)P(N+2,t) -$$
$$- \frac{k}{2}(N-1)NP(N,t). \qquad (2.2.53)$$

From a formal point of view, (2.2.53) describes random walks on a one-dimensional lattice of enumerated sites. Unlike standard problems with constant transition probabilities between sites, in (2.2.53) these probabilities depend on a site number and are essentially non-linear. Figure 2.11 shows possible transitions in the model under consideration and the relevant transition rates.

To derive an equation for the generation function, one has to multiply (2.2.53) by ζ^N and to sum over N which yields

$$\frac{\partial F(\zeta,t)}{\partial t} = k_0(\zeta-1)F(\zeta,t) + \frac{k}{2}(1-\zeta^2)\frac{\partial^2 F(\zeta,t)}{\partial \zeta^2}. \tag{2.2.54}$$

This is a partial differential equation of the second order. The stationary solution $F_0(\zeta)$ of (2.2.54) satisfies the equation

$$(1+\zeta)\frac{d^2 F_0(\zeta)}{d\zeta^2} - 2N_0^2 F_0(\zeta) = 0, \tag{2.2.55}$$

where N_0 is defined by (2.2.52). To solve (2.2.55), two boundary conditions should be specified. One of them comes from (2.2.46): $F_0(\zeta=1)=1$. The second one, especially for complicated processes in which the generation function is a solution of higher-order partial equation, is less evident. When defining the generation function (2.2.44), $|\zeta|$ cannot exceed unity, otherwise convergence of a series for $F(\zeta,t)$ becomes questionable. This is why the boundary conditions should be defined at points satisfying $|\zeta|=1$. That at $\zeta=1$ is found less evident treatment [66] for $\zeta=-1$ leads to $F_0(\zeta=-1)=0$.

A solution of (2.2.55) subject to these boundary conditions is

$$F_0(\zeta) = \vartheta \frac{I_1(4N_0\vartheta)}{I_1(4N_0)}, \tag{2.2.56}$$

where I_1 is the Bessel function of imaginary argument, $\vartheta = (\zeta+1)/2$. Using (2.2.47), (2.2.48), one can calculate mean value and dispersion, also taking into account that in a microscopic system $N_0 \gg 1$. Employing asymptotic expansion of the Bessel functions for large arguments, one gets

$$\langle N \rangle = N_0 + \frac{1}{8} + O(1/N_0). \tag{2.2.57}$$

In other words, description of kinetics in terms of a simpler equation (2.2.51) in the stationary state turns out to be correct with great accuracy. However, the fluctuation dispersion

$$\sigma_N^2 = \frac{3N_0}{4} + O(1) \approx \frac{3\langle N \rangle}{4} + O(1) \tag{2.2.58}$$

is not consistent with the Poisson distribution. It demonstrates that even simple models based on the basic kinetic equation (2.2.37) reveal a non-Poisson distribution function of fluctuations of random values N.

2.2.2.3 The Lotka–Volterra model

More interesting aspects of stochastic problems are observed when passing to systems with unstable stationary points. Since we restrict ourselves to mono- and bimolecular reactions with a maximum of two intermediate products (freedom degrees), $s = 2$, only the Lotka–Volterra model by reasons discussed in Section 2.1.1 can serve as the analog of unstable systems.

This model is based on the reaction scheme (2.1.27) and the kinetic equations (2.1.28) and (2.1.29) which in variables $N_A(t) = n_A(t)V$, $N_B(t) = n_B(t)V$ read

$$\frac{dN_A(t)}{dt} = \alpha N_A(t) - kN_A(t)N_B(t), \tag{2.2.59}$$

$$\frac{dN_B(t)}{dt} = kN_A(t)N_B(t) - \beta N_B(t), \tag{2.2.60}$$

where $k = K/V$. Stationary and periodic solutions of (2.2.59), (2.2.60) are not asymptotically stable. The stationary point of this system is a *centre* (see Section 2.1.1), a set of equations is not rough. This is why one can expect that passage from the deterministic treatment of (2.2.59), (2.2.60) to stochastic one, based on (2.2.37), can cardinaly affect the system's stability.

In the probabilistic description of this model [16, 67–69] the distribution function $P(N,t)$; $N = N_A, N_B$, is used. To perform the transition to (2.2.37), the transition probabilities $W(N \mid N')$ should be specified. In the case of $A + E \to 2A$ reaction, this probability is

$$W(N_A + 1, N_B \mid N) = \alpha N_A. \tag{2.2.61}$$

For the $A + B \to 2B$ reaction, we have

$$W(N_A - 1, N_B + 1 \mid N) = kN_A N_B. \tag{2.2.62}$$

Lastly, for the $B \to P$ reaction

$$W(N_A, N_B - 1 \mid N) = \beta N_B. \tag{2.2.63}$$

Summing up (2.2.61) to (2.2.63) we arrive at the basic kinetic equation

$$\frac{dP(\mathbf{N},t)}{dt} = \alpha(N_A - 1)P(N_A - 1, N_B, t) - \alpha N_A P(\mathbf{N}, t) +$$
$$+ k(N_A + 1)(N_B - 1)P(N_A + 1, N_B - 1, t) -$$
$$- kN_A N_B P(\mathbf{N}, t) +$$
$$+ \beta(N_B + 1)P(N_A, N_B + 1, t) - \beta N_B P(\mathbf{N}, t). \quad (2.2.64)$$

From a formal point of view it describes random walks on a square lattice with the non-linear transition probabilities. All possible kinds of transitions and their probabilities are given in Fig. 2.12.

Using (2.2.64), we obtain equations for mean values $N_A(t) = \langle N_A \rangle$ and $N_B(t) = \langle N_B \rangle$

$$\frac{dN_A(t)}{dt} = \alpha N_A(t) - k\langle N_A N_B \rangle, \qquad (2.2.65)$$

$$\frac{dN_B(t)}{dt} = k\langle N_A N_B \rangle - \beta N_B(t). \qquad (2.2.66)$$

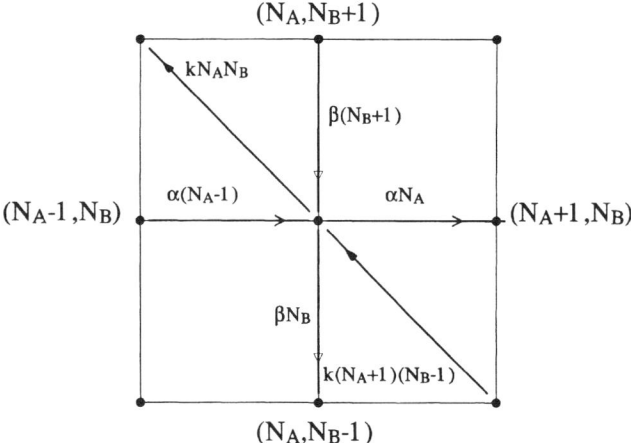

Fig. 2.12. Possible transitions in the Lotka–Volterra model.

Since random values N_A and N_B are not statistically independent, the condition $\langle N_A N_B \rangle = \langle N_A \rangle \langle N_B \rangle$ does not hold and hence (2.2.65) and (2.2.60) do not coincide with the deterministic equations (2.2.59), (2.2.60).

Let us define the correlation function $\langle \delta N_A \delta N_B \rangle$, where $\delta N_i = N_i - \langle N_i \rangle$, and new time-dependent *reaction rate*

$$k^* = k^*(t) = k\left(1 + \frac{\langle \delta N_A \delta N_B \rangle}{N_A(t) N_B(t)}\right). \tag{2.2.67}$$

With its help a set (2.2.65), (2.2.66) could be rewritten in a form similar to (2.2.59), (2.2.65)

$$\frac{dN_A(t)}{dt} = \alpha N_A(t) - k^* N_A(t) N_B(t), \tag{2.2.68}$$

$$\frac{dN_B(t)}{dt} = k^* N_A(t) N_B(t) - \beta N_B(t). \tag{2.2.69}$$

As was noted in Section 2.1.1, the concentration oscillations observed in the Lotka–Volterra model based on kinetic equations (2.1.28), (2.1.29) (or (2.2.59), (2.2.60)) are formally undamped. Perturbation of the model parameters, in particular constant k, leads to transitions between different orbits. However, the stability of solutions requires special analysis. Assume that in a given model relation between averages and fluctuations is very simple, e.g., $\langle \delta N_A \delta N_B \rangle = f(\langle N_A \rangle, \langle N_B \rangle)$, where f is an arbitrary function. Therefore k^* in (2.2.67) is also a function of the mean values $N_A(t)$ and $N_B(t)$. Models of this kind are well developed in population dynamics in biophysics [70]. Since non-linearity of kinetic equations is no longer quadratic, limitations of the Hanusse theorem [23] are lifted. Depending on the actual expression for f both stable and unstable stationary points could be obtained. Unstable stationary points are associated with such solutions as the limiting cycle; in particular, solutions which are interpreted in biophysics as *catastrophes* (population death). Unlike phenomenological models treated in biophysics [70], in the Lotka–Volterra stochastic model the relation between fluctuations and mean values could be indeed *calculated* rather than postulated.

Simulation of the random walks on a site lattice is presented in Figs 2.13 and 2.14; they show that stochastic trajectories deviate systematically from the stationary solution [16]. Alongside those which correspond to the damping oscillations, above mentioned catastrophes are also observed and characterized by $N_B = 0$, and $N_A \to \infty$. These results demonstrate indirectly

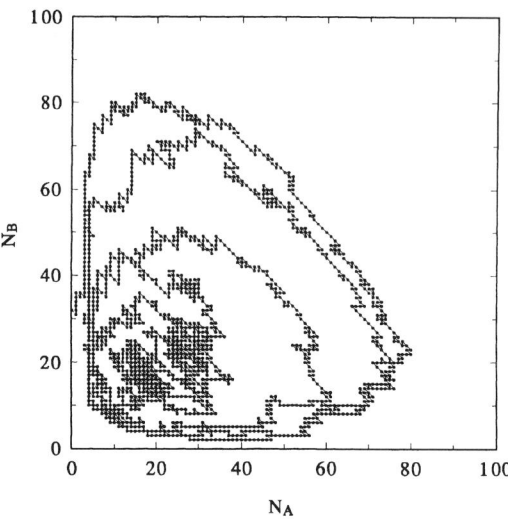

Fig. 2.13. The random trajectory in the stochastic Lotka–Volterra model, equation (2.2.64). Parameters are: $\alpha/k = \beta/k = 20$, the initial values $N_A = N_B = 20$. When the trajectory coincides with the N_B axis, prey animals A are dying out first and predators second.

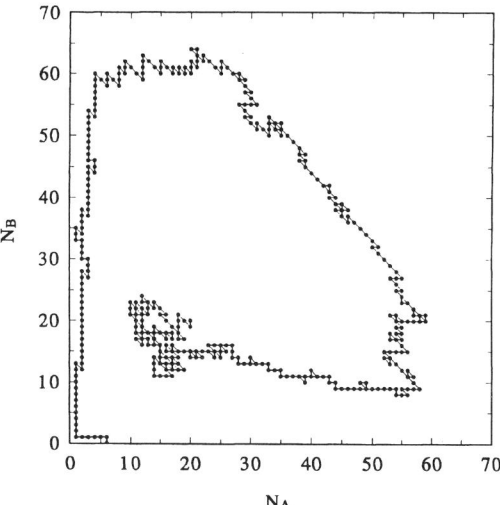

Fig. 2.14. As in Fig. 2.13 but the trajectory touches the N_A axis (death of predators and the infinite increase in the population of the prey animals).

considerable importance of fluctuations. It is shown analytically [69] that (2.2.64) has only a single and trivial solution: $P(N_A, N_B) = \delta_{N_A,0} \delta_{N_B,0}$.

Analysis of equations for second momenta like $\langle \delta N_A \delta N_B \rangle$, $\langle (\delta N_A)^2 \rangle$ and $\langle (\delta N_B)^2 \rangle$ shows that all their solutions are time-dependent. In the Lotka–Volterra model second momenta are oscillating with frequencies *larger* than that of macroscopic motion without fluctuations (2.2.59), (2.2.60). Oscillations of k^* produce respectively *noise* in (2.2.68), (2.2.69). Fluctuations in the Lotka–Volterra model are anomalous; second momenta are not expressed through mean values. Since this situation reminds the turbulence in hydrodynamics, the fluctuation regime in this model is called also *generalized turbulence* [68]. The above noted increase in fluctuations makes doubtful the standard procedure of the cut off of a set of equations for random values momenta.

The generating function of (2.2.64) could be constructed analogously to the definition (2.2.44):

$$F(\zeta_A, \zeta_B, t) = \sum_{N_A, N_B=0}^{\infty} \zeta_A^{N_A} \zeta_B^{N_B} P(N_A, N_B, t). \qquad (2.2.70)$$

The corresponding equation for the generating function

$$\frac{\partial F(\zeta_A, \zeta_B, t)}{\partial t} = \alpha \zeta_A (\zeta_A - 1) \frac{\partial}{\partial \zeta_A} F(\zeta_A, \zeta_B, t) +$$

$$+ k \zeta_B (\zeta_B - \zeta_A) \frac{\partial^2}{\partial \zeta_A \partial \zeta_B} F(\zeta_A, \zeta_B, t) -$$

$$- \beta (\zeta_B - 1) \frac{\partial}{\partial \zeta_B} F(\zeta_A, \zeta_B, t). \qquad (2.2.71)$$

Since variables ζ_A and ζ_B are not separated in (2.2.71), random values N_A and N_B are always correlated. On the other hand, this peculiarity of the equation does not permit to solve it exactly and thus asymptotic expansion has to be used. Equation (2.2.71) has no other stationary solution except trivial $F(\zeta_A, \zeta_B) = 1$, corresponding to $P(N_A, N_B) = \delta_{N_A,0} \delta_{N_B,0}$. An asymptotic solution (2.2.71) is sought in the $V \to \infty$ limit (system's volume is a large parameter), when one can assume [16], that

$$F(\zeta_A, \zeta_B, t) = \exp(V \Psi(\zeta_A, \zeta_B, t)), \qquad (2.2.72)$$

where $\Psi(\zeta_A, \zeta_B, t)$ is a value of the order of unity.

2.2.2.4 The Lotka model

In terms of the deterministic approach the Lotka model (2.1.21) is described by a set of equations

$$\frac{dN_A(t)}{dt} = k_0 - kN_A(t)N_B(t), \tag{2.2.73}$$

$$\frac{dN_B(t)}{dt} = kN_A(t)N_B(t) - \beta N_B(t), \tag{2.2.74}$$

where $k = K/V$, $k_0 = pV$. When passing to the stochastic language, one can use results obtained earlier for the Lotka–Volterra model. Here the $A + E \to 2A$ reaction is replaced by $E \to A$, the corresponding transition probability is

$$W(N_A + 1, N_B \mid \mathbf{N}) = k_0. \tag{2.2.75}$$

Figure 2.15 demonstrates a scheme of transitions. The corresponding basic kinetic equation is

$$\frac{dP(\mathbf{N},t)}{dt} = k_0 P(N_A - 1, N_B, t) - k_0 P(\mathbf{N},t) +$$
$$+ k(N_A + 1)(N_B - 1)P(N_A + 1, N_B - 1, t) -$$
$$- kN_A N_B P(\mathbf{N},t) +$$
$$+ \beta(N_B + 1)P(N_A, N_B + 1, t) - \beta N_B P(\mathbf{N},t). \tag{2.2.76}$$

Results of the stochastic simulations for the Lotka model are presented in Fig. 2.16.

The equation for generating function reads

$$\frac{\partial F(\zeta_A, \zeta_B, t)}{\partial t} = k_0(\zeta_A - 1)F(\zeta_A, \zeta_B, t) +$$
$$+ k\zeta_B(\zeta_B - \zeta_A)\frac{\partial^2}{\partial \zeta_A \partial \zeta_B} F(\zeta_A, \zeta_B, t) -$$
$$- \beta(\zeta_B - 1)\frac{\partial}{\partial \zeta_B} F(\zeta_A, \zeta_B, t). \tag{2.2.77}$$

Equations (2.2.76), (2.2.77) of the Lotka model are not analyzed so far. We suggest the readers to solve this problem as a home exercise. Despite

The treatment of stochasticity on a mesoscopic level

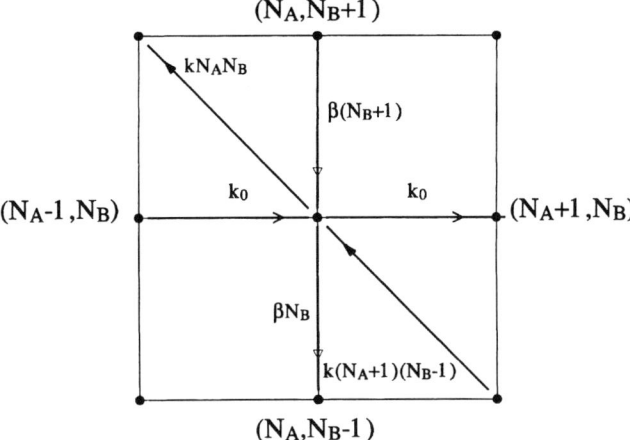

Fig. 2.15. Possible transitions in the Lotka model.

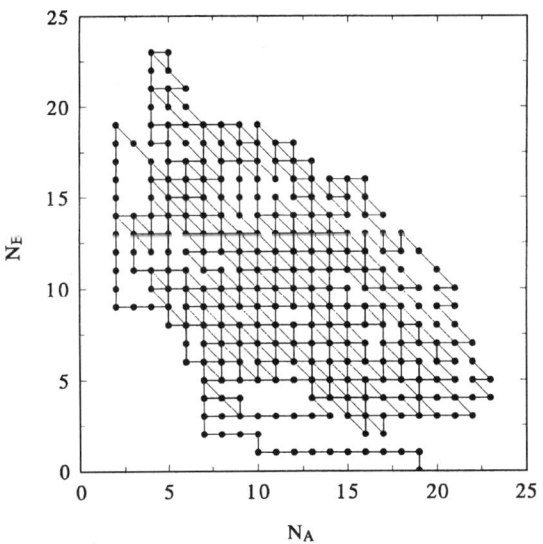

Fig. 2.16. The random trajectory in the stochastic Lotka model, equation (2.2.76). Parameters are $k_0/\beta = \beta/k = 10$, the initial values $N_A = N_B = 10$. When the trajectory touches the N_A axis, the predators B are dying out and the population of the prey animals A infinitely increases.

the fact that in terms of a formal kinetics approach the Lotka model is an example of a *rough* system (fluctuations are damped), as we noted in Section 2.1.1, for large β parameters damping could be done as small as one desires. When taking into account the fluctuation effects, the damping of macroscopic variables could turn out to be not enough to suppress "noise" emerging due to a motion of second momenta of random values, i.e., k oscillations in a set of the equations

$$\frac{dN_A(t)}{dt} = k_0 - k^* N_A(t) N_B(t), \qquad (2.2.78)$$

$$\frac{dN_B(t)}{dt} = k^* N_A(t) N_B(t) - \beta N_B(t). \qquad (2.2.79)$$

The variable k^* is defined by (2.2.67). Therefore one can expect that depending on actual parameters k_0, k and β, both stationary and non-stationary fluctuation distribution could exist.

2.2.2.5 Non-equilibrium critical phenomena

These examples clearly demonstrate the role of non-Poisson fluctuations of a number of reactants in a system. In the vicinity of the bifurcation point of equations (2.1.2), the asymptotical instability of their periodic and stationary solutions is observed and the fluctuation dispersion can infinitely increase in time deviating from Poisson regime [27]. Therefore, after prolonged time the fluctuations govern time behaviour of average quantities (concentrations). The role of perturbation here is related to the non-Poisson part of the fluctuation spectrum: one cannot neglect the higher-order momenta (dispersions) in systems with asymptotically unstable solutions. The cut-off of the infinite set of coupled equations, typical for the problem with non-linear transition probabilities $W(N \mid N')$, leads to approximate solutions which are valid at short times only. One can see here the extremely interesting analog between instabilities observed in non-equilibrium kinetic processes (reactions) and the equilibrium critical phenomena [16, 71, 72]. For systems far from the critical point the macroscopic averages and fluctuations are incomparable in their magnitudes, but near the critical point the large-number law fails: it is fluctuations which now govern the behaviour of average quantities (concentrations $n_i(t)$) and thus a system can no longer be treated in terms of these average quantities only [16, 20].

Since a number of particles involved in any reaction event are small, a change in concentration is of the order of $1/V$. Therefore, we can use for the system with complete particle mixing the asymptotic expansion in this small parameter $1/V$. The corresponding van Kampen [73, 74] procedure (see also [27, 75]) permits us to formulate simple rules for deriving the Fokker–Planck or stochastic differential equations, asymptotically equivalent to the initial master equation (2.2.37). It allows us also to obtain coefficients G_{ij} in the stochastic differential equation (2.2.2) thus liquidating their uncertainty and strengthening the relation between the deterministic description of motion and density fluctuations.

The bottleneck of this approach is obvious: the expression for transition probabilities through collective variables of a whole system (total number of particles) means that rather rare fluctuations are taken into account only, whereas their spatial correlations are neglected (i.e., different parts of a system interact being separated by a distinctive distance – the correlation length).

2.2.2.6 Spatially-extended systems

A more refined approach is based on the *local* description of fluctuations in non-equilibrium systems, which permits us to treat fluctuations of all spatial scales as well as their correlations. The birth–death formalism is applied here to the physically infinitesimal volume v_0, which is related to the rest of a system due to the diffusion process. To describe fluctuations in spatially extended systems, the whole volume is divided into blocks having distinctive sizes λ_0 ($v_0 = \lambda^{\bar{d}}$, $\bar{d} = 1, 2, 3$ is the space dimension). Enumerating these cells with the discrete variable $\vec{r}$ and defining the number of particles $N_i(\vec{r})$ therein, we can introduce the joint probability of arbitrary particle distribution over cells. Particle diffusion is also considered in terms of particle death in a given cell accompanied with particle birth in the nearest cell.

The cell sizes are expected to exceed any molecular (atomic) scale so that a number of particles therein are large, $N_i(\vec{r}) \gg 1$. The transition probabilities within cells are defined by reaction rates entering (2.1.2), whereas the hopping probabilities between close cells could easily be expressed through diffusion coefficients. This approach was successfully applied to the non-linear systems characterized by a loss of stability of macroscopic structures and the very important effect of a qualitative change of fluctuation dispersion as the fluctuation length increases has also been observed [16, 27]. In particular cases the correlation length ξ could be the introduced. The fluctuations in volumes having size less than ξ obey the Poisson distribution. To characterize this situation, Nicolis and Prigogine [16] introduced the concept of *local equilibrium*. The large-scale fluctuations demonstrate well pronounced non-equilibrium character quite different from the Poisson distribution. It is the emergence of fluctuations which makes a system to change its macroscopic state which has clear analogy with the equilibrium critical phenomena.

Applying these methods to systems in the vicinity of the non-equilibrium critical points, the conclusion was drawn [72] that the mesoscopic approach contains excess information about spatial particle distribution: the details of how the whole system's volume is divided into cells become unimportant as $\xi \to \infty$. The possibility to employ expansion in inverse powers of v_0 – similarly to a complete mixing case – was also discussed. Asymptotically it leads to the Focker–Planck equation equivalent to the Langevin-like equation.

Summing this Section up, we would like to note that in the approach discussed here the introduction of stochasticity on a mesoscopic level restricts the applicability of a method by such statements of a problem where subtle details of particle interaction become unimportant. First of all, we mean that kinetic processes with non-equilibrium critical points, when at long reaction time the correlation length exceeds all other spatial dimensions. This limitation makes us consider in the next Section 2.3 the *microscopic* level of the kinetic description.

2.3 MICROSCOPIC TREATMENT OF STOCHASTICITY

> Everything should be done as simple as possible but not simpler.
> A. Einstein

2.3.1 Many-point densities

2.3.1.1 Statistics of many-particle systems

To describe quantitatively the spatial structure of many-particle systems, the statistical physics methods widely used for condensed systems could naturally be considered as an adequate tool for reactions. An analog of a homogeneous stem of the interacting particles are condensed media without long-range order (e.g., dense gases, liquids, glasses) [76, 77]. Their only principal difference is that in a system with reaction a number of particles is varying.

For the preliminary treatment, let us consider the simplest one-component system, e.g., thermodynamically equilibrium liquid, having a fixed number of particles. Introduce the microscopic density of a particle number defining it with the help of the Dirac δ-function:

$$\hat{n}(\vec{r}) = \sum_i \delta(\vec{r} - \vec{r}_i), \tag{2.3.1}$$

where $\vec{r}_i$ are particle coordinates. The Gibbs distribution is used in an ensemble averaging $\langle \ldots \rangle$. In a homogeneous equilibrium system the mean value

$$\langle \hat{n}(\vec{r}) \rangle = n \tag{2.3.2}$$

is just macroscopic particle density. Take some arbitrary volume V and calculate the dispersion of the density fluctuations. The mean number of particles

$$\langle N \rangle = \left\langle \int \hat{n}(\vec{r}) \, d\vec{r} \right\rangle = \int \langle \hat{n}(\vec{r}) \rangle d\vec{r} = nV. \tag{2.3.3}$$

Fluctuation dispersion $\sigma_N^2 = \langle (N - \langle N \rangle)^2 \rangle \equiv \langle N^2 \rangle - \langle N \rangle^2$ could be expressed through mean squared of N:

$$\langle N^2 \rangle = \left\langle \int \hat{n}(\vec{r}) \, d\vec{r} \int \hat{n}(\vec{r}') \, d\vec{r}' \right\rangle$$

$$- \iint \langle \hat{n}(\vec{r}) \hat{n}(\vec{r}') \rangle \, d\vec{r} \, d\vec{r}'. \tag{2.3.4}$$

Let us consider mean value $\langle \hat{n}(\vec{r}) \hat{n}(\vec{r}') \rangle$. Multiplying $\hat{n}(\vec{r})$ and $\hat{n}(\vec{r}')$, we can present a product of sums like (2.3.1) in a form

$$\hat{n}(\vec{r}) \hat{n}(\vec{r}') = \sum_i \delta(\vec{r} - \vec{r}_i) \delta(\vec{r}' - \vec{r}_i) +$$

$$+ \sum_{i \neq j} \delta(\vec{r} - \vec{r}_i) \delta(\vec{r}' - \vec{r}_j). \tag{2.3.5}$$

Using the well-known δ-function property:

$$\delta(\vec{r} - \vec{r}_i) \delta(\vec{r}' - \vec{r}_i) = \delta(\vec{r} - \vec{r}') \delta(\vec{r} - \vec{r}_i), \tag{2.3.6}$$

we can express the mean of (2.3.5) as

$$\langle \hat{n}(\vec{r})\hat{n}(\vec{r}')\rangle = n\delta(\vec{r}-\vec{r}') + \rho_2(|\vec{r}-\vec{r}'|). \tag{2.3.7}$$

As it is seen from the derivation, only the second, non-singular term

$$\rho_2(|\vec{r}-\vec{r}'|) = \left\langle \sum_{i\neq j} \delta(\vec{r}-\vec{r}_i)\delta(\vec{r}'-\vec{r}_j) \right\rangle, \tag{2.3.8}$$

is related to the spatial particle correlation. The first term emerges due to repeated registration of the same particles. Since physical effects arise due to particle interaction, a singular "self-action" term enters mean values. In particular, the internal system's energy is a functional of $\rho_2(r)$. As $|\vec{r}-\vec{r}'| \to \infty$, microscopic densities become statistically independent (correlation weakening) and thus

$$\langle \hat{n}(\vec{r})\hat{n}(\vec{r}')\rangle = \langle \hat{n}(\vec{r})\rangle\langle \hat{n}(\vec{r}')\rangle = n^2. \tag{2.3.9}$$

Therefore $\rho_2(\infty) = n^2$. It is convenient to eliminate dimension-dependent concentration co-factor n^2, defining the *joint correlation function*

$$\chi^{(2)}(r) = \frac{\rho_2(r)}{n^2}. \tag{2.3.10}$$

Returning now to (2.3.4), after substitution (2.3.7) into (2.3.10) we get

$$\langle N^2 \rangle = nV + n^2 \iint \chi^{(2)}(|\vec{r}-\vec{r}'|)\,d\vec{r}\,d\vec{r}'. \tag{2.3.11}$$

The fluctuation dispersion reads

$$\sigma_N^2 = nV + n^2 \iint \left(\chi^{(2)}(|\vec{r}-\vec{r}'|) - 1\right) d\vec{r}\,d\vec{r}'. \tag{2.3.12}$$

For large volumes $V \to \infty$, (2.3.12) can be simplified

$$\frac{\sigma_N^2}{\langle N \rangle} = 1 + n \int \left(\chi^{(2)}(r) - 1\right) d\vec{r}. \tag{2.3.13}$$

Since the function $\chi^{(2)}(r)$ emerges (in calculations) usually in a form of $\chi^{(2)}(r) - 1$, in the condensed system's statistics under the correlation function

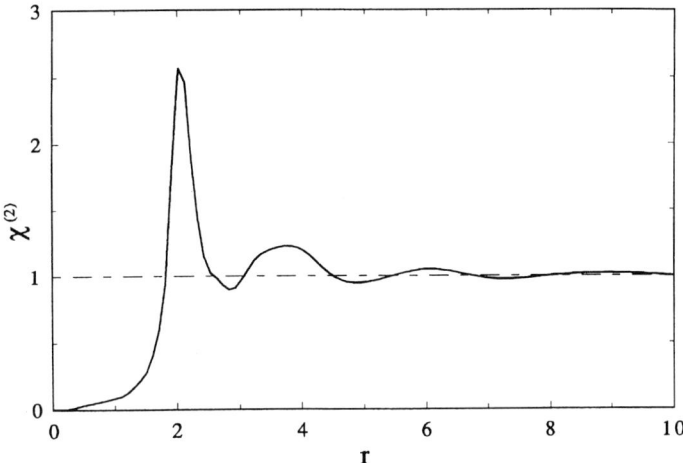

Fig. 2.17. A pattern of joint correlation function in a homogeneous condensed system.

one often means $h(r) = \chi^{(2)}(r) - 1$ with the asymptotics $h(\infty) = 0$. The qualitative behaviour of the correlation function of dense gases and liquids is given in Fig. 2.17.

The joint correlation function $\chi^{(2)}(r)$ characterizes short and intermediate orders in particle spatial relative distribution. However, what is observed in diffraction experiments is not the joint correlation function but the corresponding *structure factor* [77]

$$S(k) = S(\vec{k}) = 1 + n \int h(r) \exp\left(-i\vec{k}\vec{r}\right) d\vec{r}. \qquad (2.3.14)$$

Relation (2.3.14) could be inverted

$$\chi^{(2)}(r) = 1 + \frac{1}{2\pi^2 nr} \int_0^\infty \{S(k) - 1\} \sin(kr) k \, dk. \qquad (2.3.15)$$

Taking into account (2.3.14), (2.3.13) can be rewritten as

$$\frac{\sigma_N^2}{\langle N \rangle} = S(0). \qquad (2.3.16)$$

Since the Poisson fluctuation spectrum results in $\sigma_N^2 = \langle N \rangle$, the second term in r.h.s. of (2.3.13) defines the deviation of fluctuations from the Poisson one. This deviation is more pronounced near *the critical points* [76, 78] characterized by anomalous fluctuations. The contribution of integral into (2.3.13) depends not on a deviation of the correlation function $\chi^{(2)}(r)$ from its asymptotical limit (unity) which for large r is small but mainly in the way $\chi^{(2)}(r)$ approaches to the asymptotic limit. An estimate of its behaviour near the critical point after Ornstein and Zernike [79] is

$$S(k) = \frac{S(0)}{1 + \xi^2 k^2}, \qquad (2.3.17)$$

which corresponds to the correlation function at large r

$$h(r) \propto \frac{1}{r} \exp\left(-\frac{r}{\xi}\right). \qquad (2.3.18)$$

The variable ξ is called the *correlation radius*. Near the critical point $\xi \to \infty$ and $h(r)$ decays according to the power law.

In systems with variable number of particles, where random process of migration and recombination is Markov process, it is convenient to develop another scheme of introducing the correlation functions, avoiding singular expressions like (2.3.1) and (2.3.7).

To do it for the statistical description of a system of reactants, let us consider as an example a mixture of two kinds of particles, A's and B's, and two non-overlapping volumes $dV_1 = dV_2 = dV$, $dV \to 0$, centered at $\vec{r}_1$ and $\vec{r}_2$ respectively – Fig. 2.18. Numbers of particles $N_\nu(\vec{r}_i)$ ($\nu = A, B$) within dV_1, dV_2 are stochastic variables, $N_\nu(\vec{r}_i) = 0, 1, 2, \ldots$. The ensemble average at a given moment t is

$$\langle N_\nu(\vec{r}_i) \rangle = dN_\nu. \qquad (2.3.19)$$

Due to the spatial homogeneity it is independent of $\vec{r}_i$ allowing us to introduce the simplest spatial characteristics – macroscopic densities of particles (concentrations)

$$n_i(t) = \frac{dN_\nu}{dV}. \qquad (2.3.20)$$

Since reactant concentrations are observable, they define what is called *the temporal structure* of a process [4, 16].

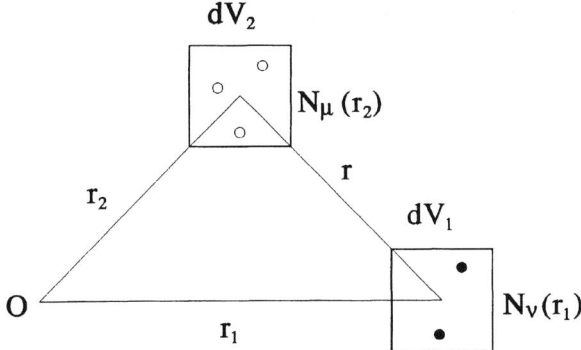

Fig. 2.18. A scheme of the statistical description of the spatially distributed system.

Since concentrations do not characterize the relative spatial distribution of particles (spatial correlations), they yield a structureless description of a system. If there were no spatial correlations, we would arrive at the relation

$$\langle N_\nu(\vec{r}_1)N_\mu(\vec{r}_2)\rangle = \langle N_\nu(\vec{r}_1)\rangle\langle N_\mu(\vec{r}_2)\rangle = \mathrm{d}N_\nu\,\mathrm{d}N_\mu. \qquad (2.3.21)$$

This is why the quantity

$$\chi^{(2)}_{\nu\mu}(r,t) = \frac{\langle N_\nu(\vec{r}_1)N_\mu(\vec{r}_2)\rangle}{\mathrm{d}N_\nu \mathrm{d}N_\mu} \qquad (2.3.22)$$

can serve as a measure of the joint particle correlation. We call the quantity $\chi^{(2)}_{\nu\mu}(r,t)$ the joint correlation functions – similar ($\nu = \mu$) or dissimilar ($\nu \neq \mu$), respectively. These functions depend on a relative distance $r = |\vec{r}| = |\vec{r}_1 - \vec{r}_2|$ only due to space homogeneity and isotropy. The natural boundary condition we put on them means neglect of correlations at large distances

$$\chi^{(2)}_{\nu\mu}(\infty,t) = 1. \qquad (2.3.23)$$

These joint correlation functions have transparent physical interpretation [78]: mean density of a number of ν-kind particles at a distance r from a given μ-kind particle (placed at origin of coordinates) is nothing but

$$C_\nu(r,t) = n_\nu(t)\chi^{(2)}_{\nu\mu}(r,t). \qquad (2.3.24)$$

In the same way triple and higher-order densities could be defined, e.g.,

$$\chi^{(3)}_{\nu\mu\lambda} = \frac{\langle N_\nu(\vec{r}_1)N_\mu(\vec{r}_2)N_\lambda(\vec{r}_3)\rangle}{dN_\nu dN_\mu dN_\lambda}. \qquad (2.3.25)$$

The infinite set of the correlation functions $\chi^{(m)}$, $m = 1, 2, \ldots$ gives a complete statistical description of the spatial structure of a system consisting of interacting particles.

It was Ziman [77] who has noted that there is little hope, at least at present, to develop an experimental technique permitting the direct measurement of these correlation functions. The only exception are the joint densities $\chi^{(2)}_{\nu\mu}(r,t)$, information about which could be learned from the diffraction structural factors of inhomogeneous systems. On the other hand, optical spectroscopy allows estimation of concentrations of such aggregate defects in alkali halide crystals as F_n ($n = 1, 2, 3, 4$) centres, i.e., n nearest anion vacancies trapped n electrons [80]. That is, we can find $\chi^{(m)}$, $m = 1$ to 4, but at small r only. Along with the difficulties known in interpretating structure factors of binary equilibrium systems (gases or liquids), obvious specific complications arise for a system of recombining particles in condensed media which, in its turn, are characterized by their own structure factors.

Therefore, the joint correlation functions $\chi^{(2)}_{\nu\mu}(r,t)$, being at least potentially observable, are more a theoretical than an experimental tool for the description of interacting particles in condensed media. Both these joint functions and macroscopic concentrations $n_\nu(t)$ determine the lowest level to characterize the *spatio-temporal* structure of a system.

Sometimes it is useful to introduce *many-particle* densities. For example, *two-point* density is

$$\rho_{\nu\mu} = \frac{\langle N_\nu(\vec{r}_1)N_\mu(\vec{r}_2)\rangle}{dV_1 dV_2}. \qquad (2.3.26)$$

Taking into account the definition (2.3.22), we obtain

$$\rho_{\nu\mu} = n_\nu(t)n_\mu(t)\chi^{(2)}_{\nu\mu}(r,t). \qquad (2.3.27)$$

These two languages differ by concentration co-factors only, the advantage of the many-particle densities is the compactness of the kinetic equations derived below. To reduce a number of indices, let us denote the joint densities of similar particles as

$$X_\nu(r,t) = \chi^{(2)}_{\nu\nu}(r,t), \qquad (2.3.28)$$

whereas those for dissimilar particles

$$Y(r,t) = \chi^{(2)}_{AB}(r,t) = \chi^{(2)}_{BA}(r,t). \qquad (2.3.29)$$

The correlation function of similar particles, X_ν, could be easily related to the simplest characteristics of spatial particle fluctuations, namely, dispersion of a number of particles within arbitrary volume V. The mean number of ν-kind particles inside V is just

$$\langle N_\nu \rangle = n_\nu(t)V, \qquad (2.3.30)$$

whereas the quantity

$$\frac{\langle (N_\nu - \langle N_\nu \rangle)^2 \rangle}{\langle N_\nu \rangle}$$
$$= 1 + \frac{n_\nu(t)}{V} \iint_V \left[X_\nu(|\vec{r}_1 - \vec{r}_2|) - 1 \right] d\vec{r}_1\, d\vec{r}_2 \qquad (2.3.31)$$

yields a measure of the particle density fluctuations. If there were no correlation of the similar particles, $X_\nu(r,t) - 1$, and r.h.s. of equation (2.3.31) is also a unity, we have the Poisson fluctuation spectrum [78]. On the contrary, any deviation from it results in the *non-Poisson spectrum*. The conclusion suggests that even the lowest level for describing spatio-temporal structure of the bimolecular recombination kinetics we mentioned above (use of concentrations, $n_A(t)$, $n_B(t)$, and joint densities $X_\nu(r,t)$, $Y(r,t)$) also takes into account the fluctuation effects we are interested in.

Since these characteristics are time-dependent, let us assume particle birth–death and migration to be the *Markov stochastic processes*. Note that making use of the stochastic models, we discuss below in detail, does not contradict the deterministic equations employed for these processes. Say, the equations for $n_\nu(t)$, $X_\nu(r,t)$, $Y(r,t)$ given in Section 2.3.1 are deterministic since both the concentrations and joint correlation functions are defined by equations (2.3.2), (2.3.4) just as ensemble average quantities. Note that the complete set of coupled equations for correlation functions is infinite. To handle it one has to cut it off restricting it by a reduced description of a system, e.g., $n_\nu(t), X_\nu(r,t), Y(r,t)$. In those cases when the *microscopic* level turns out to be too complicated, the *mesoscopic* language could be used (see Sections 2.1.2 and 2.2).

Note here that the relation between mesoscopic and microscopic approaches is not trivial. In fact, the former is closer to the *macroscopic* treatment (Section 2.1.1) which neglects the structural characteristics of a system. Passing from the micro- to meso- and, finally, to macroscopic level we loose also the initial statement of a stochastic model of the Markov process. Indeed, the disadvantages of deterministic equations used for rather simplified treatment of bimolecular kinetics (Section 2.1) lead to the macro- and mesoscopic models (Section 2.2) where the stochasticity is kept either by adding the stochastic external forces (Section 2.2.1) or by postulating the master equation itself for the relevant Markov process (Section 2.2.2). In the former case the fluctuation source is assumed to be external, whereas in the latter kinetics of bimolecular reaction and fluctuations are coupled and mutually related. Section 2.3.1.2 is aimed to consider the relation between these three levels as well as to discuss problem of how determinicity and stochasticity can coexist.

2.3.1.2 Cell formalism

Following the approach discussed in Section 2.2.2, let us divide the whole reaction volume V of the spatially extended system into N equivalent cells (domains) [81]. However, there is an essential difference with the mesoscopic level of treatment: in Section 2.2.2 a number of particles in cells were expected to be much greater than unity. Note that this restriction is not imposed on the microscopic level of system's treatment. Their volumes are chosen to be so small that each cell can be occupied by a single particle only. (There is an analogy with the lattice gas model in the theory of phase transitions [76].) Despite the finiteness of v_0 coming from atomistic reasons or lattice discreteness, at the very end we make the limiting transition $v_0 \to 0$, $N \to \infty$, $v_0 N = V$, to the continuous pattern of point dimensionless particles.

Any cell centered at some $\vec{r}$ is characterized by its occupancy number $\nu(\vec{r})$ depending on the actual reaction under study: for $A + B \to 0$, $\nu(\vec{r}) = A$, B and 0 – Fig. 2.19. Now any diffusion or reaction event could be described in terms of time-development of these occupancy numbers. Say, the diffusion motion results in replacement of the configuration $A(\vec{r})0(\vec{r}')$ for $0(\vec{r})A(\vec{r}')$,

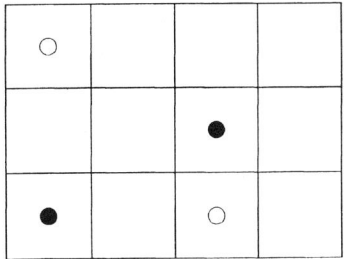

Fig. 2.19. A scheme illustrating the cell formalism. Symbols of empty cells (0) are omitted.

where $\vec{r}$, $\vec{r}'$ are the coordinates of the nearest cells. The recombination event similarly means that $A(\vec{r})B(\vec{r}') \to 0(\vec{r})0(\vec{r}')$. The new point here is that – unlike Section 2.2.2 – *non-local* character of elementary events is taken into account since the relative distances in real space become important here.

The physical state of a system with a varying number of particles is defined uniquely by a set of the population numbers $\nu(\vec{r}_1), \ldots, \nu(\vec{r}_N) = \{\nu(\vec{r})\}_N$. Assuming the reaction is the Markov process, let us introduce the distribution functions (DF's) $P(\{\nu(\vec{r})\}_N; t)$ yielding a complete probabilistic description of the problem. The recurrent relation

$$\sum_{\nu(\vec{r}_m)} P\left(\{\nu(\vec{r})\}_m; t\right) = P(\{\nu(\vec{r})\}_{m-1}; t) \tag{2.3.32}$$

permits us to find the DF's of lower order. The master equation of the Markov process could be presented either through equation for the N-cell DF $P(\{\nu(\vec{r})\}_N; t)$ or via the set of equations for partial DF's $P(\{\nu(\vec{r})\}_m; t), m = 1, 2, \ldots$ [81]. The passage to the continuous model is quite trivial. Consider m points with coordinates $\{\vec{r}\}_m = \vec{r}_1, \ldots, \vec{r}_m$ and m' points with coordinates $\{\vec{r}'\}_{m'} = \vec{r}', \ldots, \vec{r}'_{m'}$. Define now for the reactions $A + B \to 0$, $A + B \to B$ the many-particle densities $\rho_{m,m'}(\{\vec{r}\}_m; \{\vec{r}\}_{m'}; t)$ through the limiting transition $v_0 \to 0$ for an average

$$\frac{\left\langle \prod_{i=1}^{m} N_A(\vec{r}_i) \prod_{j=1}^{m'} N_B(\vec{r}'_j) \right\rangle}{v_0^{(m+m')}} \tag{2.3.33}$$

which can be calculated by means of the DF just given above. All coordinates in equation (2.3.33) are different. Since each cell contains no more than a single particle, A or B, it is easy to check that $\rho_{m,m'}(\{\vec{r}\}_m;\{\vec{r}'\}_{m'};t)$ is indeed the limit of expression

$$\rho_{m,m'}(\{\vec{r}\}_m;\{\vec{r}'\}_{m'};t)$$
$$= \lim_{v_0 \to 0} v_0^{-(m+m')} P(\{A(\vec{r})\}_m;\{B(\vec{r}')\}_{m'};t). \quad (2.3.34)$$

Therefore, the limiting transition $v_0 \to 0$ permits us to transfer the set of equations for DF's into that for the many-particle densities. The many-particle densities containing coinciding coordinates, say, $\vec{r}$ and $\vec{r}'$, are considered as the non-singular limit $|\vec{r} - \vec{r}'| \to 0$ of equation (2.3.34), carried after the limit $v_0 \to 0$. This is due the definition of the initial DF's which do not contain coinciding coordinates. This definition permits us to exclude singular terms, similar to $n\delta(\vec{r} - \vec{r}')$ in (2.3.7). Introducing, by analogy with (2.3.1), microscopic particle densities $\hat{n}_A(\vec{r},t)$ and $\hat{n}_B(\vec{r},t)$, many-point density could be defined as the mean value

$$\left\langle \prod_{i=1}^{m} \hat{n}_A(\vec{r}_i,t) \prod_{j=1}^{m'} \hat{n}_B(\vec{r}_j',t) \right\rangle, \quad (2.3.35)$$

where self-action singular terms are excluded.

2.3.1.3 Master equation
Let us consider now other terms of the kinetic equations.
(i) *Particle recombination*. For the $A + B \to 0$ reaction one gets

$$\frac{\partial}{\partial t} P(\{A(\vec{r})\}_m;\{B(\vec{r}')\}_{m'};t)$$
$$= -\sum_{i=1}^{m}\sum_{j=1}^{m'} \sigma(\vec{r}_i - \vec{r}_j') P(\{A(\vec{r})\}_m;\{B(\vec{r}')\}_{m'};t) -$$
$$- \sum_{\vec{r}'_{m'+1}}\sum_{i=1}^{m} \sigma(\vec{r}_i - \vec{r}'_{m'+1}) P(\{A(\vec{r})\}_m;\{B(\vec{r}')\}_{m'+1};t) -$$
$$- \sum_{\vec{r}_{m+1}}\sum_{j=1}^{m'} \sigma(\vec{r}_{m+1} - \vec{r}_j') P(\{A(\vec{r})\}_{m+1};\{B(\vec{r}')\}_{m'};t), \quad (2.3.36)$$

where $\sigma(\vec{r} - \vec{r}') \equiv \sigma(|\vec{r} - \vec{r}'|)$ is the recombination rate of particles A with B at $\vec{r}$ and $\vec{r}'$. The first term in equation (2.3.36) describes all the possible ways in which m particles A can recombine with m' particles B, whereas the other terms describe the recombination of particles B with A (and vice versa) not belonging to a set of $(m+m')$ particles. When dividing equation (2.3.36) by $v_0^{(m+m')}$, we arrive at

$$\frac{\partial \rho_{m,m'}}{\partial t} = -\sum_{i=1}^{m}\sum_{j=1}^{m'} \sigma(\vec{r}_i - \vec{r}'_j)\rho_{m,m'} -$$

$$-\sum_{i=1}^{m} \int \sigma(\vec{r}_i - \vec{r}'_{m'+1})\rho_{m,m'+1}\, d\vec{r}'_{m'+1} -$$

$$-\sum_{j=1}^{m'} \int \sigma(\vec{r}_{m+1} - \vec{r}'_j)\rho_{m+1,m'}\, d\vec{r}_{m+1}. \qquad (2.3.37)$$

The physical sense of individual contributions into (2.3.37) using the joint density $\rho_{2,1}$ is illustrated in Fig. 2.20.

The limitation in equation (2.3.36) that vectors $\vec{r}_{m+1}$ and $\vec{r}'_{m'+1}$ cannot coincide with $\{\vec{r}\}_m$, $\{\vec{r}'\}_{m'}$ becomes unimportant in the continuous approximation containing integrals instead of cell sums. The above mentioned changes in the DFs due to recombination correspond exactly to those done by Dettmann [82]. In the case of the A + B → B reaction one must omit the last

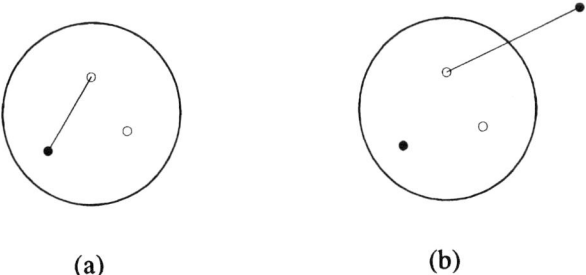

(a)   (b)

Fig. 2.20. Change $\rho_{m,m'}$ due to particle recombination. (a) Reaction of two dissimilar particles from the (m, m')-group. (b) Reaction of one of the particles from the (m, m')-group with the surrounding particles.

term in equations (2.3.36) and (2.3.37) since particles B do not disappear:

$$\frac{\partial \rho_{m,m'}}{\partial t} = - \sum_{i=1}^{m} \sum_{j=1}^{m'} \sigma(\vec{r}_i - \vec{r}_j')\rho_{m,m'} -$$

$$- \sum_{i=1}^{m} \int \sigma(\vec{r}_i - \vec{r}_{m'+1}')\rho_{m,m'+1} \, d\vec{r}_{m'+1}'. \qquad (2.3.38)$$

(ii) *Particle creation.* When considering creation of single particle A or B as well as pairs AB by irradiation in unoccupied cells, one gets for the reaction $0 \to A + B$

$$\frac{\partial}{\partial t} P(\{A(\vec{r})\}_m; \{B(\vec{r}')\}_{m'}; t)$$

$$= \sum_{i=1}^{m} \sum_{j=1}^{m'} \varphi(\vec{r}_i - \vec{r}_j') P(\{A(\vec{r})\}_m^i; \{B(\vec{r}')\}_{m'}^j; t) +$$

$$+ \sum_{\vec{r}} \sum_{i=1}^{m} \varphi(\vec{r}_i - \vec{r}) P(\{A(\vec{r})\}_m^i; \{B(\vec{r}')\}_{m'}; 0(\vec{r}); t) +$$

$$+ \sum_{\vec{r}} \sum_{j=1}^{m'} \varphi(\vec{r} - \vec{r}_j') P(\{A(\vec{r})\}_m; \{B(\vec{r}')\}_{m'}^j; 0(\vec{r}); t). \qquad (2.3.39)$$

Here $\varphi(\vec{r} - \vec{r}') = \varphi(|\vec{r} - \vec{r}'|)$ is the probability to create a particle A at $\vec{r}$ and B at $\vec{r}'$ at time units, $\{A(\vec{r})\}_m^i$ indicates that $A(\vec{r}_i)$ should be replaced by $0(\vec{r}_i)$ in the population numbers $\{A(\vec{r})\}_m$. In the limiting case $v_0 \to 0$ we assume $\varphi(\vec{r} - \vec{r}') = v_0^2 p f(\vec{r} - \vec{r}')$, where p is the irradiation intensity, $f(\vec{r}) \equiv f(r)$ is the initial distribution function of particles in just created geminate pairs,

$$\int f(r) \, d\vec{r} = 1. \qquad (2.3.40)$$

For the reaction $0 \to A$ one gets

$$\frac{\partial}{\partial t} P(\{A(\vec{r})\}_m; B(\vec{r}')\}_{m'}; t)$$

$$= \sum_{i=1}^{m} \phi(\vec{r}_i) P(\{A(\vec{r})\}_m^i; \{B(\vec{r}')\}_{m'}; t), \qquad (2.3.41)$$

with $\phi(\vec{r}) \equiv v_0 p$. Equations (2.3.39) and (2.3.41) contain the DFs where one or two population numbers $\nu(\vec{r}) = 0(\vec{r})$.

Let us rewrite equation (2.3.32) in the form

$$P(\{A(\vec{r})\}_m; \{B(\vec{r}')\}_{m'}; 0(\vec{r}); t)$$
$$= P(\{A(\vec{r})\}_m; \{B(\vec{r}')\}_{m'}; t) -$$
$$- \sum_{\nu(\vec{r})=A,B} P(\{A(\vec{r})\}_m; \{B(\vec{r}')\}_{m'}; \nu(\vec{r}); t) \qquad (2.3.42)$$

When dividing equation (2.3.42) by $v_0^{(m+m')}$ and taking into account equation (2.3.34), we can rewrite its right-hand side in the form $\rho_{m,m'} - O(v_0)$, where $O(v_0)$ tends to zero when $v_0 \to 0$. In other words, the population number $\nu(\vec{r}) = 0(\vec{r})$ in equations (2.3.39) and (2.3.41) could be omitted. Equation (2.3.39) now reads

$$\frac{\partial \rho_{m,m'}}{\partial t} = \sum_{i=1}^{m} \sum_{j=1}^{m'} pf(\vec{r}_i - \vec{r}_j')\rho_{m-1,m'-1}(\{\vec{r}\}_m^i; \{\vec{r}'\}_{m'}^j; t) +$$
$$+ p \sum_{i=1}^{m} \rho_{m-1,m'}(\{\vec{r}\}_m^i; \{\vec{r}'\}_{m'}; t) +$$
$$+ p \sum_{i=1}^{m} \rho_{m,m'-1}(\{\vec{r}\}_m; \{\vec{r}'\}_{m'}^j; t), \qquad (2.3.43)$$

where $\{\vec{r}\}_m^i$ indicates that the vector $\vec{r}_i$ is omitted in this vector set. The first term on r.h.s. of equation (2.3.43) takes into account all the ways to form a group of $(m + m')$ particles, whereas the second and third terms arise if one of reactants does not enter this group. Equation (2.3.41) becomes

$$\frac{\partial \rho_{m,m'}}{\partial t} = p \sum_{i=1}^{m} \rho_{m-1,m'}(\{\vec{r}\}_m^i; \{\vec{r}'\}_{m'}; t). \qquad (2.3.44)$$

Equation (2.3.43) corresponds also to that in [82].

(iii) *Particle motion.* Diffusion is described by

$$\frac{\partial \rho_{m,m'}}{\partial t} = D_A \sum_{i=1}^{m} \nabla_i^2 \rho_{m,m'} + D_B \sum_{j=1}^{m'} \nabla_j'^2 \rho_{m,m'}, \qquad (2.3.45)$$

where D_A, D_B are diffusion coefficients. Equation (2.3.45) can also be obtained when describing diffusive motion as a random walk in a lattice in the limiting case $v_0 \to 0$. Many-particle densities change due to diffusion as a product of concentrations taken at the corresponding spatial points.

In the case of dynamical interaction the pair potentials $U_{AA}(r)$, $U_{BB}(r)$ and $U_{AB}(r)$ should be incorporated into equation (2.3.45). It could be done using the Smoluchowski equation [27, 83, 84] for a particle drift in the external potential $W(r)$ and expressed in terms of single particle DF (or concentration of such non-interacting particles)

$$\frac{\partial g(\vec{r},t)}{\partial t} = D \nabla \left(\nabla g(\vec{r},t) + \frac{g(\vec{r},t)}{k_B T} \nabla W(r) \right), \qquad (2.3.46)$$

where k_B is the Boltzmann constant, T is temperature. The natural generalisation of equation (2.3.46) in the case of many-particle system under our study is

$$\frac{\partial \rho_{m,m'}}{\partial t} = -\sum_{i=1}^{m} \nabla_i \vec{J}^{Ai}_{m,m'} - \sum_{j=1}^{m'} \nabla'_j \vec{J}^{Bj}_{m,m'}, \qquad (2.3.47)$$

where the diffusion fluxes are defined similarly to equation (2.3.46) [85, 86]

$$\vec{J}^{Ai}_{m,m'} = -D_A \left(\nabla_i \rho_{m,m'} + \frac{\rho_{m,m'}}{k_B T} \nabla_i W^i_{m,m'} \right), \qquad (2.3.48)$$

$$\vec{J}^{Bj}_{m,m'} = -D_B \left(\nabla'_j \rho_{m,m'} + \frac{\rho_{m,m'}}{k_B T} \nabla'_j W^j_{m,m'} \right). \qquad (2.3.49)$$

In two last equations the mean force potential $W^i_{m,m'}$ is introduced, for which

$$\nabla_i W^i_{m,m'} = \sum_{i' \neq i}^{m} \nabla_i U_{AA}(|\vec{r}_i - \vec{r}_{i'}|) + \sum_{j=1}^{m'} \nabla_i U_{AB}(|\vec{r}_i - \vec{r}'_j|) +$$

$$+ \int \frac{\rho_{m+1,m'}}{\rho_{m,m'}} \nabla_i U_{AA}(|\vec{r}_i - \vec{r}_{m+1}|) \, d\vec{r}_{m+1} +$$

$$+ \int \frac{\rho_{m,m'+1}}{\rho_{m,m'}} \nabla_i U_{AB}(|\vec{r}_i - \vec{r}'_{m'+1}|) \, d\vec{r}'_{m'+1} \qquad (2.3.50)$$

The potential $W^j_{m,m'}$ is defined similarly. As it is seen from equation (2.3.50), the mean force acting on a particle A at coordinate $\vec{r}$ has both the contribution from direct interactions within a group of $(m + m')$-particles and indirect interactions (integral terms). If the particle creation and recombination terms were absent, the steady-state solution of equation (2.3.47) would correspond to putting the fluxes $\vec{J}^{Ai}_{m,m'}$ and $\vec{J}^{Bj}_{m,m'}$ equal zero. The corresponding set of the integro-differential equations

$$\nabla_i \rho_{m,m'} + \frac{\rho_{m,m'}}{k_B T} \nabla_i W^i_{m,m'} = 0, \qquad (2.3.51)$$

$$\nabla'_j \rho_{m,m'} + \frac{\rho_{m,m'}}{k_B T} \nabla'_j W^j_{m,m'} = 0 \qquad (2.3.52)$$

coincide exactly with the Ivon's equations [76] well known in the statistical physics of dense two-component gases and liquids. Therefore, equations (2.3.47) allow us to describe correctly the formation of the equilibrium state in a system consisting of dynamically interacting but nonreacting particles.

The complete set of equations for many-particle densities is nothing but a sum of contributions due to the three kinds of processes described above:

$$\frac{\partial \rho_{m,m'}}{\partial t} = \left.\frac{\partial \rho_{m,m'}}{\partial t}\right|_{\text{rec}} + \left.\frac{\partial \rho_{m,m'}}{\partial t}\right|_{\text{cr}} + \left.\frac{\partial \rho_{m,m'}}{\partial t}\right|_{\text{diff}}. \qquad (2.3.53)$$

A structure of the obtained set of equations derived by us in [81, 86] is very close to the famous BBGKI set of equations widely used in the statistical physics of dense gases and liquids [76]. Therefore, we presented the master equation of the Markov process in a form of the infinite set of deterministic coupled equations for averages (equation (2.3.34)). Practical use of these equations requires us to reduce them, retaining the joint correlation functions only.

2.3.1.4 Superposition approximation

The analogy just mentioned with the BBGKI set of equations being quite prominent still needs more detailed specification. To cut off an infinite hierarchy of coupled equations for many-particle densities, methods developed in the statistical theory of dense gases and liquids could be good candidates to be applied. However, one has to take into account that a number of the standard approximate methods, e.g., the Percus–Yevick or hyper-chain approximations, are applicable for systems with the Gibbs distribution and are based on the distinctive Boltzmann factor like $\exp(-U(r)/(k_B T))$, where $U(r)$ is the potential energy of interacting particles. The basic kinetic equation (2.3.53) has nothing to do with the Gibbs distribution. The only approximate method "neutral" with respect to the ensemble averaging is the *Kirkwood approximation* [76, 77, 87].

Despite Kirkwood's superposition approximation is widely used, the range of its applicability established earlier in statistical physics from a comparison with the molecular dynamics [76, 77] should be checked *anew* before it is applied to the kinetics of reactions. Making use of Kirkwood's superposition approximation leads to a closed set of several integro-differential equations not containing any small parameters [76]. It is clear that an accuracy of the approximations employed cannot be found in the framework of the same method, but only when comparing the results obtained with computer simulations or seldomly available exact solutions of particular problems. For instance, in the statistical physics of dense gases and liquids the diagrammatic approach argues for the so-called hyperchain approximation whereas the best results have been obtained by means of the Percus–Yevick approximation [76]. It once more clearly demonstrates that results are not necessarily better if we use more terms in corresponding diagrammatic expansion which is typical for non-convergent or semi-convergent series.

All superposition approximations mentioned above are based on the idea of multiplicative expansion, when m-reactant (m-point) distribution functions $\rho_m(1,\ldots,m)$, with arguments being the generalised coordinates, are expressed through the *correlation forms* $a^{(m)}$:

$$\ln \rho^1(1) = a^{(1)}(1),$$

$$\ln \rho_2(1,2) = a^{(1)}(1) + a^{(1)}(2) + a^{(2)}(1,2),$$

$$\ln \rho_3(1,2,3) = a^{(1)}(1) + a^{(1)}(2) + a^{(1)}(3) +$$
$$+ a^{(2)}(1,2) + a^{(2)}(2,3) + a^{(2)}(3,1) +$$
$$+ a^{(3)}(1,2,3),$$
$$\vdots \qquad (2.3.54)$$

where each of these expressions defines a new correlation form $a^{(m)}$. By omitting the correlation forms $a^{(m)}$ with $m > m_0$, all distribution functions

ρ_m with any m could be expressed through $\rho_{m'}$, $m' \leqslant m_0$ and we thus arrive at the superposition approximations. It is assumed that the correlation forms $a^{(m+1)}$ are small as compared to the $a^{(m)}$, but no distinctive small parameters are employed here. The Kirkwood superposition approximation corresponds to the choice

$$\rho_3(1,2,3) \Rightarrow \frac{\rho_2(1,2)\rho_2(2,3)\rho_2(3,1)}{\rho_1(1)\rho_1(2)\rho_1(3)}. \tag{2.3.55}$$

In its turn, the choice of $m_0 = 3$ leads to the approximation (Ziman 1979, [77])

$$\rho_4(1,2,3,4) \Rightarrow \rho_1(1)\rho_1(2)\rho_1(3)\rho_1(4) \times \tag{2.3.56}$$

$$\times \frac{\rho_3(1,2,3)\rho_3(1,2,4)\rho_3(1,3,4)\rho_3(2,3,4)}{\rho_2(1,2)\rho_2(1,3)\rho_2(1,4)\rho_2(2,3)\rho_2(2,4)\rho_2(3,4)}.$$

As was demonstrated by Kikuchi and Brush [88], using the Ising model as an example, an increase of m_0 in the expansion in the $a^{(m)}$ form secures the monotonic approach of the calculated critical parameters to exact results, except for the critical exponents which cannot be reproduced by algebraic expressions. It is important to note here that the superposition approximation permits exact (or asymptotically exact) solutions to be obtained for models revealing the critical point but not the phase transition. This should be kept in mind when interpreting the results of the bimolecular reaction kinetics obtained using approximate methods.

An alternative way to study many-particle effects is based on the correlation forms in the additive expansion [76]

$$\rho_1(1) = b^{(1)}(1),$$

$$\rho_2(1,2) = b^{(1)}(1)b^{(1)}(2) + b^{(2)}(1,2),$$

$$\rho_3(1,2,3) = b^{(1)}(1)b^{(1)}(2)b^{(1)}(3) +$$

$$+ b^{(1)}(3)b^{(2)}(1,2) + b^{(1)}(2)b^{(2)}(3,1) + b^{(1)}(1)b^{(2)}(2,3) +$$

$$+ b^{(3)}(1,2,3),$$

$$\vdots \tag{2.3.57}$$

Unfortunately, this expansion cannot be used as a basis for the development of approximate methods since – unlike the superposition approximation – in the case of considerable spatial correlation, neglect of the forms $b^{(m)}$, $m > m_0$ leads to the correlation functions *not* satisfying the proper boundary conditions and increase of m_0 does not lead to the convergence of results. A comparison of the two kinds of expansion of the many-particle distribution function demonstrates that the superposition approximation even for small m_0 corresponds to the choice in the additive expansion of $b^{(m)} \neq 0$ with any m. Therefore, in terms of the latter expansion the many-particle correlation forms $b^{(m)}$ are not neglected in the superposition approximations but are no longer independent.

Incorporation of the superposition approximation leads inevitably to a closed set of several non-linear integro-differential equations. Their non-linearity excludes the use of analytical methods, except for several cases of asymptotical automodel-like solutions at long reaction time. The kinetic equations derived are solved mainly by means of computers and this imposes limits on the approximations used. For instance, we could derive the kinetic equations for the $A + B \to C$ reaction employing the higher-order superposition approximation with $m_0 = 3, 4, \ldots$ rather than $m_0 = 2$ for the Kirkwood one. (How to realize this for the simple reaction $A + B \to B$ will be shown in Chapter 6.) However, even computer calculations involve great practical difficulties due to numerous coordinate variables entering these non-linear partial equations.

A general expression for the superposition approximation (2.3.55) has to be specified for a reaction under study. For instance, let us do it for the actual case of the bimolecular reaction employing many-particle densities $\rho_{m,m'}$. Single-particle densities are nothing but macroscopic concentrations (particle densities):

$$\rho_{1,0}(\vec{r}_1; t) = n_A(t), \qquad \rho_{0,1}(\vec{r}'_1; t) = n_B(t). \tag{2.3.58}$$

Taking into account (2.3.27)–(2.3.29), two-particle densities could be expressed easily through the joint correlation functions

$$\rho_{2,0}(\vec{r}_1, \vec{r}_2; t) = n_A^2(t) X_A(|\vec{r}_1 - \vec{r}_2|, t), \tag{2.3.59}$$

$$\rho_{0,2}(\vec{r}'_1, \vec{r}'_2; t) = n_B^2(t) X_B(|\vec{r}'_1 - \vec{r}'_2|, t), \tag{2.3.60}$$

$$\rho_{1,1}(\vec{r}_1; \vec{r}'_1; t) = n_A(t) n_B(t) Y(|\vec{r}_1 - \vec{r}'_1|, t). \tag{2.3.61}$$

Three-particle densities $\rho_{m,m'}$ with $(m+m') = 3$ could be expressed through the Kirkwood approximation as products of single-particle (2.3.58) and two-particle (2.3.59)–(2.3.61) densities:

$$\rho_{2,1}(\vec{r}_1, \vec{r}_2; \vec{r}_1'; t) \Rightarrow n_A^2(t) n_B(t) X_A(|\vec{r}_1 - \vec{r}_2|, t) \times$$
$$\times Y(|\vec{r}_1 - \vec{r}_1'|, t) Y(|\vec{r}_2 - \vec{r}_1'|, t). \qquad (2.3.62)$$

Equation (2.3.62) corresponds to the three-particle correlation function (2.3.25)

$$\chi_{AAB}^{(3)} \Rightarrow X_A(|\vec{r}_1 - \vec{r}_2|, t) Y(|\vec{r}_1 - \vec{r}_1'|, t) Y(|\vec{r}_2 - \vec{r}_1'|, t). \qquad (2.3.63)$$

Kuzovkov and Kotomin [89–91] (see also [92, 93]) were the first to use the complete Kirkwood superposition approximation (2.3.62) in the kinetic calculations for bimolecular reaction in condensed media. This approximation allows us to cut off the infinite hierarchy of equations for the correlation functions describing spatial distribution of particles of the two kinds and to restrict ourselves to the treatment of minimal set of the kinetic equations which realistically could be handled (Fig. 2.21). In earlier studies [82, 84, 91, 94–97] a shortened superposition approximation was widely used

$$\rho_{2,1}(\vec{r}_1, \vec{r}_2; \vec{r}_1'; t)$$
$$\Rightarrow n_A^2(t) n_B(t) Y(|\vec{r}_1 - \vec{r}_1'|, t) Y(|\vec{r}_2 - \vec{r}_1'|, t), \qquad (2.3.64)$$

or

$$\chi_{AAB}^{(3)} \Rightarrow Y(|\vec{r}_1 - \vec{r}_1'|, t) Y(|\vec{r}_2 - \vec{r}_1'|, t). \qquad (2.3.65)$$

In equations (2.3.64) and (2.3.65) – unlike (2.3.62) and (2.3.63) – spatial correlations of similar particles are neglected:

$$X_A(r, t) = X_B(r, t) \equiv 1. \qquad (2.3.66)$$

Use of (2.3.64) for diffusion problems without particle generation leads to the *linear* equation for the correlation function of dissimilar reactants $Y(r, t)$, which greatly simplifies the solution of kinetic equations (Chapter 6). The approximation (2.3.56) was applied for the first time to the study of kinetic processes in [98, 99].

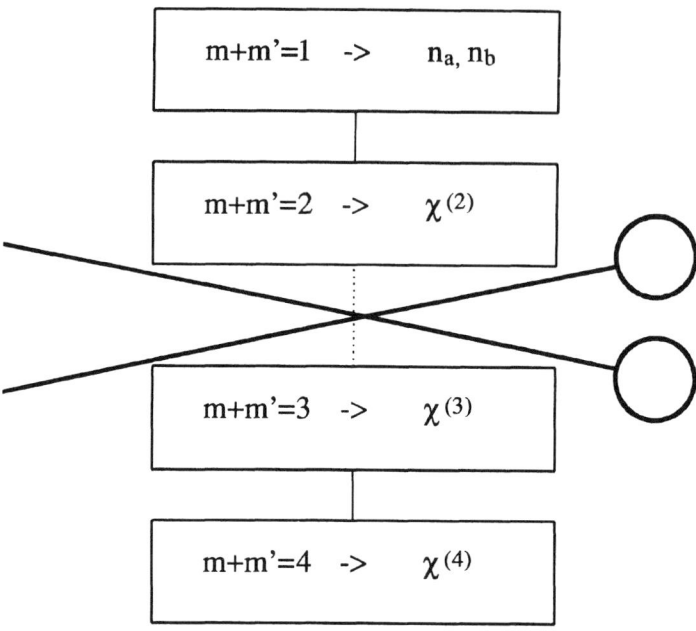

Fig. 2.21. The idea of the cut off of the infinite hierarchy of equations for the correlation functions by means of the Kirkwood superposition approximation.

The applications of the many-particle densities will be demonstrated on a full scale in further Chapters. It should be only said here that the many-particle density formalism being combined with the *shortened* Kirkwood superposition approximation, equation (2.3.64), results in the well-known equations of the standard kinetics for both neutral [83] and charged particles [100] giving just another way of their derivation. On the other hand, the use of the "full-scale" (complete) Kirkwood's approximation, equation (2.3.62), permits us to take into account the *many-particle (cooperative) effects* [81, 91, 99–102] we are studying in this book.

2.3.1.5 Many-point densities and probability densities

Note that willing to stress the relation between many-particle densities and master equation for the Markov process, we followed the formalism presented by us [81] rather than that used in the pioneering papers by Waite [84, 94, 95] and Leibfried [96], as well as in more recent studies [82, 97, 103–105] where equations similar to our equations (2.3.37), (2.3.43) were derived. They used the probability densities $w_{m,m'}(\{\vec{r}\}_m;\{\vec{r}'\}_{m'};t)$ to find at time t m particles of the A kind at points $\{\vec{r}\}_m$ and m' B particles at $\{\vec{r}'\}_{m'}$. By means of some mathematical manipulations based on combinations one can construct from $w_{m,m'}$ physically observable particle densities $\rho_{m,m'}$ [82].

A method to derive kinetic similar to (2.3.37), (2.3.43) has certain shortcomings. Say, Waite [84, 94, 95] has neglected the undistinguishability of similar reactants and related symmetry of the probability densities $w_{m,m'}(\{\vec{r}\}_m;\{\vec{r}'\}_{m'};t)$ with respect to the permutation of coordinates in $\{\vec{r}\}_m$ (or $\{\vec{r}'\}_{m'}$). Accordingly, his approach is often called "asymmetric" unlike its "symmetric" counter-partner presented by Leibfried [96] who has demonstrated that the incorporation of the fact that similar particles are undistinguishable in the case of initially correlated (geminate) pairs change both the relevant kinetic equations and their solution. The symmetric approach does not use an additional equation for time development of geminate pairs as the asymmetric one does. (For more details and discussion of this point see [105] and [81, 91, 106].) Note that the described formalism of the master equation for the Markov process argues for the symmetric approach. When deriving equation (2.3.43), Dettmann [82] enumerated particles at their birth moments which gives excess information and, in principle, can lead to some paradoxes (see for details [81]).

2.3.2 The field-theoretical formalism

2.3.2.1 Probability densities and quantum-mechanical analogy

The probability densities $w_{m,m'}(\{\vec{r}\}_m;\{\vec{r}'\}_{m'};t)$ were used in Section 2.3.1 in the many-particle density formalism just as an intermediate and, as follows from Section 2.3.1, *not necessary* step to derive equations (2.3.53). However, it was learned in the last years that a form of the corresponding master equation for $w_{m,m'}$ suggests an idea how to use its analogy with the *field-theoretical formalism* developed in quantum mechanics and to derive a novel approach to the recombination kinetics of classical particles [35, 43, 107–114].

Let us consider a mixture of classical mobile particles A and B participating in the A + B → 0 reaction occurring in a continuous neutral medium. Assume also that uncorrelated particles are created with rate p and recombine with probability $\sigma(r)$. Thus one arrives at equation [113]

$$\frac{\partial w_{m,m'}(\{\vec{r}\}_m; \{\vec{r}'\}_{m'}; t)}{\partial t}$$

$$= \sum_{i=1}^{m} D_A \nabla_i \left\{ \nabla_i w_{m,m'} + \right.$$

$$\left. + \frac{w_{m,m'}}{k_B T} \nabla_i \left[\sum_{i' \neq 1}^{m} U_{AA}(|\vec{r}_i - \vec{r}_{i'}|) + \sum_{j=1}^{m'} U_{AB}(|\vec{r}_i - \vec{r}'_j|) \right] \right\} +$$

$$+ \sum_{j=1}^{m'} D_B \nabla'_j \left\{ \nabla'_j w_{m,m'} + \right.$$

$$\left. + \frac{w_{m,m'}}{k_B T} \nabla'_j \left[\sum_{j' \neq j}^{m'} U_{BB}(|\vec{r}'_j - \vec{r}'_{j'}|) + \sum_{i=1}^{m} U_{AB}(|\vec{r}_i - \vec{r}'_j|) \right] \right\} -$$

$$- \sum_{i=1}^{m} \sum_{j=1}^{m'} \sigma(|\vec{r}_i - \vec{r}'_j|) w_{m,m'} +$$

$$+ \iint \sigma(|\vec{r}_{m+1} - \vec{r}'_{m'+1}|) w_{m+1,m'+1} \, d\vec{r}_{m+1} \, d\vec{r}'_{m'+1} +$$

$$+ \frac{p}{V} \sum_{i=1}^{m} w_{m-1,m'} + \frac{p}{V} \sum_{j=1}^{m'} w_{m,m'-1} - 2p w_{m,m'}. \quad (2.3.67)$$

The set of functions $w_{m,m'}$ ($m, m' = 0, 1, \ldots$) yields the complete probabilistic description of a system. The variation of a number of particles of both kinds is taken into account by the normalization condition

$$\sum_{m=0}^{\infty} \sum_{m'=0}^{\infty} \int \frac{d\vec{r}_1 \ldots d\vec{r}_m d\vec{r}'_1 \ldots d\vec{r}'_{m'}}{m! m'!} \times$$

$$\times w_{m,m'}(\{\vec{r}\}_m; \{\vec{r}'\}_{m'}; t) = 1. \quad (2.3.68)$$

Before analysing (2.3.67), let us consider the preliminary simple statement of the problem. For the A + B → B reaction without source ($p = 0$) the number of B particles is unchanged. Assuming also that particles do not interact, the reaction kinetics becomes linear in concentration of A's: it is enough to solve the problem of a single particle A survival surrounded by N particles B. As a consequence, an infinite set of equation (2.3.67) is reduced to a *single* equation for the probability density $w_{1,N}(\vec{r}_1; \{\vec{r}''\}_N; t)$:

$$\frac{\partial w_{1,N}(\vec{r}_1; \{\vec{r}''\}_N; t)}{\partial t}$$

$$= D_A \nabla_1^2 w_{1,N} + D_B \sum_{j=1}^{N} \nabla_j'^2 w_{1,N} - \sum_{j=1}^{N} \sigma(|\vec{r}_1 - \vec{r}_j'|) w_{1,N}. \quad (2.3.69)$$

To solve (2.3.69), the initial condition $w_{1,N}(\vec{r}_1; \{\vec{r}\}_N; 0)$ (initial distribution of traps w) has to be given. Unlike the many-particle formalism based on $\rho_{m,m'}$ with equations defining averaged values (macroscopic densities or correlation functions), in the probability density approach these averages are calculated with the help of a set of functions $w_{m,m'}$. To do so multipole integration over all particle coordinates should be performed. Even in a simplified problem statement (2.3.69) averaging over the initial trap configuration is necessary. The solution of equation (2.3.69) is in fact a many-dimensional diffusion equation with particle sources which is not a trivial problem. Due to homogeneity of a set of equations in (2.3.69) the coordinate difference, $\vec{r}_j = \vec{r}_j' - \vec{r}_1$ enters only. It permits rewriting (2.3.69) in a form

$$\frac{\partial w_N(\{\vec{r}\}_N; t)}{\partial t}$$

$$= (D_A + D_B) \sum_{j=1}^{N} \nabla_j^2 w_N + D_A \sum_{i \ne j=1}^{N} \nabla_i \nabla_j w_N - \sum_{j=1}^{N} \sigma(r_j) w_N. \quad (2.3.70)$$

As follows from (2.3.70), it is impossible to separate variables $\vec{r}_j$ unless $D_A = 0$, and mixed derivatives prevent finding of the exact solution. In the particular case $D_A = 0$ an explicit solution of (2.3.70) is known.

The solution of (2.3.69) is a purely mathematical problem well known in the theory of diffusion-controlled processes of classical particles. However, a particular form of writing down (2.3.69) allows us to use a certain mathematical analogy of this equation with quantum mechanics. Say, many-dimensional diffusion equation (2.3.69) is an analog to the Schrödinger equation for a system of N spinless particles B, interacting with the central particle A placed at the origin:

$$i\hbar \frac{\partial \Psi(\vec{r}_1; \{\vec{r}'\}_N; t)}{\partial t}$$

$$= -\frac{\hbar^2}{2m_A} \nabla_1^2 \Psi - \sum_{j=1}^{N} \frac{\hbar^2}{2m_B} \nabla_j'^2 \Psi + \sum_{j=1}^{N} U(|\vec{r}_j' - \vec{r}_1|)\Psi. \qquad (2.3.71)$$

Further development of this analogy leads to the *non-Hermitian Hamiltonian* problem describing the Bose particles. Proceeding in this way, the classical diffusion problem could be related to *quantum theory* of multiple scattering [115–118].

The use of the same analogy for the A + B → C reaction, described by a set of (2.3.67) is more problematic: coupling of these equations results in a non-conserving number of particles in a system. This problem could be much easier treated in terms of the *field-theoretical formalism*.

2.3.2.2 Secondary quantization

The formal secondary quantisation procedure presented first by Doi [107], Zeldovich and Ovchinnikov [35] reduces equations (2.3.67) into

$$\frac{\partial}{\partial t}|\Phi(t)\rangle = \widehat{H}|\Phi(t)\rangle. \qquad (2.3.72)$$

The state of a system in this equation is given by "wave function"

$$|\Phi(t)\rangle = \sum_{m=0}^{\infty} \sum_{m'=0}^{\infty} \frac{1}{m!m'!} V^{-(m+m')} \int w_{m,m'}(\{\vec{r}\}_m; \{\vec{r}'\}_{m'}; t) \times$$

$$\times \prod_{i=1}^{m} \prod_{j=1}^{m'} \widehat{\Psi}_A^+(\vec{r}_i) \widehat{\Psi}_B^+(\vec{r}_j') \, d\vec{r}_i \, d\vec{r}_j' |0\rangle. \qquad (2.3.73)$$

Taking into account the symmetry of $w_{m,m'}(\{\vec{r}\}_m; \{\vec{r}'\}_{m'}; t)$ with respect to the coordinate permutations, the commutation condition arises

$$\left[\widehat{\Psi}_\nu(\vec{r})\widehat{\Psi}_\mu^+(\vec{r}')\right] = \delta_{\nu\mu}\delta(\vec{r} - \vec{r}'), \qquad \nu, \mu = A, B, \qquad (2.3.74)$$

completed with

$$\widehat{\Psi}_\nu |0\rangle = 0 = \langle 0|\widehat{\Psi}_\nu^+. \tag{2.3.75}$$

The following relation takes place

$$\left\langle 0 \left| V^{-(m+m')/2} \prod_{i=1}^{m} \prod_{j=1}^{m'} \widehat{\Psi}_A(\vec{r}_i) \widehat{\Psi}_B(\vec{r}_j') \, d\vec{r}_i \, d\vec{r}_j' \right| \Phi(t) \right\rangle$$
$$= w_{m,m'}(\{\vec{r}\}_m; \{\vec{r}'\}_{m'}; t). \tag{2.3.76}$$

The standard expansion of operators $\widehat{\Psi}_\nu, \widehat{\Psi}_\nu^+$ into plane waves

$$\widehat{\Psi}_A(\vec{r}) = V^{-1/2} \sum_{\vec{k}} \widehat{a}_{\vec{k}} \, e^{-i\vec{k}\vec{r}},$$
$$\widehat{\Psi}_B(\vec{r}) = V^{-1/2} \sum_{\vec{k}} \widehat{b}_{\vec{k}} \, e^{-i\vec{k}\vec{r}} \tag{2.3.77}$$

permits us to express the linear operator $\widehat{H}$ in equation (2.3.72) through the Bose birth–death (creation–annihilation) operators $\widehat{a}_{\vec{k}}, \widehat{a}_{\vec{k}}^+, \widehat{b}_{\vec{k}}, \widehat{b}_{\vec{k}}^+$:

$$\widehat{H} = -\sum_{\vec{k}} D_A k^2 \widehat{a}_{\vec{k}}^+ \widehat{a}_{\vec{k}} - \sum_{\vec{k}} D_B k^2 \widehat{b}_{\vec{k}}^+ \widehat{b}_{\vec{k}} + V^{-1} \sum_{\vec{k}} \sigma(k) \widehat{a}_{\vec{k}} \widehat{b}_{-\vec{k}} -$$
$$- D_A V_A^{-1} \sum_{\vec{q},\vec{k}_1,\dots,\vec{k}_4} (\vec{q}, \vec{q}+\vec{k}) \frac{U_{AA}(q)}{k_B T} \widehat{a}_{\vec{k}_4}^+ \widehat{a}_{\vec{k}_3}^+ \widehat{a}_{\vec{k}_2} \widehat{a}_{\vec{k}_1} \times$$
$$\times \Delta(\vec{k}_4 - \vec{k}_2 - \vec{q}) \Delta(\vec{k}_1 - \vec{k}_3 - \vec{q}) -$$
$$- D_B V^{-1} \sum_{\vec{q},\vec{k}_1,\dots,\vec{k}_4} (\vec{q}, \vec{q}+\vec{k}) \frac{U_{BB}(q)}{k_B T} \widehat{b}_{\vec{k}_4}^+ \widehat{b}_{\vec{k}_3}^+ \widehat{b}_{\vec{k}_2} \times$$
$$\times \Delta(\vec{k}_4 - \vec{k}_2 - \vec{q}) \Delta(\vec{k}_1 - \vec{k}_3 - \vec{q}) -$$
$$- V^{-1} \sum_{\vec{q},\vec{k}_1,\vec{k},\vec{p}_1,\vec{p}} \left\{ \sigma(q) + (\vec{q}, D_A[\vec{q}+\vec{k}] + D_B[\vec{q}-\vec{k}]) \frac{U_{AB}(q)}{k_B T} \right\} \widehat{b}_{\vec{k}_1} \times$$
$$\times \widehat{a}_{\vec{k}_1}^+ \widehat{b}_{\vec{p}_1}^+ \widehat{a}_{\vec{k}} \widehat{b}_{\vec{p}} \, \Delta(\vec{k}_1 - \vec{k} - \vec{q}) \Delta(\vec{p}_1 - \vec{p} - \vec{q}) +$$
$$+ V p(\widehat{a}_0 - 1) + V p(\widehat{b}_0 - 1). \tag{2.3.78}$$

In equation (2.3.78) $\sigma(k)$, $U_{\nu\mu}(k)$ are the Fourier transforms of functions introduced in equation (2.3.67), $\Delta(\vec{k}) = \delta_{\vec{k},0}$ is a Kronecker delta.

According to Doi [107], Zeldovich and Ovchinnikov [35], the evolution of the state of a system given by the vector $|\Phi(t)\rangle$ obeys the Schrödinger equation with imaginary time and non-Hermitian Hamiltonian. The averaging procedure also differs from that generally-accepted in quantum mechanics.

2.3.2.3 Diagrammatic formalism

The presented form of the master equation (2.3.67) permits us to employ the *diagrammatic technique* of the perturbation theory [44, 108–110, 113]. The "free Hamiltonian" could be written as

$$\widehat{H}_0 = -\sum_{\vec{k}} D_A k^2 \widehat{a}_{\vec{k}}^+ \widehat{a}_{\vec{k}} - \sum_{\vec{k}} D_B k^2 \widehat{b}_{\vec{k}}^+ \widehat{b}_{\vec{k}}, \qquad (2.3.79)$$

a role of perturbation is played by the rest terms of equation (2.3.78). The particular choice of $\widehat{H}_0$ in equation (2.3.79) restricts problems under study to the diffusion-controlled reactions only; D_A, $D_B \neq 0$. Besides, the diagrammatic technique could be applied to the *stationary* state or not very far from it. In this case, neglecting the fluctuations in a number of particles, we conclude that the operators $\widehat{a}_0^+, \widehat{a}_0, \widehat{b}_0^+, \widehat{b}_0$ could be considered as c numbers corresponding to $\widehat{a}_0 \to n_A V$, $\widehat{b}_0 \to n_B V$, $\widehat{a}_0 \to 1$, $\widehat{b}_0 \to 1$.

A procedure similar to the condensate separation in the imperfect Bose gas was employed by Lifshitz and Pitaevski [78]. The diagrammatic technique allows us to calculate the reaction rate and steady-state joint correlation functions. A separation of a condensate from terms with $\vec{k} = 0$ cannot be done without particle production ($p = 0$), in which case $n_A, n_B \to 0$ as $t \to \infty$. In this respect the formalism presented by Lushnikov [111] for the *non*-stationary processes is of certain interest.

References

[1] H. Haken, Advanced Synergetics (Springer, Berlin, 1983).
[2] S.W. Benson, The Foundations of Chemical Kinetics (McGraw-Hill, New York, 1960).
[3] D.A. Franck-Kamenetskii, Diffusion and Thermal Conduction in Chemical Kinetics (Nauka, Moscow, 1967).
[4] W. Ebeling, Structurbildung bei irresiblen Prozessen (Teubner, Leipzig, 1976).

References

[5] L.S. Polak, Non-Equilibrium Chemical Kinetics and its Applications (Nauka, Moscow, 1979).
[6] L.S. Polak and A.S. Mikhailov, Self-Organization in Non-Equilibrium Physico-Chemical Systems (Nauka, Moscow, 1983).
[7] A.M. Zhabotinsky, Concentration Auto-Oscillations (Nauka, Moscow, 1974).
[8] C. Vidal, Chaos and Order in Nature, in: Proc. Int. Symp. on Synergetics (Springer, Berlin, 1981) p. 68.
[9] Ch. Lushchik and A. Lushchik, Decay of Electronic Excitations into Defects in Solids (Nauka, Moscow, 1989); K. S. Song and R. T. Williams, Self-Trapped Excitons (Springer, Berlin, 1993).
[10] V.A. Benderskii, V.Kh. Brikenshtein and P.G. Pilippov, Phys. Status Solidi B: 117 (1983) 9.
[11] R. Noks, Theory of Excitons (Academic Press, New York, 1966).
[12] V.L. Ermolaev, E.N. Bodunov, E.B. Sveshnikova and T.A. Shahverdov, Non-Radiative Energy Transfer (Nauka, Moscow, 1977).
[13] H. Eyring, S.H. Lin and S.M. Lin, Basic Chemical Kinetics (Wiley, New York, 1980).
[14] A. Andronov, E. Leontovich, I. Gordon and A. Mayer, The Qualitative Theory of the Second Order Dynamical Systems (Nauka, Moscow, 1966).
[15] A. Andronov, E. Leontovich, I. Gordon and A. Mayer, Bifurcation Theory of Planar Dynamical Systems (Nauka, Moscow, 1967).
[16] G. Nicolis and I. Prigogine, Self-Organization in Non-Equilibrium Systems (Wiley, New York, 1977).
[17] A.J. Lotka, J. Phys. Chem. 14 (1910) 271.
[18] A.J. Lotka, J. Am. Chem. Soc. 27 (1920) 1595.
[19] V. Volterra, Leçons sur la Theorie Mathematique de la Lutte Pour la Vie (Paris, 1931).
[20] I. Prigogine, From Being to Becoming: Time and Complexity in the Physical Sciences (Freeman, San Francisco, 1980).
[21] H. Haken, Synergetics (Springer, Berlin, 1978).
[22] H. Haken, Rep. Prog. Phys. 52 (1989) 515.
[23] P. Hanusse, C. R. Acad. Sci. Ser. C: 274 (1972) 1245.
[24] I. Prigogine and R. Lefever, J. Chem. Phys. 48 (1968) 1695.
[25] E.N. Lorenz, J. Atmos. Sci. 20 (1963) 130.
[26] R.Z. Sagdeev, D.A. Usikov and G.M. Zaslavsky, Nonlinear Physics. From the Pendulum to Turbulence and Chaos (Harwood Academic Publishers, London, 1988).
[27] C.W. Gardiner, Handbook of Stochastic Methods for Physics, Chemistry and the Natural Sciences (Springer, Berlin, 1983).
[28] U. Gösele, Chem. Phys. Lett. 43 (1976) 61.
[29] U. Gösele and A. Seeger, Philos. Mag. 34 (1976) 177.
[30] U. Gösele, Prog. React. Kinet. 13 (1984) 63.
[31] A. Mozumder, S.M. Pimbott, P. Clifford and N.J.B. Green, Chem. Phys. Lett. 142 (1987) 385.
[32] J. Tyson and J. Light, J. Chem. Phys. 59 (1973) 4164.
[33] Ya.B. Zeldovich, Sov. Electochem. 13 (1977) 677.
[34] Ya.B. Zeldovich and A.A. Ovchinnikov, Sov. Phys. JETP Lett. 26 (1977) 588.

[35] Ya.B. Zeldovich and A.A. Ovchinnikov, Sov. Phys. JETP 74 (1978) 1588.
[36] S.F. Burlatsky, Teor. Eksp. Khim. 14 (1978) 483 (Sov. Theor. Exp. Chem., in Russian).
[37] S.F. Burlatsky and A.A. Ovchinnikov, Zh. Fiz. Khim. 52 (1978) 2847 (Sov. J. Phys. Chem., in Russian).
[38] A.A. Belyi, A.A. Ovchinnikov and S.F. Timashev, Teor. Eksp. Khim. 18 (1982) 269 (Sov. Theor. Exp. Chem., in Russian).
[39] I.M. Sokolov, Sov. Phys. JETP Lett. 44 (1986) 53.
[40] I.M. Sokolov, Sov. Phys. JETP 94 (1988) 199.
[41] Ya.B. Zeldovich and A.S. Mikhailov, Sov. Phys. Usp. 30 (1987) 977.
[42] A.G. Vitukhnovsky, B.L. Pyttel and I.M. Sokolov, Phys. Lett. A: 128 (1988) 161.
[43] A.A. Ovchinnikov, S.F. Timashev and A.A. Belyi, Kinetics of Diffusion-Controlled Chemical Processes (Nuova Science, New York, 1988).
[44] A.S. Mikhailov, Phys. Rep. 184 (1989) 308.
[45] D. Toussaint and F. Wilczek, J. Chem. Phys. 78 (1983) 2642.
[46] K. Kang and S. Redner, Phys. Rev. Lett. 52 (1984) 955.
[47] K. Kang and S. Redner, Phys. Rev. A: 32 (1985) 435.
[48] H. Schnörer, I. Sokolov and A. Blumen, Phys. Rev. A: 42 (1990) 7075.
[49] B. Balagurov and V. Vax, Sov. Phys. JETP 38 (1974) 968.
[50] A.A. Ovchinnikov and Ya.B. Zeldovich, Chem. Phys. 28 (1978) 215.
[51] P. Grassberger and I. Procaccia, J. Chem. Phys. 77 (1982) 6291.
[52] A.M. Berezhkovskii, Ya.A. Makhnovskii and R.A. Suris, Chem. Phys. 137 (1989) 41.
[53] R. Stratonovich, Selected Problems of the Fluctuational Theory in Radio (Radio, Moscow, 1961).
[54] A.S. Mikhailov and I.V. Uporov, Sov. Phys. Usp. 144 (1984) 79.
[55] Ya.B. Zeldovich, S.A. Molchanov, A.A. Ruzmaikin and D.D. Sokolov, Sov. Phys. Usp. 30 (1987) 353.
[56] S. Ma, Modern Theory of Critical Phenomena (Benjamin, London, 1976).
[57] A. Patashinskii and V. Pokrovskii, Fluctuational Theory of Phase Transitions (Nauka, Moscow, 1982).
[58] G. Dewel, D. Walgraef and P. Borckmans, Zs. Phys. Abt. B: 28 (1977) 235.
[59] D. Walgraef, G. Dewel and P. Borckmans, Phys. Rev. A 21 (1980) 397.
[60] A.A. Ovchinnikov and S.F. Burlatsky, Sov. Phys. JETP Lett. 43 (1986) 494.
[61] G.S. Oshanin, S.F. Burlatsky and A.A. Ovchinnikov, Phys. Lett. A 139 (1989) 245.
[62] K. Lindenberg, B.J. West and R. Kopelman, Phys. Rev. Lett. 60 (1988) 1777.
[63] Z. Yi-Cheng, Phys. Rev. Lett. 59 (1987) 1725.
[64] G. Nicolis and A. Babloyantz, J. Chem. Phys. 51 (1969) 2632.
[65] G. Nicolis, M. Malek-Mansour, K. Kitshara and A. van Nypelseer, Phys. Lett. A 48 (1974) 217.
[66] R. Mazo, J. Chem. Phys. 62 (1975) 4244.
[67] G. Nicolis, J. Stat. Phys. 6 (1971) 195.
[68] G. Nicolis and I. Prigogine, Proc. Nat. Acad. Sci. (USA) 68 (1971) 2102.
[69] V.T.N. Reddy, J. Stat. Phys. 13 (1974) 61.
[70] A.D. Bazikin, Mathematical Biophysics of Interacting Populations (Nauka, Moscow, 1985).

[71] G. Nicolis, in: Non-Linear Phenomena in Physics and Biology, eds R.H. Enns and B.L. Jones (Plenum, New York, 1981) p. 185.
[72] G. Nicolis, in: Stochastic Nonlinear Systems in Physics, Chemistry and Biology, eds L. Arnolds and R. Lefever (Springer, Berlin, 1981) p. 44.
[73] N.C. van Kempen, Can. J. Phys. 39 (1961) 551.
[74] N.C. van Kempen, Adv. Chem. Phys. 34 (1976) 245.
[75] S. Grossmann, in: Stochastic Nonlinear Systems in Physics, Chemistry and Biology, eds L. Arnold and R. Lefever (Springer, Berlin, 1983).
[76] R. Balescu, Equilibrium and Non-Equilibrium Statistical Mechanics (Wiley, New York, 1975).
[77] J.M. Ziman, Models of Disorder (Cambridge Univ. Press, London, 1979).
[78] E.M. Lifshitz and L.P. Pitaevskii, Statistical Physics, Vol. 2: Condensed Matter Theory (Nauka, Moscow, 1987).
[79] L. Ornstein and F. Zernike, Proc. Acad. Sci. (Amsterdam) 17 (1914) 793.
[80] A.M. Stoneham, Theory of Defects in Solids (Clarendon Press, Oxford, 1975).
[81] E.A. Kotomin and V.N. Kuzovkov, Chem. Phys. 76 (1983) 479.
[82] K. Dettmann, Phys. Status Solidi 10 (1965) 269.
[83] M. Smoluchowski, Zs. Phys. Chem. Abt. B: 92 (1917) 129.
[84] T.R. Waite, J. Chem. Phys. 28 (1958) 103.
[85] V.N. Kuzovkov, Teor. Eksp. Khim. 21 (1985) 33 (Sov. Theor. Exp. Chem., in Russian).
[86] V.N. Kuzovkov and E.A. Kotomin, Czech. J. Phys. B 35 (1985) 541.
[87] J.G. Kirkwood, J. Chem. Phys. 76 (1935) 479.
[88] R. Kikuchi and S.G. Brush, J. Chem. Phys. 47 (1967) 195.
[89] V.N. Kuzovkov and E.A. Kotomin, J. Phys. C: 13 (1980) L499.
[90] V.N. Kuzovkov and E.A. Kotomin, Phys. Status Solidi B: 105 (1981) 789.
[91] E.A. Kotomin and V.N. Kuzovkov, Rep. Prog. Phys. 55 (1992) 2079.
[92] E.A. Kotomin and V.N. Kuzovkov, Phys. Status Solidi B: 108 (1981) 37.
[93] Yu.B. Gaididei, A.I. Onipko and I.V. Zozulenko, Chem. Phys. 117 (1987) 367.
[94] T.R. Waite, Phys. Rev. 107 (1957) 463.
[95] T.R. Waite, J. Chem. Phys. 32 (1960) 21.
[96] G. Leibfried, Bestrahlungseffekte in Festkrpern (Teubner, Stuttgart, 1965).
[97] A.I. Onipko, Physics of Many-Body Systems, Vol. 2 (Naukova Dumka, Kiev, 1982) p. 60.
[98] V.N. Kuzovkov, Latv. PSR Zinat. Akad. Vestis Fiz. Teh. Zinat. Ser. 5 (1986) 46 (Proc. Latv. Acad. Sci. Phys. Technol. Ser., in Russian).
[99] V.N. Kuzovkov and E.A. Kotomin, Rep. Prog. Phys. 51 (1988) 1479.
[100] P. Debye, J. Electrochem. Soc. 32 (1942) 265.
[101] V.N. Kuzovkov and E.A. Kotomin, Chem. Phys. 81 (1983) 335.
[102] V.N. Kuzovkov and E.A. Kotomin, Chem. Phys. 98 (1985) 351.
[103] K. Dettmann, G. Leibfried and K. Schröder, Phys. Status Solidi 22 (1967) 423.
[104] A. Suna, Phys. Rev. B: 1 (1970) 1716.
[105] Yu. Kalnin, Latv. PSR Zinat. Akad. Vestis Fiz. Teh. Zinat. Ser. 5 (1982) 3 (Proc. Latv. Acad. Sci. Phys. Technol. Ser., in Russian).

[106] E.A. Kotomin and V.N. Kuzovkov, Teor. Eksp. Khim. 18 (1982) 274 (Sov. Theor. Exp. Chem., in Russian).
[107] M. Doi, J. Phys. A: 9 (1976) 1479.
[108] A.S. Mikhailov, Phys. Lett. A 85 (1981) 214.
[109] A.S. Mikhailov, Phys. Lett. A 85 (1981) 427.
[110] A.S. Mikhailov and V.V. Yashin, J. Stat. Phys. 38 (1985) 347.
[111] A.A. Lushnikov, Sov. Phys. JETP 91 (1986) 1376.
[112] S.F. Burlatsky, A.A. Ovchinnikov and K.A. Pronin, Sov. Phys. JETP 92 (1986) 625.
[113] A.M. Gutin, A.S. Mikhailov and V.V. Yashin, Sov. Phys. JETP 92 (1987) 941.
[114] G.S. Oshanin, S.F. Burlatsky and A.A. Ovchinnikov, Khim. Fiz. 8 (1989) 372 (Sov. J. Chem. Phys., in Russian).
[115] Yu.B. Gaididei and A.I. Onipko, Mol. Cryst. Liq. Cryst. 62 (1980) 213.
[116] M. Bixon and R. Zwanzig, J. Chem. Phys. 75 (1981) 2354.
[117] M. Muthukumar, J. Chem. Phys. 76 (1982) 2667.
[118] D.F. Calef and J.M. Deutch, J. Chem. Phys. 79 (1983) 203.

Chapter 3

From Pair Kinetics Toward the Many-Reactant Problem

3.1 BASIC DEFECTS AND PROCESSES IN SOLIDS

> If you found a truth in the nature, put it into a book where it will be served even worth.
> (Wenn sie die Wahrheit in der Natur gefunden haben, so schmeißen sie sie wieder in ein Buch, wo sie noch schlechter aufgehoben ist)
>
> Lichtenberg

3.1.1 Mechanisms of defect creation in solids

Irradiation of all kinds of solids (metals, semiconductors, insulators) is known to produce pairs of the point *Frenkel defects* – vacancies, v, and interstitial atoms, i, which are most often spatially well-correlated [1–9]. In many ionic crystals these Frenkel defects form the so-called F and H *centres* (anion vacancy with trapped electron and interstitial halide atom X^0 forming the chemical bonding in a form of quasimolecule X_2^- with some of the nearest regular anions, X) – Fig. 3.1. In metals the analog of the latter is called the *dumbbell interstitial*.

Under moderate energies, the primary event of incident particle interaction with crystals of any nature is very simple: it is an *elastic pair collision* resulting in displacement of atoms (ions) into an interstitial position [1, 7, 10] provided they received energy exceeding a *threshold value* [11] which is typically 10–20 eV.

However, in the last two decades it has been shown experimentally [1, 7, 8,12–14] and theoretically [15–18] that in many wide-gap insulators including alkali halides the primary mechanism of the Frenkel defect formation is *sub-threshold*, i.e., lattice defects arise from the non-radiative decay of excitons whose formation energy is less than the forbidden gap of solids, typically $\leqslant 10$ eV. These excitons are created easily by X-rays and UV light. Under ionic or electron beam irradiations the main portion of the incident particle

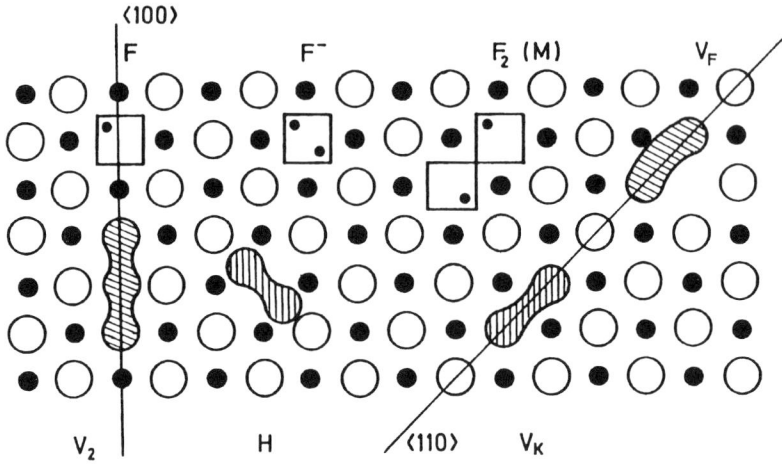

Fig. 3.1. Models of basic radiation defects in ionic solids.

energy is also utilized for excitation of the *electronic subsystem* of solids (i.e., for the creation of electrons, holes and excitons).

The function of the initial distribution of complementary defects – v, i pairs (also called *geminate*) – over relative distances depends strongly not only on the material but on whether impact or sub-threshold mechanism occurs and also on the particular kind of irradiation (e.g., X-rays or photons [8]). On creation of a v, i pair, the interstitial atom has the excess kinetic energy due to which it can be displaced, e.g., by a chain of *focusing collisions* (usually along an axis of close-packed rows of atoms). For example, such displacements for 50-eV crowdions in α-Fe are about $20a_0$ [11], where a_0 is the lattice spacing. In contrast, in alkali halides these distances are much shorter, and for most geminate partners, they lie in an interval from the 1st to the 5th nearest neighbours (nn) in the crystalline lattice.

3.1.2 Mechanisms of defect recombination

An initial distribution function within geminate pairs directly defines their stability. In terms of the *black sphere* model, dissimilar defects (v, i) disappear instantly when approaching to within, or when just created by irradiation at the critical relative distance r_0 (called also *clear-cut* radius) – see

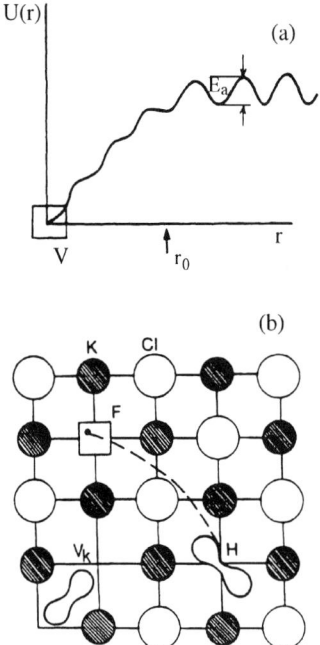

Fig. 3.2. Two principal mechanisms of defect recombination in solids. (a) Complementary defect annihilation, r_0 is the clear-cut (black sphere) radius, (b) distant tunnelling recombination due to overlap of wave functions of defects. Two principal kinds of hole centres – H and V_k are shown.

Fig. 3.2(a). Its typical value varies between 3–5 a_0 for metals [3, 4] and nearest neighbours for alkali halides. In some semiconductors, e.g., In_2Te_3, the radius of the instability zone could be very large (e.g., [19–22]). The relevant physical mechanism is annihilation of interstitial atoms with their "own" vacancies, which occurs in the time interval of several lattice vibrations, $\sim 10^{-13}$ s, and results in the restored perfect crystalline lattice. This mechanism takes place for all kinds of solids. Thus we can write down phenomenologicaly for the *recombination probability* (per unit time)

$$\sigma(r) = \sigma_0 \theta(r_0 - r), \qquad \sigma_0 \to \infty, \qquad (3.1.1)$$

where $\theta(x)$ is the Heaviside step-function.

Another, alternative mechanism of defect recombination is spontaneous *electron tunnelling* from an electron centre to its hole partner (or in terms of semiconductors, from *donor* to *acceptor*) as shown in Fig. 3.2(b). Being of a quantum-mechanical nature, it depends on the overlap of the wave functions of the two defects and is characterized by the exponential law [21, 23–27]

$$\sigma(r) = \sigma_0 \exp(-r/r_0). \tag{3.1.2}$$

The parameter σ_0 depends on the recombination mechanism (radiative vs. non-radiative) and is typically of the order 10^7 or 10^{15} s^{-1}, respectively [21] (see, however, Zamaraev et al. [27], who observed σ_0 up to 10^{21} s^{-1}). The other recombination parameter, r_0, is nothing but half the Bohr radius of the wave-function of an electron centre and is, for example, about 0.5 Å for F and Ag^0 centres and 1 Å for shallower Tl^0 centre in KCl. For paramagnetic defects this parameter could be found by means of EPR and ENDOR [28–30].

Tunnelling recombination of primary F, H pairs can result either in closely spaced v$^+$, i$^-$ pairs (the so-called α, I centres) which annihilate immediately due to Coulomb interaction and a consequently large instability radius. However some i$^-$ ions occur in crowdion configurations, and leave vacancy moving away up to 4–5 a_0 even at 4 K [31]. The distinctive feature of tunnelling recombination is its temperature independence, which makes it one of the major low-temperature secondary processes in insulating solids with defects.

3.1.3 Defect dynamic interaction

It is clear that *defect interactions* must play an important role in their stability and transformations. Along with the generally-known *Coulomb attraction*

$$U(r) = \frac{e_1 e_2}{\varepsilon r}, \tag{3.1.3}$$

observed for oppositely charged defects (e.g., v$^+$ and i$^-$) in ionic solids, the *elastic interaction* is present even for neutral defects in both metals and insulators [32–34]. Its interaction energy has the following asymptotic form at large distances r of

$$U(\vec{r}) = -\frac{\alpha(\vartheta, \varphi)}{r^3}. \tag{3.1.4}$$

This interaction arises from the overlap of the deformation fields around both defects. For weakly anisotropic cubic crystals and isotropic point defects, the long-range (dipole–dipole) contribution obeys equation (3.1.4) with $\alpha(\vartheta, \varphi) \propto [\alpha_4]$ (i.e., the cubic harmonic with $l = 4$). In other words, the elastic interaction is *anisotropic*. If defects are also anisotropic, which is the case for an H centre (X_2^- molecule), in alkali halides or crowdions in metals, there is little hope of getting an analytical expression for α [35]. The calculation of $U(r)$ for F, H pairs in a KBr crystal has demonstrated [36] that their attraction energy has a maximum along an $\langle 001 \rangle$ axis with $\langle 110 \rangle$ orientation of the H centre; reaching for 1 nn the value -0.043 eV. However, in other directions their elastic interaction was found to be repulsive.

3.1.4 Mechanisms of defect migration

At temperatures when interstitial atoms and ions become *mobile* (typically, 20–30 K in both metals and insulators), they perform thermally-activated incoherent hops, the frequencies of which are exponentially dependent on reciprocal temperature:

$$\nu = \nu_0 \exp\left(-\frac{E_a}{k_B T}\right), \tag{3.1.5}$$

where E_a is the hopping activation energy. The relevant diffusion coefficients in case $\bar{d} = 3$ can be shown to be

$$D = D_0 \exp\left(-\frac{E_a}{k_B T}\right), \qquad D_0 = \frac{\lambda^2 \nu_0 f}{6}, \tag{3.1.6}$$

where λ is a hopping length and f is a correlation factor [37–39]. The inverse of the ν, $\tau_0 = \nu^{-1}$, is called a *waiting time* which defect spends on a lattice site before it hops instantaneously into another equivalent site. (In anisotropic media D is no longer a scalar but a tensor [40, 41].)

In alkali halides the basic mobile defects participating in diffusion-controlled recombination processes are H and V_k centres [42]. Their thermally-activated motion in KBr begins at $T \geqslant 35$ and 150 K respectively. The relevant hopping activation energies are 0.09 eV and 0.47 eV, respectively. Taking into account that H centres begin to reorient on their sites already at $T \geqslant 15$ K [42], one can assume that an approaching F centre would react with H centre in the energetically most favourable orientation

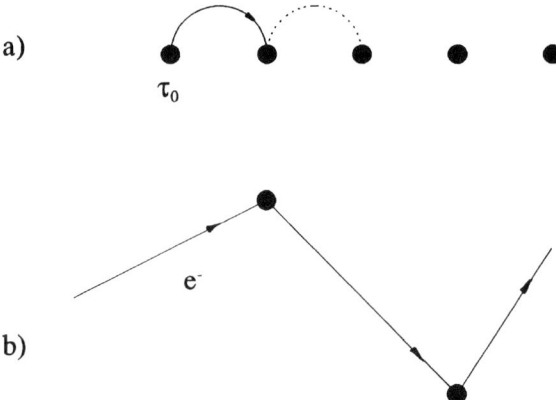

Fig. 3.3. Schematic classification of processes in solid with defects. (a) Defect diffusion; (b) carrier flight.

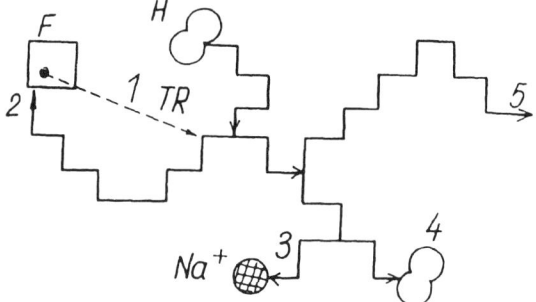

Fig. 3.4. Processes defending the survival probability of F centres in alkali halide crystals: 1 – tunnelling recombination of close F, H defects, 2 – their annihilation, 3 – trapping of mobile H centre at impurity, 4 – formation of immobile dimer centre, 5 – H-centre leaves its geminate partner in random walks on a lattice.

and thus we can still treat their effective interaction as isotropic one. Another example of mobile species is I centre (interstitial anion i$^-$) which starts to migrate above 30 K. On the other hand, α and F centres become mobile above the room temperature only.

Many-phonon hopping processes could be treated as defect localization during some waiting time τ_0 with further sudden and instant hop and new localization in a nearest lattice site for τ_0 etc. The same situation takes place for solvated electrons in solutions and electron migration over impurities in doped semiconductors. In these cases hopping lengths are stochastic variables.

Another kind of *flight motion* [43, 44] occurs for free carriers in gases and heavily doped semiconductors; it is almost temperature independent. Here most of their time particles find themselves in the flight motion with flight lengths greatly exceeding a lattice constant interrupted by rare scatterings on phonons (Fig. 3.3).

During their diffusive walks, H centres can either approach their "own" F centres to within the distance $r \leqslant r_0$ and recombine with them in the course of the so-called *geminate* (monomolecular) reaction or leave them behind in their random walks. Some of these H centres recombine with "foreign" F centres, thus participating in *bimolecular* reactions. The rest of the H centres become trapped by impurities, dislocations, or aggregate in the form of *immobile* dimer H_2 centres thus going out of the secondary reactions as shown in Fig. 3.4. In other words, the *survival probability* of the geminate pairs (F centres) directly defines the defect *accumulation efficiency* and thus, a material's sensitivity to radiation.

3.1.5 Classification of processes in solids

Defect diffusion traditionally is treated as a process in *continuum* medium. However, discreteness of the crystalline lattice becomes important in particular situations, e.g., when defect recombination occurs in several hops (nearest neighbour recombination) [3, 4] or even for nearest-site hops of defects if their recombination is controlled by the tunnelling whose probability greatly changes on a scale of lattice constant [45, 46].

Figures 3.5 and 3.6 present schematic classification of regimes observable for the $A + B \to 0$ reaction. We will concentrate in further Chapters of the book mainly on diffusion-controlled kinetics and will discuss very shortly an idea of *trap-controlled kinetics* [47–49]. Any solids contain pre-radiation defects which are called *electron traps and recombination centres* – Fig. 3.7. Under irradiation these traps and centres are filled by electrons and holes respectively. The probability of the electron thermal ionization from a trap obeys the usual Arrhenius law: $\gamma = s \exp(-E/(k_B T))$, where s is the so-called *frequency factor* and E thermal ionization energy. When the temperature is increased, electrons become delocalized, flight over the conduction band and recombine with holes on the recombination centres. Such

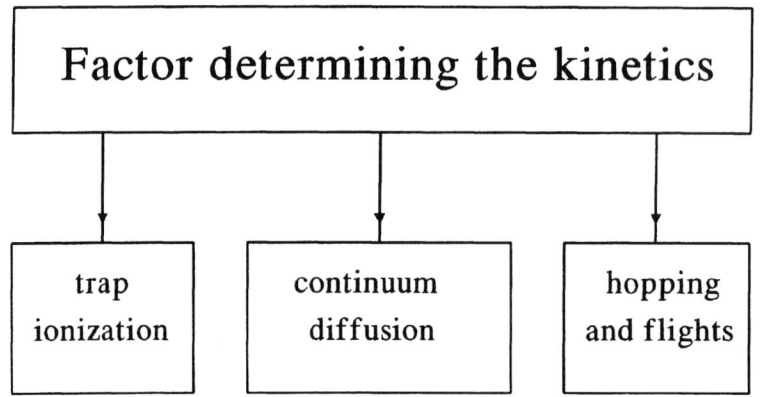

Fig. 3.5. Different factors determining the kinetics of the A + B → 0 reaction.

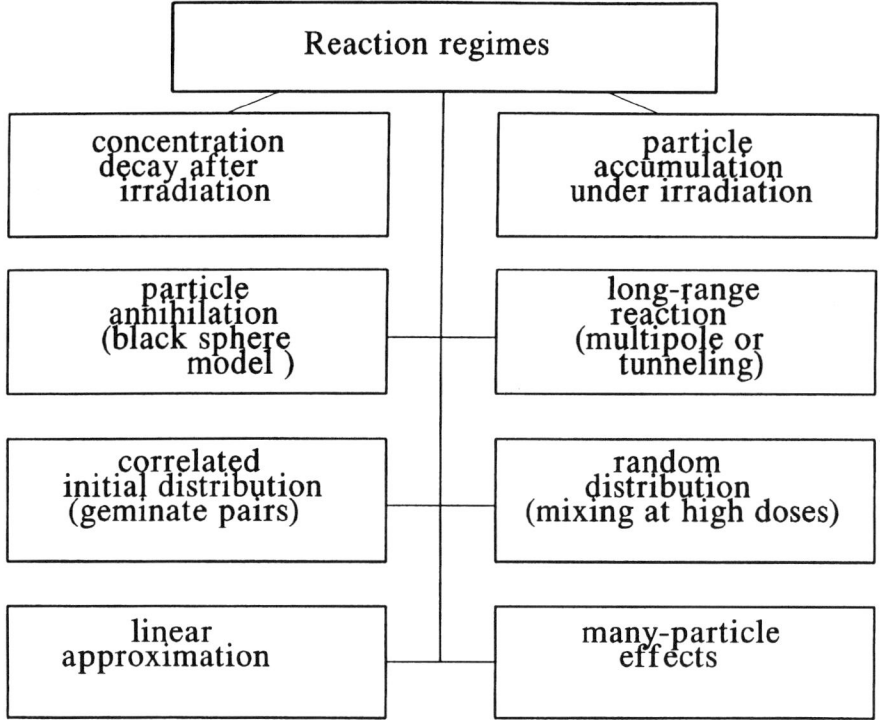

Fig. 3.6. Different regimes of the A + B → 0 reaction.

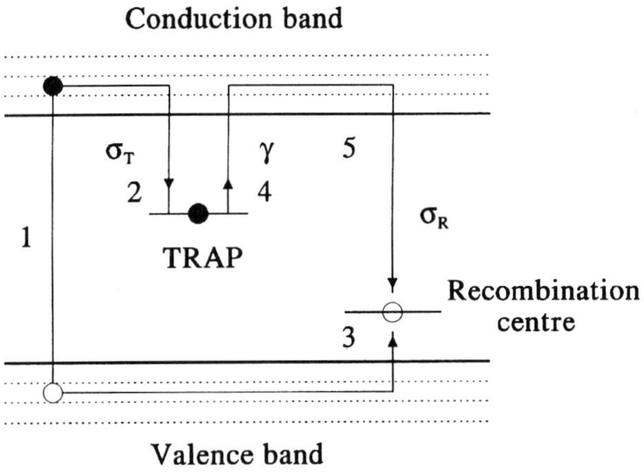

Basic defects and processes in solids

Conduction band

σ_T γ 5
2   4
                    σ_R
1
      TRAP
                      Recombination
                3   centre

Valence band

Fig. 3.7. Trap-controlled carrier recombination: 1 – excitation of solid with creation of electron (•) and hole (◯) pair; 2, 3 – their localization (trapping) by defects; 4 – thermal ionization of electron from a trap; 5 – its recombination with the recombination centre.

kinetics is considered usually in terms of balance equations:

$$\frac{dn_T}{dt} = -\gamma n_T + \sigma_T(\nu_T - n_T)n,$$

$$\frac{dn_R}{dt} = -\sigma_R n_R n,$$

$$\frac{dn}{dt} = \gamma n_T - \sigma_T(\nu_T - n_T)n - \sigma_R n_R n,$$

$$\nu_T = n_T + n_{T^*}, \qquad (3.1.7)$$

where ν_T, n_T and n_{T^*} are total concentration of traps for filled and ionized traps respectively; σ_T and σ_R are cross-sections of carrier trapping and recombination respectively; n is mobile electron concentration. Since the thermal ionization probability for electrons has the temperature dependence formally coinciding with that for defect hopping in the diffusion-controlled processes, sometimes it is not easy to identify which of two processes takes place [46, 50].

In the last decades, both experimental data [2, 51] and theoretical studies [52–56] revealed the effect of the statistical *similar defect aggregation* under defect accumulation (permanent particle source). It means that the initial random mixture of defects of two kinds A and B during bimolecular reaction is

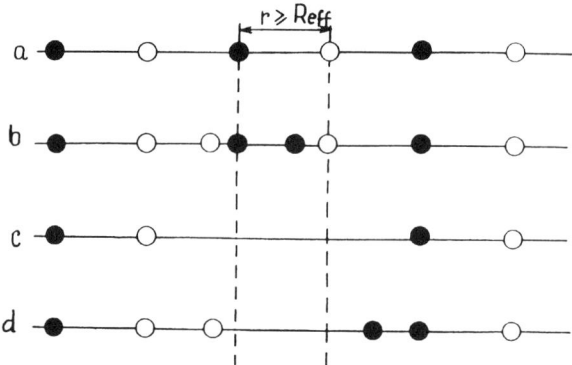

Fig. 3.8. Qualitative illustration of the similar defect aggregation under irradiation. (a) Random distribution, mean distance between defects r exceeds the recombination radius R. (b) New pair is created near the pre-existing defects. (c) It disappears. (d) Similar defect aggregates could be established, if newly-created defects find themselves near defects of the same kind.

replaced by *loose aggregates* (domains) containing predominantly defects of the *same* kind, i.e., A or B only (Fig. 1.18). The greater are the initial concentrations and/or the longer is reaction time, the more this effect is pronounced. The defect segregation into aggregates is the *universal phenomenon* having common features with self-organization observed earlier for complex chemical systems [57, 58]. It takes place *irrespective* of the particular reaction regime, i.e., it is observed for both mobile and immobile defects, for annihilation and tunnelling recombination, in defect decay after the irradiation is switched off, and with the defect source (irradiation) [53]. Radiation-induced aggregation of similar defects was observed experimentally in the electronic microscope by Hobbs et al. [59] and after prolonged X-ray irradiation by Faraday and Compton [51] via concentration of dimer F_2 centres which is *anomalously high* for random distribution even at 4 K. This effect permits to accumulate defect concentrations with the saturation value which exceeds several times that of the Poisson (random) distribution of defects.

This similar defect segregation arises due to the *local fluctuations of particle (defect) densities* which is illustrated qualitatively in Fig. 3.8: production by a change of two or more similar defects (say, A) nearby creates a *germ* of their aggregate which increases in size with time since it is more stable and has a greater chance to survive than several isolated defects A; the probability that the same number of defects B will be created statistically in the same region to destroy this A-germ is very small. On the other hand, diffusive approach of some defect B to the A-rich aggregate results in disappearance of a single peripheral defect A, but doesn't affect defects in a centre of the aggregate. Due to this effect of *statistical defect screening* an *effective recombination* volume per vacancy in an aggregate is much smaller than for the isolated vacancy $v_0 = (4/3)\pi r_0^3$ [60–63]. In order to study theoretically defect aggregation, several new methods of physical and chemical kinetics were developed in recent years; they are discussed in detail below. The concentration of accumulated stable defects depends also on the *temperature* (for a given irradiation dose, i.e., a total number of created defects). If the Frenkel defects are correlated into geminate pairs, this temperature dependence reflects their survival probability. In alkali halides, η is known to obey the *Arrhenius law* [14]

$$\eta \propto \exp\left(-\frac{E^*}{k_\mathrm{B} T}\right), \qquad (3.1.8)$$

where E^* is effective accumulation energy. On the other hand, another relation is widely used in metals to approximate the dependence $\eta = \eta(T)$, namely [3, 4]

$$\eta \propto 1 - cT^{-1/3}, \qquad (3.1.9)$$

c is constant. We will discuss these relations below and demonstrate their relationship.

3.2 ANNEALING OF GEMINATE PAIRS OF DEFECTS

> This is a theory explaining the northern lights by the brilliance of the herring scales.
> (Dieses ist eine Theorie, die [...] das Nordlicht durch den Glanz der Heringe erklärt)
>
> Lichtenberg

3.2.1 Smoluchowski equation

Before discussing the statistics and kinetics of many-particle systems, which requires the use of rather refined and complicated formalism, let us start first of all from a *two-particle* problem. Although it is presented from the methodological point of view, it is realistic. For example, it can describe reaction kinetics for complementary radiation Frenkel defects in solids which are known to be created in pairs of closely spaced defects [5, 8]. It is a common phenomenon in both metals and insulators.

At small radiation doses (the number of radiation-produced defects), the mean distance l between components of such geminate pairs (the vacancy and an interstitial atom) is much *less* than the mean distance between different pairs $l_0 = n^{-1/3}$, where n is defect concentration. The initial defect distribution is described by the distribution function $f(r)$. Below a certain temperature (typically < 30 K for interstitial atoms and 200 K for vacancies in alkali halides), defects are immobile. With a temperature increase, the defects perform thermally activated random hops between the nearest lattice sites. This is usually considered to be *continuous diffusion*.

Due to the existence of two quite different distinctive distances (scale factors) – l_0 and l – the recombination kinetics also reveals two stages called *monomolecular* and *bimolecular* respectively. The defects survived in their geminate pairs go away, separate and start to mix and recombine with dissimilar components from other pairs. It is clear that the problem of kinetics of the monomolecular process is reduced to the time development of the probability $w(t)$ to find any *single* geminate pair AB as a function of the initial spatial distribution of the pair components $f(r)$, recombination law $\sigma(r)$ and interaction $U_{AB}(r)$. The smaller the initial concentration of defects, $n(0) \to 0$, as $l_0 \to \infty$, the more correct is the separation of the kinetics into two substages, whereas the treatment of the case of semi-mixed geminate pairs is a very difficult problem discussed below.

Until the geminate pairs start to mix, i.e., at relatively short times $\tau \ll l_0^2/D$, where $D = D_A + D_B$ is a relative diffusion coefficient, the monomolecular kinetics reads $n(t) = n(0)w(t)$, with $n(0) = n_A(0) = n_B(0)$ being initial particle concentration. The distinctive feature of this stage is the linearity of the recombination kinetics $n(t)$ with respect to the irradiation dose $n(0)$.

To describe the *geminate kinetics*, let us start with the motion equations for the probability density $W(\vec{r}_A, \vec{r}_B, t)$, where $W(\vec{r}_A, \vec{r}_B, t)d\vec{r}_A d\vec{r}_B$ gives the probability to find at the moment t particle A at volume $d\vec{r}_A$ centred at $\vec{r}_A$ and particle B at $d\vec{r}_B$ centred at $\vec{r}_B$. The relevant motion equation reads [64]

$$\frac{\partial W(\vec{r}_A, \vec{r}_B, t)}{\partial t} = \left(D_A \nabla_A^2 + D_B \nabla_B^2 - \sigma(r)\right) W(\vec{r}_A, \vec{r}_B, t), \qquad (3.2.1)$$

D_A and D_B are diffusion coefficients and $r = |\vec{r}_A - \vec{r}_B|$. The *survival probability* of a pair $\omega(t)$ comes from integration over the particle coordinates

$$\omega(t) = \iint W(\vec{r}_A, \vec{r}_B, t) \, d\vec{r}_A \, d\vec{r}_B. \tag{3.2.2}$$

Since in the isotropic medium the recombination kinetics is governed by the *relative* distance $\vec{r} = \vec{r}_A - \vec{r}_B$, it is convenient to introduce new coordinates $\vec{r}, \vec{R}$. The latter,

$$\vec{R} = \frac{\vec{r}_A/D_A + \vec{r}_B/D_B}{1/D_A + 1/D_B}, \tag{3.2.3}$$

is a centre-of-mass coordinate. Now we can rewrite equation (3.2.1) in the form

$$\frac{\partial W(\vec{r}, \vec{R}, t)}{\partial t} = \left(D^* \nabla_R^2 + D \nabla_r^2 - \sigma(r)\right) W(\vec{r}, \vec{R}, t), \tag{3.2.4}$$

where $D^* = D_A D_B / D$ is centre-of-mass diffusion coefficients. In a homogeneous medium the information about the centre-of-mass could be easily eliminated by integrating

$$w(r, t) = \int W(\vec{r}, \vec{R}, t) \, d\vec{R}, \tag{3.2.5}$$

which yields the geminate partners relative spatial distribution $w(r,t)$ at any moment t provided the initial distribution $f(r)$ is spherically symmetric. Therefore, the probability to find a geminate pair is

$$\omega(t) = \int w(r, t) \, d\vec{r}. \tag{3.2.6}$$

Differentiating equation (3.2.5) and making use of the motion equation (3.2.4), we arrive at the well-known *Smoluchowski equation* [65–70]

$$\frac{\partial w(r, t)}{\partial t} = D \Delta w(r, t) - \sigma(r) w(r, t), \tag{3.2.7}$$

where the continuous motion operator is just the Laplace operator

$$\Delta = \frac{\partial^2}{\partial r^2} + \frac{(\bar{d}-1)}{r}\frac{\partial}{\partial r}, \qquad (3.2.8)$$

where $\bar{d}$ is the space dimensionality.

The distinctive feature of the kinetics sought for is its dependence on the *relative* diffusion coefficient $D = D_A + D_B$ only. In turn, the time derivative of equation (3.2.6) reads (using also equation (3.2.7))

$$\frac{dw(t)}{dt} = -P(t), \qquad (3.2.9)$$

where

$$P(t) = \int \sigma(r)w(r,t)\,d\vec{r}. \qquad (3.2.10)$$

When deriving equation (3.2.10), it was also assumed that particles disappear only due to recombination.

Equation (3.2.7) could be generalized for the case of particle *interaction* characterized by the *pair potential* $U(r) = U_{AB}(r)$ [66, 69]:

$$\frac{\partial w(r,t)}{\partial t} = D\nabla\left(\nabla w(r,t) + \frac{w(r,t)}{k_B T}\nabla U(r)\right) - \sigma(r)w(r,t). \qquad (3.2.11)$$

As compared with equation (3.2.7), equation (3.2.11) has an additional term due to a particle drift in the external potential $U(r)$ which is assumed to be isotropic.

For $\bar{d} = 3$, the solution of equation (3.2.9) could easily be found to be

$$w(t) = 4\pi \int_0^\infty w(l\mid t)f(l)l^2\,dl, \qquad (3.2.12)$$

where $f(r) = w(r,0)$ is the normalized initial distribution.

$$\int f(r)\,d\vec{r} = 1, \qquad (3.2.13)$$

but $w(l \mid t)$ is the conditional survival probability (provided initial ($t = 0$) separation within pair is l):

$$w(r, 0) = \frac{\delta(r - l)}{4\pi l^2}. \tag{3.2.14}$$

For the *static recombination*, $D = 0$, equation (3.2.7) reads

$$w(r, t) = f(r)e^{-\sigma(r)t}. \tag{3.2.15}$$

It is clear from equation (3.2.15) that $\tau(r) = 1/\sigma(r)$ is nothing but the *lifetime* of a pair separated by a distance r.

It is convenient to analyze the general equations (3.2.7) and (3.2.11) in terms of the simplest model of the recombination event called *clear-cut radius* or *black reaction sphere* – equation (3.1.1):

$$\begin{aligned}\sigma(r) &= \sigma_0, \quad \sigma_0 \to \infty \quad (r \leqslant r_0), \\ \sigma(r) &= 0, \quad\quad\quad\quad\quad\; (r > r_0).\end{aligned} \tag{3.2.16}$$

This means that *no* recombination occurs unless the particles approach each other to within the critical distance r_0 which is associated with the unavoidable fall into the recombination sphere. It can be shown that for instant recombination, $\sigma_0 \to \infty$, $w(r < r_0, t) = 0$, which results in the generally accepted *Smoluchowski boundary condition*

$$w(r_0, t) = 0. \tag{3.2.17}$$

For the black sphere model under consideration the integration in equations (3.2.7) and (3.2.11) over the spherical layer $r > r_0$ gives equation (3.2.10) with the flux of particles over the recombination sphere surface

$$P(t) = 4\pi D r_0^2 \left. \frac{\partial w(r, t)}{\partial r} \right|_{r=r_0}. \tag{3.2.18}$$

Therefore, making use of this simple model one can solve and qualitatively analyze different peculiarities of the monomolecular recombination. Partially reflecting boundary conditions (grey recombination sphere) are studied in [50, 68]. For the effects of reaction sphere anisotropy, see Chapter 5.

3.2.2 Noninteracting particles

As an example, let us consider now the simplest case of non-interacting particles, $U(r) = 0$, all particles are created with the *same* separation l (3.2.14). The solution of equation (3.2.7) is [66]:

$$w(r,t) = \frac{1}{4\pi l r (4\pi Dt)^{1/2}} \left\{ \exp\left[-\frac{(r-l)^2}{4Dt}\right] - \exp\left[-\frac{(r+l-2r_0)^2}{4Dt}\right] \right\}. \quad (3.2.19)$$

The integration of equation (3.2.9) yields the time development of the conditional probability to find a pair at time t:

$$w(l \mid t) = 1 - \frac{r_0}{l} \frac{2}{\pi^{1/2}} \int_{\frac{(l-r_0)}{2(Dt)^{1/2}}}^{\infty} e^{-\zeta^2} d\zeta. \quad (3.2.20)$$

For a short reaction time, $2(Dt)^{1/2} \ll (l-r_0)$, the asymptotic expansion of equation (3.2.20) gives the estimate

$$w(l \mid t) \approx 1 - \frac{r_0}{l} F(t), \quad (3.2.21)$$

where

$$F(t) = \frac{2(Dt)^{1/2}}{\pi^{1/2}(l-r_0)} \exp\left[-\frac{(r-l)^2}{4Dt}\right]. \quad (3.2.22)$$

This expression turns out to be surprisingly close to the asymptotic survival probability seen from equation (3.2.20)

$$w(l \mid \infty) = 1 - \frac{r_0}{l} \quad (\bar{d} = 3), \quad (3.2.23)$$

even for the intermediate time, $t \approx (l-r_0)^2/(4D)$. This is even more so for a long time when $(l-r_0) \ll 2(Dt)^{1/2}$. Equation (3.2.23) demonstrates clearly that a certain portion of pairs has separated due to diffusive walks into infinity. For $\bar{d} = 3$, the complete recombination, $w(l \mid \infty) = 0$, occurs only if one partner is created at the very edge of the recombination sphere or inside it, $l \to r_0$.

In the two-dimensional case (e.g., on a surface or for highly anisotropic layered solids), we get in equation (3.2.7) the modified Laplace operator (3.2.8) and instead equation (3.2.18) we have

$$P(t) = 2\pi D r_0 \frac{\partial w(r,t)}{\partial r}\bigg|_{r=r_0}, \qquad (3.2.24)$$

which results in the modification of equation (3.2.21) at $2(Dt)^{1/2} \ll (l-r_0)$:

$$w(l \mid t) \approx 1 - \left(\frac{r_0}{l}\right)^{1/2} F(t). \qquad (3.2.25)$$

At first glance, it seems quite similar to equation (3.2.21). However, they drastically differ as $t \to \infty$:

$$w(l \mid \infty) = 0 \quad (\bar{d} = 1, 2). \qquad (3.2.26)$$

Another way to derive the expression for survival probability $w(l \mid t)$ is to use the *Laplace transform*

$$\widehat{w}(l \mid s) = \int_0^\infty w(l \mid t) e^{-st}\, dt. \qquad (3.2.27)$$

From equations (3.2.7) and (3.2.9) modified for $\bar{d} = 2$ one obtains

$$s\widehat{w}(l \mid s) = 1 - \frac{K_0\left((s/D)^{1/2} l\right)}{K_0\left((s/D)^{1/2} r_0\right)}, \qquad (3.2.28)$$

where K_0 is the modified Bessel function of zero order.

Using the limiting relation

$$w(l \mid \infty) = \lim_{s \to 0} s\widehat{w}(l \mid s) \qquad (3.2.29)$$

and the asymptotic behaviour $K_0(z) \approx -\ln(z/2)$ as $z \to 0$, we again get equation (3.2.26).

3.2.3 Drift in the potential field

In the case of particle interaction described by their drift in potential, the black sphere model again allows us to obtain a simple solution of the kinetics sought for. Imposing the boundary conditions, equation (3.2.17) and $w(\infty, t) = 0$, as well as the initial condition (3.2.14) on the motion equation

$$\frac{\partial w(r,t)}{\partial t} = D\nabla\left(\nabla w(r,t) + \frac{w(r,t)}{k_B T}\nabla U(r)\right) \qquad (3.2.30)$$

and using the Laplace transform

$$\widehat{w}(r,s) = \int_0^\infty w(r,t)e^{-st}\,dt, \qquad (3.2.31)$$

we arrive at ($\bar{d} = 3$)

$$s\widehat{w}(r,s) - \frac{\delta(r-l)}{4\pi l^2} = D\nabla\left(\nabla \widehat{w}(r,s) + \frac{\widehat{w}(r,s)}{k_B T}\nabla U(r)\right). \qquad (3.2.32)$$

Integration over the spherical layer $r > r_0$ and use of equation (3.2.6),

$$\widehat{\omega}(l \mid s) = \int \widehat{w}(r,s)\,d\vec{r}, \qquad (3.2.33)$$

lead to

$$s\widehat{\omega}(l \mid s) = 1 - 4\pi Dr_0^2 \frac{\partial \widehat{w}(r,s)}{\partial r}\bigg|_{r=r_0}. \qquad (3.2.34)$$

If we now put $s \to 0$, we finally get

$$\omega(l \mid \infty) = 1 - \left(r^2 \frac{\partial Z(r)}{\partial r}\right)_{r=r_0}. \qquad (3.2.35)$$

Here $Z(r) = 4\pi D\widehat{w}(r,0)$ is a solution of the following inhomogeneous equation:

$$\nabla\left(\nabla Z(r) + \frac{Z(r)}{k_B T}\nabla U(r)\right) = -\frac{\delta(r-l)}{l^2} \qquad (3.2.36)$$

with the boundary conditions

$$Z(r_0) = Z(\infty) = 0. \qquad (3.2.37)$$

Equation (3.2.35) could be transformed into a form more convenient for further analysis [71]:

$$w(l \mid \infty) = -\lim_{r \to \infty} r^2 \frac{\partial Z(r)}{\partial r}. \qquad (3.2.38)$$

Let $y(r)$ be a solution of the homogeneous equation obtained putting the r.h.s. of equation (3.2.36) to zero,

$$\nabla \left(\nabla y(r) + \frac{y(r)}{k_B T} \nabla U(r) \right) = 0, \qquad (3.2.39)$$

subject to the boundary conditions

$$y(r_0) = 0, \qquad y(\infty) = 1. \qquad (3.2.40)$$

It is shown below (Chapter 4) that its solution is also important for the bimolecular stage of a reaction defining a stationary recombination profile. The second linearly independent solution of this equation could be expressed through $y(r)$ as

$$y^*(r) = y(r) \int_r^\infty \exp\left[-\frac{U(r')}{k_B T}\right] \frac{dr'}{[r' y(r')]^2}. \qquad (3.2.41)$$

Besides for $r \to \infty$

$$y^*(r) \propto \frac{1}{r}. \qquad (3.2.42)$$

Their Wronskian is

$$W[y, y^*] = y(r) \frac{\partial y^*(r)}{\partial r} - y^*(r) \frac{\partial y(r)}{\partial r}$$

$$= \frac{1}{r^2} \exp\left[-\frac{U(r)}{k_B T}\right]. \qquad (3.2.43)$$

The solution of equation (3.2.36) subject to the corresponding boundary conditions (3.2.37) and the continuity condition at $r = l$ could be sought for in a form

$$Z(r) = \begin{cases} cy(r)y^*(l), & r < l, \\ cy^*(r)y(l), & r > l. \end{cases} \quad (3.2.44)$$

The constant c here could be found from the solution at $r < l$ and $r > l$

$$\left. \frac{\partial Z(r)}{\partial r} \right|_{l-0}^{l+0} = -\frac{1}{l^2}. \quad (3.2.45)$$

Equation (3.2.45) arises from integration of equation (3.2.36) in the interval $l-\delta$ to $l+\delta$, where δ is an infinitesimal parameter. With the help of equations (3.2.44), (3.2.45) one gets

$$cW[y, y^*]\big|_{r=l} = -\frac{1}{l^2}, \quad (3.2.46)$$

and therefore

$$c = \exp\left[\frac{U(l)}{k_B T}\right]. \quad (3.2.47)$$

At last, from equation (3.2.38) we obtain very simple result [72]: the survival probability

$$w(l \mid \infty) = y(l) \exp\left[\frac{U(l)}{k_B T}\right], \quad (3.2.48)$$

which is independent of the diffusion coefficient D. Similarly to equation (3.2.23), the *effective reaction radius* R_{eff} could be defined as a coefficient in the asymptotic expansion

$$w(l \mid \infty) \approx 1 - R_{\text{eff}}/l + \cdots, \quad (3.2.49)$$

as $l \to \infty$ [50, 71], which is equivalent to

$$R_{\text{eff}} = \lim_{l \to \infty} l\big(1 - w(l \mid \infty)\big) \quad (3.2.50)$$

or

$$R_{\text{eff}} = b - \lim_{l \to \infty} \frac{lU(l)}{k_B T}, \qquad (3.2.51)$$

where b is the expansion coefficient

$$y(l) \approx 1 - b/l + \cdots, \qquad (3.2.52)$$

as $l \to \infty$.

The second term in expansion (3.2.51) is non-zero for slowly decreasing interactions, e.g., Coulomb [71]. For charged particles obeying the Coulomb law

$$U(r) = \frac{e_A e_B}{\varepsilon r} \qquad (3.2.53)$$

it is convenient to introduce a new parameter L via

$$\frac{U(r)}{k_B T} = \pm L/r, \qquad (3.2.54)$$

so that

$$L = \frac{|e_A e_B|}{\varepsilon k_B T}. \qquad (3.2.55)$$

(The signs '+' and '−' are for repulsion and attraction, respectively.) L is called *Onsager radius* [73] at which the Coulomb interaction energy equals the thermal energy $k_B T$. Finally, for the Coulomb attraction (see more details in Chapter 4), the effective reaction radius is [67]

$$R_{\text{eff}}^{(-)} = \frac{L}{1 - \exp[-L/r_0]}. \qquad (3.2.56)$$

When $L \gg r_0$, the effective reaction radius coincides practically with the Onsager radius, $R_{\text{eff}}^{(-)} \approx L$.

On the other hand, for repulsion [67] one gets

$$R_{\text{eff}}^{(+)} = \frac{L}{\exp[L/r_0] - 1}. \qquad (3.2.57)$$

As $L \gg r_0$ the effective radius could be exponentially small, $R_{\text{eff}}^{(+)} \approx L\exp[-L/r_0]$. For the weak Coulomb interaction, as $L \to 0$, we naturally obtain from equations (3.2.56) and (3.2.57) that $R_{\text{eff}}^{(\pm)} \approx r_0$; the reaction rate again is controlled by the recombination at the black sphere radius. At last, both effective radii – for repulsion and attraction – are trivially related [50, 71]:

$$R_{\text{eff}}^{(-)} - R_{\text{eff}}^{(+)} = L. \qquad (3.2.58)$$

Strictly speaking, it is correct in the case of *complete* particle recombination at the black sphere only; partial particle reflection is discussed by Doktorov and Kotomin [50]. Incorporation of the *back reactions* into the kinetics of geminate recombination has been presented quite recently by [74, 75]. The effective radius for an elastic interaction of defects in crystals, (3.1.4), was calculated by Schröder [3], Kotomin and Fabrikant [76].

As it was said in Section 3.1, the survival probability of the geminate pairs $\omega(l \mid \infty)$ directly defines the efficiency of the Frenkel defect accumulation in solids. Let us assume that the concentration of radiation-created defects is n. Then after a transient period during which mobile interstitials recombine with their geminate partners or leave them thus surviving, we have n_{acc} of *stable* defects obeying the relation $n_{\text{acc}} = \eta n$, where $\eta = \omega(\infty)$ is *accumulation efficiency*, which could be identified with the survival probability of a *single* mobile interstitial (provided that different geminate pairs do not mix).

In the case of the Coulomb attraction of defects, which is true for α, I pairs, equation (3.2.48) yields (the solution for $y(r)$ is given in Chapter 4):

$$\omega(l \mid \infty) = \frac{\exp\left(\frac{L}{r_0} - \frac{L}{l}\right) - 1}{\exp\left(\frac{L}{r_0}\right) - 1}. \qquad (3.2.59)$$

If both $L \gg r_0$ and $l \gg r_0$, the Arrhenius law takes place with the effective activation energy E^* equal to that of the Coulomb attraction just after the creation of charged pair, similar to the case of elastic interaction. Note that since the actual value of the Onsager radius L in alkali halides could be as large as $L \approx 100$ Å ($T = 100$ K, $\varepsilon \approx 5$, $e_A = -e_B = e$), only a small fraction of pairs with a large enough initial separation, $l \approx L$, has a chance to be separated [77].

Let us consider now annealing of *neutral* defects whose annihilation is stimulated by elastic interaction, characterized by the energy $U(r) = -\alpha/r^3$ ($\alpha > 0$) – equation (3.1.4). In the case where all geminate pairs are created at the same distance l, we have [71, 76]

$$\omega(l \mid \infty) = 1 - \frac{r_0}{l} \frac{\exp\left(-\frac{g}{l^3}\right) \Phi\left(1, \frac{4}{3}, \frac{g}{l^3}\right)}{\exp\left(-\frac{g}{r_0^3}\right) \Phi\left(1, \frac{4}{3}, \frac{g}{r_0^3}\right)}, \qquad (3.2.60)$$

where $g = \alpha/k_B T$, and Φ is the degenerate hypergeometric function.

At high temperatures, when $g \ll l^3$, but not r_0^3, one gets the relation widely used for metals (e.g., [78, 79]): equation (3.2.49), with $R_{\text{eff}} = r_0$. (It could be shown that for the *general* interaction law $U(r) = -\alpha/r^m$ equation (3.2.49) also holds provided $g \ll l^m$.) At low temperatures, when $g \gg l^3$, one gets quite another expression for the survival probability:

$$\omega(l \mid \infty) = \frac{l^2 \exp\left(-\frac{g}{l^3}\right) - r_0^2 \exp\left(-\frac{g}{r_0^3}\right)}{\Gamma\left(\frac{1}{3}\right) g^{2/3}}, \qquad (3.2.61)$$

where Γ is Euler's gamma function. If $l \gg r_0$, equation (3.2.61) reads

$$\omega(l \mid \infty) \approx \frac{l^2 \exp\left(-\frac{g}{l^3}\right)}{\Gamma\left(\frac{1}{3}\right) g^{2/3}} \propto \exp\left(-\frac{g}{l^3}\right), \qquad (3.2.62)$$

i.e., it obeys the Arrhenius relation with the effective energy equal to the elastic interaction energy of defects at the moment of their creation.

To illustrate the general theory developed above, let us briefly consider here as an example the recombination of the geminate F, H pairs in alkali halide crystals. Steady-state experiments on irradiated alkali halides show that during the linear heating of a sample $T = T_0 + \beta t$ (β is a heating rate, T_0 is initial temperature), *several* recombination (annealing) stages are observed e.g., in KBr separated by several K [8, 80]. This process was simulated theoretically by Kotomin et al. [77, 81].

They have calculated the continuous diffusion equation (3.2.30) with $U(r) = -\alpha/r^3$ for several kinds of nn F, H centres in the crystalline lattice. Figure 3.9 demonstrates well that both defect initial separation and an elastic interaction are of primary importance for geminate pair recombination kinetics. The 3nn defects are only expected to have noticeable survival probability. Its magnitude agrees well with equation (3.2.60).

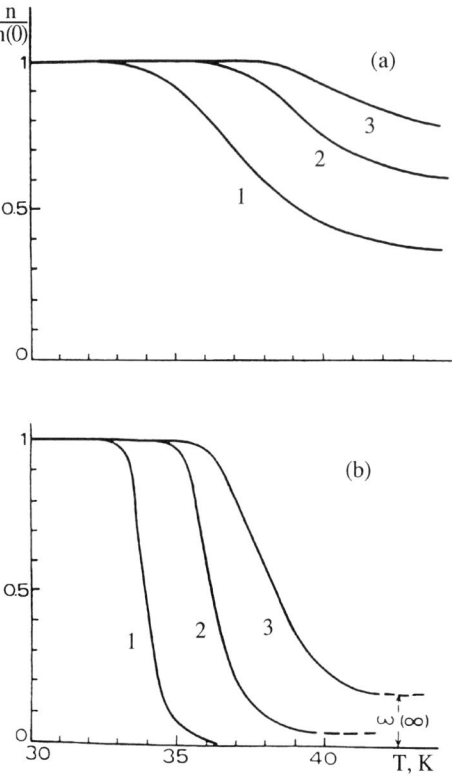

Fig. 3.9. The thermostimulated correlated annealing of F, H defects in KBr. Curves 1 to 3 – H-centres are in relative positions $(1,1,0)$ nearest neighbours (nn), $(2,0,0)$ (2nn), $(2,2,0)$ (3nn) respectively (in units of $a_0/2$, a_0 is the lattice constant, a vacancy is at the origin of coordinates). An elastic interaction is neglected (a) and taken into account (b). Parameters used: $r_0 = 3$ Å, $\alpha = 3$ eV Å^3, $E_a = 0.09$ eV, $D_0 = 1.5 \times 10^{11}$ Å^2 s^{-1}, $\beta = 0.1$ K s^{-1}. Survival probably $w(\infty)$ is shown by arrow.

In a more realistic and complicated case, defects at *several* relative distances are produced. As is easily seen from Fig. 3.10, the step-structure on the annealing curve is pronounced to be worse, the greater the distance of an H centre from a vacancy; it is no longer easily seen even for 2nn and 3nn, despite their elastic interaction. A comparison of theoretical calculations with actual experimental data as well as a more detailed treatment of the F, H annealing kinetics is given in [77].

Annealing of geminate pairs of defects

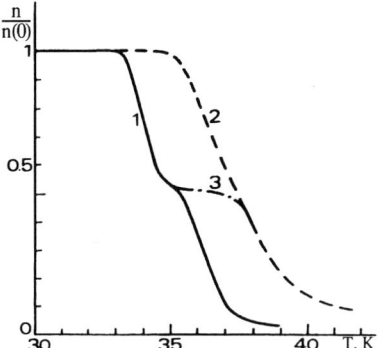

Fig. 3.10. Same as in Fig. 3.9 for the initial defect distributions consisting of several kinds of pairs of close defects taken in equal concentrations. Curve 1 – 1nn and 2nn, 2 – 2nn and 3nn, 3 – 1nn and 3nn.

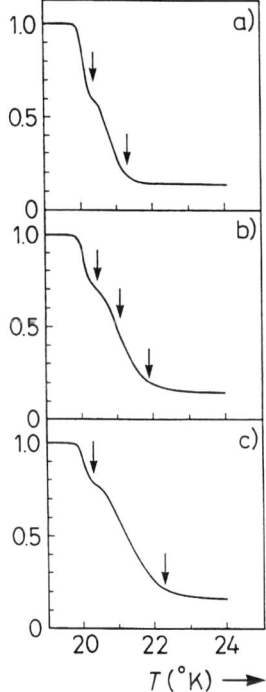

Fig. 3.11. Same as in Fig. 3.9 for simultaneous annealing of 1nn, 2nn and 3nn of F, I centres in KBr crystals.

As it is known, I centres are the most mobile radiation-induced radiation defects in alkali halides and therefore they play an essential role in low-temperature defect annealing. It is known, in particular, from thermally-stimulated conductivity and thermally-stimulated luminescence measurements, that these centres recombine with the F and F' electron centres which results in an electron release from anion vacancy. This electron participates in a number of secondary reactions, e.g., in recombination with hole (H, V_k) centres. Results of the calculations of the correlated annealing of the close pairs of I, F centres are presented in Fig. 3.11. The conclusion could be drawn that even simultaneous annealing of three kinds of pairs (1nn, 2nn and 3nn in equal concentrations) results in the step-structure of concentration decay in complete agreement with the experimental data [82].

3.2.4 Recombination on a discrete crystalline lattice

Up to now we have been considering defect diffusion in *continuous approximation*, despite the fact that crystalline lattice discreteness was explicitly taken into account defining the initial distribution for geminate pairs. Note, however, that such continuous diffusion approximation is valid only *asymptotically* when defects (particles) before recombination made large number of hops (see Kotomin and Doktorov [50]). This condition could be violated for recombination of *very close* defects which can happen in several hops. The *lattice* statement of the annihilation kinetics has been discussed in detail by Schröder et al. [3, 4, 83], Dederichs and Deutz [34]. Let us consider here just the most important points of this problem.

The *lattice* diffusion equation reads

$$\frac{dw(\vec{n},t)}{dt} = \sum_{\vec{m}} \left[H(\vec{n}-\vec{m})\, w(\vec{m},t) - H(\vec{m}-\vec{n})\, w(\vec{n},t) \right] +$$

$$+ p(\vec{n},t). \tag{3.2.63}$$

Here $w(\vec{n},t)$ is the probability to find an interstitial atom in a site $\vec{n}$ at time t, $H(\vec{n}-\vec{m})$ the hopping rate and $p(\vec{n},t)$ the production rate. For the nearest neighbour hopping we have

$$H(\vec{n}-\vec{m}) = S(\vec{n}-\vec{m})\nu_0 \exp\left(-\frac{E_a}{k_B T}\right), \tag{3.2.64}$$

$$S(\vec{n} - \vec{m}) = \begin{cases} 1/z & \text{if } \vec{n} \text{ and } \vec{m} \text{ are nearest sites,} \\ 0 & \text{otherwise,} \end{cases} \qquad (3.2.65)$$

z is a number of the nearest neighbours (say, 6 for a cubic lattice). It is convenient to introduce the dimensionless time $t' = t/\tau$, $\tau = \nu_0^{-1} \exp(E_a/(k_B T))$, and $p'(\vec{n}, t) = p(\vec{n}, t)\tau$, which permits to simplify equation (3.2.63)

$$H(\vec{n} - \vec{m}) \equiv S(\vec{n} - \vec{m}), \qquad \sum_{\vec{n} \text{ (or } \vec{m})} H(\vec{n} - \vec{m}) = 1. \qquad (3.2.66)$$

The boundary condition of instant annihilation when an interstitial atom finds itself in one of the unstable sites around vacancy reads $w(\vec{\mu}) = 0$, $\vec{\mu} \in \{\vec{\nu}\}$. (A shape and size of the instability region are discussed, e.g., by Schröder [3], Dzhumanov and Khabibullaev [16]).

Now the *effective* recombination (annihilation) radius could be defined similarly to that in the continuous approximation

$$\eta_R = \frac{R_{\text{eff}}}{l}, \qquad (3.2.67)$$

where η_R is recombination probability, l – distance between defects at the moment of their creation. On the other hand, η_R could be explicitly calculated as a fraction of interstitial atoms entering the recombination region $\{\vec{\nu}\}$. Let us calculate the absorption rate $a(\vec{m}) \geqslant 0$ satisfying the boundary condition and the steady-state diffusion equation

$$\sum_{\vec{m}} [H(\vec{n} - \vec{m})w(\vec{m}) - H(\vec{m} - \vec{n})w(\vec{n})] +$$
$$+ p(\vec{n}) - a(\vec{n}) = 0. \qquad (3.2.68)$$

In a model of nearest-neighbour hopping $a(\vec{m})$ are nonzero on the surface sites of the recombination sphere only. The recombination probability sought for is

$$\eta_R = \sum_{\vec{\nu}} a(\vec{\nu}), \qquad \text{provided} \quad \sum_{\vec{n}} p(\vec{n}) = 1. \qquad (3.2.69)$$

Solution of (3.2.68) could be presented via the Green's function G in the form

$$w(\vec{n}) = \sum_{\vec{m}} G(\vec{n} - \vec{m})(p(\vec{m}) - a(\vec{m})), \qquad (3.2.70)$$

where G is defined through equations

$$\sum_{\vec{m}} F(\vec{n} - \vec{m})G(\vec{m} - \vec{k}) = \delta_{\vec{n},\vec{k}}, \qquad (3.2.71)$$

$$F(\vec{n} - \vec{m}) = \delta_{\vec{n},\vec{m}} - H(\vec{n} - \vec{m}),$$

$$\sum_{\vec{m} \text{ (or } \vec{n})} F(\vec{n} - \vec{m}) = 0. \qquad (3.2.72)$$

The Green's function G is calculated, in particular, in [3, 4] – analytically for b.c.c. lattice and numerically for s.c. and f.c.c. lattices. It could be calculated as a particular case of the static lattice Green's function widely used in the solid state physics.

Imposing on equation (3.2.70) the boundary condition $w(\vec{\nu}) = 0$, one arrives at a set of linear equations for $a(\vec{\nu})$:

$$\sum_{\vec{\mu}} G(\vec{\nu} - \vec{\mu})a(\vec{\mu}) = \sum_{\vec{m}} G(\vec{\nu} - \vec{m})p(\vec{m}). \qquad (3.2.73)$$

Summation on the r.h.s. runs over all site on the surface of the recombination region. Taking into account the normalisation condition of the defect production, the recombination probability reads

$$\eta_R = \sum_{\vec{\mu}} a(\vec{\mu}) = \sum_{\vec{\mu},\vec{\nu}} \widetilde{G}^{-1}(\vec{\mu} - \vec{\nu}) \sum_{\vec{m}} G(\vec{\nu} - \vec{m})p(\vec{m}), \qquad (3.2.74)$$

where $\widetilde{G}$ is the Green's function projection onto the recombination surface. Now the problem is reduced to the inversion of a single finite matrix. In the case of a *single* recombination site we have

$$\eta_R(l) = G(\vec{l})/G(\vec{0}), \qquad (3.2.75)$$

provided atoms are produced in the site $\vec{l}$. To reduce number of unknown parameters $a(\vec{\nu})$, recombination region symmetry could be used. For large l we can replace $G(\vec{\nu} - \vec{m})$ the r.h.s. of equation (3.2.73) by its *asymptotic* value

$$G(\vec{\nu} - \vec{m}) \approx G_{\text{as}}(\vec{\nu} - \vec{m}) = \frac{V_{\text{at}}}{4\pi D R_{\text{eff}}}, \qquad (3.2.76)$$

where V_{at} is atomic volume (a_0^3 for s.c. lattice), D is a diffusion coefficient, R_{eff} is an effective radius.

G_{as} is equivalent to the continuous Green's function and yields a good approximation for $G(\vec{m})$ even for small R_{eff}. For $p(\vec{m}) = \delta_{\vec{m},\vec{l}}$ one gets the asymptotical expression for the recombination probability $\eta_{\text{as}}(l) = \eta_R = R_{\text{eff}}/l$, with

$$R_{\text{eff}} = \sum_{\vec{\mu},\vec{\nu}} \widetilde{G}^{-1}(\vec{\mu} - \vec{\nu}) \frac{V_{\text{at}}}{4\pi D}. \qquad (3.2.77)$$

Calculation done by Schröder and Eberlein [3, 83], has shown than even for small recombination regions, e.g., that containing only nearest to a vacancy sites in the b.c.c. lattice, $R_{\text{eff}} \approx 0.81 a_0$ which very close to the linear size of the recombination region, $R_{\text{max}} = a_0\sqrt{3}/4 \approx 0.866 a_0$.

These results yield an impressive justification for the use of the continuous diffusion approximation even for short distances and small recombination regions.

References

[1] E. Sonder and W.A. Sibley, in: Point Defect in Solids, Vol. 1, eds J.H. Crawford and L.M. Slifkin (Plenum, New York, 1972) p. 201.
[2] W. Schilling and K. Sonnenberg, J. Phys. F: 3 (1973) 322.
[3] K. Schröder, Diffusion-Controlled Reactions of Point Defects, Report 1083-FF, Jülich, 1974.
[4] K. Schröder, Point Defects in Metals (2), Springer Tracts Mod. Phys. 87 (1980) 71.
[5] N. Itoh, Adv. Phys. 31 (1982) 49.
[6] N. Itoh, Semicond. Insul. 5 (1983) 165.
[7] M. Ueta, H. Kanzaki, K. Kobayashi, Y. Toyozawa and E. Manamura, Excitonic Processes in Solids (Springer, Berlin, 1986).

[8] Ch. Lushchik and A. Lushchik, Decay of Electronic Excitations into Defects in Solids (Nauka, Moscow, 1989).
[9] V.V. Emtsev, T.V. Machovets, V.V. Mikhnovich and N.A. Vitkovskii, Radiat. Eff. Defects Solids 111/112 (1989) 99.
[10] M.A. Elango, Elementary Inelastic Radiation Processes (Nauka, Moscow, 1988).
[11] V. Kirsanov, A. Suvorov and Yu. Trushin, Processes of Radiation-Induced Defect Production in Metals (Energoatomisdat, Moscow, 1985).
[12] M. Klinger, Ch. Lushchik, T. Mashovets, Sov. Phys. Usp. 147 (1985) 523.
[13] E.D. Aluker, V.V. Gavrilov, R.G. Deich and S.A. Chernov, Transient Radiation-Stimulated Processes in Alkali Halide Crystals (Zinatne, Riga, 1987).
[14] N. Itoh and K. Tanimura, J. Phys. Chem. Solids 51 (1990) 717.
[15] Y. Toyozawa, in: Excitonic Processes in Solids (Springer, Berlin, 1986) p. 203.
[16] S. Dzhumanov and P.K. Khabibulaev, Phys. Status Solidi B: 152 (1989) 395.
[17] A. Shluger and E. Stefanovich, Phys. Rev. B: 42 (1990) 9664.
[18] R.T. Williams and K.S. Song, J. Phys. Chem. Solids 51 (1990) 679; K.S. Song and R.T. Williams, Self-Trapped Excitons (Springer, Berlin, 1993).
[19] V.M. Koshkin and Yu.P. Zabrodskii, Bull. Acad. Sci. USSR Phys. Ser. 227 (1976) 1323.
[20] V.M. Koshkin and Yu.P. Zabrodskii, Sov. Phys. Solid State 18 (1976) 2857.
[21] P. Dean, in: Progr. in Solid State Chemistry, Vol. 8 (Pergamon Press, Oxford, New York, 1973) p. 1.
[22] Yu.P. Zabrodskii, Yu.B. Reshetnyak and V.M. Koshkin, Sov. Phys. Solid State 32 (1990) 69.
[23] A.T. Vink, J. Lumin. 9 (1974) 159.
[24] E.A. Kotomin and I.I. Fabrikant, J. Lumin. 9 (1975) 502.
[25] S.A. Rice and M.J. Pilling, Progr. React. Kinet. 9 (1978) 93.
[26] A.B. Doktorov and E.A. Kotomin, Phys. Status Solidi B: 114 (1982) 9.
[27] K.I. Zamaraev, R.F. Khairutdinov and V.P. Zhdanov, Long-Range Electron Tunneling in Chemistry, Vol. 30 (North-Holland, Amsterdam, 1990) (Russian edn: Nauka, Novosibirsk, 1985).
[28] V.F. Grachev and M.F. Deigen, Sov. Phys. Usp. 125 (1978) 631.
[29] M. Moreno, J. Phys. Chem. Solids 52 (1991) 835.
[30] J.M. Spaeth and F.K. Koschnick, J. Phys. Chem. Solids 52 (1991) 1; J.M. Spaeth, J.R. Niklas and R.H. Bartram, Structural Analysis of Point Defects in Solids (Springer, Berlin, 1993).
[31] V.M. Lisitsyn, Latv. PSR Zinat. Akad. Vestis Fiz. Teh. Zinat. Ser. 3 (1990) 59 (Proc. Latv. Acad. Sci. Phys. Technol. Ser., in Russian).
[32] I.D. Eshelby, Solid State Phys. 3 (1956) 79.
[33] A.M. Kosevich, Foundations of the Crystalline Lattice Mechanics (Nauka, Moscow, 1972).
[34] P.H. Dederichs and J. Deutz, Gitter- und Kontinuumstheorie der Wechselwirkung von Punktdefekten, Report-1600, Jülich, 1979.
[35] W. Scheu, W. Frank and H. Kronmüller, Phys. Status Solidi B: 82 (1977) 523.
[36] K. Bachmann and H. Peisl, J. Phys. Chem. Solids 31 (1980) 1525.

References

[37] J.R. Manning, Diffusion Kinetics for Atoms in Crystals (Van Nostrand, Princeton, New York, 1968).
[38] S. Mrowec, Defects and Diffusion in Solids (Elsevier, Amsterdam, 1980).
[39] A.S. Laskar, G. Brebec and C. Monty (eds), Diffusion in Materials, NATO ASI Series in Appl. Sciences, Vol. 179 (1990).
[40] U. Gösele, Progr. React. Kinet. 13 (1984) 63.
[41] A. Mozumder, S.M. Pimbott, P. Clifford and N.J.B. Green, Chem. Phys. Lett. 142 (1987) 385.
[42] M.N. Kabler, in: Point Defects in Solids, ed. J. Crawford (Plenum, New York, 1972) Ch. 4.
[43] N.F. Mott and N.A. Davis, Electronic Processes in Non-Crystalline Solids (Clarendon, Oxford, 1979).
[44] B. Shklovskii and A. Efros, Electronic Properties of Doped Semiconductors (Nauka, Moscow, 1979).
[45] E.A. Kotomin, I.A. Tale, V.G. Tale, P. Kulis and P. Butlers, J. Phys. C: 1 (1989) 6777.
[46] E.A. Kotomin, L.N. Kantorovich, I.A. Tale and V.G. Tale, J. Phys. C: 4 (1991) 7429.
[47] V.V. Antonov-Romanovskii, Kinetics of Photoluminescence of Crystal Phosphors (Nauka, Moscow, 1966).
[48] R. Chen, J. Mater. Sci. 11 (1976) 1521.
[49] L. Kantorovich, E. Kotomin, V. Kuzovkov, I. Tale, A. Shluger and Yu. Zakis, Models of Processes in Wide-Gap Solids with Defects (Zinatne, Riga, 1991).
[50] E.A. Kotomin and A.B. Doktorov, Phys. Status Solidi B: 114 (1982) 287.
[51] B.F. Faraday and W.D. Compton, Phys. Rev. Sect. A 138 (1965) 893.
[52] Ya.B. Zeldovich and A.S. Mikhailov, Sov. Phys. Usp. 153 (1987) 469.
[53] V.N. Kuzovkov and E.A. Kotomin, Rep. Prog. Phys. 51 (1988) 1479.
[54] A.S. Mikhailov, Phys. Rep. 184 (1989) 308.
[55] A. Seeger, Radiat. Eff. Defects Solids 111/112 (1989) 355.
[56] V.L. Vinetsky, Yu.H. Kalnin, E.A. Kotomin and A.A. Ovchinnikov, Sov. Phys. Usp. 33 (1990) 793.
[57] H. Haken, Ed. Chaos and Order in Nature (Springer, Berlin, 1984).
[58] C. Vidal and H. Lemarchard, La Réaction Créatrice Dynamique des Systemes Chimiques (Hermann, Paris, 1988).
[59] L.W. Hobbs, A.E. Hughes and D. Pooley, Proc. R. Soc. London Ser. A: 332 (1973) 167.
[60] G. Lück and R. Sizmann, Phys. Status Solidi 5 (1964) 683.
[61] G. Lück and R. Sizmann, Phys. Status Solidi 17 (1966) K61.
[62] G. Lück and R. Sizmann, Nucleonik 8 (1966) 256.
[63] V.L. Vinetsky, Sov. Phys. Solid State 25 (1983) 1159.
[64] G. Leibfried, in: Bestrahlungseffekte in Festkrpern (Teubner, Stuttgart, 1965) p. 266.
[65] M. Smoluchowski, Zs. Phys. Chem. 92 (1917) 129.
[66] T.R. Waite, Phys. Rev. 107 (1957) 463; 107 (1957) 471.
[67] T.R. Waite, J. Chem. Phys. 28 (1958) 103.
[68] D. Peak and J.W. Corbett, Phys. Rev. B: 5 (1972) 1226.
[69] C.W. Gardiner, Handbook of Stochastic Methods for Physics, Chemistry and the Natural Sciences (Springer, Berlin, 1983).

[70] A.A. Ovchinnikov, S.F. Timashev and A.A. Belyi, Kinetics of Diffusion-Controlled Chemical Processes (Nova Science, New York, 1988).
[71] E.A. Kotomin and I.I. Fabrikant, J. Phys. C: 10 (1977) 4931.
[72] E.A. Kotomin, I.I. Fabrikant and I.A. Tale, J. Phys. C: 10 (1977) 2903.
[73] L. Onsager, J. Chem. Phys. 2 (1934) 599.
[74] N.J. Agmon, Chem. Phys. 88 (1988) 5639.
[75] E. Pines, D. Huppert and N. Agmon, J. Chem. Phys. 88 (1988) 5620.
[76] E.A. Kotomin and I.I. Fabrikant, Radiat. Eff. 46 (1980) 85.
[77] E.A. Kotomin, A.I. Popov and R. Eglitis, J. Phys. Cond. Matter, 4 (1992) 5901; E.A. Kotomin, A.I. Popov and M. Hirai, J. Phys. Soc. Jpn 63 (1994) 2602.
[78] D.E. Becker, F. Dvorschak and H. Wollenberger, Phys. Status Solidi B: 54 (1972) 455.
[79] K. Schröder and K. Dettmann, Z. Phys. B 22 (1975) 343.
[80] N. Itoh and M. Saidoh, J. Phys. (Paris) 34 (1973) C9-101.
[81] E.A. Kotomin, Latv. PSR Zinat. Akad. Vestis Fiz. Teh. Zinat. Ser. 5 (1985) 122 (Proc. Latv. Acad. Sci. Phys. Technol. Ser., in Russian).
[82] A.I. Popov, E.A. Kotomin and R. Eglitis, Phys. Status Solidi B: 175 (1993) K39.
[83] K. Schröder and E. Eberlein, Z. Phys. B 22 (1975) 181.

Chapter 4

The Linear Approximation in Bimolecular Reaction Kinetics

4.1 THE SHORTENED SUPERPOSITION APPROXIMATION (A + B → 0 REACTION)

> The fact that they preach in churches does not free them from the necessity of a lightning rod.
> (Da in den Kirchen gepredigt wird, macht deswegen die Blitzableiter auf ihnen nicht unnötig)
>
> Lichtenberg

4.1.1 Initial conditions

The kinetics of the diffusion-controlled reaction $A + B \to 0$ under study is defined by the *initial conditions* imposed on the kinetic equations. Let us discuss this point using the production of geminate particles (defects) as an example. Neglecting for the sake of simplicity diffusion and recombination (note that even the kinetics of immobile particle accumulation under steady-state source is not a simple problem – see Chapter 7), let us consider several equations from the *infinite* hierarchy of equations (2.3.43):

$$\frac{\partial \rho_{1,0}(\vec{r}_1;t)}{\partial t} = p,$$

$$\frac{\partial \rho_{0,1}(\vec{r}_1';t)}{\partial t} = p. \tag{4.1.1}$$

$$\frac{\partial \rho_{2,0}(\vec{r}_1,\vec{r}_2;t)}{\partial t} = p\rho_{1,0}(\vec{r}_1;t) + p\rho_{1,0}(\vec{r}_2;t), \tag{4.1.2}$$

$$\frac{\partial \rho_{0,2}(\vec{r}_1',\vec{r}_2';t)}{\partial t} = p\rho_{0,1}(\vec{r}_1';t) + p\rho_{0,1}(\vec{r}_2';t), \tag{4.1.3}$$

$$\frac{\partial \rho_{1,1}(\vec{r}_1; \vec{r}_1'; t)}{\partial t} = p\rho_{1,0}(\vec{r}_1; t) + p\rho_{0,1}(\vec{r}_1'; t) + pf(|\vec{r}_1 - \vec{r}_1'|), \qquad (4.1.4)$$

$$\frac{\partial \rho_{2,1}(\vec{r}_1, \vec{r}_2; \vec{r}_1'; t)}{\partial t}$$
$$= p\rho_{1,1}(\vec{r}_1; \vec{r}_1'; t) + p\rho_{1,1}(\vec{r}_2; \vec{r}_1'; t) + p\rho_{2,0}(\vec{r}_1, \vec{r}_2; t) +$$
$$+ pf(|\vec{r}_1 - \vec{r}_1'|)\rho_{1,0}(\vec{r}_2; t) + pf(|\vec{r}_2 - \vec{r}_1'|)\rho_{1,0}(\vec{r}_1; t). \qquad (4.1.5)$$

Since at the very beginning, $t = 0$, there were no particles, we can put all many-particle densities to zero, $\rho_{m,m'} = 0$. As follows from equations (4.1.1), at time t_0 concentrations are $n_A(t_0) = n_B(t_0) = n(t_0) = n_0 = pt_0$. Integrating equations (4.1.2) to (4.1.4), we arrive at

$$\rho_{2,0} = \rho_{0,2} = n_0^2, \qquad (4.1.6)$$

$$\rho_{1,1} = n_0^2 + n_0 f(|\vec{r}_1 - \vec{r}_1'|), \qquad (4.1.7)$$

whereas the three-particle density is

$$\rho_{2,1} = n_0^3 + n_0^2 f(|\vec{r}_1 - \vec{r}_1'|) + n_0^2 f(|\vec{r}_2 - \vec{r}_1'|). \qquad (4.1.8)$$

These two- and three-particle densities could be used as the initial conditions for the diffusion-controlled particle recombination, when the particle source is switched off. In terms of the correlation functions these initial conditions are $n(0) = n_0$,

$$X_\nu(r, 0) = 1, \qquad (4.1.9)$$

$$Y(r, 0) = 1 + f(r)/n_0, \qquad (4.1.10)$$

$$\chi_{AAB}^{(3)} = 1 + f(|\vec{r}_1 - \vec{r}_1'|)/n_0 + f(|\vec{r}_2 - \vec{r}_1'|)/n_0. \qquad (4.1.11)$$

Strictly speaking, particle diffusion and recombination occurring in a course of irradiation might affect equations (4.1.9) to (4.1.11), but for a short time (small doses) we can neglect it. Equations (4.1.9) and (4.1.10) describe the random distribution of similar particles entering different geminate pairs and the correlation of dissimilar particles within these pairs. If there were no correlation within geminate pairs, $f(r)/n_0 \ll 1$,

$$X_\nu(r,0) = 1 \quad \text{and} \quad Y(r,0) = 1. \tag{4.1.12}$$

In turn, equations (4.1.9) and (4.1.10) yield a general initial condition necessary for an analysis of the transition from the monomolecular to bimolecular kinetics. When the initial particle distribution is unknown (which is usually the case, e.g., after prolonged irradiation), it is assumed, as a rule, to be the simplest, i.e., the *Poisson* one, equation (4.1.12).

4.1.2 The shortened superposition approximation

Turning back now to the kinetics of the diffusion-controlled $A + B \to 0$ recombination, and neglecting reactant interaction, let us write down several equations from an infinite hierarchy of equations (2.3.37), (2.3.45), and (2.3.53):

$$\frac{\partial \rho_{1,0}}{\partial t} = D_A \nabla_1^2 \rho_{1,0} - \int \sigma(|\vec{r}_1 - \vec{r}_1'|) \rho_{1,1} \, d\vec{r}_1',$$

$$\frac{\partial \rho_{0,1}}{\partial t} = D_B \nabla_1'^2 \rho_{0,1} - \int \sigma(|\vec{r}_1 - \vec{r}_1'|) \rho_{1,1} \, d\vec{r}_1, \tag{4.1.13}$$

$$\frac{\partial \rho_{2,0}}{\partial t} = D_A \nabla_1^2 \rho_{2,0} + D_A \nabla_2^2 \rho_{2,0} -$$
$$- \int \sigma(|\vec{r}_1 - \vec{r}_1'|) \rho_{2,1} \, d\vec{r}_1' - \int \sigma(|\vec{r}_2 - \vec{r}_1'|) \rho_{2,1} \, d\vec{r}_1', \tag{4.1.14}$$

$$\frac{\partial \rho_{0,2}}{\partial t} = D_B \nabla_1'^2 \rho_{2,0} + D_B \nabla_2'^2 \rho_{0,2} -$$
$$- \int \sigma(|\vec{r}_1 - \vec{r}_1'|) \rho_{1,2} \, d\vec{r}_1 - \int \sigma(|\vec{r}_1 - \vec{r}_2'|) \rho_{1,2} \, d\vec{r}_1, \tag{4.1.15}$$

$$\frac{\partial \rho_{1,1}}{\partial t} = D_A \nabla_1^2 \rho_{1,1} + D_B \nabla_1'^2 \rho_{1,1} - \sigma(|\vec{r}_1 - \vec{r}_1'|) \rho_{1,1} -$$
$$- \int \sigma(|\vec{r}_2 - \vec{r}_1'|) \rho_{2,1} \, d\vec{r}_2 - \int \sigma(|\vec{r}_1 - \vec{r}_2'|) \rho_{1,2} \, d\vec{r}_2'. \tag{4.1.16}$$

Keeping in mind the physical meaning of single- and two-particle densities, equations (2.3.58) to (2.3.61), we can rewrite now equations (4.1.13) in a form similar to the standard chemical kinetics (Section 2.1). To do it, let us first transform the integral expression

$$\int \sigma(|\vec{r}_1 - \vec{r}_1'|)\rho_{1,1}\,d\vec{r}_1'$$

$$= n_A(t)n_B(t)\int \sigma(|\vec{r}_1 - \vec{r}_1'|)Y(|\vec{r}_1 - \vec{r}_1'|,t)\,d\vec{r}_1'. \tag{4.1.17}$$

Passing from the integration over $\vec{r}_1'$ to that over the variable $\vec{r} = \vec{r}_1 - \vec{r}_1'$, we arrive at the equation

$$\frac{dn_\nu(t)}{dt} = -K(t)n_A(t)n_B(t), \quad \nu = A, B, \tag{4.1.18}$$

similar to that found in formal kinetics formalism but with the *time-dependent* reaction rate

$$K(t) = \int \sigma(r)Y(r,t)\,d\vec{r}. \tag{4.1.19}$$

In fact, the latter is a *functional* of the correlation function of dissimilar particles, i.e., to calculate $K(t)$ we need to know either $Y(r,t)$ or $\rho_{1,1}$. In its turn, equation (4.1.16) demonstrates that these latter are coupled with three-point densities etc. Therefore, to solve the problem, we have to cut off the infinite equation hierarchy, thus only approximately describing the fluctuation spectrum. Usually it is done by means of the complete Kirkwood superposition approximation, equations (2.3.62) and (2.3.63), or the shortened approximation, equations (2.3.64) and (2.3.65).

In this Section we consider namely the latter. Under the shortened Kirkwood's approximation equation (2.3.64) is expressed through $Y(r,t)$ only, therefore the substitution of equation (2.3.64) into (2.3.16) results in a closed equation uniquely defining $Y(r,t)$. This equation is *linear* in Y (see the title of this Section) and equations (4.1.14), (4.1.15) are not used any longer. Substituting equation (2.3.64) into (4.1.16), using the relative coordinate $\vec{r} = \vec{r}_1 - \vec{r}_1'$ and after some manipulations

$$\frac{\partial \rho_{1,1}}{\partial t} = \frac{\partial}{\partial t}\left(n_A(t)n_B(t)Y(r,t)\right)$$

$$= n_A(t)n_B(t)\frac{\partial Y(r,t)}{\partial t} +$$

$$+ \left(n_B(t)\frac{dn_A(t)}{dt} + n_A(t)\frac{dn_B(t)}{dt}\right)Y(r,t)$$

$$= n_A(t)n_B(t)\frac{\partial Y(r,t)}{\partial} -$$

$$- n_A(t)n_B(t)\bigl(n_A(t) + n_B(t)\bigr)K(t)Y(r,t), \tag{4.1.20}$$

$$D_A \nabla_1^2 \rho_{1,1} + D_B \nabla_1'^2 \rho_{1,1} = n_A(t)n_B(t)(D_A + D_B)\nabla^2 Y(r,t), \tag{4.1.21}$$

$$\int \sigma\bigl(|\vec{r}_2 - \vec{r}_1'|\bigr)\rho_{2,1}\,d\vec{r}_2$$

$$= n_A^2(t)n_B(t)Y(r,t)\int \sigma\bigl(|\vec{r}_2 - \vec{r}_1'|\bigr)Y\bigl(|\vec{r}_2 - \vec{r}_1'|,t\bigr)d\vec{r}_2$$

$$= n_A^2(t)n_B(t)K(t)Y(r,t), \tag{4.1.22}$$

we arrive at equation as follows

$$\frac{\partial Y(r,t)}{\partial t} = D\nabla^2 Y(r,t) - \sigma(r)Y(r,t). \tag{4.1.23}$$

Incorporation of the reactant interaction, equations (2.3.47) to (2.3.50), leads to the *non-linear equations*, both in $Y(r,t)$ and in $n_A(t)$ and $n_B(t)$. Non-linear terms arise due to the integral terms in equation (2.3.50) describing the effective interaction of particles with their environment. The linearization of this equation, justified if the concentrations are small, yields

$$\frac{\partial Y(r,t)}{\partial t} = D\nabla\left(\nabla Y(r,t) + \frac{Y(r,t)}{k_B T}\nabla U(r)\right) - \sigma(r)Y(r,t) \tag{4.1.24}$$

where $U(r) = U_{AB}(r)$ is two-particle interaction potential. They say that the non-linear equations (4.1.18) define the *concentration dynamics* whereas through equations (4.1.23) and (4.1.24) the *correlation dynamics* is introduced. The distinctive feature of this linear approximation [1–3] is that it neglects *back-coupling* of these two dynamics since the correlation dynamics through the reaction rate $K(t)$ determines the concentration time-development, but still is independent of it (Fig. 4.1).

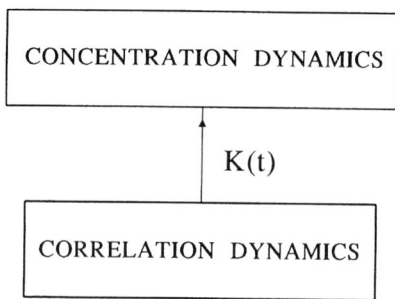

Fig. 4.1. The schematic relation between concentration dynamics and correlation dynamics in the linear approximation.

Since in the linear approximation under study there is no back coupling between dynamics of concentrations and correlations, the central point of the kinetics is a solution of the partial differential equations (4.1.23) and (4.1.24). Obtaining the time development of the concentrations becomes here quite trivial; from equation (4.1.18) we easily obtain for unequal concentrations, $n_B > n_A$, $\delta n = n_B(0) - n_A(0) = \text{const}$, that

$$n_A(t) = \frac{\delta n}{\{1 + \delta n/n_A(0)\} \exp\left[\delta n \int_0^t K(t')\,dt'\right] - 1}. \qquad (4.1.25)$$

For large difference in concentrations, $n_B(0) \gg n_A(0)$, $n_B \approx \text{const}$, this equation is reduced to

$$n_A(t) \approx n_A(0) \exp\left[-n_B(0) \int_0^t K(t')\,dt'\right]. \qquad (4.1.26)$$

If particle concentrations are equal, $n(t) = n_A(t) = n_B(t)$, one gets

$$\frac{1}{n(t)} = \frac{1}{n(0)} + \int_0^t K(t')\,dt'. \qquad (4.1.27)$$

Therefore, if the reaction rate $K(t)$ is known, the A+B $\to$ 0 reaction kinetics is defined uniquely.

Equation (4.1.23) for the correlation function describing a linearized many-particle problem coincides, formally speaking, with equation (3.2.7) obtained for the two-particle problem. It is also true for equations (3.2.11) and (4.1.24), (3.2.10) and (4.1.19). In terms of mathematics, they differ by the boundary conditions only: $Y(\infty,t) = 1$ comes from the weakening of correlations as $r \to \infty$, whereas $w(r,t)$ yields the non-normalized particle distribution over distances and its finite volume integral is secured by the condition more rigid than $w(\infty,t) = 0$.

4.1.3 The infinitely diluted system

As we have mentioned in Chapter 2, the accuracy of the kinetic equations derived using the superposition approximation cannot be checked up in the framework of the *same* theory. It is the analysis of the limiting case of the infinitely diluted system, $n_0 \to 0$, which nevertheless permits us to compare approximate results obtained in the linearized approximation with the exact solution of the two-particle problem (Chapter 3).

For equal particle concentrations, where $n_A(t) = n_B(t) = n(t)$,

$$\frac{dn(t)}{dt} = -K(t)n^2(t). \tag{4.1.28}$$

Making use now of the initial conditions, equation (4.1.10), describing the correlated particle distribution, we could expect that in the limiting case of infinitely separated geminate pairs, as $n_0 \to 0$, the solution of the set of equations (4.1.19), (4.1.23) and (4.1.28) should coincide with the result obtained in Chapter 3 for geminate pairs. It is clear from equation (4.1.10) that the initial distribution in pairs $Y(r,0)$ is defined by the normalized function $f(r)$ multiplied by the inverse particle concentration, n_0^{-1}. Since $n_0 \to 0$, this term $f(r)/n_0$ becomes predominant, therefore for infinitely diluted pairs we can neglect the bimolecular contribution into kinetics. At the moment t when some of the geminate pairs already have recombined, the spatial distribution of survived particles is described by the non-normalized $w(r,t)$ introduced in Chapter 3. Taking into account equation (3.2.6), we can redefine the normalized distribution function as

$$\tilde{f}(r,t) = w(r,t)/\omega(t). \tag{4.1.29}$$

At the same time, the concentrations can be easily expressed through the survival probability, $n(t) = n_0\omega(t)$. We could expect a solution in the form

$$Y(r,t) \approx \frac{\tilde{f}(r,t)}{n(t)} + O(1) = \frac{w(r,t)}{n_0 w^2(t)} + O(1), \qquad (4.1.30)$$

where $O(1)$ are terms of the order of unity and thus unimportant as $n_0 \to 0$. Substituting equation (4.1.30) into equation (4.1.19) and putting $n(t) = n_0 w(t)$ into equation (4.1.28), after limiting transition $n_0 \to 0$ we arrive at the finally familiar equations (3.2.9) and (3.2.10). However, after the same limiting transition equation (4.1.23) becomes transformed into something essentially different from equation (3.2.7), namely

$$\frac{\partial w(r,t)}{\partial t} = D\Delta w(r,t) - \sigma(r) w(r,t) + 2w(r,t) \frac{d \ln w(t)}{dt}. \qquad (4.1.31)$$

The use of equation (4.1.31) instead of equation (3.2.7) gives the survival probability (for the initial condition (3.2.14)) $w(l \mid \infty) = 0.5$ [4] as $l \to r_0$ – rather than the exact result $w(r_0 \mid \infty) = 0$ coming from equation (3.2.23)! Therefore, the set of equations (4.1.19), (4.1.23) and (4.1.28) *fails* to reproduce the limiting case of the infinitely diluted system. Let us clarify the origin of this effect. The conclusion can be drawn from equation (4.1.11) that as $n_0 \to 0$, the second and third terms are of the order of n_0^{-1}. Instead, when deriving equation (4.1.23), the use of both the complete, equation (2.3.63), and shortened, equation (2.3.65), Kirkwood's superposition approximations result in the three-particle density

$$\chi_{AAB}^{(3)} = 1 + \frac{1}{n_0} f(|\vec{r}_1 - \vec{r}_1'|) + \frac{1}{n_0} f(|\vec{r}_2 - \vec{r}_1'|) +$$
$$+ \frac{1}{n_0^2} f(|\vec{r}_1 - \vec{r}_1'|) f(|\vec{r}_2 - \vec{r}_1'|). \qquad (4.1.32)$$

As $n_0 \to 0$, the predominant term here is of the order of n_0^{-2}. Therefore, the shortcoming of equation (4.1.23) arises due to the incorrect use of the superposition approximation in a situation of very particular (strongly correlated) particle distribution. Strictly speaking, the correct treatment of the recombination process with arbitrary initial distribution requires the usage of the *complete* set of correlation functions. At the joint correlation level, such description yields reasonable results only for the particle distribution close to a random equation (4.1.12). For the infinitely diluted system the correlation function (4.1.10) of dissimilar closely spaced particles reveals the strong correlation at short distances, $Y(r,0) \gg 1$, which should affect the higher-order correlation functions too. This fact is obviously neglected in Kirkwood's superposition approximation and this point needs more detailed treatment [5].

The bottleneck of equations (4.1.19), (4.1.23) and (4.1.28) (derived for the first time by Leibfried [4], see also [6]) has stimulated development of particular *interpolating schemes* (in terms of Antonov-Romanovskii [7]). This point has been discussed more than once ([7] and references therein, as well as [8–15]).

4.1.4 Waite's interpolation equations

In particular, in the widely-spread Waite's approach the reaction rate on intuitive grounds is represented by a sum of the mono- and bimolecular contributions

$$\frac{dn(t)}{dt} = \left.\frac{dn(t)}{dt}\right|_{bi} + \left.\frac{dn(t)}{dt}\right|_{mono}, \qquad (4.1.33)$$

arising due to the recombination within geminate pairs and between components of different pairs during their diffusion-controlled mixing, respectively. In terms of mathematics it corresponds to the *split* of the correlation function of dissimilar reactants, equation (4.1.10), into geminate and bimolecular parts: $Y(r,0) = Y_{bi}(r,0) + Y_g(r,0)$, where $Y_g(r,0) = f(r)/n_0$ and $Y_{bi}(r,0) = 1$. The latter condition arises from the random distribution of different pairs. The bimolecular recombination is described by equation (4.1.18) but with the obtained reaction rate $K(t)$ making use of equations (4.1.19) and (4.1.23), provided $Y(r,0) = Y_{bi}(r,0) = 1$. Equation (4.1.18) could be used to estimate the bimolecular kinetics

$$\varphi_A(t) = \frac{n_A(t)}{n_A(0)} = \exp\left[-\int_0^t K(t')n_B(t')\,dt'\right],$$

$$\varphi_B(t) = \frac{n_B(t)}{n_B(0)} = \exp\left[-\int_0^t K(t')n_A(t')\,dt'\right]. \qquad (4.1.34)$$

For equal concentrations we get

$$\varphi_A(t) = \varphi_B(t) = \varphi(t) = \exp\left[-\int_0^t K(t')n(t')\,dt'\right]. \qquad (4.1.35)$$

If there were no bimolecular recombination, we would get from equation (3.2.9) that

$$\left.\frac{dn(t)}{dt}\right|_{mono} = -n_0 P(t), \qquad (4.1.36)$$

where $P(t)$ is defined by equations (3.2.7) and (3.2.10). The bimolecular recombination has to modify equation (4.1.36), taking into account that components of geminate pairs can recombine "foreign" particles too. Taking this into account, we arrive at the equation

$$\left.\frac{dn(t)}{dt}\right|_{mono} = -n_0 P(t)\varphi^2(t). \qquad (4.1.37)$$

Therefore, whole kinetics reads as a sum of both contributions:

$$\frac{dn(t)}{dt} = -K(t)n^2(t) - n_0 P(t)\varphi^2(t). \qquad (4.1.38)$$

Equation (4.1.38) is obviously constructed in such a way that it correctly reproduces the solution of a pair problem.

However, Waite's approach has several shortcomings (first discussed by Kotomin and Kuzovkov [14, 15]). First of all, it contradicts a universal principle of statistical description itself; the particle distribution functions (in particular, many-particle densities) have to be defined *independently* of the kinetic process, but it is only the physical process which determines the actual form of kinetic equations which are aimed to describe the system's time development. This means that when considering the diffusion-controlled particle recombination (there is no source), the actual mechanism of how particles were created – whether or not correlated in geminate pairs – is not important; these are concentrations and joint densities which uniquely determine the decay kinetics. Moreover, even the knowledge of the coordinates of all the particles involved in the reaction (which permits us to find an *infinite* hierarchy of correlation functions $\chi^{(m)}, m = 2, \ldots, \infty$, and thus is more than enough to describe statistically the process under study *does not allow* Waite to find the initial distribution function of geminate pairs $w(r,t)$ in order to identify it with $Y(r,0)$: due to the undistinguishability of all A (and B) particles there is *no* criterium except intuition how to separate whole particle distribution into pairs created either genetically or not.

In fact, Waite's approach needs a kind of *demon* to mark (enumerate) pairs of particles at moments of their birth and then following the time development $w(r,t)$ of all such geminate pairs – even when these pairs already completely have mixed at the bimolecular reaction stage! What is said above demonstrates quite well how the violation of statistical principles in deriving kinetic equations can lead to unphysical paradoxes (discussed in detail in [14, 15]).

Hereafter, let us assume a random initial distribution of particles, as in equation (4.1.12). In this case the three-particle densities $\chi^{(3)}$ suit well their approximation at the beginning stage of recombination, which, however, could not be the case after a *long* reaction time. This is why we should not overestimate the results of the linear approximation discussed below.

4.1.5 Recombination of immobile reactants

In the case of long-range recombination of *immobile* particles ($D = 0$) occurring at a low temperature, we can use equation (4.1.23) to obtain

$$\frac{\partial Y(r,t)}{\partial t} = -\sigma(r) Y(r,t). \tag{4.1.39}$$

Its solution for a random particle distribution, $Y(r,0) = 1$, is

$$Y(r,t) = \exp\bigl[-\sigma(r)t\bigr]. \tag{4.1.40}$$

The integral entering equations (4.1.25) and (4.1.27) could easily be expressed as

$$\begin{aligned} F(t) &= \int_0^t K(t')\,dt' \\ &= \int \bigl\{1 - \exp\bigl[-\sigma(r)t\bigr]\bigr\}\,d\vec{r} \\ &= \int \bigl\{1 - Y(r,t)\bigr\}\,d\vec{r}. \end{aligned} \tag{4.1.41}$$

Therefore, the universal asymptotic decay kinetics are: for equal concentrations

$$\frac{1}{n(t)} = \frac{1}{n(0)} + F(t) \tag{4.1.42}$$

and if the concentrations differ greatly, $n_B \gg n_A$, we have

$$n_A(t) \approx n_A(0) \exp\bigl[-n_B(0)F(t)\bigr]. \tag{4.1.43}$$

The actual time behaviour of $F(t)$, shown in equation (4.1.41), depends on the choice of $\sigma(r)$. This quantity $F(t)$ could be presented in a more universal and transparent form in terms of the *correlation length* ξ.

Let us consider as an illustration the multipole interaction which is observed in many solids and liquids due to the interaction of the electronically excited donors D^* with acceptors A resulting in an energy transfer: $D^* + A \to D + A^* \to D + A + h\nu$ [16, 17]. Its probability (per unit time) is

$$\sigma(r) = \sigma_0/r^m, \quad m = 6, 8, 10, \ldots. \tag{4.1.44}$$

The case $m = 6$ corresponds to the most long-range dipole–dipole interaction. Define how the time-dependent correlation length $\xi = \xi_0$ (subscript '0' indicates that the linear approximation is employed):

$$\xi_0 = (\sigma_0 t)^{1/m}. \tag{4.1.45}$$

As could easily be demonstrated, the correlation function (4.1.40) is stationary if the auto-model variable $\eta = r/\xi_0$ is used instead of r:

$$Y(r,t) = Y(\eta) = \exp[-\eta^{-m}]. \tag{4.1.46}$$

Another example is the *tunnelling recombination* [18] of defects discussed in the Section 3.1, equation (3.1.2). Note that this equation holds also for an exchange *triplet–triplet energy transfer* which makes the phenomenological theories of energy transfer and tunnelling recombination quite similar [18]. Let us define at long t the *correlation length*

$$\xi_0 = r_0 \ln(\sigma_0 t). \tag{4.1.47}$$

If $\xi_0 \gg 1$, the correlation function $Y(r,t)$ becomes practically stationary. Indeed, in terms of the auto-model (scaled) variable $\eta = r/\xi_0$ we get

$$Y(r,t) = \exp\bigl[-e^{\xi_0(\eta-1)}\bigr] \approx \begin{cases} 0, & \eta < 1, \\ 1, & \eta > 1. \end{cases} \tag{4.1.48}$$

In a general case the correlation length ξ_0 could be defined through the expression

$$\sigma(\xi_0)t = 1. \tag{4.1.49}$$

As $r < \xi_0$, $Y(r,t) \approx 0$ (AB pairs practically do not exist at relative distances $r < \xi_0$). On the other hand, at $r > \xi_0$, $Y(r,t) \approx 1$, i.e., the joint correlation function could be approximated by the Heaviside step function (see equation (4.1.48)). In doing so, the integral in equation (4.1.47) could be estimated as

$$F(t) \approx V(\xi_0) = \int_{V(\xi_0)} d\vec{r} = \int_0^{\xi_0} r^{\bar{d}-1}\,dr \int d\Omega_{\bar{d}} = \frac{\gamma_{\bar{d}}}{\bar{d}} \xi_0^{\bar{d}}. \tag{4.1.50}$$

Here $V(\xi_0)$ is the volume of $\bar{d}$-dimension sphere with a radius ξ_0. $\gamma_{\bar{d}} = \int d\Omega_{\bar{d}}$ is a factor arising due to the angle integration in $\bar{d}$-dimensional space. For $\bar{d} = 1, 2$ or 3, $\gamma_{\bar{d}} = 2, 2\pi$, or 4π, respectively. For the multipole interaction the integral in equation (4.1.47) could be explicitly calculated

$$F(t) = k_0 \xi_0^{\bar{d}}, \qquad k_0 = \frac{\gamma_{\bar{d}}}{\bar{d}} \Gamma(1 - \bar{d}/m), \tag{4.1.51}$$

where $\Gamma(x)$ is Euler's gamma-function. The conclusion itself suggests that in equation (4.1.50) the factor $\xi_0^{\bar{d}}$ is reproduced correctly, unlike k_0 where a deviation from an explicit result comes from the step-like approximation used. For tunnelling recombination, say, for $\bar{d} = 3$, we get the expansion at $\xi_0 \gtrsim r_0$

$$F(t) = \frac{4\pi}{3}\left(\xi_0^3 + c_1 \xi_0^2 r_0 + c_2 \xi_0 r_0^2 + c_3 r_0^3\right), \tag{4.1.52}$$

These constants are $c_1 = 1.732$, $c_2 = 5.934$, $c_3 = 5.445$. As $\xi_0 \gg r_0$, equation (4.1.52) transforms into equation (4.1.50).

Therefore, as $\xi_0 \to \infty$, we obtain asymptotic reaction laws as follows: for equal concentrations

$$n(t) \propto 1/\left(c\xi_0^{\bar{d}}\right), \tag{4.1.53}$$

where $c \approx \gamma_{\bar{d}}/\bar{d}$, and

$$n_A(t) \propto \exp\left(-cn_B(0)\xi_0^{\bar{d}}\right), \tag{4.1.54}$$

for unequal concentrations. Note that in the case of the multipole interaction, equation (4.1.44), the only distinctive length scale is the correlation length ξ_0. In contrast, for tunnelling recombination there are *two* such scales (r_0 and ξ_0). Therefore, in the latter case reaction kinetics is defined by a *competition* of these scales: (4.1.52) gives different time behaviours at $\xi_0 \sim r_0$ and $\xi_0 \gg r_0$.

4.1.6 Diffusion-controlled reactions

The black sphere approximation permits us to obtain the most simple and physically transparent results for the kinetics of diffusion-controlled reactions. We should remind that this approximation involves a strong negative correlation of dissimilar particles at $r \leqslant r_0$, where $Y(r,t) \equiv 0$, described by the Smoluchowski boundary condition

$$Y(r_0, t) = 0. \tag{4.1.55}$$

In this case equations (4.1.23) and (4.1.24) are simplified:

$$\frac{\partial Y(r,t)}{\partial t} = D\nabla^2 Y(r,t), \tag{4.1.56}$$

$$\frac{\partial Y(r,t)}{\partial t} = D\nabla \left(\nabla Y(r,t) + \frac{Y(r,t)}{k_B T} \nabla U(r) \right). \tag{4.1.57}$$

Similarly to equations (3.2.10) and (3.2.18), the reaction rate for $\bar{d} = 3$ is

$$K(t) = 4\pi D r_0^2 \left. \frac{\partial Y(r,t)}{\partial r} \right|_{r=r_0}. \tag{4.1.58}$$

Let us first consider equation (4.1.56). For the uncorrelated initial distribution, $Y(r > r_0, 0) = 1$, its solution is [8]

$$Y(r,t) = 1 - \frac{r_0}{r} + \frac{2r_0}{r\pi^{1/2}} \int_0^{(r-r_0)/2l_D} \exp(-\zeta^2)\,d\zeta, \tag{4.1.59}$$

where $l_D = \sqrt{Dt}$ is diffusion length, whereas in the general case of an arbitrary initial distribution

$$Y(r,t) = \frac{1}{2\pi^{1/2} l_D r} \int_0^\infty Y(r',0) \left\{ \exp\left[-\frac{(r-r')^2}{4l_D^2}\right] - \right.$$

$$-\exp\left[-\frac{(r+r'-2r_0)^2}{4l_D^2}\right]\right\}r'\,dr'. \tag{4.1.60}$$

Defining $K(t)$ through equation (4.1.58), we can obtain with the help of equation (4.1.59) that

$$K(t) = 4\pi D r_0 \left(1 + \frac{r_0}{\sqrt{\pi Dt}}\right) = 4\pi D r_0 \left(1 + \frac{r_0}{\pi^{1/2} l_D}\right). \tag{4.1.61}$$

Transient kinetics with the time-dependent reaction rate $K(t)$, equation (4.1.61), have been observed experimentally more than once (e.g., by Tanimura and Itoh [19]). In the steady-state case, $t \to \infty$, we get finally

$$y(r) \equiv Y(r, \infty) = 1 - \frac{r_0}{r}, \tag{4.1.62}$$

and the stationary reaction rate

$$K_0 \equiv K(\infty) = 4\pi D r_0. \tag{4.1.63}$$

The formation of the steady-state recombination profile (4.1.62) occurs for the space dimension $\bar{d} = 3$ only. For instance, if $\bar{d} = 2$, taking into account the change in the diffusion operator Δ and the expression for the two-dimensional reaction rate

$$K(t) = 2\pi D r_0 \left.\frac{\partial Y(r,t)}{\partial r}\right|_{r=r_0}, \tag{4.1.64}$$

we get finally [20]

$$K(t) = \frac{8D}{\pi} \int_0^\infty \exp\left[-\frac{l_D^2}{r_0^2}\zeta^2\right] \frac{d\zeta}{\{J_0(\zeta)+Y_0(\zeta)\}\zeta}, \tag{4.1.65}$$

where $J_0(x)$, and $Y_0(x)$ are the zero-order Bessel and Neuman functions. At short time, $l_D < r_0$, equation (4.1.65) could be approximated as

$$K(t) \approx \pi D \left(1 + \frac{2r_0}{\sqrt{\pi Dt}}\right) = \pi D \left(1 + \frac{2r_0}{\pi^{1/2} l_D}\right). \tag{4.1.66}$$

At a great time, $l_D > r_0$,

$$K(t) \approx \frac{2\pi D}{\ln(cl_D/r_0)}, \quad c \approx 1.125, \tag{4.1.67}$$

i.e., $K(t) \propto 1/\ln t$ as $t \to \infty$.

In other words, the peculiarity of the *two*-dimensional motion which has led to the zero survival probability of correlated pairs, equation (3.2.26), for randomly distributed particles consists of the complete zerofication of the reaction rate at a great time, $K(\infty) = 0$. The logarithmic dependence of the reaction rate on time does not considerably affect the asymptotic behaviour of macroscopic concentrations. Introducing the *critical exponent* α

$$\alpha = \lim_{t \to \infty} \alpha(t), \quad \alpha(t) = -\frac{d \ln n_A(t)}{d \ln t}, \tag{4.1.68}$$

we obtain the *same* critical exponent $\alpha = 1$ in both cases – $\bar{d} = 2$ and $\bar{d} = 3$, for $n_A(t) = n_B(t)$. The peculiarity of the case $\bar{d} = 2$ results only in the *logarithmic* time correction to the decay law $n(t) \propto t^{-1}$ if $n_A(t) = n_B(t)$ or in power time corrections to $n_A(t) \propto \exp(-\text{const} \cdot t)$, if $n_A(t) \ll n_B(t)$.

Equations (4.1.61) and (4.1.66) demonstrate well the above-mentioned competition of the two scales l_D and r_0. In the linear approximation the solution of the problem for $\bar{d} = 1$ is expressed via l_D only:

$$K(t) = \frac{2D}{\sqrt{\pi D t}} = \frac{2D}{\pi^{1/2} l_D}, \tag{4.1.69}$$

It corresponds to $\alpha = 1/2$, i.e., $n(t) \propto t^{-1/2}$. The value l_D could be identified with the correlation length ξ_0: for $\bar{d} = 1$ and 2, strong negative correlations are observed: at $r < \xi_0 = l_D$ and $l_D \gg r_0$ the correlation function $Y(r,t) \ll 1$. Observation of the stationary solution (4.1.62) makes the existence of the correlation length ξ_0 less evident; deviation of $Y(r,t)$ from its asymptotics $Y(\infty, t) = 1$ at all r is not well pronounced. However, this deviation has a long-range character. So, at $\xi_0 = l_D \to \infty$ we get from equation (4.1.62)

$$y(r) - 1 = -\frac{r_0}{r} \propto r^{-1}. \tag{4.1.70}$$

As it was noted in Section 2.3 in connection with equation (2.3.18), a power approximation of the correlation function corresponds to the *critical point*.

Incorporation of particle drift due to their elastic or the Coulomb interaction, accompanied by the replacement of equation (4.1.56) by equation (4.1.57), essentially complicates the problem. In the simplest case we can restrict ourselves to the quasi-steady state and black sphere model for $\bar{d} = 3$.

That is, we are interested in $y(r) = Y(r, \infty)$ and the corresponding reaction rate, $K_0 = K(\infty)$. The former satisfies the equation (3.2.39) with the boundary conditions (3.2.40). As it is clear from Chapter 3, the solution of equation (3.2.39) defines uniquely the survival probability $w(l \mid \infty)$ of geminate pairs. In a particular case of the Coulomb interaction, the solution of the steady-state equation (3.2.39) is simplified since for the unscreened Coulomb potential the relation $\nabla^2 U(r) = 0$ holds. Integrating the differential equation

$$\frac{d^2 y(r)}{dr^2} + \left(\frac{2}{r} + \frac{1}{k_B T} \frac{\partial U(r)}{\partial r} \right) \frac{dy(r)}{dr} = 0 \qquad (4.1.71)$$

we obtain the recombination profiles for the *attraction* $(U < 0)$

$$y^{(-)}(r) = \frac{\exp(L/r_0) - \exp(L/r)}{\exp(L/r_0) - 1}, \qquad (4.1.72)$$

and *repulsion*, respectively

$$y^{(+)}(r) = \frac{\exp(-L/r) - \exp(-L/r_0)}{1 - \exp(-L/r_0)}, \qquad (4.1.73)$$

where $U(r) = \pm L/r$, L is the Onsager radius [21]. Defining the reaction rate through equation (4.1.58), we obtain respectively [10, 22]

$$K_0^{(-)} = 4\pi D \frac{L}{1 - \exp(-L/r_0)}, \qquad (4.1.74)$$

$$K_0^{(+)} = 4\pi D \frac{L}{\exp(L/r_0) - 1}. \qquad (4.1.75)$$

Similarly to the black sphere model, equation (4.1.63), the effective reaction radius R_{eff} could be defined through $K_0 = 4\pi D R_{\text{eff}}$. Comparing the R_{eff} obtained in such a way, with the results of Chapter 3, the conclusion suggests that they coincide, i.e., both definitions of the effective radii turn out to be equivalent. Different aspects of the solution of the kinetic equations (4.1.57) for charged particles were discussed more than once [2, 3, 23–30]. In particular, Hong and Noolandi [31] presented a *time-dependent* solution of the problem.

4.2 TUNNELLING RECOMBINATION

> If I correctly imagine a genealogy of the lady called Science, her elder sister's name is Ignorance.
> (Wenn ich die Genealogie der Dame Wissenschaft recht kenne, so ist die Unwissenheit ihre ältere Schwester)
>
> Lichtenberg

4.2.1 Static reactions

Considering the reaction kinetics in the preceding Sections of Chapter 4, we have restricted ourselves to the simplest case of the recombination rate $\sigma(r)$ corresponding to the black sphere approximation, equation (3.2.16). However, if recombination is long-range, like that described by equations (4.1.44) or (3.1.2), one has to use equations (4.1.23) and (4.1.24), which yield essentially more complicated kinetics, especially for the transient period. Let us discuss briefly the main features of the diffusion-controlled kinetics controlled by *tunnelling* recombination, equation (3.1.2) (see also [32–34]).

Let us consider first the case of low temperatures and immobile defects (*static reaction*). In this situation electron tunnelling is the *only* channel of defect recombination and concentration decay. Depending on the defect relative spatial distribution, two recombination regimes can arise.

For the exciton mechanism of defect production in alkali halides the Frenkel *pairs of well correlated defects* are known to be created [35], the mean distance between defects inside these pairs is much smaller than that between different pairs. The geminate pair distribution function could often be approximated as

$$f(r) = \frac{1}{R} \exp\left(-\frac{r}{R}\right). \tag{4.2.1}$$

Tunnelling recombination

Making use of equations (3.2.9), (3.2.10), and equations (3.2.15), which are valid for static recombination, we arrive at (provided $\sigma_0 t \gg 1$)

$$n(t) = n_0 \Gamma(1 + r_0/R)(\sigma_0 t)^{-r_0/R}, \qquad (4.2.2)$$

where Γ is the Euler gamma function.

The intensity of any bimolecular, and in particular, tunnelling recombination, defined as

$$I \propto \left| \frac{\mathrm{d}n(t)}{\mathrm{d}t} \right|, \qquad (4.2.3)$$

could easily be calculated by means of equation (4.2.2):

$$I \propto t^{-(1+r_0/R)}. \qquad (4.2.4)$$

Since the Bohr radius of the electron centres in ionic crystals is typically rather small, ($r_0 \leqslant 1$ Å), usually the ratio $r_0/R = 0.01$–0.05 and equation (4.2.4) reveals the algebraic decay law of intensity. This decay kinetics has been observed more than once (see [18, 34] for more details where other kinds of spatial distributions are also discussed).

Equations (3.2.9), (3.2.10) and (3.2.15) yield an explicit time decay of concentrations:

$$n(t) = n_0 \int f(r) \exp\left[-\sigma(r)t\right] \mathrm{d}\vec{r}. \qquad (4.2.5)$$

For large correlation lengths ξ_0 defined by equation (4.1.47) we can estimate equation (4.2.5) in the spirit of equation (3.1.2), thus getting the simple equation

$$n(t) \approx n_0 \int_{\xi_0}^{\infty} f(r) \mathrm{d}\vec{r}. \qquad (4.2.6)$$

Equation (4.2.6) yields a good approximation of (4.2.5) if the relative change of $f(r)$ is small as coordinate r varies by r_0. Differentiating equation (4.2.4) with respect to time and keeping in mind the dependence of ξ_0 on time, one can restore the initial defect distribution $f(r)$ from the recombination intensity, equation (4.2.3), see below equation (4.3.31). Therefore, measurements of $n(t)$ or $I(t)$ in sufficiently wide time interval allow us to get physical information about defect creation mechanism. In particular, it was employed to estimate the effect of I impurity upon the primary Frenkel defect separation in KBr [36, 37].

For the *random* initial distribution equations (4.1.53), and (4.1.54) of the bimolecular kinetics should be used. It results from equation (4.1.53) ($\bar{d} = 3$) that the recombination intensity is

$$I \propto \frac{\ln^2(\sigma_0 t)}{t}. \tag{4.2.7}$$

Comparison of equation (4.2.7) with (4.2.4) (or other expressions based on equation (4.2.5)) and particular choice of the function $f(r)$ are widely used to distinguish qualitatively between two limiting cases of random and well-correlated defect distributions.

As it will be shown in Chapter 6, for large reaction depths (times), many-particle effects begin to play an important role. This means that our simple estimates presented here are no longer valid. Since our derivation assumed that $\xi_0 \gg 1$, the question of the range of applicability of these estimates will be discussed anew.

4.2.2 Diffusion-controlled reactions

To describe quantitatively the diffusion-controlled tunnelling process, let us start from equation (4.1.23). Restricting ourselves to the tunnelling mechanism of defect recombination only (without annihilation), the boundary condition should be imposed on $Y(r,t)$ in equation (4.1.23) at $r = 0$ meaning *no* particle flux through the coordinate origin. Another kind of boundary conditions widely used in radiation physics is the so-called *radiation boundary condition* (which however is not well justified theoretically) [33, 38]. The idea is to solve equation (4.1.23) in the interval $r > R$ with the partial *reflection* of the particle flux from the sphere of radius R:

$$4\pi D R^2 \left.\frac{\partial Y(r,t)}{\partial r}\right|_{r=R} = kY(R,t). \tag{4.2.8}$$

The boundary condition (4.2.8) is also called the *grey sphere* – in contrast to the earlier-considered black sphere, equation (3.2.16), which is its limiting case as $k \to \infty$ and is characterized by complete particle absorption at contact.

Let us introduce the *efficiency* of a contact reaction

$$\eta = \frac{k}{k + K_D}, \qquad K_D = 4\pi DR, \tag{4.2.9}$$

as well as the distinctive dimensionless parameter

$$\chi^2 = \frac{r_0^2}{D}\sigma(R), \tag{4.2.10}$$

characterizing the relative contribution of tunnelling into reaction rate. When defect mobility is small ($D \to 0$) and tunnelling recombination is strong, this parameter is $\chi \gg 1$. In the opposite case of highly mobile defects and weak tunnelling, $\chi \leqslant 1$.

The steady-state solution of equations (4.1.23) and (4.2.8) for $\bar{d} = 3$ is ([39], see also [33]):

$$y(r) = \frac{2r_0}{r}\left\{K_0\left(2\chi e^{-\frac{r-R}{2r_0}}\right) - \vartheta I_0\left(2\chi e^{-\frac{r-R}{2r_0}}\right)\right\},$$

$$y(\infty) = 1, \tag{4.2.11}$$

where

$$\vartheta = \frac{K_0(2\chi) - (R/r_0)\chi(1-\eta)K_1(2\chi)}{I_0(2\chi) + (R/r_0)\chi(1-\eta)I_1(2\chi)}, \tag{4.2.12}$$

I_ν and K_ν are modified Bessel's functions of the order ν ($\nu = 0, 1$) of the first and second kind, respectively. In the limiting case of a total annihilation (absorption) at contact ($\eta = 1$) we have

$$\vartheta = \frac{K_0(2\chi)}{I_0(2\chi)}. \tag{4.2.13}$$

The conclusion could be drawn from Fig. 4.2 that the steady-state profile depends essentially on the defect mobility or temperature – unlike the black sphere model, equation (4.1.70). The steady-state solution $y(r)$ defines the stationary reaction rate $K(\infty)$ through the *effective radius* of reaction R_{eff}:

$$K(\infty) = 4\pi DR_{\text{eff}}. \tag{4.2.14}$$

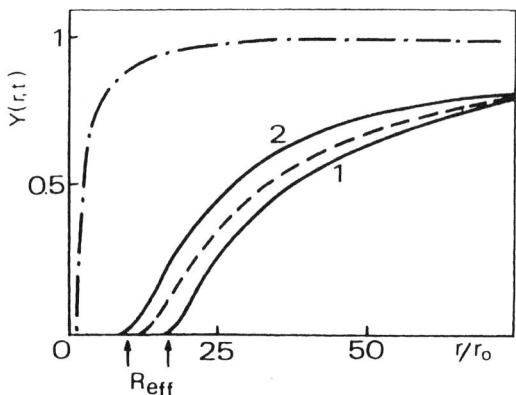

Fig. 4.2. Quasi-steady-state tunnelling recombination profile for V_k centres in KCl calculated at 180 K (curve 1) and 190 K (curve 2) (equation (4.2.11)). The relevant effective radii, equation (4.2.15), are shown by arrows. Dotted line gives the black sphere approximation, broken line – the transient process (discussed in Section 4.2).

The simplest method to find R_{eff} is to analyze the asymptotics of the solution (4.2.11) as $r \to \infty$

$$y(r) \approx 1 - \frac{R_{\text{eff}}}{r}. \tag{4.2.15}$$

There is a complete analogy here with what has been said about equation (4.1.63). Let us analyze now the effective radius of the diffusion-controlled tunnelling recombination in more detail. From equations (4.2.12) and (4.2.15), one gets

$$R_{\text{eff}} = R + 2r_0(\ln \chi + C + \vartheta) = R + 2r_0\big(\ln(\gamma\chi) + \vartheta\big), \tag{4.2.16}$$

Here $C = 0.57721$ is the Euler's constant, $\gamma = \exp(C)$. In the case of the black sphere ($\eta = 1$) we obtain for weak ($\chi \ll 1$) and strong ($\chi \gg 1$) tunnelling respectively

$$R_{\text{eff}} \approx R + 2r_0\chi^2, \quad \chi \ll 1, \tag{4.2.17}$$

$$R_{\text{eff}} \approx r_0 \ln\left(\gamma^2 \frac{\sigma_0 r_0^2}{D}\right), \quad \chi \gg 1. \tag{4.2.18}$$

Note that the effective radius (4.2.18) of "strong" tunnelling takes place under the condition $(\sigma(R)r_0^2/D) \gg 1$, but not $(\sigma_0 r_0^2/D) \gg 1$, which means that strong tunnelling occurs at the contact R. In this case, the effective radius is independent of R and deviates from the exact value by the exponentially small quantity $\exp(-2\chi)$ only. It should be mentioned that equation (4.2.18) does not necessarily mean that $R_{\text{eff}} \gg R$. All what is said above is valid if the so-called *binarity collision* condition holds [33]. It states that linear approximation holds provided a number of particles within a reaction volume with the radius R_{eff} is small:

$$\frac{4\pi}{3} R_{\text{eff}}^3 n_0 \ll 1. \qquad (4.2.19)$$

For ordinary defect concentrations $n < 10^{18}\,\text{cm}^{-3}$, and $r_0 < 1\,\text{Å}$, $\sigma_0 \sim 10^7\,\text{s}^{-1}$ the condition (4.2.19) indeed takes place down to the very small diffusion coefficients, $D \approx 10^{-10}\,\text{cm}^2\,\text{s}^{-1}$.

The behaviour of the steady-state function $y(r)$ is simplest for weak tunnelling recombination (almost pure contact recombination). In this case $y(r) = 0$ when $r \leqslant R$, and when $r > R$, equation (4.2.15) holds with $R_{\text{eff}} = R$.

In another extreme case of the very strong tunnelling $(\sigma_0 r_0^2/D) \gg \ln(\sigma(R)r_0^2/D) \gg 1$, we have $R_{\text{eff}} \gg r_0$ (Fig. 4.2). In this case the $y(r)$ is exponentially small, $y(r) \propto \exp(-2\chi)$, at $r < R_{\text{eff}}$ and follows equation (4.2.15) as $r > R_{\text{eff}}$. The increase of $y(r)$ from a very small value up to its asymptotic value occurs so sharply that one can approximate $y(r)$ in the following way: $y(r)$ is zero as $r < R_{\text{eff}}$ and equation (4.2.15) holds at $r > R_{\text{eff}}$. Comparison with the case of weak tunnelling (contact reaction) shows that strong tunnelling to a good approximation is characterized by the black sphere model, but now with R_{eff} *dependent on the diffusion coefficient*. Readers can find many more details, in particular about the effects of the particle reflection on the reaction sphere, in the review article [33]. Let us estimate here the magnitude of R_{eff} for the actual defects participating in the diffusion-controlled reactions in alkali halides: the electron (Ag^0, Tl^0, F) and hole (H, V_k) pairs in KBr. Put $r_0 \leqslant 1\,\text{Å}$ for electronic centres and $\sigma_0 = 10^{15}\,\text{s}^{-1}$, $10^7\,\text{s}^{-1}$ for V_k and H centres respectively. (The first figure is typical for non-radiative tunnelling transitions.) The diffusion coefficient entering equation (4.2.18) is

$$D = \frac{\lambda^2 \nu_0}{6} \exp\left(-E_a/(k_B T)\right), \qquad (4.2.20)$$

Here λ is defect hopping length, ν_0 is a hopping frequency and E_a is activation energy for a hop. These latter are $E_a = 0.09$ eV, $\nu_0 = 10^{10}$ s^{-1} (H centre) and $E_a = 0.53$ eV, $\nu_0 = 10^{13}$ s^{-1} (V$_k$ centre) [40, 41]. Equations (4.2.17) and (4.2.18) give R_{eff} falling from 20 Å down to 5 Å (which is close to the contact reaction radius R) within the temperature intervals T 50–80 K and 160–240 K, respectively. This demonstrates that tunnelling recombination is important in a relatively wide temperature interval (which is the greater, the larger is E_a) and even for rather mobile defects. Note that when at high temperatures defects become very mobile, they approach each other within contact radius R so quickly that electron tunnelling has no time to occur.

4.2.3 Non-stationary process

Let us briefly consider now *non-steady-state* effects in the kinetics of tunnelling recombination. Time development of the correlation function $Y(r,t)$ could be found either in form of the Laplace transform [42–45] or an infinite series expansion [39, 46]. The most transparent (in terms of physics) method to study the time development of $Y(r,t)$ and of the relevant $n_\nu(t)$ is direct *computer calculations* of the kinetic equation (see also [47, 48]. These computer calculations [49, 50] have demonstrated that an approach to the steady-state $Y(r,\infty) = y(r)$ is, at least, the two-stage process (Fig. 4.3). At the very beginning when $t \simeq \sigma(R)^{-1}$, concentration sharply decreases due to the rapid tunnelling recombination of close dissimilar defects. It results in the sharp recombination profile of $Y(r,t)$ which is characterized by the distinctive radius $R(t)$, equation (4.1.47), increasing in time. Since at this stage the recombination process does not depend on defect diffusion, the latter only weakly affects the time development of the recombination profile. After $R(t)$ reached the value of the effective diffusion-controlled radius R_{eff}, equation (4.2.18), the second, diffusion-controlled stage begins in which the slope of the recombination profile decreases, $Y(r,t)$ becomes much more smooth and the steady-state $Y(r,t)$ is formed finally as $t \gg R_{\text{eff}}^2/D$.

A comparison of the time developments of $Y(r,t)$ in two limiting cases (pure contact reaction and strong tunnelling recombination) demonstrates their *qualitative difference*. In the latter case, the first stage is very short and is finished already at $t \simeq \sigma(R)^{-1}$; the further change of $Y(r,t)$ is defined here entirely by the non-stationary diffusion. The relevant reaction rate for

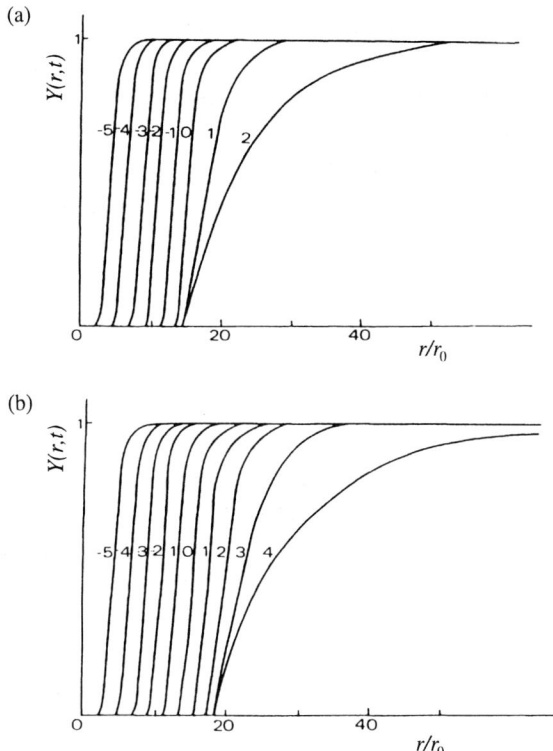

Fig. 4.3. The process of steady-state recombination profile formation for the strong tunnelling case; parameter $\chi^2 = (r_0^2 \sigma_0/D) \times 10^7$ (a) and 2×10^8 (b). Figures in curves show $\log t$.

the initial random distribution is well known (equation (4.1.61), see, e.g., Waite [10]):

$$K(t) = 4\pi D R \left(1 + \frac{R}{\sqrt{\pi D t}}\right). \tag{4.2.21}$$

Therefore, equation (4.2.21) with the substitution of R_{eff} for R cannot describe correctly the process of the steady-state formation if the diffusion process is controlled by the strong tunnelling ($\chi \gg 1$). In other words, strong tunnelling could be described in terms of the effective recombination radius R_{eff} analogous to the black sphere in the *steady-state* reaction stage only.

Another distinctive feature of strong tunnelling recombination could be seen after a step-like (sudden) increase (decrease) of temperature (or diffusion coefficient – see equation (4.2.20)) when the steady-state profile has already been reached. Such *mobility stimulation* leads to the prolonged transient stage from one steady-state $y(r, T_1)$ to another $y(r, T_2)$, corresponding to the diffusion coefficients $D(T_1)$ and $D(T_2)$ respectively. This process is shown schematicaly in Fig. 4.2 by a broken curve. It should be stressed that if tunnelling recombination is not involved, there is *no* transient stage at all since the relevant steady state profile $y(r) = 1 - R/r$, equation (4.1.62), doesn't depend on $D(T)$.

The effect of an inertial increase in recombination intensity $I(t)$ was observed experimentally in many insulating crystals including alkali halides [51, 52], $Ba_3(PO_4)_2$ [53], α-Al_2O_3 [54], Na-salt of DNA (Fig. 4.4) [52]. The advantage of this technique which is efficient for the identification of tunnelling stages of reactions in insulators is that the initial and final steady-state recombination profiles are known and could be calculated by means of equation (4.2.11) unlike the *initial* distribution which is usually believed to be *random* but in fact is unknown.

The treatment of the time-dependent equation (4.1.23) has shown [55] that the transient kinetics is controlled by *three* parameters: the ratio of the diffusion coefficients, $D^* = D(T_2)/D(T_1) = \exp(-E_a \delta T/T_1^2)$ ($\delta T = T_2 - T_1$ is temperature increment), $\sigma_0 r_0^2/D$ and r_0^2/D. The first parameter, D^*, defines an increase in recombination intensity $I(T_2)/I(T_1)$ (vertical scale) and thus permits us to get the hopping activation energy E_a. The parameter r_0^2/D could be found by fitting the calculated transient time to the experimentally observed one (horizontal scale).

If half the Bohr radius of the electron centre r_0 is known (e.g., for F-centres in alkali halides from ESR experiments, see, e.g., [56]), the pre-exponential diffusion coefficient D_0 could be found.

A theoretical analysis of the experimental kinetics for V_k centres in KCl–Tl, as well as for self-trapped holes in α-Al_2O_3 and Na-salt of DNA, is presented in [55]. The fitting of theory to the experimental curves is shown in Fig. 4.4. Partial agreement of theory and experiment observed in the particular case of V_k centres was attributed to the violation of the *continuous* approximation in the diffusion description. This point is discussed in detail below in Section 4.3. Note in conclusion that the fact of the observation of prolonged increase in recombination intensity itself demonstrated slow mobility of defects. In the case of pure irradiated crystals, it is a strong

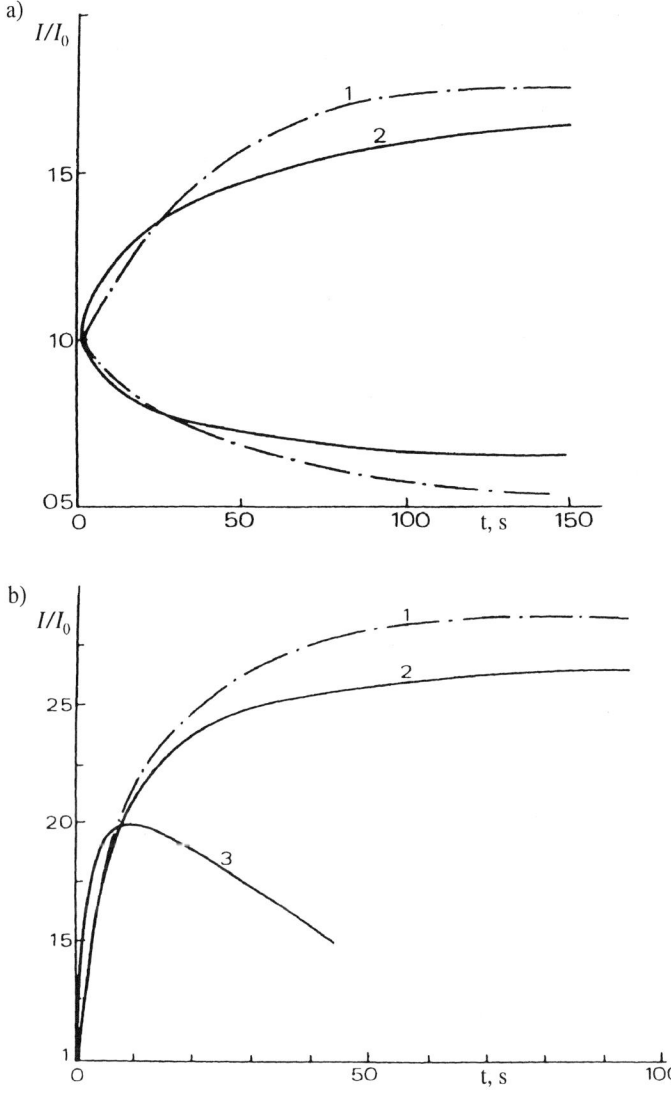

Fig. 4.4. (a) The transient kinetics of tunnelling luminescence intensity increase due to hypothetical self-trapped holes in α-Al$_2$O$_3$ [53, 55]. Dash-dotted line 1 – experimental, full line 2 – theoretical. Temperature increment: 198.2 K $\to$ 201.5 K (two curves above), and temperature decrease 204.1 K $\to$ 201.5 K (two curves below). (b) Same for the Na-salt of DNA. Dash-dotted line 1 – experimental, full lines – theoretical: for three-dimensional recombination (curve 2) and one-dimensional (curve 3). Temperature increase 141 $\to$ 146 K.

argument for hole *self-trapping* (small-radius polaron formation). The non-steady-state effects are also discussed in Section 4.4 with special emphasis on large hopping lengths.

4.2.4 Effects of defect interaction

In many cases of interest tunnelling recombination of defects is accompanied by their elastic or Coulomb interaction, which is actual, e.g., for F, H and V_k, A^0 pairs of the Frenkel defects in alkali halides, respectively. In these cases the equation defining the steady-state recombination profile is

$$D\nabla \left(\nabla y(r) + \frac{y(r)}{k_B T} \nabla U(r) \right) - \sigma(r) y(r) = 0, \qquad (4.2.22)$$

with $\sigma(r) = \sigma_0 \exp(-r/r_0)$ and $y(\infty) = 1$.
It defines the reaction rate

$$K(\infty) = \int \sigma(r) y(r) \, d\vec{r} \equiv 4\pi D R_{\text{eff}}, \qquad (4.2.23)$$

entering the equation (4.1.28) for concentration decay. Substituting

$$y(r) = \frac{z(r)}{r} \exp\left[-\beta U(r)/2 \right], \qquad \beta = \frac{1}{k_B T}, \qquad (4.2.24)$$

we obtain

$$\frac{d^2 z(r)}{dr^2} + z(r) \left(\frac{\beta \nabla^2 U(r)}{2} - \frac{\beta^2 [\nabla U(r)]^2}{4} - \frac{\sigma(r)}{D} \right) = 0, \qquad (4.2.25)$$

with the boundary conditions imposed $z(R) = 0$, $z(r \to \infty) = r - b$, where b is an asymptotic coefficient of the steady-state at $r \to \infty$.

As it was said above (Section 3.2), for the elastic interaction this coefficient coincides with the effective radius of recombination, $R_{\text{eff}} = b$, whereas for the Coulomb interaction R_{eff} is defined in equation (3.2.51). Therefore the problem of obtaining the steady-state reaction rate is reduced to the finding the asymptotic coefficient b of the solution of equation (4.2.25). Formally it coincides with the quantum-mechanical *scattering length* on the potential which is half the value in brackets in equation (4.2.25). In real cases of the elastic and Coulomb interactions equation (4.2.25) could be rewritten as

$$\frac{d^2 z(r)}{dr^2} - z(r)\left(\frac{L^2}{4r^4} + \frac{\sigma_0}{D}\exp(-r/r_0)\right) = 0, \quad (4.2.26)$$

$$\frac{d^2 z(r)}{dr^2} - z(r)\left(\frac{9g^2}{4r^8} + \frac{3g}{r^5} + \frac{\sigma_0}{D}\exp(-r/r_0)\right) = 0, \quad (4.2.27)$$

where the *Onsager radius* L, equation (3.2.55), is introduced for the Coulomb interaction and the parameter $g = \frac{\alpha}{k_B T} = \alpha\beta$ (with the dimension m^3) for the elastic one.

The solution of the equation (4.2.26) cannot be found in an analytical form and thus some approximations have to be used, e.g., *variational principle*. Its formalism is described in detail [33, 57, 58] for both lower bound estimates and upper bound estimates. Note here only that there are *two* extreme cases: when $\sigma(r)/D$ term is small compared to the drift term, reaction is controlled by defect interaction, in the opposite case it is controlled by tunnelling recombination. The first case takes place, e.g., at high temperatures (or small solution viscosities if solvated electron is considered).

The steady-state for diffusion-controlled recombination *without* interaction is given by equation (4.2.15), whereas its analog for the *isotropic elastic attraction* ($g > 0$) without tunnelling reads [59, 60]

$$z(r) = \exp\left[-\frac{g}{2r^3}\right]\left(r\exp\left[\frac{g}{r^3}\right]b\Phi\left(1, \frac{4}{3}, \frac{g}{r^3}\right)\right), \quad (4.2.28)$$

where Φ is a degenerate hypergeometric function, from which one gets

$$R^{el}_{eff} = \frac{R\exp[g/R^3]}{\Phi(1, 4/3, g/R^3)}$$

$$\approx \begin{cases} \frac{3g^{1/3}}{\Gamma(1/3)} = 1.12(\alpha\beta)^{1/3}, & g \gg R^3, \\ R, & g \ll R^3. \end{cases} \quad (4.2.29)$$

As it follows from equation (4.2.29), at *low* temperatures R^{el}_{eff} decreases with temperature as $T^{-1/3}$ [61–63] and approaches an annihilation radius at high temperatures.

On the other hand, the effective radius of the annihilation stimulated by the Coulomb interaction is well known to be [8, 10, 22, 64]

$$R_{\text{eff}}^{\mp} = \frac{1}{\int_R^\infty \exp[\beta U(r)] \frac{dr}{r^2}} = \frac{L}{2}\left(\coth\left[\frac{L}{2R}\right] \pm 1\right), \quad (4.2.30)$$

where the signs '−' and '+' correspond to the attraction ($U < 0$) and repulsion ($U > 0$) respectively, L is the Onsager radius, equation (3.2.55).

One can argue that for the strong tunnelling, $\chi \gg 1$, equation (4.2.10), when the diffusion-controlled tunnelling recombination is characterized by the effective radius

$$R_D = r_0 \ln\left(\frac{\sigma_0 r_0^2}{D}\right) \gg R,$$

equation (4.2.30) could be generalized by replacing R for R_D [39, 46]

$$R_{\text{eff}}^{c\mp} \approx \frac{L}{2}\left(\coth\left[\frac{L}{2R_D}\right] \pm 1\right). \quad (4.2.31)$$

For $R_D > L$ one gets $R_{\text{eff}}^c \approx R_D$, while if $R_D < L$, $R_{\text{eff}}^{c-} \approx L$. Accuracy of this formula when $R_D \approx L$ is not clear. To check it up, calculations were done for two typical cases corresponding to shallow donors in semiconductors ($r_0 = 20$ Å) and deep centres in ionic reaction ($r_0 = 2$ Å) [65]. In the first case the reaction is controlled by a drift in the Coulomb field, when $L > R_D$ within *all* the intervals of the diffusion coefficients considered (Fig. 4.5(a)) whereas in the second case, quite on the contrary, the recombination is controlled by tunnelling (Fig. 4.5(b)). What is surprising, that in both cases equation (4.2.31) describes the explicit result very well even if $R_D \approx L$! It could be shown that R_{eff}^{c-} has to exceed the L by the value $L/2$ which is also seen in Fig. 4.5 (in the interval $R_D^{(2)} > L^{(2)}$).

Equation (4.2.31) works well also for the case of the Coulomb repulsion, shown in Fig. 4.5(b). The effective radius decreases with increase of D since $R_{\text{eff}}^+ = R_{\text{eff}}^- - L$ and $R_{\text{eff}}^- \to L$ as R decreases (and D increases). As $L > R_D$, equation (4.2.31) yields $R_{\text{eff}}^+ \approx L\exp(-L/R_D)$ which really holds for the curve 1.

It should be stressed that R_{eff}^c means the effective radius at which mobile defect is trapped by the Coulomb field of its partner but the electron tun-

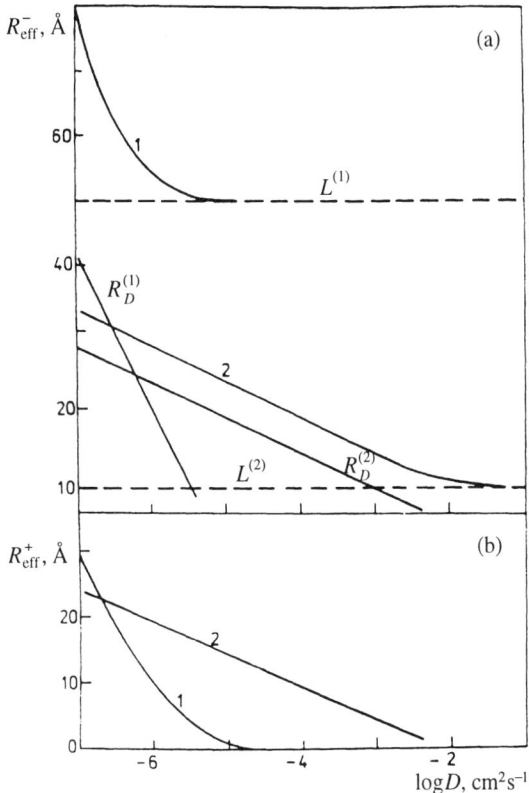

Fig. 4.5. Effective radii of the diffusion-controlled tunnelling recombination for the Coulomb attraction (a) and repulsion (b) (after [65]). Curve 1 and 2 are results of computer calculations with parameters: $\sigma_0 = 10^7\,\text{s}^{-1}$, $r_0 = 20$ Å, $R_D = 4$ Å and $\sigma_0 = 10^{14}\,\text{s}^{-1}$, $r_0 = 2$ Å, $R_D = 4$ Å respectively. L – the Onsager radius, equation (3.2.55), R_D – radius of strong tunnelling recombination, equation (4.3.7).

nelling takes place at a much shorter distance (where the correlation function of dissimilar reactants tends to zero) – see Fig. 4.6.

We do not discuss here variational estimates of the R_{eff}^c; for details see the review article [33]. Note here only that radius estimates show that for a typical small-radius electron defect, the activator atom A^0 and R_{eff}^{c-} begin to deviate from the Onsager radius L at very small D values only (low temperature) when the applicability of binary approximation itself is questionable. It comes from the fact that due to small values of $r_0 \approx 0.5$ Å and $\varepsilon \approx 5$ the Onsager

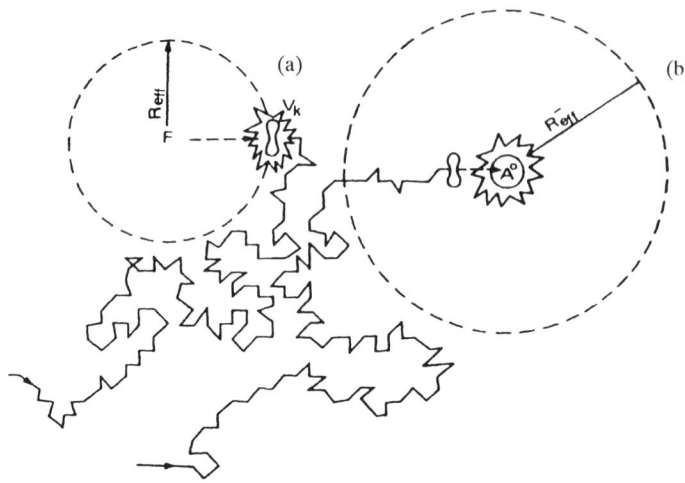

Fig. 4.6. Schematic pattern of the tunnelling recombination of positively changed V_k centre with neutral electron centre (e.g., F centre) (a) and with oppositely charged activator atom (e.g., Tl^0) (b). In the second case the Coulomb field traps V_k at long distance R_{eff}^-, but electron transfer itself occurs at much shorter distance.

radius exceeds greatly that for tunnelling recombination, R_{eff}. In other words, recombination kinetics is practically controlled by a drift in the Coulomb field. Another situation is expected to happen in semiconductors and polar liquids, where r_0 and ε are much greater. There exist a lot of experimental evidences of such effective tunnelling [32, 34, 66].

Let us consider now several variational estimates of the effective radius taking into account annihilation, tunnelling and an *elastic interaction*. If tunnelling term in equation (4.2.25) is large in comparison with others in brackets, we can use for the upper estimate equation (4.2.15) as a trial function $y(r)$ which leads to

$$R_{\text{el}}^{\text{up}} \leqslant R_{\text{eff}} + \int_R^\infty \left(\frac{9g^2}{4r^8} + \frac{3g}{r^5}\right) y(r) \, dr, \qquad (4.2.32)$$

where R_{eff} is given by equation (4.2.78). (This integral can be calculated numerically.)

In the opposite case of *large D* (high temperature) equation (4.2.28) could be used as a trial function $z(r)$ thus obtaining

$$R_{\text{el}}^{\text{up}} \leqslant R_{\text{el}} + \frac{\sigma_0}{D} \int_R^\infty z^2(r) \exp(-r/r_0) \, dr, \qquad (4.2.33)$$

R_{el} is given by equation (4.2.29). Since $z(r)$ contains a rapidly decreasing exponential, e^{-g/r^3}, and usually $g \gg r_0^3$ we can use the *saddle-point method* to estimate this integral. Thus, if $g \gg R^3$, we obtain

$$R_{\text{el}}^{\text{up}} \leqslant R_{\text{el}} + \frac{c_1 \sigma_0}{D} g^{7/24} r_0^{7/8} \exp\left(-4(3r_0)^{-3/4} g^{1/4}\right), \qquad (4.2.34)$$

$$c_1 = \frac{(\pi/2)^{1/2} 3^{13/8}}{\Gamma^2(1/3)}. \qquad (4.2.35)$$

On the other hand, employing the Kohn variational principle [67] two *lower bound* estimates may be obtained [59, 60]

$$R_1^{\text{low}} \geqslant R_{\text{el}}^{\text{up}} - \left(\frac{\sigma_0}{D}\right)^2 \frac{1}{3|g|} \int_0^\infty \left(z(r) \exp(-r/r_0)\right)^2 r^5 \, dr$$

$$\approx R_{\text{el}}^{\text{up}} - c_2 \left(\frac{\sigma_0}{D}\right)^2 |g|^{13/24} \left(\frac{r_0}{2}\right)^{27/8} \times$$

$$\times \exp\left\{-4|g|^{1/4} \left(\frac{3r_0}{2}\right)^{-3/4}\right\}, \qquad (4.2.36)$$

$$c_2 = \frac{(\pi/2)^{1/2} 3^{15/8}}{\Gamma^2(1/3)}. \qquad (4.2.37)$$

Another estimate is

$$R_2^{\text{low}} \geqslant R_{\text{el}}^{\text{up}} - \frac{c_1}{\kappa} \left(\frac{\sigma_0}{D}\right)^2 g^{7/24} r_0^{17/8} \exp\left(-4(3r_0)^{-3/4} g^{1/4}\right),$$

$$\kappa = (3eg/5r_0)^5. \qquad (4.2.38)$$

One can conclude from equation (4.2.33) that the correction δR to the effective radius due to tunnelling (the second term) decreases with both the decrease of parameters σ_0, r_0 characterizing tunnelling strength, and the increase of temperature (diffusion coefficient). Since this correction makes the contribution to the rate constant, $\delta K = 4\pi D \delta R$, independent on D, the

Arrhenius law, $K \propto \exp(-E_{\text{eff}}/(k_B T))$, *does not* hold here. (The same is true for the Coulomb interaction [39, 46, 68].) Variational estimates of the effective radius at high temperatures, when the recombination is controlled predominantly by *annihilation*, are discussed in [60]. Variational estimates of the effective radius taking into account annihilation, tunnelling and an *elastic interaction* were discussed in detail in [33].

Making use of the elastic constant entering equation (3.1.4) for F, H centres in KBr $\alpha = 3$ eV Å^3 [69], one can estimate easily that the effective radius of annihilation stimulated by elastic interaction, R_{el} (equation (4.2.29)) varies from 11 Å down to 7 Å as the temperature increases from 40 K to 200 K (and then is independent of the annihilation radius $R \approx 4$ Å). On the other hand, the effective radius of tunnelling recombination, equation (4.2.17), decreases from 10 Å (at 40 K) down to 5 Å (60 K). It coincides with the elastic radius, R_{el}, at ≈ 37 K, where diffusion is very slow and the binary approximation, equation (4.2.19), does not hold any longer.

In Fig. 4.7, another case of A^0, H pairs is shown when half the Bohr radius of the electron activator centre is typically about 1 Å (e.g., for Tl^0 in KCl). Here a sum of two upper estimates (curves 5 and 6) practically *coincide* at all temperatures with the exact radius (curve 1) and with lower-bound estimates (curve 7 and 8) above 70 K.

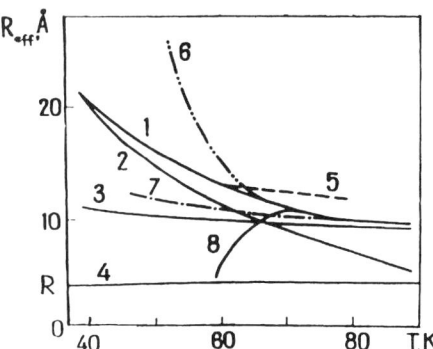

Fig. 4.7. Temperature dependence of the effective radius of H, A^0 recombination in KBr controlled by an elastic interaction, diffusion and tunnelling. Curve 1 – exact result, 2 – effect of tunnelling and annihilation, 3 – isotropic attraction and annihilation, 4 – pure annihilation. Variational estimates: upper bound when (i) tunnelling dominates (equation (4.2.32) – curve 5) or an elastic interaction dominates (equation (4.2.34) – curve 6). Curve 7 – lower bound estimate, equation (4.2.36), when an elastic interaction is a predominant factor.

4.2.5 The anisotropic effects

As it is shown in Fig. 3.1(b) in Chapter 3, typical hole centres in alkali halides – H and V_k centres being X_2^- – quasi-molecules oriented along the $\langle 110 \rangle$ axis are rather anisotropic which is observed experimentally, e.g., via *polarized recombination luminescence* [70, 71]. Their analog is a dumb-bell interstitial in many metals.

It should be remembered that even for isotropic defects their elastic interaction is *anisotropic*, due to crystalline anisotropy, equation (3.1.4). A pair of the simplest Frenkel defects – vacancy and an interstitial atom – attract each other in the direction $\langle 100 \rangle$, but their interaction becomes repulsive, e.g., along $\langle 111 \rangle$ and $\langle 101 \rangle$ axes.

For an anisotropic defect, like crowdions or di-atomic quasi-molecules (H and V_k centres), the problem becomes much more complicated and often permits only a numerical solution (e.g., [72]). For example, an estimate of the interaction energy of a crowdion with a vacancy in Cu in the direction perpendicular to the crowdion axis is $\leqslant 0.1$ eV at the relative distance $\sqrt{2}\, a_0$ (a_0 is a lattice constant) if both are in the same plane, but this energy becomes 0.02 eV only for a distance twice as large ($\varphi = 0$ in Fig. 4.8(a)). Increase of the angle φ from $0°$ up to $45°$ (crowdion lies in the next plane) can change a sign of their interaction.

In the case of F, H centres in alkali halides their maximum attraction and repulsion correspond to the configurations 1 and 2 in Fig. 4.8(b). Even for the nearest F, H pair in KBr, the attraction energy is rather small: ≈ 0.04 eV.

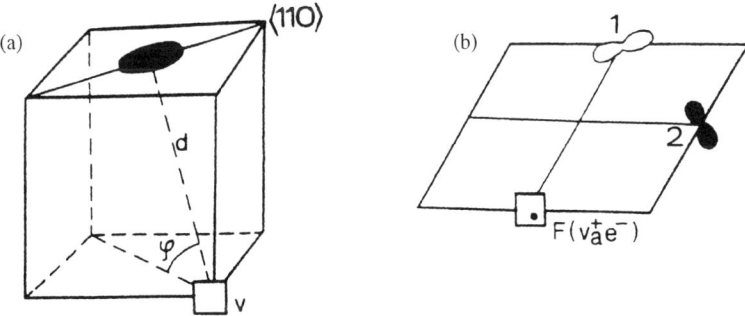

Fig. 4.8. Elastic interaction of of crowdion with vacancy in metals (a) and F, H centres in alkali halides (b). Configuration 1 is energetically the most favourable with $E_{\text{int}} = -0.043$ eV.

Another effect of crystal anisotropy is that the diffusion coefficient D turns out to become the *diffusion tensor* $\widehat{D}$ [73–75]. It results in the asymmetric escape probabilities of geminate pairs in different directions.

To consider defect recombination controlled by the anisotropic potential, equation (3.1.4), it was suggested [76] to rewrite the quasi-steady-state equation (4.2.22) in a form

$$\nabla^2 z(\vec{r}) + V(\vec{r}) z(\vec{r}) = 0, \qquad (4.2.39)$$

where the substitution (4.2.24) is used and

$$V(\vec{r}) = \left(\frac{\beta \nabla^2 U(r)}{2} - \frac{\beta^2 [\nabla U(r)]^2}{4} \right). \qquad (4.2.40)$$

Further, both $f(\vec{r})$ and $V(\vec{r})$ are expanded in the spherical harmonics and thus the problem is reduced to the solution of a finite set of coupled ordinary differential equations.

The variational procedure was developed in [61, 63] for calculating R_{eff} for anisotropic potentials. Employing different trial functions, it is shown that for a strong elastic interaction the effective radius is

$$R_{\text{el}} \approx (0.52 \div 0.54) \left(\frac{\alpha}{k_B T} \right)^{1/3}, \qquad (4.2.41)$$

i.e., it is proportional to $T^{-1/3}$ (cf. equation (4.2.29)). A comparison with an isotropic attraction demonstrates that roughly *half* of the recombination sphere is efficient now only for the approaching defects whereas defects avoid their other half due to the strong repulsion.

It is convenient to consider a model of an anisotropic recombination region: the *reflecting* recombination sphere (*white sphere*) with *black* reaction spots on its surface [77, 78]. The measure of the reaction anisotropy here is the geometrical *steric factor* Ω which is a ratio of a black spot square to a total surface square. Such a model could be actual for reactions of complex biologically active molecules and tunnelling recombination when the donor electron has an asymmetric (e.g., p-like) wavefunction. Note the non-trivial result that at small Ω, due to the partial averaging of the reaction anisotropy by rotational motion arising due to numerous repeated contacts of reactants before the reaction, the reaction rate is $K_0 \propto \Omega^{1/2}$ rather than the intuitive estimate $K_0 \propto \Omega$.

4.3 NON-DIFFUSION DEFECT RECOMBINATION AND NON-STATIONARY KINETICS

> What we have today are books about books and depictions about depictions.
> (Heutzutage haben wir schon Bücher von Büchern und Beschreibungen von Beschreibungen)
>
> Lichtenberg

4.3.1 The hopping recombination

Defect recombination is treated traditionally as a diffusion-controlled process, usually without discussing whether stochastic defect hops *always* result in the diffusion equation like (4.1.23). On the one hand, *both* stochastic flights and short-range hopping lead to a macroscopic defect *diffusion motion* after a sufficient number of random walks (cf. Section 3.2). On the other hand, the *recombination mechanism* can be either a flight or a hopping (or related one) respectively, but it can be regarded as *diffusion* under conditions essentially different: the *kinematics of defect encounter* is of principal importance here. It is shown [18, 27–29, 79] that the hopping motion itself can result in two quite different (hopping and diffusion) *recombination* mechanisms. The latter one holds only if the relative defect motion in the scale of an essential change of the tunnelling probability $\sigma(r)$, i.e., r_0, consists of a *large number of hops*. This is a very strong restriction which does not take place, e.g., for electrons in slightly doped glasses and semiconductors. In this case alternative *hopping* mechanism holds, where particles enter and leave the recombination sphere in a *single* hop. Note that limits of the continuous diffusion description were first discussed for the *energy transfer* in doped crystals and liquids [27–29, 80].

In this Section we describe briefly the principal effects arising during replacement of the continuous diffusion, equation (4.1.23), by its more general analog

$$\frac{\partial Y(r,t)}{\partial t} = \widehat{L} Y(r,t) - \sigma(r) Y(r,t). \qquad (4.3.1)$$

Here the diffusion operator $D\Delta$ is replaced by the operator $\widehat{L}$ describing motion as stochastic hops in the continuous coordinate space. Let $\varphi(r)$ be the distribution function of hop lengths normalized as

$$\int \varphi(r)\,\mathrm{d}\vec{r} = 1. \qquad (4.3.2)$$

The mean square hop length is

$$\lambda^2 = \langle r^2 \rangle = \int r^2 \varphi(r) \, d\vec{r}. \tag{4.3.3}$$

Defect *motion* at a large relative distance could always be treated as diffusion with the coefficient $D = \frac{\lambda^2}{6\tau}$, where τ is the mean (waiting) time between successive hops. Thus we arrive at

$$\widehat{L}Y(r,t) = \frac{1}{\tau}\left(\int \varphi(\vec{r}')Y(|\vec{r}-\vec{r}'|,t)\, d\vec{r}' - Y(r,t)\right). \tag{4.3.4}$$

In view of the spherical symmetry of the problem, this expression could be simplified and reduced to the one-dimensional integral

$$\widehat{L}Y(r,t) = \frac{1}{\tau}\left(\int_0^\infty \Phi(r,r')Y(r',t)\, dr' - Y(r,t)\right), \tag{4.3.5}$$

where $\Phi(r,r')$ is an integral kernel defined by the distribution $\varphi(r)$, and conditional probability density of obtaining the particle at r after its hop from r'.

The quasi-steady-state hopping recombination rate $K(\infty) = K_0$ is related to the coefficient R_{eff} via equation (4.2.14) as in the diffusion-controlled case. As in equation (4.2.15), this R_{eff} is defined by the asymptotics of the solution, $Y(r,\infty) = y(r)$, as $r \to \infty$. It is important, however, that R_{eff} cannot generally be treated as the *effective recombination radius*. It holds provided that the hop length is much smaller than the distinctive scale r_0 of tunnelling recombination

$$\lambda \ll r_0. \tag{4.3.6}$$

If the condition (4.3.6) holds, in the integrand in equation (4.3.5) the function $Y(|\vec{r}-\vec{r}'|,t)$ could be expanded in the Taylor series in powers of $\vec{r}'$ up to second-order terms since the function $Y(r,t)$ is a smoother function than $\varphi(r)$. In doing so, the operator (4.3.5) is reduced to the ordinary diffusion $D\Delta$ one and $R_{\text{eff}} = R_{\text{D}}$, where

$$R_{\text{D}} = r_0 \ln \frac{\sigma_0 r_0^2}{D}. \tag{4.3.7}$$

Note that for the strong tunnelling, $\ln \chi \gg 1$, $R_D \approx R_{\text{eff}}$, equation (4.2.18).

Since typically $r_0 \simeq 1$ Å, and λ is restricted in a crystal or liquid by a distance of several Å, the condition (4.3.7) can be violated which makes the diffusion description inadequate and an alternative *hopping recombination* formalism should be used.

4.3.2 Estimates of the effective reaction radius and reaction rate

If hop length is large enough, the predominant contribution into integral (4.3.5) comes from large distances r' lying *outside* the effective recombination sphere, where $Y(r',t) \approx 1$. Therefore we can estimate (4.3.5) as

$$\widehat{L}Y(r,t) = \frac{1}{\tau}(1 - Y(r,t)). \tag{4.3.8}$$

Solution of equation (4.3.1) with the operator (4.3.8) could be easily found

$$Y(r,t) = \frac{1 + \sigma(r)\tau \exp[-\sigma(r)t - t/\tau]}{1 + \sigma(r)\tau}. \tag{4.3.9}$$

As $t \to \infty$, we get the steady-state $y(r) = \frac{1}{1+\sigma(r)\tau}$. The hopping reaction rate K_0 is respectively

$$K_0 = \frac{1}{\tau} \int \frac{\sigma(r)\tau}{1 + \sigma(r)\tau} d\vec{r}. \tag{4.3.10}$$

In the particular case of tunnelling recombination

$$\frac{\sigma(r)\tau}{1 + \sigma(r)\tau} = \frac{1}{\exp\left[-\frac{r-R}{r_0}\right] + 1}. \tag{4.3.11}$$

Here the quantity

$$R_H = r_0 \ln(\sigma_0 \tau) \tag{4.3.12}$$

is nothing but the *effective recombination radius* [81] in the case of long hops. Equation (4.3.12) differs from its diffusion analog, equation (4.3.7), since instead of diffusion time r_0^2/D necessary to cross recombination layer of the thickness r_0, we have now waiting time τ. It is easy to show that hopping kinetics hold if

$$\lambda \gg R_\text{H}. \tag{4.3.13}$$

This means that defects enter the recombination sphere and leave it in a single hop.

Since equation (4.3.11) is quite similar to the Fermi–Dirac distribution, we can approximate it by step-like function which is zero as $r \leqslant R_\text{H}$ and unity if $r > R_\text{H}$. When doing so, we obtain from equation (4.3.10) the steady-state reaction rate

$$K_0 = \frac{4\pi}{3\tau} R_\text{H}^3. \tag{4.3.14}$$

Note that the steady-state $y(r)$ obtained from equation (4.3.9) as $t \to \infty$ does not completely agree with the asymptotic behaviour of the quantity established earlier due to absence here of the term R_eff/r. This term comes from the diffusion character of distant tunnelling recombination. The term R_eff/r arises from the next after equation (4.3.8) term in the expansion of the integral (4.3.5) in parameter $(R/\lambda)^2 \ll 1$. Nevertheless, the use of such a function as (4.3.9) to calculate K_0 does not affect results since in deriving equation (4.3.10) the *asymptotic* value of $y(r)$ is only used rather than its behaviour at large r. A comparison of hopping recombination with equation (4.2.18) yields

$$R_\text{H} = 2R(R/\lambda)^2. \tag{4.3.15}$$

In view of equation (4.3.13) R_eff characterizing hopping recombination is less than R. Determined experimentally via $K_0/(4\pi D)$, this quantity may turn out to be even less than the annihilation radius R (atomic scale or lattice parameter), but it increases with growing viscosity, which is characteristic of the diffusion stage of a strong tunnelling. A combination of these contradicting properties indicates that we face hopping control of the defect recombination [27–29].

To give a transparent physical interpretation to both hopping and diffusion kinetics controlled by strong tunnelling recombination, the reaction rate could be presented in the universal form as

$$K_0 = \frac{\widehat{v}}{\widehat{\tau}}, \tag{4.3.16}$$

where $\widehat{v}$ is volume of the recombination sphere, and $\widehat{\tau}$ is time the defect stays inside it.

In the hopping mechanism $\widehat{v} = \frac{4\pi}{3} R_H^3$, to be compared with a quite similar expression $\widehat{v} = \frac{4\pi}{3} R_D^3$ for the diffusion. The time required to cross recombination sphere is $\widehat{\tau} = R_D^2/(3D)$ for diffusion and $\widehat{\tau} = \tau$ for hopping mechanism.

4.3.3 General solution for the arbitrary hopping length

To treat the problem of recombination kinetics for *arbitrary* hopping lengths $0 < \lambda < \infty$, it is convenient to restrict ourselves to the *exponential* hopping length distribution known as the Torrey model [82]. In this particular case the integral kernel $\Phi(r, r')$ which in equation (4.3.5) is defined by distribution results in

$$\varphi(l) = \frac{l}{\lambda_0} \exp(-l/\lambda_0), \tag{4.3.17}$$

$$\Phi(r, r') = \frac{r'}{r} \left(\varphi(|r - r'|) - \varphi(r + r') \right), \tag{4.3.18}$$

where λ_0 is the most probable hopping length and mean square displacement is $\lambda^2 = 6\lambda_0^2$. Although this model is adequate for an actual physical situation (motion of a solvated electron), it turned out to be very useful for the analytical treatment of the problem under study since it permits us to reduce the integral *stationary* equation:

$$\frac{1}{\tau} \left(\int_0^\infty \Phi(r, r') y(r') \, dr' - y(r) \right) - \sigma(r) y(r) = 0 \tag{4.3.19}$$

to the *differential* one [27–29, 81]

$$\lambda_0^2 \frac{d^2}{dr^2} (r\eta(r)) - \frac{\sigma(r)\tau}{1 + \sigma(r)\tau} (r\eta(r)) = 0, \tag{4.3.20}$$

where $\eta(r) = y(r)(1 + \sigma(r)\tau)$. The effective radius is defined as usual through its asymptotic behaviour, $R_{\text{eff}} = \lim_{r \to \infty} d\eta(r)/dr$. It is found to be [83]

$$R_{\text{eff}} = r_0 \left(\ln \gamma^2 + \zeta + 2\Psi(1/x) + x + G(\zeta, x) \right), \tag{4.3.21}$$

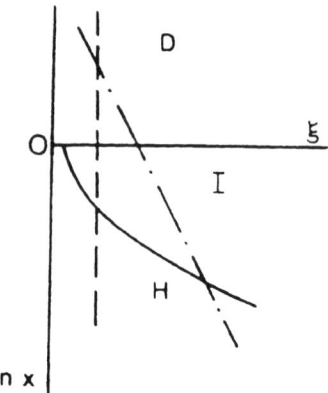

Fig. 4.9. Discrimination of migration-controlled reactions [81]. D – diffusion ($\chi < 1$), I – intermediate regime (lies between abscissa axis and full line), and H – hopping reaction (below this line). The cases $\tau =$ const and $D =$ const are marked by lines – – – and — · — · — respectively.

where Ψ is the Euler psi function. Here two dimensionless parameters, $\zeta = \ln(\sigma_0 \tau)$ and $x = \lambda_0/r_0$, are introduced, and $G(\zeta, x)$ is a complicated function containing the hypergeometric functions of the first order. Equation (4.3.21) is universal and covers *all range* of hopping lengths; moreover, it involves both the *kinetic* stage (immobile defects) and the migration-controlled stage. It was treated in full scale in [81]. Since we are interested in the latter regime now, let us restrict ourselves to the right half plane $\zeta > 0$ (Fig. 4.9).

Depending on λ_0, this half plane is divided vertically into three regions – diffusion, intermediate and hopping. For the first two regimes one can simplify equation (4.3.21) putting there $G = 0$, thus getting

$$R_{\text{eff}} = r_0 \left(\ln \gamma^2 + \zeta + 2\Psi(1/x) + x \right). \tag{4.3.22}$$

On the other hand, when approaching the intermediate region from the hopping side (large x), the following estimate was derived:

$$R_{\text{eff}} \approx r_0 \left(\zeta - x \operatorname{th}(\zeta/x) \right). \tag{4.3.23}$$

To understand an effect of λ upon the kinetics, it is useful to consider the cross-section $D = \lambda_0^2 \tau_0^{-1} =$ const. It could be shown that at large ζ the approximations (4.3.22) and (4.3.23) overlap in the intermediate region. Con-

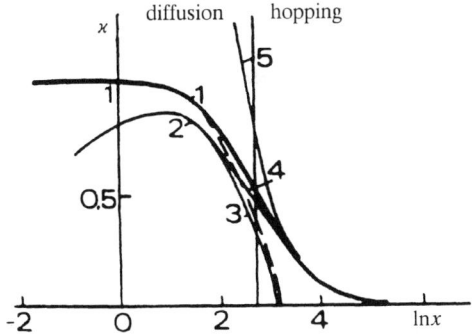

Fig. 4.10. The ratio of effective hopping reaction radius to the diffusion one vs. hop length [81]. 1 – exact calculation. Different approximations: curve 2 – equation (4.3.26), 3 – (4.3.22), 4 – (4.3.23), 5 – (4.3.27). Diffusion mechanism holds between the axis $x = 0$ and the vertical line, hopping on the right hand side of it.

sidering the asymptotic behaviour of equation (4.3.22) at $x \ll 1$ and $x \gg 1$, one gets two corrections to the diffusion-controlled radius:

$$R_{\text{eff}}^{\text{d}} = r_0 \ln \frac{\sigma_0 \tau r_0^2 \gamma^2}{\lambda_0^2}, \qquad (4.3.24)$$

namely:

$$R_{\text{eff}}^{(1)} \approx R_{\text{eff}}^{\text{d}} - \frac{\lambda_0^2}{6 r_0} + O\left(\frac{\lambda_0^3}{r_0^2}\right) \qquad (4.3.25)$$

and

$$R_{\text{eff}}^{(2)} \approx R_{\text{H}} - \lambda_0, \qquad r_0 \ll \lambda_0 \ll R_{\text{H}}, \qquad (4.3.26)$$

respectively. R_{H} is given by equation (4.3.12). As $\zeta/x \ll 1$, equation (4.3.23) yields hopping-controlled radius

$$R_{\text{eff}}^{(3)} \approx \frac{R_{\text{H}}^3}{3 \lambda_0^2} \left(1 - 0.6(R_{\text{H}}/\lambda_0)^2\right), \qquad R_{\text{H}} \ll \lambda_0. \qquad (4.3.27)$$

Figure 4.10 shows that two approximations, equations (4.3.22) and (4.3.23), become close in the middle of the intermediate region and both are approaching asymptotically equation (4.3.26). Note that the case of multipole reactions, equation (4.1.44), was considered in detail in [81]. The case of *combined* defect hopping with two different lengths has been treated by Kipriyanov and Karpushin [84].

4.3.4 Non-stationary hopping kinetics

The analytical formalism just discussed has two shortcomings: first, the usage of quite particular hop length distribution and, secondly, the restriction to the steady-state properties. The Torrey model becomes inadequate for point defects in crystals, where single hop lengths λ between the nearest lattice sites takes place, $\varphi(r) = \frac{1}{4\pi\lambda^2} \delta(r-\lambda)$ in equation (4.3.4). This results in the integral equation (4.3.1) for the correlation function of dissimilar defects,

$$\widehat{L} Y(r,t) = \frac{1}{\tau}\left(\frac{1}{2r\lambda}\int_{|r-\lambda|}^{r+\lambda} Y(r',t) r' \, dr' - Y(r,t)\right). \qquad (4.3.28)$$

In order to analyze the *temporal* process of steady-state formation in a course of tunnelling reactions in crystals, Kotomin [85] solved numerically equation (4.3.28) and the relevant equation (4.1.19) for the reaction rate $K(t)$. It is clearly seen in Fig. 4.11 that the steady-states formed after the transient

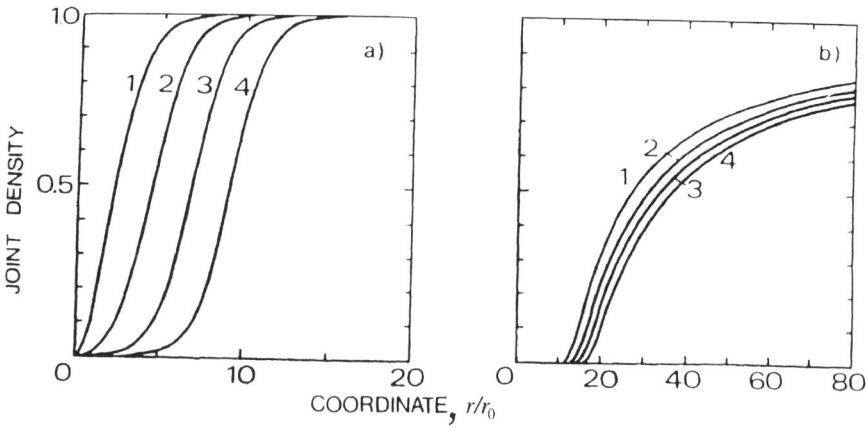

Fig. 4.11. The stationary correlation functions of dissimilar reactants for long hops (a), equation (4.3.9) (curves 1 to 4: $\sigma_0 = 10^1, 10^2, 10^3, 10^4$) and short hops (b), equation (4.2.11) (curves 1 to 4: $\lambda/r_0 = 0.5, 0.2, 0.1, 0.05$, $\sigma_0 = 10^4$).

period in two limiting cases of long hop lengths (a) and small hop lengths (b) are quite different.

Figure 4.12 shows how the steady-state is formed from the random initial distribution for *large* hop lengths. For $\lambda = 20r_0$ the recombination profile is reached very quickly and coincides with the asymptotic hopping solution. As λ decreases, the profile becomes more and more smooth, with a tail extending to greater distances. When a hop length reaches a value close to $\lambda \approx r_0$ (Fig. 4.12(d)), the profile differs greatly from one typical for the hopping regime (Fig. 4.11(a)), being similar to the case of continuous

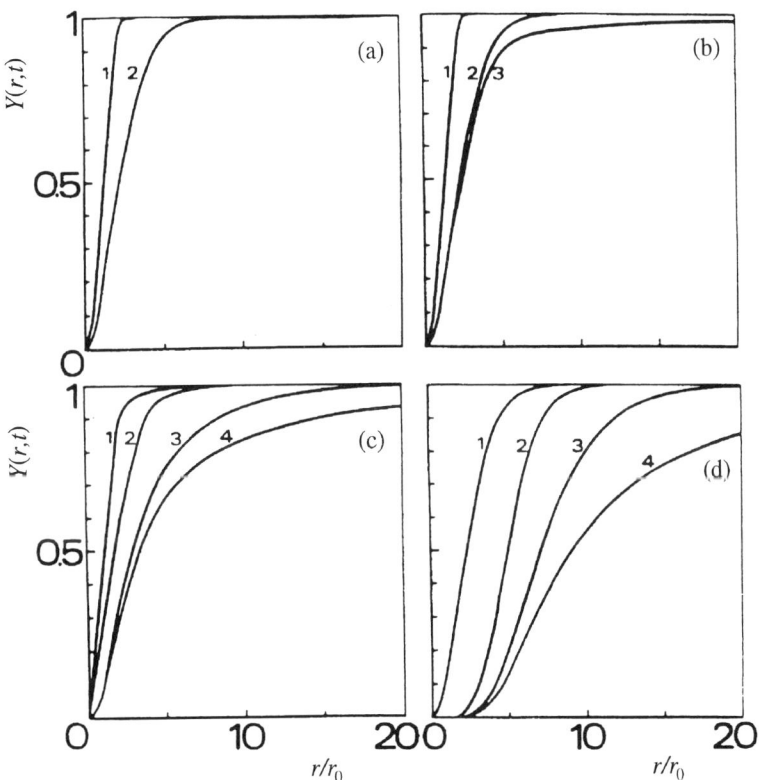

Fig. 4.12. The time development of the correlation function for large hop lengths. (a) $\lambda = 20r_0$, curves 1, 2: $t/\tau = 0.1, 10$. Curve 2 gives equation (4.3.9). (b) $\lambda = 10r_0$, curve 1, 3: $t/\tau = 0, 1, 100$. Curve 2 – equation (4.3.9). (c) $\lambda = 5r_0$, curves 1 to 4: $t/\tau = 0.1, 0.5, 10, 500$. (d) $\lambda = r_0$, curve 2 to 4: $t/\tau = 10, 100, 2000$. Curve 1 – equation (4.3.9).

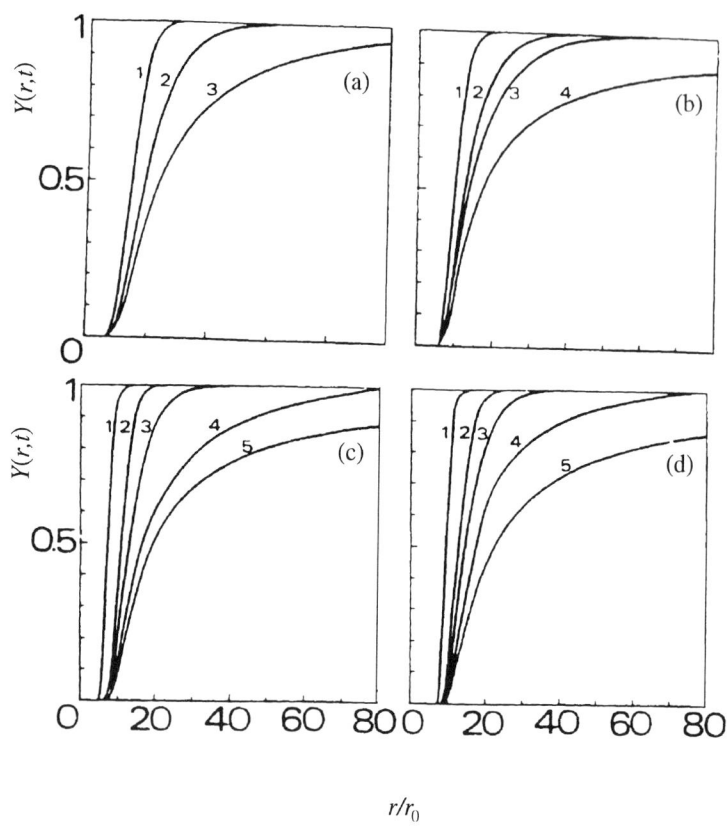

Fig. 4.13. Same as Fig. 4.12 for short hop lengths, $\sigma_0 = 10$. (a) $\lambda = 0.5r_0$. Curves 1 to 3: $t/\tau = 10^1, 10^2, 10^4$; (b) $\lambda = 0.2r_0$. Curves 1 to 3: $t/\tau = 10^3, 10^4, 3 \times 10^4$. Curve 4 – the continuous diffusion approximation, equation (4.2.11). (c) $\lambda = 0.1r_0$. Curves 1 to 4: $t/\tau = 100, 500, 2 \times 10^4, 8 \times 10^4$. Curve 5 – equation (4.2.11). (d) $\lambda = 0.05r_0$. Curves 1 to 4: $t/\tau = 10^3, 2 \times 10^4, 9 \times 10^4, 10^6$. Curve 5 (steady-state) is reached at $t \approx 5 \times 10^6$ only.

diffusion, plotted in Fig. 4.11(b). In particular, the reaction profile is shifted to larger distances and Y tends to zero at $r \leqslant 4r_0$.

This trend continues as λ decreases further (Fig. 4.13). At last, for small $\lambda < 0.05r_0$ the reaction profile practically coincides with the limiting case of *continuous diffusion*, equation (4.3.11). It is seen in Fig. 4.14 how fastly the transient time required to reach the steady-state increases with decreasing λ. For $\lambda > r_0$ this happens very rapidly, whereas for small λ (continuous diffusion) it goes very slowly.

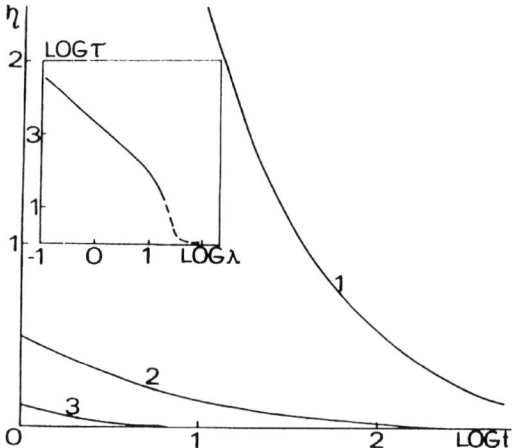

Fig. 4.14. The relative deviation from the steady-state, $\eta = K(t)/K_0 - 1$ vs time [85]. The curves 1 to 3 correspond to $\lambda = 1r_0, 5r_0$ and $10r_0$, respectively. The insert represents the transient time τ vs. λ.

Qualitatively speaking, as the hop length tends to zero, the transient time can be estimated in the spirit of equation (4.2.21) as

$$t_{tr} \approx \frac{5R_{\text{eff}}^2}{D} = \frac{30\ln^2\left(6\sigma_0 r_0^2 \tau/\lambda^2\right)\tau}{\lambda^2} \tag{4.3.29}$$

which would yield a straight line with a slope of -2 in the coordinates $\log t_{tr}$ vs. $\log \lambda$. In fact, the calculated slope (see insert in Fig. 4.14) is smaller than predicted. For $\lambda < r_0$ they differ by a factor of 2 but for $\lambda = 5r_0$ the calculated $t_{tr} = 200\tau$ already greatly exceeds $t_{tr} = 5\tau$ coming from equation (4.3.29).

As it follows from equation (4.3.29), in another extreme case, $\lambda \to \infty$, t_{tr} tends to τ, as it is indicated by a broken line in the insert of Fig. 4.14. It is also demonstrated in [85] that the continuous diffusion approximation, equation (4.1.63), gives quite reasonable reaction rates up to $\lambda \leqslant 5r_0$. This comes from the fact that $K(t)$ is a convolution of the correlation function and the exponentially decaying reaction probability $\sigma(r)$, that is, the essential deviation of the reaction profile from the diffusion limit does not affect the reaction rate considerably.

4.3.5 Illustrative examples: doped alkali halides and silica glasses

The results obtained in these computer simulations of the hopping reactions were applied in [86] to the V_k centre recombination in KCl–Tl^0 stimulated by step-like temperature increase. As it is clearly seen from Fig. 4.15, taking directly into account *finite* hop lengths ($\lambda = a_0/(2\sqrt{2}) \approx 2.2$ Å to be compared with $r_0 \lesssim 1$ Å of the electron Tl^0 centre) permits us to obtain a much better agreement with the experimental data than the standard continuous diffusion approximation (curves 1 and 3, respectively).

As it was shown above, the static kinetics of tunnelling luminescence is much simpler than the diffusion-controlled kinetics. The static tunnelling luminescence intensity often obeys the empirical *Becquerel's law* [87] $I(t) \propto t^{-\alpha}$, where the distinctive decay parameter $\alpha = -d\log I(t)/(d\log t)$ is defined by the spatial distribution of defects only, usually considered to be either the isolated pairs of spatially well-correlated dissimilar defects (low dose irradiation) or the random mixture of dissimilar defects (high doses and/or high temperatures) [88]. Moreover, in the case of pairwise distribution, the *partial lightsum method* has been presented [88–91] in order to restorate the defect initial spatial distribution $f(r)$ within geminate pairs – see equation (4.2.1) and below. What we have discussed here are the *tran-*

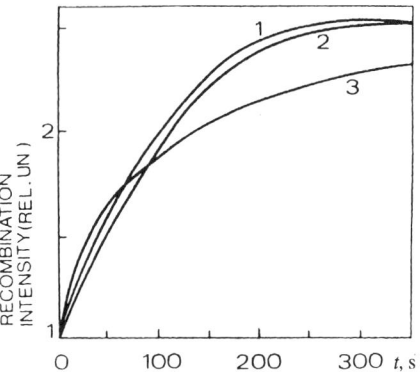

Fig. 4.15. A comparison of experimental delayed kinetics of an increase of V_k tunnelling luminescence intensity after sudden change of their mobility (temperature increase from 175 to 180 K) in KCl with theory [86]. 1 – hopping kinetics for $\lambda = 2r_0$ obtained by means of equation (4.4.1), 2 – experimental curve, 3 – results of continuous diffusion approximation ($\lambda \approx 0$), equation (4.2.11).

sient effects in the kinetics of diffusion-controlled tunnelling recombination. However, only few papers have dealt with the *transient* recombination kinetics when the *static* regime is changed for the diffusion-controlled regime [51] and vice versa [36, 92, 93] after step-like temperature increase (decrease) or photo-excitation switch-on (-off).

The influence of excitation temperature on the the α-value parameter for tunnelling luminescence kinetics is described in [94].

It has first been observed by Tale and Gailitis [51] that after a sudden temperature change the tunnelling luminescence increase is *delayed* due to an inertial diffusive motion of defects; the relevant theory was developed quite recently [55] and discussed above. On the other hand, when the defects are again frozen, a partial lightsum method can be used demonstrating clearly [95] that the concentration of close dissimilar defects has been *enriched* due to their preceding diffusive approach.

The experimental kinetics of the transient tunnelling luminescence observed for alkali halides and sodium-silicate glasses in a temperature increase-decrease cycle have been presented and discussed recently [92, 93] with a special emphasis to distinguish *two alternative cases* – when the tunnelling luminescence kinetics is controlled by *anisotropic defect* rotation and by a thermally-stimulated *motion*, starting from the tunnelling luminescence experimental peculiarities. The conclusion was drawn from their analysis [93] that in the particular case of the sodium-silicate glasses the electrons undergo thermally stimulated hops below the mobility edge characterized by the activation energy $\sim k_B T$.

The end of the present Section aims both to summarize the just mentioned peculiarities of the non-steady-state transient kinetics of the tunnelling luminescence due to step-wise temperature changes, and to develop the theoretical basis for distinguishing two alternative reasons for the tunnelling luminescence temperature dependence: thermally activated *defect diffusion or rotation*.

There are *three* reasons why the temperature change can affect the tunnelling luminescence of radiation defects in wide-gap insulators characterized by a strong electron–phonon interaction:

i) The higher-lying vibrational states become populated with the temperature increase which is known to result in broadening and shift of the luminescence peak as well as in its thermal quenching;

ii) The above-mentioned population of vibration levels of electron and hole defects involved in the recombination can increase the *probability* of tunnelling an electron transfer, as in equation (3.1.2) which, generally speaking, is temperature dependent;

iii) If one of the defects is *anisotropic* (e.g., V_k or Ag^{2+} centres in alkali halides), its *rotation* can increase the recombination probability. Another reason discussed above is the defect diffusion which stimulates their mutual approach.

If the first reason is the case, the slope of the curve $\log I(t)$ vs. $\log t$ must remain unchanged after the temperature increase, and when the temperature is decreased, the curve stands back in line again with that seen before the temperature change, as in Fig. 4.16(a); the defect distribution function is *not* affected due to the temperature cycle. On the other hand, the tunnelling luminescence spectrum is changed, even if the total lightsum remains the same, and the wings of the luminescence peak increase whereas the intensity near its maximum decreases with the temperature, respectively.

In a case when either the second or the third reason takes place, the spatial distribution function will be *irreversibly* changed. The F', V_k pair can be considered as an example in $KCl-SO_4$ crystals (F' is F centre with a trapped electron), which reveals the tunnelling luminescence peak at 3.9 eV [96].

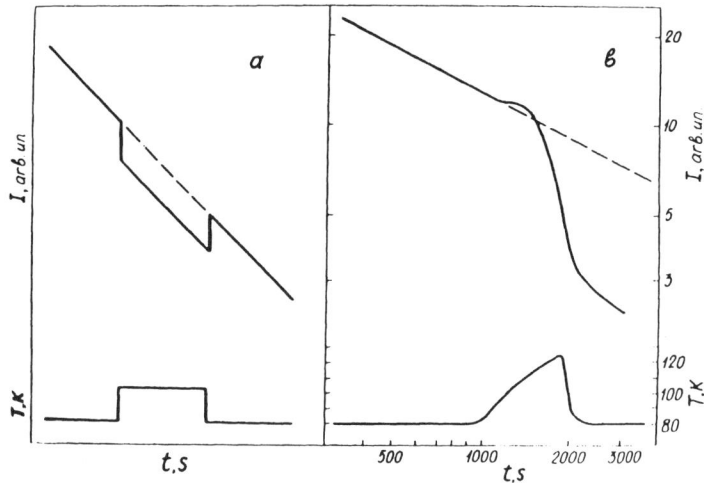

Fig. 4.16. (a) Reversible change of the tunnelling luminescence kinetics due to the step-like temperature stimulation (schematicaly). (b) The tunnelling luminescence kinetics for F, V_k defects in $KCl-SO_4$ due to the temperature stimulation after X-ray excitation at 125 K [106].

Here, heating the samples above 120 K leads to the release of H centres earlier trapped by the substitutional Na^+ ions (H_A centres [97]). These mobile H centres are believed to destroy F' centres which leads to the irreversible decrease of the I starting from the very moment of the temperature increase (Fig. 4.16(b)): after the temperature again decreases, the luminescence intensity *does not* stand back in line with the basic *reference dependence* of $I(t)$ vs. t observed before the temperature cycle. The influence of excitation time τ duration upon the tunnelling luminescence decay kinetics has been analysed in [88]. It is shown that after $t \geqslant 3\tau$, $I(t) \propto t^{-\alpha}$ holds quite well. This is why in actual experiments the temperature stimulation was only imposed only at $t \geqslant 800$ s.

As an example of the second effect – the anisotropic defect *rotation* – the Ag^{2+} centre in KCl-Ag crystal could be used. The tunnelling luminescence of Ag^0, Ag^{2+} pairs peaking at 2.2 eV is well known [98, 99] and this centre below 50 K possesses the tetragonal distortion due to the Jahn–Teller effect [70, 98]. Above 50 K the dynamical Jahn–Teller effect occurs and effect for Ag^+ centre between three equivalent site positions. It is known [98] that the Ag^{2+} symmetry axis reorientation with respect to the Ag^0, Ag^{2+} direction leads to the considerable change of their recombination probability, which is accompanied by the change of a *sign of the tunnelling luminescence polarization*. Note here that such polarization has also been observed for the quasi-molecular V_k centres (Fig. 3.1) entering Tl^0, V_k pairs [70], Ag^0, V_k [70, 100], Na^0, V_k [71] and F, V_k [96] pairs (see also [101]). However, in that case – unlike Ag^{2+} – we *cannot* distinguish between V_k rotation and hopping.

The effect of a sudden temperature change upon the tunnelling luminescence kinetics for Ag^0, Ag^{2+} pairs in KCl is plotted in Fig. 4.17. At a low temperature (8 K), the pairs where the Ag^{2+} symmetry axis is close to being perpendicular to the inter-defect axis recombine predominantly (an initial stage of the curve 1a). After they have been recombined, close "wrongly-oriented" Ag^0, Ag^{2+} pairs with low tunnelling probability still exist. As the temperature increases up to 60 K, Ag^{2+} centres begin to reorient themselves and the surviving close Ag^0, Ag^{2+} pairs disappear in their turn. It is important to note that the observed intensity increase is practically *inertial* less, i.e., $I(t)$ follows *immediately* the temperature change. Quite on the contrary, the temperature decrease to 8 K is accompanied by a *smooth* $I(t)$ decrease. The $I(t)$ now lies below the reference line extrapolated from the initial stage of the static tunnelling luminescence decay (the broken line in Fig. 4.17)

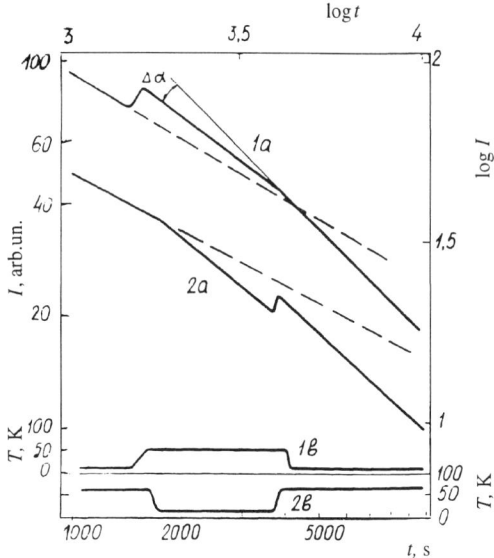

Fig. 4.17. The tunnelling luminescence kinetics for Ag^0, Ag^{2+} defects in KCl-Ag (curves 1a, 2a) due to the temperature stimulation cycle shown in curves 1b, 2b [106]. The excitation temperature is 280 K.

since a fraction of Ag^{2+} centres is frozen in the configuration preventing their further tunnelling recombination.

Let us now analyze, following Rogulis and Kotomin [102], the experimental non-stationary tunnelling luminescence kinetics observed after a sudden temperature change for V_k, F pairs in KBr, Ag^0 and V_k in KCl-Ag as well as for an electron thermal release from Ag^0 (Ag^0 and Ag^{2+} in NaCl-Ag).

Figure 4.18 shows the kinetics of the tunnelling luminescence growth after the step-like temperature increase observed for F, V_k centres in pure KBr crystal (for the static decay the kinetics is described well by the Becquerel's law with $\alpha \approx 0.8$). Unlike the case of rotation, Fig. 4.17, it is obviously *delayed* with the distinctive τ which becomes shorter, as temperature is raised (for more details see [18, 51]). This delay arose directly due to the transient period resulting in the formation of the quasi-steady-state recombination profile. (The same behaviour has been observed for Ag^0, V_k centres in KCl-Ag).

This is not the case for Ag^0 centre *ionization* in NaCl-Ag where no *delay* was observed since the thermally released electron practically *instantly* recombines the Ag^{2+} centre. The decay kinetics after temperature decreases

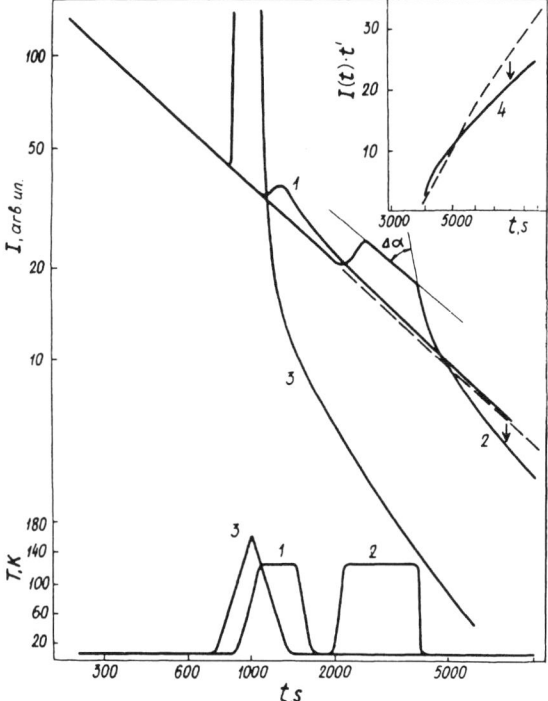

Fig. 4.18. The tunnelling luminescence kinetics for F, V_k pairs in KBr (upper curves 1, 2, 3) vs. temperature stimulation (lower curves 1, 2, 3 respectively). α is the slope change after the stimulation is switched off. According to the partial lightsum method [37, 87–90] the insert (curve 4) yields the relative dissimilar defect distribution occurring after the temperature stimulation. Arrows show the change of both the tunnelling luminescence decay intensity $I(t)$ and the spatial defect distribution caused by the stimulation; a close defect concentration increases whereas that for the distant pairs decreases.

is also inertial, which permits us to distinguish thermal ionization from the rotation. The relevant $I(t)$ slope change during stimulation as compared with the static decay law could be *both positive and negative* depending on the number of factors – e.g., the defect Coulomb interaction, the initial spatial distribution, the diffusion coefficient, etc.

As the temperature decreases suddenly, the $I(t)$ falls below the reference straight line observed for the diffusion-controlled recombination. The distinctive feature of the transient kinetics is the great slope change: $\Delta\alpha \geqslant 0.5$ (see Fig. 4.18, curve 2).

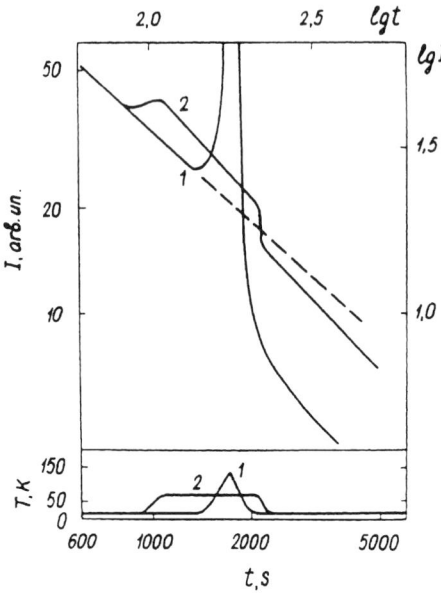

Fig. 4.19. The tunnelling luminescence kinetics for the E_1^-, Tb^{3+} pairs in $Na_2O \cdot 3SiO_2\text{-}Tb^{3+}$ glass. The ultraviolet excitation was at 120 K with further short thermal excitation of electrons above the mobility edge (curve 1) and prolonged heating up to 70 K below the excitation threshold (curve 2) [93].

The influence of the temperature stimulation upon the tunnelling luminescence of the sodium-silicate glasses has also been analyzed in [93] in order to distinguish between two alternative models of the thermostimulated tunnelling luminescence: the direct tunnelling electron transfer between two defects, and electron intermediate hops over intrinsic glass traps. To answer this question, the tunnelling kinetics for $Na_2O \cdot 3SiO_2\text{-}Tb^{3+}$ was measured using the temperature stimulation cycle as described above. The tunnelling recombination of the intrinsic E_1^- centre with the $(Tb^{3+})^+$ impurity results in the distinctive Tb^{3+} luminescence narrow lines [103]. The conclusion can be drawn from Fig. 4.19 that for both cases – short thermal excitation of electrons above the mobility edge (curve 1) and the prolonged sample maintaining at 70 K when thermal excitation above the mobility edge is suppressed (curve 2) – the slope of the tunnelling luminescence decay curves (being plotted in the $\log I(t)$ vs. $\log t$ coordinates) *increases* similarly with respect to the reference line. An estimate of the spatial defect distribution by means of the partial lightsum method (see below) indicates a *decrease* in the number of far E_1^-, $(Tb^{3+})^+$ pairs in relative comparison with the close pairs. Therefore, two similar kinetics argue that the electron thermally-activated hopping over traps characterized by the activation energy $\sim k_B T$ takes place during sample heating. This conclusion is also supported by the experimental fact that after the sample was shortly heated up to 90 K, the tunnelling luminescence intensity exceeded (during several hundred of seconds) the $I(t)$ magnitude extrapolated from the initial stage of the decay law $I(t)$, which signifies an increase of the concentration of *close* E_1^-, $(Tb^{3+})^+$ pairs [93].

4.3.6 Effects of defect anisotropy

The influence of *anisotropy* of acceptor wavefunction upon tunnelling luminescence kinetics was treated in [104]. The conclusion was drawn that for the *static* tunnelling luminescence it just results in the redefinition of the σ_0 parameter. However, we are interested here in the *non*-steady-state kinetics and shall demonstrate below that, particularly at this stage, anisotropic recombination reveals distinctive behaviour which allows us to identify it.

Since the actual angular dependence of the $\sigma(r)$ is rather complicated (e.g., [105]), Rogulis [106] has used the following simple approximation

$$\sigma(r) = \sigma_0(1 + b\cos\varphi)\exp(-r/r_0), \qquad (4.3.30)$$

where $b \geqslant 0$. φ is an angle between the inter-defect direction and the axis perpendicular to the acceptor symmetry axis (e.g., Ag^{2+} centre).

Assuming that the *initial* dissimilar defect distribution over distances r and angles φ for isolated pairs of correlated defects is random, authors [102, 106] have presented kinetics of luminescence decay controlled by *anisotropic tunnelling* recombination due to defect rotation. Curves 1 and 2 in Fig. 4.20 demonstrate the calculated effects of the thermally-stimulated reorientation – the temperature switching-on and -off respectively. The magnitude of the parameter b entering equation (4.3.30) was optimised in curve 1, Fig. 4.20, to get the best fitting to the experimental kinetics during and after thermal stimulation (both the $I(t)$ increase and the slope α). The typical magnitude of the slope change after the stimulation is switched off is $\Delta\alpha \leqslant 0.2$.

Kinetics of the tunnelling recombination depends greatly upon the defect *mobility* (whether a static tunnelling luminescence regime at low temperatures or the diffusion-controlled regime arising at higher temperatures when defects become mobile) and their spatial distribution.

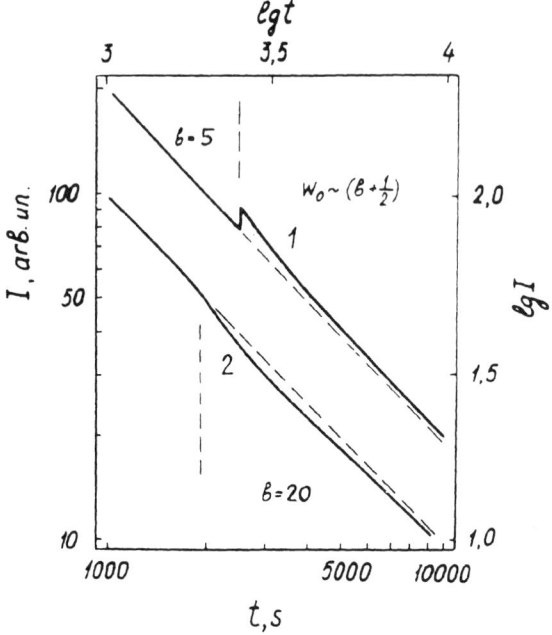

Fig. 4.20. The influence of the tunnelling recombination anisotropy upon the non-steady-state luminescence kinetics. Curves 1 and 2 correspond to the defect reorientation switching-on and -off, respectively at the moments given by broken lines [106].

For the static tunnelling luminescence decay and correlated defect distribution the joint density of dissimilar defects is governed by equations (4.1.40) to (4.1.42).

The conclusion was drawn above that for the isotropic case the *exponentially falling defect* distribution results in $I(t) \propto t^{-(1+r_0/R)}$, as in equation (4.2.4). Thus, for the distribution *increasing* with r (at least within the limits of short distances), $I(t)$ is expected to decrease more slowly than $I(t) \propto t^{-1}$. Both cases more than once were observed experimentally [18, 32, 34, 39, 43, 87].

Let us consider now the tunnelling luminescence kinetics for the *transient* processes when the temperature stimulation is suddenly switched on and off, respectively. This is different from the transient kinetics treated above, which arises due to an *inertial* passage between two quasi-steady states corresponding to different temperatures.

Non-diffusion defect recombination and non-stationary kinetics 227

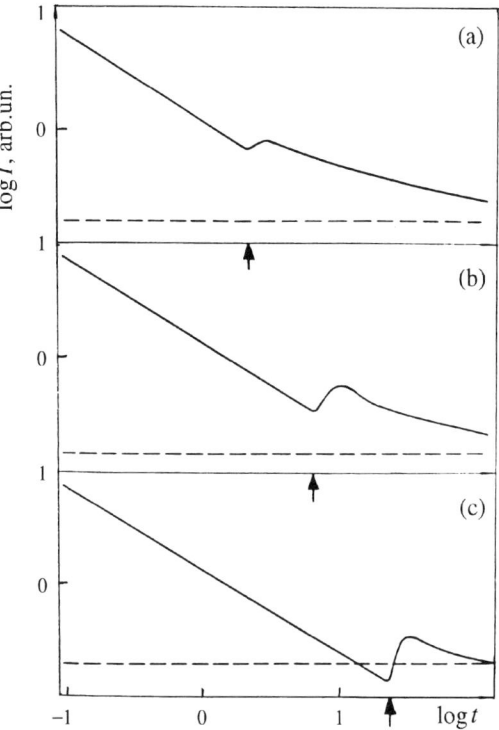

Fig. 4.21. The computer simulation [102] demonstrating how the temperature stimulation affects the initial stage of the diffusion-controlled tunnelling luminescence. The cases a, b, c, correspond to three subsequent moments $t_1 < t_2 < t_3$ of the switching-on the stimulation (diffusion). The dashed lines are corresponding $I(t)$ for the quasi-stationary recombination, provided the defect concentration changes negligibly.

Figure 4.21 shows the diffusion kinetics (equations (4.1.18), (4.1.19) and (4.1.23)) of an initial stage of the thermal stimulation made after a static tunnelling decay for a certain time; in all cases one can see the luminescence *increase*. It is greater, the closer the luminescence intensity at the instant stimulation moment to its quasi-steady value (broken line). Calculations were carried out assuming that the temperature stimulation yields relatively weak perturbation, i.e., the defect concentration $n(t)$ in equation (4.1.18) changes much more slowly than the reaction rate $K(t)$ and therefore $I(t) \propto K(t)$.

This tunnelling luminescence increase (well pronounced on the experimental curves – Figs 4.18, 4.19) originates from the peculiarities of the

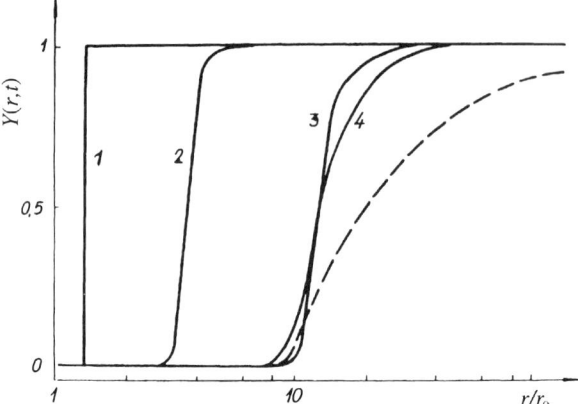

Fig. 4.22. The time-development of the recombination profile of dissimilar defects [102]. Curves 1, 2, 3 show the effect of static tunneling recombination; 4 – a short time after the diffusion has started; broken curve – the profile as $t \to \infty$.

time-development of the recombination profile shown in Fig. 4.22. The monotonous shift with time of the static recombination profile to greater distances (curves 1 and 3) is *hampered* when reactants begin to diffuse. $Y(r,t)$ even *increases* at short distances where $Y(r,t) \leqslant 0.2$ for a certain time (curve 4) then striving to the steady-state profile (the broken curve). If the defect concentration should fall slowly, $I(t)$ tends with time to the constant magnitude, $I(t) \propto K_0 \cong 4\pi D R_{\text{eff}}$. However, this *is not* the case for the experiments shown in Fig. 4.19, likely either due to a considerable decrease of defect concentrations or due to recombination within correlated pairs. (It should be reminded that in the case $\bar{d} = 1$, $Y(r,t)$ is *never* constant, but *always* shifts with time.)

As the temperature stimulation is switched off, the static kinetics is governed by equation (4.1.40) with the initial distribution function $y(r)$ from equation (4.2.11). However, all attempts [102] to describe in such a way the experimental tunnelling luminescence decay for F and V_k in KBr (Fig. 4.18) were unsuccessful. Both this observation and the absence of the plateau of $I(t)$ during the temperature stimulation, characteristic for the quasi-steady states, argue that the tunnelling *recombination takes place in correlated pairs*. This is in line with the conclusion [107] that for ordinary defect concentrations $\leqslant 10^{16}$ cm^{-3} (X-ray sample excitation for minutes) and the time $\leqslant 10^5$ s the slope is dose-independent but $I(t) \propto$ dose [95].

In the *method of partial lightsums* the lightsums $S(t)$ of the tunnelling luminescence emitted during logarithmically increasing time intervals are measured instead of the ordinary luminescence intensity $I(t)$ which allows us to estimate with the help of equations (4.2.3) and (4.2.6) the initial defect distributions

$$S(t) \propto I(t)t \propto f\left[r_0 \ln(\sigma_0 t)\right]. \qquad (4.3.31)$$

That is, an observation of the emitted luminescence $I(t)$ in a wide time interval permits us to restore the defect distribution $f(r)$. Equation (4.3.31) has been used for finding the distribution within F, V_k pairs in KBr when the temperature stimulation is finished [102, 106]. At the beginning the $I(t)$ exceeds the reference straight line (curves 2 and 4 in Fig. 4.18), which corresponds to the *enriched* concentration of closely spaced defects, whereas the further $I(t)$ dip below the reference line means a *decrease* of the number of well-separated defects. As the temperature is increased up to to 135 K for a short time (curve 1 in Fig. 4.18)), the experimental curve lies above the reference line at all times. For a strong temperature stimulation up to 170 K, even for a short time, we generally observe a great reduction in a number of *all* pairs, especially distant (curve 3 in Fig. 4.18).

Therefore, the method of partial lightsums illustrates once more effects of the defect diffusion: a number of close pairs increases, whereas that of distant pairs decreases, *unlike the case of defect rotation*.

The spatial redistribution acts to increase considerably the slope in the coordinates $\log I(t)$ vs. $\log t$, leading to the typical value $\Delta\alpha \geqslant 0.5$. On the other hand, there is *no* such defect redistribution over distances if the defect reorientation is operative, which leads to the *much smaller* change of the slope, $\Delta\alpha \leqslant 0.2$.

To conclude this Section, we would like to stress that both experimental and theoretical analyses of the *non-steady-state kinetics* of the tunnelling luminescence of defects in insulators after the step-like stimulation allow us to *distinguish* the anisotropic defect rotation and diffusion. For the defect rotation, sharp increases of the $I(t)$ and its smooth decrease are observed for the temperature stimulation cycle, whereas an *opposite* effect occurs for the defect diffusion.

In the case of *thermal ionization* of the electronic centre, e.g., Ag^0, both the increase and the decrease of the $I(t)$ are very sharp due to the inertial character of the released electron motion. At last, the electron migration over traps in a glass is characterized by a sharp $I(t)$ increase and by a decay slope greater than for diffusion. An additional analysis based on the method of partial lightsums allows to conclude that the defect spatial redistribution from random to close pairs takes place here.

References

[1] V.V. Antonov-Romanovskii, Kinetics of Photoluminescence of Crystal Phosphors (Nauka, Moscow, 1966).
[2] H. Eyring, S.H. Lin and S.M. Lin, Basic Chemical Kinetics (Wiley, New York, 1980).
[3] A.I. Onipko, in: Physics of Many-Body Systems, Vol. 2 (Naukova Dumka, Kiev, 1982) p. 60.
[4] G. Leibfried, in: Bestrahlungseffekte in Festkörpern (Teubner, Stuttgart, 1965) p. 266.
[5] V.N. Kuzovkov and E.A. Kotomin, Chem. Phys. 81 (1983) 335.
[6] Yu. Kalnin and F. Pirogov, Phys. Status Solidi B: 84 (1977) 521.
[7] V.V. Antonov-Romanovskii, Tr. Fiz. Inst. Akad. Nauk SSSR 117 (1980) 55 (Proc. Lebedev Phys. Inst. Acad. Sci. USSR, in Russian).
[8] T.R. Waite, Phys. Rev. 107 (1957) 463.
[9] T.R. Waite, Phys. Rev. 107 (1957) 471.
[10] T.R. Waite, J. Chem. Phys. 28 (1958) 103.
[11] T.R. Waite, J. Chem. Phys. 32 (1960) 21.
[12] Yu.H. Kalnin, Latv. PSR Zinat. Akad. Vestis Fiz. Teh. Zinat. Ser. 3 (1977) 70 (Proc. Latv. Acad. Sci. Phys. Technol. Ser., in Russian).
[13] Yu.H. Kalnin, Latv. PSR Zinat. Akad. Vestis Fiz. Teh. Zinat. Ser. 5 (1982) 3 (Proc. Latv. Acad. Sci. Phys. Technol. Ser., in Russian).
[14] E.A. Kotomin and V.N. Kuzovkov, Teor. Eksp. Khim. 18 (1982) 274 (Sov. Theor. Exp. Chem., in Russian).
[15] E.A. Kotomin and V.N. Kuzovkov, Chem. Phys. 76 (1983) 479.
[16] R. Noks, Theory of Excitons (Academic Press, New York, 1966).
[17] V.M. Agranovich and M.D. Galanin, Modern Problems in Condensed Matter Science, Vol. 3: Electron Excitation Energy Transfer in Condensed Matter (North-Holland, Amsterdam, 1982).
[18] A.B. Doktorov and E.A. Kotomin, Phys. Status Solidi B: 114 (1982) 9.
[19] K. Tanimura and N. Itoh, J. Phys. Chem. Solids 42 (1981) 901.
[20] U. Gösele and F.A. Hantley, Phys. Lett. A 55 (1975) 291.
[21] L. Onsager, J. Chem. Phys. 2 (1934) 599.
[22] P. Debye, J. Electrochem. Soc. 32 (1942) 265.
[23] E.W. Montroll, J. Chem. Phys. 14 (1946) 202.
[24] A. Mozumder, J. Chem. Phys. 48 (1968) 1659.
[25] I.L. Magee and A.B. Tayler, J. Chem. Phys. 56 (1972) 3061.
[26] G.C. Abell and A. Mozumder, J. Chem. Phys. 56 (1972) 4079.
[27] A.B. Doktorov, A.A. Kipriyanov and A.I. Burstein, Sov. Opt. Spectr. 45 (1978) 497.

References

[28] A.B. Doktorov, A.A. Kipriyanov and A.I. Burstein, Sov. Opt. Spectr. 45 (1978) 684.
[29] A.B. Doktorov, A.A. Kipriyanov and A.I. Burstein, Sov. Phys. JETP, 47 (1978) 623.
[30] V.N. Kuzovkov and E.A. Kotomin, Chem. Phys. 98 (1985) 351.
[31] K.M. Hong and J. Noolandi, J. Chem. Phys. 69 (1987) 5026.
[32] S.A. Rice and M.J. Pilling, Prog. React. Kinet. 9 (1978) 93.
[33] E.A. Kotomin and A.B. Doktorov, Phys. Status Solidi B: 114 (1982) 287.
[34] K.I. Zamaraev, R.F. Khairutdinov and V.P. Zhdanov, The Comprehensive Chemical Kinetics, Vol. 30: Long-Range Electron Tunnelling in Chemistry (North-Holland, Amsterdam, 1989).
[35] Ch. Lushchik and A. Lushchik, Decay of Electronic Excitations into Defects in Solids (Nauka, Moscow, 1989).
[36] I.K. Vitol and V.Ya. Grabovskis, in: Abstr. Eur. Top. Conf. on Defects in Ionic Crystals, Marseille, 1973, p. 198.
[37] I.K. Vitol and V.Ya. Grabovskis, Bull. Acad. Sci. USSR Phys. Ser. 38 (1974) 1223.
[38] D. Peak and J.W. Corbett, Phys. Rev. B: 5 (1972) 1226.
[39] I. Fabrikant and E. Kotomin, J. Lumin. 9 (1975) 502.
[40] M.N. Kabler, in: Point Defects in Solids (1), ed. J. Crawford (Plenum, New York, 1972) Ch. 4.
[41] E.A. Kotomin, I.I. Fabrikant and I.A. Tale, J. Phys. C: 10 (1977) 2903.
[42] Yu.A. Berlin, Dokl. Akad. Nauk SSSR 223 (1975) 625.
[43] R.F. Khairutdinov, Dokl. Akad. Nauk SSSR 228 (1976) 149.
[44] M.J. Pilling and S.A. Rice, J. Chem. Soc. Faraday Trans. II 71 (1975) 1311; 17 (1975) 1563.
[45] M.J. Pilling and S.A. Rice, J. Chem. Soc. Faraday Trans. II 72 (1976) 792.
[46] I. Fabrikant and E. Kotomin, in: Electronic and Ionic Processes in Ionic Crystals 2 (1974) 78; 2 (1974) 93; 2 (1974) 108.
[47] J. Kuba, J. Lumin. 37 (1987) 287.
[48] J. Kuba and B. Sipp, J. Lumin. 33 (1985) 255.
[49] A.A. Gailitis and I.K. Vitol, Bull. Acad. Sci. USSR Phys. Ser. 35 (1971) 1340.
[50] A.A. Gailitis, PhD thesis (University of Latvia, Riga, 1972).
[51] I.A. Tale and A.A. Gailitis, Bull. Acad. Sci. USSR Phys. Ser. 35 (1971) 1336.
[52] E.A. Kotomin, A.I. Popov and R. Eglitis, J. Phys. C: 4 (1992) 5901.
[53] I. Tale, P. Kulis and V. Kronghaus, J. Lumin. 20 (1979) 343.
[54] P. Kulis, PhD thesis (University of Latvia, Riga, 1987).
[55] E.A. Kotomin, I.A. Tale, V.G. Tale, P. Kulis and P. Butlers, J. Phys. C: 1 (1989) 6777.
[56] V.F. Grachev and M.F. Deigen, Sov. Phys. Usp. 125 (1978) 631.
[57] E.A. Kotomin and I.I. Fabrikant, Latv. PSR Zinat. Akad. Vestis Fiz. Teh. Zinat. Ser. 1 (1979) 53 (Proc. Latv. Acad. Sci. Phys. Technol. Ser., in Russian).
[58] E.A. Kotomin and I.I. Fabrikant, Latv. PSR Zinat. Akad. Vestis Fiz. Teh. Zinat. Ser. 3 (1979) 76 (Proc. Latv. Acad. Sci. Phys. Technol. Ser., in Russian).
[59] E.A. Kotomin and I.I. Fabrikant, J. Phys. C: 10 (1977) 4931.
[60] E.A. Kotomin and I.I. Fabrikant, Radiat. Eff. 46 (1980) 85.
[61] K. Schröder, Diffusion-Controlled Reactions of Point Defects, Report 1083-FF, Jülich, 1974.

[62] K. Schröder, Point Defects in Metals (2), Springer Tracts Mod. Phys. 87 (1980) 71.
[63] K.Schröder and K. Dettmann, Z. Phys. B 22 (1975) 343.
[64] N.A. Fuks, Z. Phys. 89 (1934) 36.
[65] E.A. Kotomin, Khim. Phys. 3 (1984) 581 (Sov. Chem. Phys., in Russian).
[66] V.M. Byakov, V.L. Grishkin and A.A. Ovchinnikov, Preprint Inst. Theor. Phys. N40, Moscow, 1977.
[67] L. Spruch and L. Rosenberg, Phys. Rev. A: 12 (1975) 1297.
[68] I. Fabrikant and E. Kotomin, J. Phys. C: 10 (1977) 493.
[69] K. Bachmann and H. Peisl, J. Phys. Chem. Solids 31 (1980) 1525.
[70] C.J. Delbecq, Y. Toyozawa and P.H. Yuster, Phys. Rev. B: 9 (1974) 4497.
[71] K. Imanaka, A.H. Kayal, A.C. Mazgar and J. Rossel, Phys. Status Solidi B: 108 (1981) 449.
[72] W. Scheu, W. Frank and H. Kronmüller, Phys. Status Solidi B: 82 (1977) 523.
[73] U. Gösele, Prog. React. Kinet. 13 (1984) 63.
[74] A. Mozumder, S.M. Pimbott, P. Clifford and N.J.B. Green, Chem. Phys. Lett. 142 (1987) 385.
[75] S.M. Pimblott, A. Mozumder and N.B. Green, J. Chem. Phys. 90 (1989) 6595.
[76] M.H. Yoo and W.H. Butler, Phys. Status Solidi B: 77 (1977) 181.
[77] A.B. Doktorov and B.I. Yakobson, Chem. Phys. 60 (1981) 223.
[78] A.I. Burstein, A.B. Doktorov and V.A. Morozov, Chem. Phys. 104 (1986) 1.
[79] A.B. Doktorov and A.I. Burstein, Sov. Phys. JETP 68 (1975) 1349.
[80] A.I. Burstein, Sov. Phys. JETP 35 (1972) 882.
[81] A.I. Burstein, A.B. Doktorov, V.A. Morozov, A.A. Kipriyanov and S.G. Fedorenko, Sov. Phys. JETP 61 (1985) 516.
[82] H.S. Torrey, Phys. Rev. 52 (1953) 962.
[83] A.B. Doktorov, Physica A 90 (1978) 109.
[84] A.A. Kipriyanov and A.A. Karpushin, Khim. Phys. 7 (1988) 60 (Sov. Chem. Phys., in Russian).
[85] E.A. Kotomin, Dynamic Processes in Condensed Molecular Systems (Word Scientific, Singapore, 1990) p. 414.
[86] E.A. Kotomin, L. Kantorovich, I. Tale and V. Tale, J. Phys. Condens. Matter, 4 (1992) 7429.
[87] I.K. Vitol, A.A. Gailitis and V.Ya. Grabovskis, Uch. Zap. Latv. Gos. Univ. 208 (1974) 16 (Proc. Latv. Univ., in Russian).
[88] A.A. Gailitis, Uch. Zap. Latv. Gos. Univ. 234 (1975) 42 (Proc. Latv. Univ., in Russian).
[89] I.K. Vitol, Uch. Zap. Latv. Gos. Univ. 234 (1973) 42 (Proc. Latv. Univ., in Russian).
[90] I.K. Vitol and A.A. Gailitis, in: Abstr. Eur. Top. Conf. on Defects in Ionic Crystals, Marseille, 1973, p. 197.
[91] V.Ya. Grabovskis and I.K. Vitol, Bull. Acad. Sci. USSR Phys. Ser. 38 (1974) 1223.
[92] I.K. Vitol and U.T. Rogulis, Thermoactivated Spectroscopy of Defects in Ionic Crystals (Latv. Univ. Press, Riga, 1983) p. 83.
[93] V.I. Arbuzov, I.K. Vitol, A.R. Kangro, U.T. Rogulis and M.V. Tolstoy, Fiz. Khim. Stekla 12 (1986) 75 (Sov. Phys. Chem. Glass, in Russian).

References

[94] I.A. Tale, D.K. Millers and A.A. Gailitis, in: Abstr. XIX Sov. Conf. Luminescence, Crystal Phosphors, Riga, 1970, p. 179.

[95] I.K. Vitol, V.Ya. Grabovskis and A.R. Kangro, Electr. and Ionic Processes in Ionic Crystals, Vol. 6 (Latv. Univ. Press, Riga, 1977) p. 82.

[96] V.Ya. Grabovskis and I.K. Vitol, J. Lumin. 20 (1979) 337.

[97] D. Schoemaker, Phys. Rev. B: 3 (1971) 3516.

[98] C.J. Delbecq, D.L. Dexter and P.H. Yuster, Phys. Rev. 17 (1978) 4765.

[99] P.G. Baranov, Yu.P. Veschunov and N.G. Romanov, Sov. Phys. Solid State 22 (1980) 3732.

[100] T. Tashiro, S. Takeuchi, M. Saidoh and N. Itoh, Phys. Status Solidi B: 92 (1979) 611.

[101] D.E. Aboltin, A.U. Grinfeld, E.A. Krivads and V.G. Plekhanov, in: Abstr. 5th Sov. Conf. Rad. Phys. and Chem. Ionic Crystal, Riga, 1983, p. 229.

[102] U.T. Rogulis and E.A. Kotomin, Radiad. Eff. Defects Solids 111/112 (1989) 191.

[103] V.I. Arbuzov, I.K. Vitol, A.R. Kangro, L.B. Popova and M.N. Tolstoy, Fiz. Khim. Stekla 8 (1982) 82 (Sov. Phys. Chem. Glass, in Russian).

[104] A.B. Doktorov, R.F. Khairutdinov and K.I. Zamaraev, Chem. Phys. 61 (1981) 351.

[105] B. Brocklehurst, J. Phys. Chem. 83 (1979) 536.

[106] U.T. Rogulis, PhD Dissertation (Inst. of Physics of Latv. Acad. Sci., Salaspils, 1986).

[107] U.K. Kanders, Uch. Zap. Latv. Gos. Univ. 254 (1976) 57 (Proc. Latv. Univ., in Russian).

Chapter 5

The Fluctuation-Controlled Kinetics: The Basic Formalism of Many-Point Particle Densities

5.1 THE A + B → 0 REACTION

> A detailed study of individual organs makes us unable to understand a life of a whole organism.
>
> V. Klyuchevsky, old Russian historian

5.1.1 The non-linear correlation dynamics

5.1.1.1 Kinetic equations

In this Section we consider again the kinetics of bimolecular A + B → 0 recombination but instead of the linearized approximation discussed above, the *complete* Kirkwood superposition approximation, equation (2.3.62) is used which results in emergence of two new joint correlation functions for *similar* particles, $X_\nu(r,t)$, ν = A, B. The extended set of the correlation functions, $n_A(t), n_B(t), X_A(r,t), X_B(r,t)$ and $Y(r,t)$ is believed to be able now to describe the *intermediate* order in the particle spatial distribution.

Substitution of equation (2.3.62) into a set of equations (4.1.13) to (4.1.16) for noncharged (neutral) particles ($U_{\nu\mu}(r) = 0$) does not affect equations (4.1.18) and (4.1.19) whereas the linear equation (4.1.23) describing the correlation dynamics splits now into *three* integro-differential equations. Main stages of the passage from general equations (4.1.14)–(4.1.16) for the joint densities to those for the joint correlation functions have been demonstrated earlier, see (4.1.20) and (4.1.21). Therefore let us consider only those terms which are affected by the use of superposition approximation. Hereafter we use the relative coordinates $\vec{r} = \vec{r}_1 - \vec{r}_1'$, $\vec{r}'' = \vec{r}_2 - \vec{r}_1'$ and

$\vec{r}^* = \vec{r}_1 - \vec{r}_2 \equiv \vec{r} - \vec{r}'$. The superposition approximation yields

$$\int \sigma\bigl(|\vec{r}_2 - \vec{r}_1'|\bigr)\rho_{2,1}\,\mathrm{d}\vec{r}_2$$

$$= n_A^2(t)n_B(t)Y(r,t)\int \sigma(r')Y(r',t)X_A(r^*,t)\,\mathrm{d}\vec{r}_2$$

$$= n_A^2(t)n_B(t)Y(r,t)\int \sigma(r')Y(r',t)X_A(r^*,t)\,\mathrm{d}\vec{r}'. \tag{5.1.1}$$

When obtaining this final form, the integration variable was changed in (5.1.1).

The *correlation dynamics* could be presented in a form

$$\frac{\partial X_A(r,t)}{\partial t} = 2D_A\nabla^2 X_A(r,t) - 2X_A(r,t)n_B(t)I[Y], \tag{5.1.2}$$

$$\frac{\partial X_B(r,t)}{\partial t} = 2D_B\nabla^2 X_B(r,t) - 2X_B(r,t)n_A(t)I[Y], \tag{5.1.3}$$

$$\frac{\partial Y(r,t)}{\partial t} = D\nabla^2 Y(r,t) - \sigma(r)Y(r,t) -$$

$$- Y(r,t)\sum_{\nu=A,B} n_\nu(t)I[X_\nu]. \tag{5.1.4}$$

For the compactness of equations (5.1.2) to (5.1.4), the functional (spatial convolution) has been introduced above as

$$I[Z] = \int \sigma(r')Y(r',t)\bigl(Z(r^*,t) - 1\bigr)\,\mathrm{d}\vec{r}',$$

$$r^* = |\vec{r} - \vec{r}'|, \quad Z = X_\nu, Y, \tag{5.1.5}$$

arising directly due to Kirkwood's superposition approximation. Similarly to manipulations done in (4.1.50), let us present the unit volume in $\bar{d}$-dimensional space as

$$\mathrm{d}\vec{r}' = r'^{\bar{d}-1}\,\mathrm{d}r'\,\mathrm{d}\Omega_{\bar{d}}, \quad \int \mathrm{d}\Omega_{\bar{d}} = \gamma_{\bar{d}}. \tag{5.1.6}$$

Therefore the functional $I[Z]$ reads

$$I[Z] = \gamma_{\bar{d}} \int \sigma(r')Y(r',t)\left(\widetilde{Z}(r^*,t) - 1\right) r'^{\bar{d}-1} \, dr', \qquad (5.1.7)$$

where $\widetilde{Z}(r^*,t)$ is a function of r, r' and t, obtained from $Z(r^*,t)$ by an angle averaging,

$$\widetilde{Z}(r^*,t) = \frac{\int Z(r^*,t) \, d\Omega_{\bar{d}}}{\int d\Omega_{\bar{d}}}. \qquad (5.1.8)$$

Taking into account that $r^* = (r^2 + r'^2 - 2rr'\cos\vartheta)^{1/2}$, where ϑ is an angle between $\vec{r}$ and $\vec{r}'$, the mean value in equation (5.1.8) could be easily calculated. In doing so we arrive at:

$$\widetilde{Z}(r^*,t) = \frac{1}{2} \sum_{\cos\vartheta=\pm 1} Z(r^*,t)$$

$$= \frac{1}{2}\left(Z(|r-r'|,t) + Z(r+r',t)\right), \quad \bar{d} = 1, \qquad (5.1.9)$$

or

$$\widetilde{Z}(r^*,t) = \frac{1}{\pi} \int_{-1}^{+1} \frac{Z(\zeta,t) \, ds}{(1-s^2)^{1/2}},$$

$$\zeta = (r^2 + r'^2 + 2rr's)^{1/2}, \quad \bar{d} = 2. \qquad (5.1.10)$$

Lastly, for $d = 3$

$$\widetilde{Z}(r^*,t) = \frac{1}{2rr'} \int_{|r-r'|}^{r+r'} Z(\zeta,t) \zeta \, d\zeta. \qquad (5.1.11)$$

The shortened Kirkwood superposition approximation (2.3.64) differs from the complete one, equation (2.3.63), by the additional condition imposed on the correlation function of similar particles; $X_\nu(r,t) = 1$ at *any* time t. Its substitution into equation (5.1.4) and taking into account equation (5.1.5) leads, as one could expect, to the linearized equation (4.1.23) for the correlation dynamics. Therefore, the applicability range of the linearized kinetic

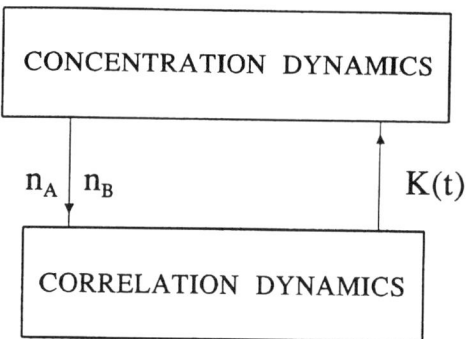

Fig. 5.1. The interrelation between the two kinetics: of concentrations and correlations treated in terms of complete Kirkwood superposition approximation.

equation as compared with the generalized equations (5.1.2) to (5.1.4) depends on the initial particle spatial distribution and violation in a course of reaction of the condition $X_\nu(r,t) = 1$ which becomes increasingly incorrect as time increases, even for the random *initial* distribution.

The non-linearity of the equations (5.1.2) to (5.1.4) prevents us from the use of analytical methods for calculating the reaction rate. These equations reveal *back-coupling of the correlation and concentration dynamics* – Fig. 5.1. Unlike equation (4.1.23), the non-linear terms of equations (5.1.2) to (5.1.4) contain the current particle concentrations $n_A(t)$, $n_B(t)$ due to which the reaction rate $K(t)$ turns out to be concentration-dependent. (In particular, it depends also on initial reactant concentration.) As it is demonstrated below, in the *fluctuation-controlled kinetics* (treated in the framework of all joint densities) such fundamental steady-state characteristics of the linear theory as a recombination profile and a reaction rate as well as an effective reaction radius are no longer useful. The purpose of this fluctuation-controlled approach is to study the general trends and kinetics peculiarities rather than to calculate more precisely just mentioned actual parameters.

5.1.1.2 Infinitely diluted system and a pair problem

It was demonstrated in [1] that for infinitely diluted system, $n(0) = n_0 \to 0$, the solution of a set (4.1.19), (4.1.28), (5.1.2) to (5.1.4) subject to the boundary conditions (4.1.9) and (4.1.10), completely coincides with the exact solution of a pair problem studied in Section 3.2. At first glance it seems to be rather unexpected, since for the problem with initial conditions (4.1.10) use of Kirkwood's superposition approximation, equations (2.3.63) or (2.3.64), leads to the approximation of three-particle correlation functions by equation (4.1.32) which differs considerably in the limit $n_0 \to 0$ from the exact solution given by equation (4.1.11) at $t = 0$. Indeed, it was shown above in Chapter 4, that in the linear approximation Kirkwood's superposition approximation shortcoming has led to incorrect equation (4.1.31) instead of (3.2.7). In the fluctuation kinetics – equations (5.1.2) to (5.1.4) – the superposition approximation affects the non-linear terms only.

It has been shown [1] that substitution of the expected solution (4.1.30) into equations for the similar particle correlation functions, (5.1.2) and (5.1.3), results in $X_\nu(r,t)$ going to zero (practically instantly, during $\tau \sim n_0^{-1}$) at distances where $w(r,t) \neq 0$. Such zerofication has no physical sense but arises just due to the incorrect terms in equation (4.1.32) being of the order of n_0^{-2}. The incorrectness in the definition of the correlation function of similar particles in equation (5.1.4) *cancels* the shortcoming of the superposition approximation, thus by a chance giving the exact solution (3.2.7) of a pair problem. The point is that the "defective" term entering equation (4.1.32) becomes predominant as $n_0 \to 0$, but at the same time the correlation function of similar particles $X_\nu(r,t)$ could be neglected when one calculates the survival probability $\omega(t)$ and the distribution function $w(r,t)$ characterising the time evolution of reactant pairs. This is why serious disadvantages of $X_\nu(r,t)$ nevertheless do not affect the concentration dynamics.

Therefore, even successful reproduction of the exact results (in our case – of a pair problem) *does not* give a final proof of the correctness of the fluctuation kinetics equations applied to the system with a strong spatial correlation. In our opinion, this problem is still far from its solution. Besides, the demonstrated disadvantages of the kinetic equations associated with Kirkwood's approximation, raise the general question about their applicability, even provided the initial conditions given by equation (4.1.12) agree well with the employed superposition approximation. The point is that in a course of bimolecular reaction three- and higher-order particle correlations appear and therefore the justification that the superposition approximation fits well the random initial distribution inevitably fails at *long* reaction times. The accuracy of kinetic equations derived cannot be checked up by their comparison with an exactly soluble model (since it is absent) but in a line with many solid-state problems, it is direct computer simulation which could give a strong argument. Due to long reaction time we are interested in, such simulations turn out to be quite time-consuming.

5.1.1.3 Small parameter concept

The correlation dynamics in terms of the linear approximation (4.1.23) and $X_\nu(r,t) = 1$ could be formally derived from the general set (5.1.2) to (5.1.4) due to its linearisation, i.e., neglecting the non-linear terms with concentrations as co-factors. At first sight, this procedure along with general neglect of the similar particle correlations could be justified by following simple arguments. Concentrations of both kinds of particles, $n(t) = n_A(t) = n_B(t)$, involved into the irreversible reaction $A + B \to 0$ strive to zero as $t \to \infty$. Multiplying these concentrations by the distinctive recombination volume $v \sim r_0^{\bar{d}}$, one indeed gets formally a *small dimensionless parameter* $n(t)r_0^{\bar{d}}$. Following a "small parameter philosophy", the non-linear terms in equations (5.1.2) to (5.1.4) could be neglected as $t \to \infty$ since $n(t)r_0^{\bar{d}} \to 0$. It means that many-particle phenomena could manifest themselves as *concentration effects* only, i.e., the competition between several B's to recombine any A (or vice versa) described by these non-linear terms seems to be important for great particle concentrations only. If it were true, one could expect for infinitely diluted systems, $n(t) \to 0$, that equations (5.1.4) are reduced to linear equation (4.1.23), whereas equations (5.1.2), (5.1.3) without non-linear terms yield uncorrelated similar particle distribution. However, the asymptotic results presented in Chapter 2 *contradict* such a small parameter concept.

The point is that this approach ignores the distinctive feature of the bimolecular process – its non-equilibrium character. The fundamental result known in the theory of non-equilibrium systems [2, 3] is that they tend to become self-organised to a degree which could be characterised by the joint correlation functions, $X_\nu(r,t)$ and $Y(r,t)$. The idea to use $n(t)r_0^{\bar{d}}$ as a small parameter were right, unless there are *no other* distinctive parameters of the same dimension as r_0.

As it was shown in Chapter 2, even the linear approximation demonstrates emergence of the distinctive scale factors increasing in time – the *correlation lengths* ξ. So, for the multipole interaction (4.1.44) there is no scale r_0 at all, whereas dimensionless parameter $n(t)\xi_0^{\bar{d}}$, where ξ_0 is defined by equations (4.1.45) and (4.1.53) remains constant at long time. The same is true for the diffusion – controlled reaction; we can use the diffusion length $l_D = \sqrt{Dt}$ as ξ_0, thus resulting in $n(t)l_D^{\bar{d}} \approx $ const as $d \leqslant 2$ (the correlation function $Y(r,t)$ is not stationary). However, the fluctuation estimate of concentrations $n(t)$ in (2.1.78), $n(t) \propto (Dt)^{-\bar{d}/4} = l_D^{-\bar{d}/2}$ leads to the *infinitely* increasing value of $n(t)l_D^{\bar{d}}$, $\bar{d} < 4$.

We can thus conclude that decrease in particle concentration during the reaction with time and existence of the dimensionless parameter $n(t)r_0^{\bar{d}}$ is *not enough* to use the latter as a small characteristic parameter governing the kinetics. The formation of the distinctive length ξ increasing in time affects also the joint correlation functions: at short distances $r < \xi$ they start to deviate from their asymptotic magnitudes $X_\nu(\infty, t) = Y(\infty, t) = 1$ thus demonstrating reaction-induced the non-Poisson fluctuations of the particle densities. Note that in functional (5.1.5) the correlation function $Z(r^*, t)$ ($Z(r, t) = X_\nu(r, t)$ and $Y(r, t)$) enters as the difference $(Z(r^*, t) - 1)$, i.e., non-linear terms in (5.1.2) to (5.1.4) are non-zero if the fluctuation spectrum is non Poissonian. It could be reasonably interpreted as a formation of *loose clusters* of similar particles, A's or B's, giving rise to these long-range correlations. These clusters are dynamical, e.g., they can dissolve in one spatial point and arise at another.

There is an analogy between these *long-range* particle correlations formed in a course of chemical reactions and equilibrium system of charged particles where the small-parameter approach also fails. Another analogy exists with equilibrium system of particles interacting by *short-range* forces and revealing phase transitions and long-range correlations of structure elements (Section 2.3.1). The physics of both equilibrium and non-equilibrium critical phenomena requires development of the adequate mathematical formalism not employing any small parameters. As it is well known [4, 5] the superposition-like approximations were successfully used earlier in describing critical phenomena in condensed matter physics but a criterium of their accuracy, as was said above lies *beyond* the formalism developed and quality of approximations made could be understood only by comparison with either exact solutions of particular problems or with direct computer simulations.

In terms of mathematics (5.1.2) to (5.1.4) are nothing but generalized diffusion equations

$$\frac{\partial g(r, t)}{\partial t} = D\nabla^2 g(r, t) - q(r, t)g(r, t) \tag{5.1.12}$$

with particle sources $q(r, t)g(r, t)$ (or sinks – dependent on a sign of $q(r, t)$).

One has to distinguish between a *direct* creation of reactant correlations due to disappearance of some AB pairs and an *indirect* channel of correlation formation when one of partners of any AA, BB or AB pair recombines with surrounding particle. The former mechanism is taken into account through the

$\sigma(r)Y(r,t)$ term in (5.1.4) whereas the latter – through non-linear terms in (5.1.2)–(5.1.4), containing the functionals $I[Z]$. Since dynamical interaction of particles is not incorporated into the kinetic equations, the *similar* particle correlation arises in the indirect way. However, in the linear approximation the direct mechanism of the correlation formation is taken into account, being defined entirely by the $\sigma(r)$ value. It is characterised by such distinctive scales as r_0, ξ_0 lengths (for reactions of immobile particles since the diffusion being described by the scale l_D *does not* produce spatial correlations, i.e., diffusion terms in the kinetic equations do not serve as sources).

The indirect mechanism of the correlation formation could be characterized by quite small $q(r,t)$ values entering equation (5.1.12). Thus in (5.1.2) and (5.1.3) we have

$$q(r,t) = 2n_\nu(t) I[Y].$$

For the reaction $A + B \to 0$ $Y(r,t) \leqslant 1$, therefore

$$|q(r,t)| \leqslant 2n_\nu(t)|I[0]| = 2n_\nu(t) K(t),$$

where the relation $I[Z] = -K(t)$ for $Z(r,t) = 0$ could be used, keeping in mind definitions (5.1.5) and (4.1.19). As one can see, $q(r,t)$ decays in time as $n_\nu(t) K(t)$. It is important to stress, however, that a role of sources in equations like (5.1.12) depends not on $q(r,t)$, but more on its integral property $Q(t) = \int q(r,t)\, d\vec{r}$. If $Y(r,t) \ll 1$ at $r < \xi$, we have $|Q(t)| \sim n(t) K(t) \xi^{\bar{d}}$ which could be rather large; i.e., an emergence of *new* correlation lengths $\xi \gg \xi_0$ is of primary importance. Unlikely the equations of linear approximation (4.1.23), in the set (5.1.2) to (5.1.4), the concentration motion affects here the correlation dynamics. Therefore, new scales emerge in the equations for the correlation dynamics having the length dimension; in particular, for equal concentrations it is l_0, with $l_0^{-\bar{d}} = n(0)$.

To illustrate this point, let us consider recombination of immobile particles with the multipole interaction. In the linear approximation there exists a single correlation length ξ_0, equation (4.1.45). From two lengths l_0 and ξ_0, we can combine a new one

$$\xi = l_0 \Psi(\xi_0/l_0), \tag{5.1.13}$$

where $\Psi(x)$ is an arbitrary function. In the region of the applicability of the linear approximation one gets $\xi \approx \xi_0$ and $\Psi(x) \approx x$.

All said above demonstrates how *new degrees of freedom* arises when more refined approximation (2.3.64) is used – the complete superposition approximation instead of the standard equation (2.3.64), but it does not prove existence of effects themselves. Solution of non-linear equations for the correlation dynamics is discussed in next Sections of this Chapter.

5.1.1.4 Reaction of immobile particles

The kinetic equations in the case of immobile particles, $D_A, D_B = 0$, come from a set (5.1.2) to (5.1.4):

$$\frac{d \ln X_A(r,t)}{dt} = -2n_B(t) I[Y], \qquad (5.1.14)$$

$$\frac{d \ln X_B(r,t)}{dt} = -2n_A(t) I[Y], \qquad (5.1.15)$$

$$\frac{d \ln Y(r,t)}{dt} = -\sigma(r) - \sum_{\nu=A,B} n_\nu(t) I[X_\nu]. \qquad (5.1.16)$$

Let us illustrate an idea of the numerical solution of the set (5.1.14) – (5.1.16) (concentration equation (4.1.18) can be integrated trivially) making use of the basic equation

$$\frac{dg(r,\tau)}{d\tau} = -q(r,\tau;\{g\}), \qquad (5.1.17)$$

to which any of above-written equations could be reduced. Here $g(r,\tau)$ is the logarithm of an arbitrary correlation function, an initial condition (4.1.12) corresponds to $g = 0$; $\{g\}$ denotes in a compact way a set of all functions g, $\tau = \tau(t)$ is convenient for numerical solution. For example, Schnörer et al. [6, 7] in their study of the tunnelling recombination used the correlation length ξ_0. Making discretization $r_m = m\Delta r$, $\tau_i = i\Delta\tau$, $g(r_m, \tau_i) = g_m^i$, let us write down the difference analog of (5.1.17) in a form

$$\frac{g_m^{i+1} - g_m^i}{\Delta \tau} = -\frac{1}{2}\big(q(r_m, \tau_i; \{g^i\}) + q(r_m, \tau_{i+1}; \{g^{i+1}\})\big). \qquad (5.1.18)$$

In the symmetric difference scheme (5.1.18) integrals in (5.1.7), (5.1.8) are calculated within accuracy of $O(\Delta r^2)$, therefore an accuracy of the difference scheme is $O(\Delta\tau^2 + \Delta r^2)$. Function $(r_m, \tau_{i+1}; \{g^{i+1}\})$ is non-linear in g, therefore the solution of (5.1.18) should be found iteratively: expression $q(r_m, \tau_{i+1}; \{g^{i+1}\})$ in (5.1.18) is replaced by $q(r_m, \tau_{i+1}; \{\widehat{g}\})$, with the initial guess $\widehat{g}(0) = g_m^i$ and using that $\widehat{g}(j+1) = g_m^{i+1}(j)$, where $g_m^{i+1}(j)$ is j-th iteration solution. Iterating is stopped provided $|g_m^{i+1}(j) - g_m^{i+1}(j+1)| < \varepsilon$ for all m, ε is the iteration accuracy.

Disadvantage of this direct method to solve (5.1.14) to (5.1.16) is that the required computational time is proportional to ξ^2 due to an increase in coordinate parameter m (functions $\{g\}$ are non-zero at $r < \xi$, where ξ increases in time). This is why another coordinate variable $\eta = r/\xi$, is more useful. Its use results in the correlation functions which are practically stationary at long times [8]. The basic equation is

$$\frac{\partial g(x,\tau)}{\partial \tau} - \frac{\partial g(x,\tau)}{\partial x} = -q(x,\tau;\{g\}), \qquad (5.1.19)$$

where $x = \ln \eta$. Using discretization $x_m = m\Delta x$, $\tau_i = i\Delta\tau$, $g(x_m, \tau_i) = g_m^i$, let us write down the difference analogs (symmetrical scheme):

$$\frac{\partial g(x,\tau)}{\partial \tau} \Rightarrow \frac{1}{2\Delta\tau}\left(g_m^{i+1} + g_{m+1}^{i+1} - g_m^i - g_{m+1}^i\right), \qquad (5.1.20)$$

$$\frac{\partial g(x,\tau)}{\partial x} \Rightarrow \frac{1}{2\Delta x}\left(g_{m+1}^{i+1} - g_m^{i+1} + g_{m+1}^i - g_m^i\right), \qquad (5.1.21)$$

$$q(x,\tau;\{g\}) \Rightarrow \frac{1}{4}\left(q_m^{i+1} + q_{m+1}^{i+1} + q_m^i + q_{m+1}^i\right). \qquad (5.1.22)$$

The accuracy of this difference scheme is $O(\Delta\tau^2 + \Delta x^2)$, provided the integrals entering (5.1.7) and (5.1.8) are calculated with accuracy of $O(\Delta x^2)$. The difference equations are solved first from the left to the right with the boundary condition $g_N = 0$ (no correlations at $r \gg \xi$). Similar to the discussed scheme (5.1.18), non-linear effects are taken here into account iteratively.

5.1.1.5 Diffusion-controlled reactions. Black sphere model

The integro-differential equations (5.1.2) to (5.1.4) could be considerably simplified (which is of great importance for actual computer calculations) in the *black sphere approximation* (3.2.16). Let us consider the particular case of *instant* recombination; $\sigma(r) = \sigma_0 \theta(r_0 - r)$, $\sigma_0 \to \infty$ ($\theta(x)$ is the Heaviside step-function) which holds, for instance, for defects in metals – interstitial atom recombines with vacancy nearby in time of the order of several lattice vibrations (10^{-13} s). Denoting $h(r,t) = \sigma(r)Y(r,t)$ and multiplying equation (5.1.4) by $\theta(r_0 - r)$, we arrive at

$$\sigma_0^{-1} \frac{\partial h(r,t)}{\partial t} = \theta(r_0 - r)D\nabla^2 Y(r,t) - h(r,t) - $$
$$ - \sigma_0^{-1} h(r,t) \sum_{\nu=A,B} n_\nu(t) I[X_\nu]. \quad (5.1.23)$$

For great values of σ_0 equation (5.1.23) is nothing but a differential equation with a small parameter multiplying the derivative, whose main term of the asymptotic expansion is the solution of the degenerate equation ($\sigma_0^{-1} = 0$)

$$h(r,t) = \theta(r_0 - r)D\nabla^2 Y(r,t). \quad (5.1.24)$$

When deriving equation (5.1.24), we took into account that the functional (5.1.7) remains finite as $\sigma_0^{-1} \to 0$.

Due to the instant recombination all the dissimilar particles with relative distances $r \leqslant r_0$ disappear, which results in the Smoluchowski boundary condition

$$Y(r,t) = 0, \quad r \leqslant r_0. \quad (5.1.25)$$

The degenerate equation could be used to simplify the integral expressions. Substitution of (5.1.24) into (4.1.19), leads to

$$K(t) = \int \sigma(r)Y(r,t)\,d\vec{r} = \int h(r,t)\,d\vec{r}$$
$$= \int_{V(r_0)} D\nabla^2 Y(r,t)\,d\vec{r} = \int D\nabla Y(r,t)\,d\vec{S}. \quad (5.1.26)$$

Using the integral theorem, a volume integral $V(r_0)$ in (5.1.26) for $\bar{d}$-dimensional sphere of the radius r_0 is transformed into a surface one, from where one gets

$$K(t) = \gamma_{\bar{d}} r_0^{\bar{d}-1} \left. \frac{\partial Y(r,t)}{\partial r} \right|_{r=r_0}. \quad (5.1.27)$$

An expression (5.1.27) coincides with equations (4.1.58) and (4.1.64) for $\bar{d} = 2$ and 3, giving thus an evidence for correctness of (5.1.24).

To calculate functionals $I[Z]$, let us use the presentation (5.1.7). Denoting for compactness $(\widetilde{Z}(r^*,t) - 1) = q = q(r, r', t)$, we arrive at

$$I[Z] = \gamma_{\bar{d}} \int h(r',t) q r'^{\bar{d}-1} dr'$$

$$= \gamma_{\bar{d}} \int_0^{r_0} Dr'^{1-\bar{d}} \frac{\partial}{\partial r'} \left(r'^{\bar{d}-1} \frac{\partial}{\partial r'} \right) Y(r',t) q r'^{\bar{d}-1} dr'$$

$$= \gamma_{\bar{d}} D \left\{ q \frac{\partial}{\partial r'} Y(r',t) - Y(r',t) \frac{\partial}{\partial r'} q \right\} r'^{\bar{d}-1} \Bigg|_{r'=r_0} +$$

$$+ \gamma_{\bar{d}} \int_0^{r_0} DY(r',t) \nabla'^2 q r'^{\bar{d}-1} dr'. \tag{5.1.28}$$

Keeping in mind (5.1.25), one gets from (5.1.28)

$$I[Z] = \gamma_{\bar{d}} D r_0^{\bar{d}-1} q(r, r_0, t) \frac{\partial Y(r',t)}{\partial r'} \Bigg|_{r'=r_0}. \tag{5.1.29}$$

The derivative in (5.1.29) could be expressed through the reaction rate $K(t)$:

$$I[Z] = K(t) q(r, r_0, t), \tag{5.1.30}$$

or

$$I[Z] = K(t) \big(\widetilde{Z}(r^*, t) - 1\big) \Big|_{r'=r_0}. \tag{5.1.31}$$

We use hereafter the dimensionless variables: $r' = r/r_0$, $t' = Dt/r_0^2$, $D'_\nu = 2D_\nu/D$, $n'_\nu = \gamma_{\bar{d}} n_\nu r_0^{\bar{d}}$, $K'(t) = K(t)/(\gamma_{\bar{d}} D r_0^{\bar{d}-2})$. Omitting below primes, we arrive at the equation for the concentrations formally coinciding with equation (4.1.18) but having the redefined reaction rate

$$K(t) = \frac{\partial Y(r,t)}{\partial r} \Bigg|_{r=1}. \tag{5.1.32}$$

The correlation dynamics is governed by equations

$$\frac{\partial X_A(r,t)}{\partial t} = D_A \Delta X_A(r,t) - 2K(t) X_A(r,t) n_B(t) J[Y], \quad (5.1.33)$$

$$\frac{\partial X_B(r,t)}{\partial t} = D_B \Delta X_B(r,t) - 2K(t) X_B(r,t) n_A(t) J[Y], \quad (5.1.34)$$

$$\frac{\partial Y(r,t)}{\partial t} = \Delta Y(r,t) - K(t) Y(r,t) \sum_{\nu=A,B} n_\nu(t) J[X_\nu]. \quad (5.1.35)$$

The diffusion operator Δ entering (5.1.33) to (5.1.35) is *dimension*-dependent (equation (3.2.8)), the functionals $J[Z]$ are:

$$J[Z] = \frac{1}{2}\bigl(Z(|r-1|,t) + Z(r+1,t) - 2\bigr), \quad \bar{d} = 1, \quad (5.1.36)$$

$$J[Z] = \frac{1}{\pi} \int_{-1}^{+1} \frac{[Z(\zeta,t) - 1]\,ds}{(1-s^2)^{1/2}},$$
$$\zeta = (r^2 + 1 + 2rs)^{1/2}, \quad \bar{d} = 2, \quad (5.1.37)$$

and

$$J[Z] = \frac{1}{2r} \int_{|r-1|}^{r+1} [Z(\zeta,t) - 1] \zeta \, d\zeta, \quad \bar{d} = 3. \quad (5.1.38)$$

The standard boundary conditions reflect weakening of the correlations at large relative distances: $X_\nu(\infty,t) = Y(\infty,t) = 1$. We impose also the Smoluchowski boundary condition,

$$Y(r,t) \equiv 0, \quad r \leqslant 1. \quad (5.1.39)$$

For the joint correlation functions of similar particles the "native" boundary condition is absence of the flux over the coordinate origin

$$\lim_{r \to 0} D_\nu r^{\bar{d}-1} \frac{\partial X_\nu(r,t)}{\partial r} = 0, \quad (5.1.40)$$

since a given spatial point could be occupied by one particle only.

Since in the chosen-above units $D_A + D_B = 2$, the partial diffusion coefficients are no longer independent but the distinctive parameter of theory is their *ratio* $\kappa = D_A/(D_A + D_B)$, $0 \leqslant \kappa \leqslant 1$.

The initial conditions of random reactant distribution in the black sphere model agree with the boundary condition given by equation (5.1.39). However, since the latter condition means that there are no dissimilar particles at the relative distances $r < r_0 = 1$, both spatial distributions, $X_\nu(r,t)$ and $Y(r,t)$, strictly speaking, are no longer random. An analysis of a set of equations (5.1.33) to (5.1.35) permits to clarify the mechanism of how the initial correlation, equation (5.1.39), develops in time. The non-linear terms of kinetic equations (5.1.33)–(5.1.35), derived above in the superposition approximation, contain the functionals $J[Z]$. At the very beginning, $t = 0$, the functionals $J[X_\nu]$ are zero resulting in the linearized equation for the correlation function of dissimilar particles $Y(r,t)$. However, because of the Smoluchowski boundary condition the third functional, $J[Y]$, is nonzero at $r < 2$. In terms of mathematics, the last terms of equations (5.1.33) to (5.1.35) are nothing but sources or sinks dependent on their signs. The emergence of correlations in equations (5.1.33), (5.1.34) through $J[Y] = 0$ means that similar reactants are also no longer characterized by the Poisson distribution, $X_\nu(r,t) = 1$. In its turn, their deviation from the unity value produces the correlation source in equation (5.1.35) – unlike initial correlation (5.1.39), this additional reaction-produced correlation is time-dependent. At short time the distinctive scale of the spatial correlations remains the same as follows from equation (5.1.39), i.e., of the order of r_0. However, the above described correlation back-coupling creates a new time-dependent scale. This point, as well as a set of kinetic equations (5.1.33) to (5.1.35) is analysed in detail below.

As a basic model of a set of equation derived above, the generalized diffusion equation with non-linear source could be used

$$\frac{\partial g(r,t)}{\partial t} = D\nabla^2 g(r,t) - q(r,t,\{g\})g(r,t). \tag{5.1.41}$$

To avoid lengthy calculations, let us confine ourselves to the implicit calculation scheme. Introducing discretization $t_i = i\Delta t$, $r_m = m\Delta r$, $g(r_m, t_i) = g_m^i$, one gets

$$\frac{g_m^{i+1} - g_m^i}{\Delta t} = D\widehat{L}g_m^{i+1} - q_m(t^{i+1}, \{g^{i+1}\})g_m^{i+1}, \tag{5.1.42}$$

with the difference operator

$$\widehat{L}g_m^{i+1} = a_m g_{m-1}^{i+1} - b_m g_m^{i+1} + c_m g_{m+1}^{i+1}, \qquad (5.1.43)$$

corresponding to the diffusion operator. The coefficients in (5.1.43) were taken for an optimal conservative scheme; the non-linear equation (5.1.42) is solved iteratively in the spirit of approach described above (equation (5.1.18)). One-dimensional linear parabolic equation is solved on each iteration step, methods of their solutions are well known. For negative sources, $q < 0$, good convergence is achieved using the equation

$$\frac{g_m^{i+1} - g_m^i}{\Delta t} = D\widehat{L}g_m^{i+1} -$$

$$- 2q_m(t^{i+1}, \{\widehat{g}\})\widehat{g}_m + q_m(t^{i+1}, \{\widehat{g}\})g_m^{i+1}. \qquad (5.1.44)$$

Accuracy of the difference scheme is $O(\Delta t + \Delta r^2)$, which could be reduced to $O(\Delta t^2 + \Delta r^2)$ by means of the symmetrical difference scheme. In practice schemes with monotonously increasing spatial and temporal steps are usually used for these purposes [1, 9–11]. As $r \sim 1$, Δr is small but increases with r whereas Δt increment is limited by the condition that the relative change of g_m at any step should not exceed a given small value. Unlike the case of immobile particle reaction, the calculation of the functionals $J[Z]$, (5.1.37) and (5.1.38), requires one-dimensional integration only which is not time-consuming.

5.1.1.6 Reactant dynamical interaction

When incorporating the dynamical interactions given by equation (2.3.47), the diffusion terms in equations (5.1.2) to (5.1.4) become [12]

$$\left.\frac{\partial X_A(r,t)}{\partial t}\right|_{\text{diff}} = 2D_A \nabla \left(\nabla X_A(r,t) + X_A(r,t)\nabla \mathfrak{U}_A(r,t)\right),$$

$$\nabla \mathfrak{U}_A(r,t) = \frac{1}{k_B T}\Bigg\{\nabla U_{AA}(r) +$$

$$+ \Bigg[n_A(t)\int \nabla U_{AA}(r^*)X_A(r^*,t)X_A(r',t)\,d\vec{r}'' +$$

$$+ n_\mathrm{B}(t) \int \nabla U_{\mathrm{AB}}(r^*) Y(r^*,t) Y(r',t) \, \mathrm{d}\vec{r}' \Big] \Big\}, \qquad (5.1.45)$$

$$\left. \frac{\partial X_\mathrm{B}(r,t)}{\partial t} \right|_\mathrm{diff} = 2 D_\mathrm{B} \nabla \left(\nabla X_\mathrm{B}(r,t) + X_\mathrm{B}(r,t) \nabla \mathfrak{U}_\mathrm{B}(r,t) \right),$$

$$\nabla \mathfrak{U}_\mathrm{B}(r,t) = \frac{1}{k_\mathrm{B} T} \Big\{ \nabla U_{\mathrm{BB}}(r) +$$

$$+ \Big[n_\mathrm{B}(t) \int \nabla U_{\mathrm{BB}}(r^*) X_\mathrm{B}(r^*,t) X_\mathrm{B}(r',t) \, \mathrm{d}\vec{r}' +$$

$$+ n_\mathrm{A}(t) \int \nabla U_{\mathrm{AB}}(r^*) Y(r^*,t) Y(r',t) \, \mathrm{d}\vec{r}' \Big] \Big\}, \qquad (5.1.46)$$

$$\left. \frac{\partial Y(r,t)}{\partial t} \right|_\mathrm{diff} = D \nabla \left(\nabla Y(r,t) + Y(r,t) \nabla \mathfrak{U}(r,t) \right),$$

$$\nabla \mathfrak{U}(r,t) = \frac{1}{k_\mathrm{B} T} \Big\{ \nabla U_{\mathrm{AB}}(r) +$$

$$+ \frac{D_\mathrm{A}}{D} \Big[n_\mathrm{A}(t) \int \nabla U_{\mathrm{AA}}(r^*) X_\mathrm{A}(r^*,t) Y(r',t) \, \mathrm{d}\vec{r}' +$$

$$+ n_\mathrm{B}(t) \int \nabla U_{\mathrm{AB}}(r^*) Y(r^*,t) X_\mathrm{B}(r',t) \, \mathrm{d}\vec{r}' \Big] +$$

$$+ \frac{D_\mathrm{B}}{D} \Big[n_\mathrm{B}(t) \int \nabla U_{\mathrm{BB}}(r^*) X_\mathrm{B}(r^*,t) Y(r',t) \, \mathrm{d}\vec{r}' +$$

$$+ n_\mathrm{A}(t) \int \nabla U_{\mathrm{AB}}(r^*) Y(r^*,t) X_\mathrm{A}(r',t) \, \mathrm{d}\vec{r}' \Big] \Big\}, \qquad (5.1.47)$$

where $\vec{r}^* = \vec{r} - \vec{r}'$.

To consider a role of dynamical reactant interactions, let us treat the Coulomb one as a distinctive example. Similarly to the case of an equilibrium system of charged particles, the superposition approximation does not permit us to describe correctly the effective reactant interaction since the long-range Coulomb forces change qualitatively the fluctuation spectrum of a system and therefore the *collective effects* (charge screening at the long relative distances) should be taken into account. This screening is *non-equilibrium* (being diffusion- and recombination-controlled) and thus *cannot* be described in a trivial way through the ordinary Debye radius, $r_D = (4\pi e^2 n/(k_B T))^{-1/2}$. Instead of equations (5.1.45) to (5.1.47), another scheme was developed in [9, 12] for the calculating the effective Coulomb potentials – which is in a line with the Debye–Hückel theory of electrolytes [4]. Its principal advantage is the consistent coupling between interaction potentials and the joint correlation functions.

Let us put particle A at the coordinate origin. Keeping in mind physical interpretation of the correlation function in (2.3.24), an expression $C_A(r,t) = n_A(t) X_A(r,t)$ defines the density of A's at the distance r from a given particle A at moment t. The charge density of these particles is

$$\tilde{\rho}_A(r,t) = e_A C_A(r,t) = e_A n_A(t) X_A(r,t), \tag{5.1.48}$$

where e_A is a charge of A. The charge density of B's is respectively

$$\tilde{\rho}_B(r,t) = e_B C_B(r,t) = e_B n_B(t) Y(r,t). \tag{5.1.49}$$

The potential $\phi_A(r,t)$ produced by these charged in a medium with dielectric constant ε is obtained from the Poisson equation

$$\nabla^2 \phi_A(r,t) = -\frac{4\pi}{\varepsilon} \left(\tilde{\rho}_A(r,t) + \tilde{\rho}_B(r,t) \right). \tag{5.1.50}$$

For the Coulomb attraction of reactants A and B ($e_A = -e_B$) present in equal concentrations $n(t)$ and the black sphere model the kinetic equations read ($\bar{d} = 3$)

$$\frac{dn(t)}{dt} = -K(t) n^2(t), \qquad K(t) = \left. \frac{\partial Y(r,t)}{\partial r} \right|_{r=1}, \tag{5.1.51}$$

$$\frac{\partial X_\nu(r,t)}{\partial t} = D_\nu \nabla \left(\nabla X_\nu(r,t) + X_\nu(r,t) \nabla \mathfrak{U}_\nu(r,t) \right) -$$
$$- 2K(t) n(t) X_\nu(r,t) J[Y], \tag{5.1.52}$$

$$\frac{\partial Y(r,t)}{\partial t} = \nabla\bigl(\nabla Y(r,t) + Y(r,t)\nabla \mathfrak{U}(r,t)\bigr) -$$
$$- K(t)n(t)Y(r,t) \sum_{\nu=A,B} J[X_\nu]. \qquad (5.1.53)$$

(The dimensionless units are again used here, as earlier in equations (5.1.33) to (5.1.35)).

The self-consistent potentials $\mathfrak{U}_\nu(r,t)$ are defined as

$$\nabla^2 \mathfrak{U}_\nu(r,t) = -n(t)L\bigl(X_\nu(r,t) - Y(r,t)\bigr) \qquad (5.1.54)$$

and are accompanied by the boundary condition

$$\lim_{r \to 0} r\mathfrak{U}_\nu(r,t) = L. \qquad (5.1.55)$$

In equations (5.1.54) and (5.1.55) the dimensionless Onsager radius

$$L = \frac{e_A^2}{\varepsilon k_B T r_0} \qquad (5.1.56)$$

is used. The average potential in (5.1.53) is nothing but

$$\mathfrak{U}(r,t) = -\frac{1}{2}\bigl(\mathfrak{U}_A(r,t) + \mathfrak{U}_B(r,t)\bigr). \qquad (5.1.57)$$

In the case of the Coulomb repulsion, it is assumed that additional charges of opposite sign exist in a system to keep its electroneutrality. If these charges are distributed at random, and $n_A(t) = n_B(t) = n(t)$, $e_A = e_B$ hold, we arrive at the same equations (5.1.51) to (5.1.53) as above, but with the *redefined potentials*

$$\nabla^2 \mathfrak{U}_\nu(r,t) = -n(t)L\bigl(X_\nu(r,t) + Y(r,t) - 2\bigr), \qquad (5.1.58)$$

$$\mathfrak{U}(r,t) = \frac{1}{2}\bigl(\mathfrak{U}_A(r,t) + \mathfrak{U}_B(r,t)\bigr). \qquad (5.1.59)$$

When making use of the standard initial conditions $X_\nu(r,0) = 1$, $Y(r > 1, 0) = 1$ for the correlation functions, the reaction kinetics is governed uniquely by the three parameters: initial concentration $n(0)$, the Onsager radius L and the ratio of diffusion coefficients $\kappa = D_A/D$. If reactant have different diffusion coefficients, the correlation functions of similar particles A–A and B–B, $X_A(r,t)$ and $X_B(r,t)$ no longer coincide and the self-consistent reaction potentials for A's and B's differ also. Despite the fact that a numerical solution of these equations requires certain modification of the difference scheme, taking into account potential singularities, the scheme itself is similar to that used for neutral particles. Original results are presented below in Chapter 6.

5.1.2 The superposition approximation

5.1.2.1 General comments

The approximations of the superposition-type like equation (2.3.54), are used in those problems of theoreticals physics when other-kind expansions (e.g., in powers of a small parameter) cannot be employed. First of all, we mean physics of phase transitions and critical phenomena [4, 13–15] where there are *no* small parameters at all. Neglect of the higher correlation forms $a^{(m)}$ in (2.3.54) introduces into solution errors which cannot be, in fact, estimated within the framework of the method used. That is, accuracy of the superposition-like approximations could be obtained by a comparison with either simplest explicitly solvable models (like the *Ising model* in the theory of phase transitions) or with results of direct computer simulations. Note, first of all, several distinctive features of the superposition approximations.

None of these approximations applied to the statistical theory of dense gases and liquids could predict existence of the melting point [4, 5]. Consequently, there is a range of particle densities and temperatures where the superposition approximations fail. Less dramatic results are obtained employing such approximations for calculating the lattice models in the physics of phase transitions [13, 14, 16]. As an illustration, let us consider spin ordering, i.e., ferromagnetic version of the Ising model [13]. A whole temperature interval is divided into two regions: high temperature (disordered) phase at temperature $T > T_0$ and low-temperature (ordered) one with $T < T_0$, where T_0 is the phase transition temperature. At $T < T_0$ spontaneous magnetisation M arises without any applied external magnetic field. Let us introduce now an *inverse* temperature $\tilde{t} = T^{-1}(t_0 = T_0^{-1})$. As $\tilde{t} = 0$ $(T \to \infty)$, a system is completely disordered. As $\tilde{t}$ increases $(\tilde{t} < \tilde{t}_0)$, it becomes partly ordered in the sense that short-order and intermediate-order emerge and they are characterized by the correlation function similar to (2.3.18). In the ferromagnetic case the correlation length ξ characterises the linear sizes of blocks constituting system's volume; in each block spins are predominately oriented in the same direction but a total magnetic moment of a whole system is zero.

Applying superposition approximations to the Ising model, one finds an evidence for the phase transition existence but the critical parameter $\tilde{t}_0$ is systematically underestimated (T_0 is overestimated respectively). Errors in calculation of $\tilde{t}_0$ are greater for *low* dimensions $\bar{d}$. Therefore, the superposition approximation is effective, first of all, for the *qualitative* description of the phase transition in a spin system. In the vicinity of phase transition a number of *critical exponents* $\alpha, \beta, \gamma, \ldots$, could be introduced, which characterize the critical point, like $\xi \propto |\tilde{t} - \tilde{t}_0|^{-\alpha}$, $M \propto (\tilde{t} - \tilde{t}_0)^\beta$, or $\chi_T \propto |\tilde{t} - \tilde{t}_0|^{-\gamma}$ for the magnetic permeability. Superposition approximations give only *classical* values of the critical exponents $\alpha = \alpha_0$, $\beta = \beta_0$, $\gamma = \gamma_0, \ldots$, obtained earlier in the classical *molecular field* theory [13, 14], say $\beta_0 = 1/2$, $\gamma_0 = 1$, whereas exact magnitudes of the critical exponents depend on the space dimension $\bar{d}$. To describe the *intermediate* order in a spin system in terms of the superposition approximation, an additional correlation length is introduced, $\xi_0 = |\tilde{t} - \tilde{t}_0|^{-\alpha_0}$, which does not coincide with the true ξ. In the phase transition vicinity, $\xi \to \infty$, the system is characterized by anomalous fluctuations leading to non-analytical behaviour of thermodynamical potentials. Therefore, simple algebraic expressions used in the formalism of superposition approximations for the correlation functions of a higher order, cannot *in principle* result in the corresponding kind of non-analyticity. At the same time, the molecular-field approach reproduces correctly the critical exponents in a system with long-range order [14].

These well-known results of the physics of phase transitions permit us to stress useful analogy of the critical phenomena and the kinetics of bimolecular reactions under study. Indeed, even the simplest linear approximation (Chapter 4) reveals the correlation length ξ_0 – see (4.1.45) and (4.1.47), or $\xi_0 = l_D$ for the diffusion-controlled processes. At $t = 0$ reactants are randomly distributed and thus there is no spatial correlation between them. These correlations arise in a course of the reaction, the correlation length ξ_0 increases monotonously in time but $\xi_0 \to \infty$ at $t \to \infty$ only. Consequently, a formal difference from statistical physics is that an approach to the critical point is one-side, $t_0 \to \infty$, and the "ordered" phase is absent here. There is also evident correspondence between the parameter $\tilde{t}$ in the theory of *equilibrium* phase transitions and time t in the *kinetics* of the bimolecular processes. By analogy with theory of phase transitions, different-kind *critical exponents* like $n(t) \propto \xi_0^{-\alpha_0}$ in the linear approximation could be introduced but with the "classical" values of $\alpha_0 = \bar{d}$ – both for reactions of immobile reactants, equation (4.1.53) and diffusion-controlled reactions, $\xi_0 = l_D$ as $\bar{d} \leqslant \bar{d}_0 = 2$. Results presented in Chapter 2, in particular equation (2.1.78), indicate clearly existence of other values of the critical exponent, $\alpha = \bar{d}/2$, and stress importance of the fluctuation effects.

Keeping in mind what has been said about the superposition approximation in physics of phase transitions, there are serious doubts whether it could be possible to reproduce correctly long-time kinetics of the bimolecular reactions ($\xi \to \infty$). However, there exists a circumstance which gives some hope: in this kinetics sought for the critical point exists at $t = \infty$ only! An analog of such a system is the one-dimensional Ising model, i.e., with the critical point at infinitely (there is no ordered phase-superposition approximations able to give explicit values of the correlation length and thermodynamic functions).

All the above-said demonstrates well that there are arguments for and against applicability of the superposition approximation in the kinetics of bimolecular reactions. Because of the absence of exactly solvable problems, it is computer simulation only which can give a final answer. Note at once some peculiarities of such computer simulations. The largest deviations from the standard chemical kinetics could be expected at long t (large ξ). Unlike computer simulations of *equilibrium* phenomena [4] where the particle density is constant, in the kinetics problems particle density $n(t)$ decays in time which puts natural limits on time of reaction. An increase of the standard deviation at small values of $N(t) = \langle N \rangle$ when calculating the mean concentration in computer simulations compel us to interrupt simulations at the *reaction depth* $\Gamma = \Gamma_0 \approx 3$, where

$$\Gamma = -\log \frac{n_A(t)}{n_A(0)}. \tag{5.1.60}$$

Nearly the same limits of Γ exist in real solid state experiments. However, the relevant maximal time t_m which could be achieved in such computer simulations (see equation (5.1.60)) for a given Γ_0, could turn out to be not long enough for determining the *asymptotic* laws under question. For instance, existence of so-called *small critical exponents* in physics of phase transitions [14] was not experimentally confirmed since to obtain these critical exponents, the process covering several orders of the parameter $|\tilde{t} - \tilde{t}_0|$ should be measured, what in fact is unrealistic. In such cases only computer simulations can permit us to estimate accuracy of the superposition approximation within quite limited time interval.

In this respect of great interest are problems in which abnormal asymptotics sought for are formed already at $\Gamma < \Gamma_0$ (in practice they are displayed as linear dependencies in the corresponding coordinates). It can be achieved in computer simulations using great initial reactant densities. Taking into account that largest errors of the superposition approximation were discovered in the statistical physics of dense gases and liquids namely for large particle densities, it should be accepted that use of such large initial densities (often exceeding those observed in actual physical experiments) is justified.

The same is true for low-dimensional systems, $\bar{d} = 1$ and 2. The point is not only that for such systems the better statistics could be achieved accompanied with reasonable computational time spent for it. Another circumstance is that we can expect here that the superposition approximation gives greater errors. For example, for one-dimensional contact recombination the so-called "bus effect" is known [17]: given particles A and B can react only after particles separating them disappear during reaction. This *topological* effect is not foreseen by the superposition approximation but can affect considerably the reaction rate.

Let us consider accuracy of the superposition approximation for the two quite different classes of problems: a long-range reaction of immobile particles and a diffusion-controlled one, where the diffusion length l_D arises.

5.1.2.2 Computer simulations of immobile particles

Computer simulations of bimolecular reactions for a system of immobile particles (incorporating their production) has a long history see, e.g., [18–22]. For the first time computer simulation as a test of analytical methods in the reaction kinetics was carried out by Zhdanov [23, 24] for $\bar{d} = 3$. Despite the fact that his simulations were performed up to rather small reaction depths, $\Gamma_0 < 1$, it was established that of all empirical equations presented for the tunnelling recombination kinetics (those of linear approximation – (4.1.42) or (4.1.43)) turned out to be mostly correct (note that equations (5.1.14) to (5.1.16) of the *complete* superposition approximation were not considered.) On the other hand, irrespective of the initial reactant densities and space dimension $\bar{d}$ for reaction depths $\Gamma \simeq \Gamma_0$ his theoretical curves deviate from those computer simulated by $\geqslant 10\%$. Accuracy of the superposition approximation in $\bar{d} = 3$ case was first questioned by Kuzovkov [25], it was also based on the computer simulations [23]. For more detailed testing of analytical methods in the low-dimensional systems, $\bar{d} = 1$ and 2 see [6, 7, 26].

These direct computer simulations of the $A + B \to 0$ reaction between immobile particles in one- and two-dimensional lattices were performed in the following way [6, 7, 26]. The initial particle distribution was random one of N_A particles A and N_B particles B occupying regular lattice sites of N-site chain or square lattice, then no site could be occupied by a single particle. The relevant *occupation numbers* (lattice densities) are $\rho_a(0) = N_A/N$ and $\rho_b(0) = N_B/N$. The periodical boundary conditions were imposed on the basic lattice fragment. As a distance between any two particles was taken their minimum separation *incorporating* periodically repeated particles. Any particle A can react with any B, the recombination rate $\sigma(r)$ depends on the relative distance $r_{ij} = |\vec{r}_i - \vec{r}_j'|$ only. Starting from $t = 0$, a single AB pair disappears on each time-step.

According to (3.2.15) the probability for a given AB pair to survive t seconds is

$$\phi_{AB}(t) = \exp\left[-\sigma(r_{ij})t\right]. \tag{5.1.61}$$

Probability $\phi(t)$ for *all* AB pairs to survive t seconds is a product of all factors $\phi_{AB}(t)$

$$\phi(t) = \prod_i \prod_j \phi_{AB}(t) = \exp\left(-\sum_i \sum_j \sigma(r_{ij})t\right). \tag{5.1.62}$$

It means that time intervals τ_m between decay of a given AB pair and another similar decay are distributed exponentially with the mean decay time $\tau = [\sum_i \sum_j \sigma(r_{ij})]^{-1}$. Probability w_{AB} that a *given* AB pair will recombine next is proportional to its recombination rate $\sigma(r_{ij})$: $w_{AB} = \sigma(r_{ij})\tau$.

Computer simulations consist of the following steps. (i) Calculating of the time-step $\tau = [\sum_i \sum_j \sigma(r_{ij})]^{-1}$. (ii) Making use of random numbers and a set of weighting factors $w_{AB} = \sigma(r_{ij})\tau$. (iii) A given AB pair recombines and reaction time t is increased by τ. (iv) Passage to (i).

Note some essential details of the simulation scheme. A choice of a particles' pair to disappear consists of *two* steps: with the help of the weighting factors $w_A = \sum_j w_{AB}$ a single A particle was chosen, whereas particle B was sought respectively amongst B's with the help of weights w_{AB}. Since recombination rates $\sigma(r_{ij})$ of different pairs can differ by several orders of magnitude, numerical errors have to be controlled. Instead of the mean decay time τ, in principle, in successive steps stochastic time decay variation τ_m could be done, assuming τ_m to be exponentially distributed with the mean value τ. This procedure does not affect averaged result, but increases the dispersion. Besides the time development of the particle densities, the joint correlation functions $X_\nu(r,t)$ and $Y(r,t)$ were calculated periodically. To do it, a number of similar (dissimilar) pairs separated by a distance varying between r and $r + \Delta r$, was calculated and normalized in order to satisfy the asymptotic behaviour, equation (2.3.23) of the correlation function as $r \to \infty$. The figures obtained were additionally averaged over different simulation realizations (as a rule, more than 10 realizations) and different initial random particle distributions provided the initial reactant densities were fixed to be the same. Note that simulations on lattices are limited in time: the correlation length, whose time development could be found through analysis of the correlation functions, has to be essentially less than the linear size of a system. Similar principles have been also used in computer simulations of immobile particle *accumulation* [20, 21].

5.1.2.3 Tunnelling recombination

In this Section different analytical approximations for the static tunnelling recombination based on the recombination rate (3.1.2) are compared with the direct statistical simulations. For dimensions $\bar{d} = 1, 2$ considered here a new spatial scale $r_0 = 5a$ was used (a is lattice spacing). All concentrations are given in dimensionless units $r_0^{-\bar{d}}$, where as time t is in units of σ_0^{-1}.

First of all, let us discuss the case of equal concentrations: $n_A(t) = n_B(t) = n(t)$ when two kinds of similar correlation functions coincide: $X_\nu(r,t) = X(r,t)$, $\nu = $ A, B. In Fig. 5.2 the concentration development in the one-dimensional case is presented [26]. The curve (a) gives averaged (over 10 simulations) computer-calculated density. Stripped lines demonstrate dispersion of results: they correspond to the curves $\langle n(t) \rangle \pm s(t)$, where $s^2(t) = \langle n^2(t) \rangle - \langle n(t) \rangle^2$ is a standard deviation. Curve (b) shows the *numerical* solution of a set (4.1.19), (4.1.28), and (5.1.14) to (5.1.16) derived in the framework of the superposition approximation. Curve (c) gives results of the linear approximation (4.1.41) and (4.1.42). At last, the additional curve (d) is drawn just to illustrate concentration behaviour at short times. In the linear approximation we neglect similar reactant correlation, $X(r,t) = 1$, whereas in curve (d) *dissimilar* (AB) reactant correlations (4.1.40) are also

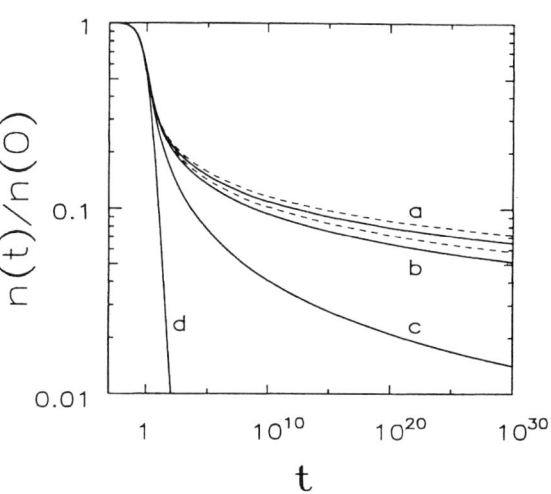

Fig. 5.2. The kinetics of tunnelling recombination $n(t)$ for $\bar{d} = 1$. The initial concentrations $n_A(0) = n_B(0) = 0.5$ ($N_A = N_B = N_0 = 10^4$ at $t = 0$) on the chain from 10^5 sites; a – computer simulations (standard deviations are indicated), b – the superposition approximation, c – linear approximation, d – the complete neglect of all correlations, see equation (5.1.64).

neglected: $X(r,t) = Y(r,t) = 1$. In such a case the reaction rate $K(t)$ turns out to be time-independent,

$$K(t) = K_0 - \int \sigma(r)\,d\vec{r} = \gamma_{\bar{d}} r_0^{\bar{d}} \sigma_0 \Gamma(\bar{d}), \qquad (5.1.63)$$

where $\Gamma(x)$ is the Euler gamma function. Thus kinetics obeys very simple relation

$$\frac{n(t)}{n(0)} = \frac{1}{1 + K_0 n(0)t}. \qquad (5.1.64)$$

In Fig. 5.3 similar curves are presented for $\bar{d} = 2$, which are averaged over 20 simulations.

Note that if primitive approach (5.1.64) is valid at $t < 1$ only (small reaction depths $\Gamma < 0.5$), the linear approximation has the greater applicability range, $\Gamma \leqslant 1$ whereas at $\Gamma > 1$ the linear approximation begins to deviate considerably from computer simulations.

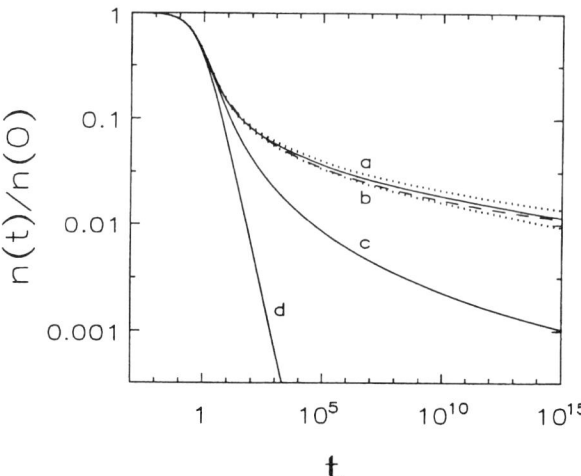

Fig. 5.3. The kinetics of the static tunnelling recombination $n(t)$ for $\bar{d} = 2$. Concentrations: $n_A(0) = n_B(0) = 0.25$ ($N_A = N_B = N_0 = 10^4$ at $t = 0$). Square lattice contained 1000×1000 sites: a – computer simulations (standard deviations are shown); b – superposition approximation; c – linear approximation neglecting similar reactant correlations, equation (5.1.64).

The conclusion which could be drawn from Fig. 5.3 is that the superposition approximation works surprisingly well for the $A + B \to 0$ reaction. As it is shown below in Chapter 6, computer simulations in Figs 5.2 and 5.3 at long time enter into the new regime where the kinetics is getting *fluctuation-controlled*. For $\bar{d} = 1$ and at long t the superposition approximation results in the kinetics (b) falling out the limits of the standard deviation (but still has correct asymptotics – see Chapter 6 below) thus giving *lower-bound* estimate. In $\bar{d} = 2$ case the curve of this approximation lies in the limits of the standard deviation, thus computer simulations coincide with results of numerical calculations of the kinetic equations.

In the $\bar{d} = 3$ case accuracy of the superposition approximation was tested for smaller reaction depths ($\Gamma < 1$) [11, 25] – see Fig. 5.4.

The correlation functions can serve as an additional test of the correctness of the superposition approximation. A comparison of computer simulations and this approximation are presented in Figs 5.5 and 5.6. Since due to insufficient statistics the correlation functions begin to oscillate at short distances, they were additionally smoothed through averaging over interval $\Delta r \sim 10 r_0$.

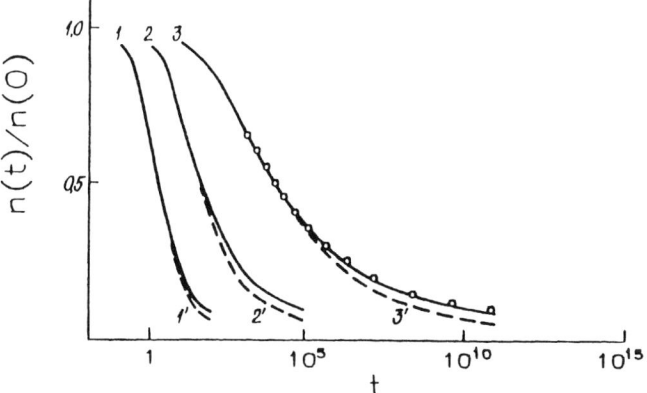

Fig. 5.4. The kinetics of tunnelling recombination $n(t)$ for $\bar{d} = 3$. The initial concentrations: $n_A = n_B = n_0 = 0.1$ (curves 1, 1'), 0.01 (curves 2, 2'), 0.001 (curves 3, 3'). Curves 1 to 3 corresponds to the superposition approximation, whereas curves 1'–3' represent the linear approximation. Computer simulations [23] are given by circles.

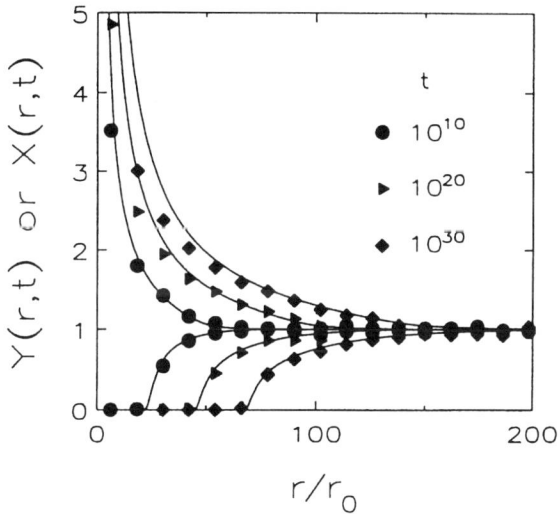

Fig. 5.5. The correlation functions for similar (X) and dissimilar (Y) particles for $\bar{d} = 1$. Full curves are result of the superposition approximation, whereas symbols – computer simulations averaged over 10 realizations and additionally over intervals $\Delta r = 5r_0$. Parameters are the same as in Fig. 5.2.

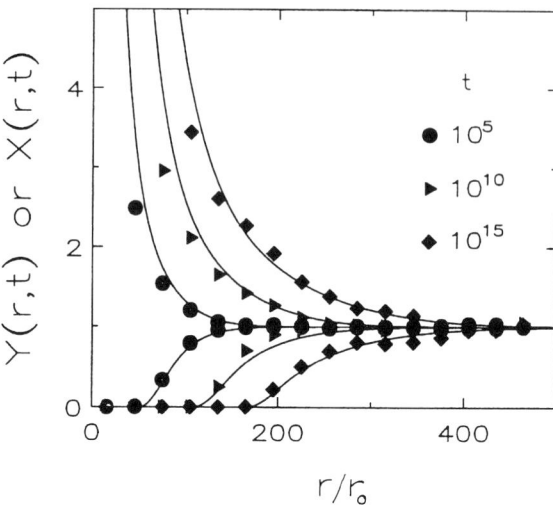

Fig. 5.6. Same as Fig. 5.5 for $\bar{d} = 2$. Computer simulations are averaged over 20 realizations and additionally over intervals $\Delta r = 6r_0$. Parameters are the same as in Fig. 5.3.

Analysis of the correlation functions demonstrates also impressive general agreement between the superposition approximation and computer simulations. Note, however certain overestimate of the similar particle correlations, $X(r,t)$, at small r, especially for $\bar{d} = 1$. In its turn the correlation function of *dissimilar* particles, $Y(r,t)$, demonstrates complete agreement with the statistical simulations. Since the time development of concentrations is defined entirely by $Y(r,t)$, Figs 5.2 and 5.3 serve as an additional evidence for the *reliability* of the superposition approximation. An estimate of the small distances here at which the function $Y(r,t)$ is no longer zero corresponds quite well to the earlier introduced correlation length ξ_0, equation (5.1.47): as one can see in fact that at moment t there are no AB pairs separated by $r < \xi_0$.

For unequal concentrations, $n_A(t) < n_B(t)$, the reaction depth $\Gamma < \Gamma_0 = 3$ reached in computer simulations is not enough for finding asymptotic laws but still permits to estimate qualitatively the accuracy of the superposition approximation. In Figs 5.7 and 5.8 numerical solution of the relevant kinetic equations is compared with computer simulations. To make situation more transparent, the linear approximation results are plotted in curves (d) for a *single* choice of initial concentrations only.

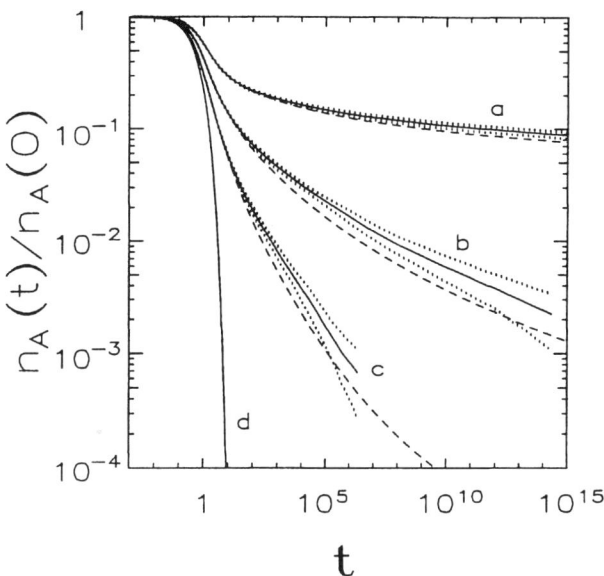

Fig. 5.7. The kinetics of the tunnelling recombination $n(t)$ for $\bar{d} = 1$. Full curves a to c are averages over 10 realisations, standard deviations are shown. Broken lines – the superposition approximation. Initial concentrations: $n_A(0) = 0.5$; whereas $n_B(0) = 0.5$ (a), 0.75 (b), 1.00 (c). Curve d is the linear approximation with parameters of c.

An increase of the standard deviation at $\Gamma \simeq 3$ due to small number of survived particles, demonstrates a limited possibility of the direct statistical simulations for a system with a variable number of particles. However, certain conclusions could be drawn even for such limited statistical information. Say, if for equal concentrations the analytical theory based on the superposition approximation seems to be quite adequate, for unequal concentrations its deviation from the computer simulations greatly increases in time. The superposition approximation gives the *lower bound* estimate of the actual kinetic curves $n_A(t)$ but if for $\bar{d} = 2$ shown in Fig. 5.8 the deviation is considerable, for $\bar{d} = 1$ (Fig. 5.7) it is not observed, at least for the reaction depths considered.

More concrete conclusions could be drawn for the linear approximation applicability: it is adequate for *small* reaction depths $\Gamma < 1$, whereas at $\Gamma > 1$ it is in serious error. In its turn, errors of the superposition approximation are essentially less, the relevant *lower bound* estimate is quite acceptable to fit theoretical parameters to the experimental curves.

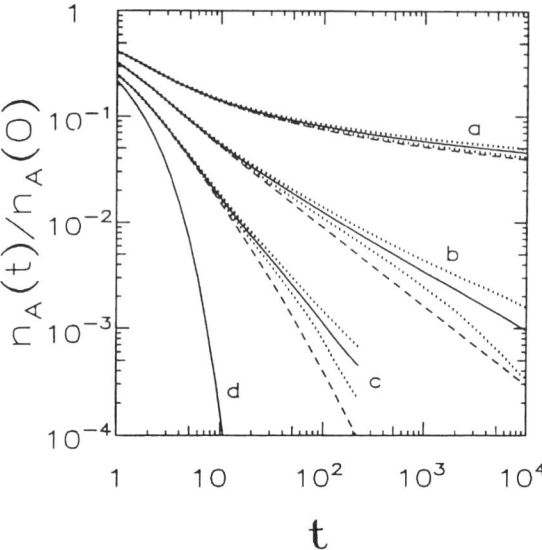

Fig. 5.8. The kinetics of the tunnelling recombination $n(t)$ for $\bar{d} = 2$. Notations are the same as in Fig. 5.7. Initial concentrations: $n_A(0) = 0.25$; $n_B(0) = 0.25$ (a), 0.3125 (b), 0.375 (c).

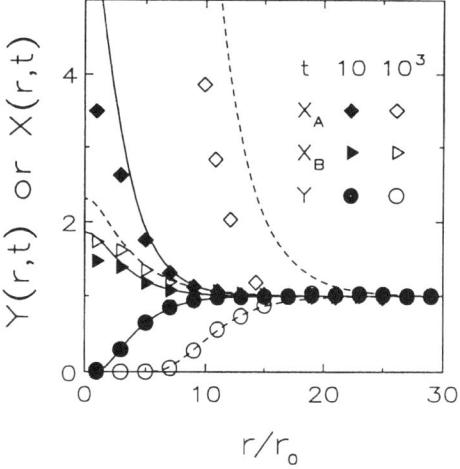

Fig. 5.9. The correlation functions of similar (X_A, X_B) and dissimilar (Y) particles for $\bar{d} = 2$. Full $(t = 10)$ and broken $(t = 10^3)$ lines are results of the superposition approximation. Symbols show computer simulations. Initial concentrations: $n_A(0) = 0.25$ and $n_B(0) = 0.3125$ as in curve b, Fig. 5.8. Curves are averaged over 10 realisations and additionally over intervals $\Delta r = 2r_0$.

Quantitative deviations are seen also from the correlation shown in Fig. 5.9. The correlation functions of dissimilar particles $Y(r,t)$ are in good agreement with simulations, which results also in a reliable reproduction of the decay kinetics for $n_A(t)$ – unlike behaviour of the correlation functions of the *similar* particles $X_\nu(r,t)$ which is very well pronounced for $X_A(r,t)$. Positive correlations, $X_\nu(r,t) > 1$ as $r < \xi$, argue for the similar particle aggregation, and the superposition approximation tends to *overestimate* their density. The obtained results permit to conclude that the approximation (2.3.63) of the three-particle correlation function could be in a serious error for the excess of one kind of reactants.

5.1.2.4 Multipole interaction

The multipole interaction of immobile particles (4.1.44) is an additional way to check up advantages of the superposition approximation [8]. The reason is that the tunnelling recombination (3.1.2) serves better as an example of *short-range* reaction. Indeed, the distinctive scale r_0 characterizing distant (non-contact) interaction could be defined as

$$r_0^{-1} = -\lim_{r \to \infty} \frac{\ln \sigma(r)}{r}. \tag{5.1.65}$$

For the tunnelling recombination r_0 is typical of the order of 1 Å. In this sense the multipole interaction is a *long-range* one since substitution of (4.1.44) into (5.1.65) yields $r_0 = \infty$. Generally speaking, the accuracy of the superposition approximation can turn out to be dependent on the actual recombination law.

To be short, in Fig. 5.10 results of the computer simulation (full curves) and the superposition approximation (broken curves) are compared for $\bar{d} = 1$ and 2 and equal particle concentrations, $n_A(t) = n_B(t) = n(t)$. Time is given is units a^m/σ_0, where a is a lattice constant.

As it took place for the tunnelling recombination, divergence in results is not large. It will be shown in Chapter 6 that the reaction depths studied here are enough to establish appearance of the new asymptotic kinetic laws. The superposition approximation giving a lower bound *estimate* of the kinetics, reproduces correctly the kinetics at long times. Results of the linear approximation are not plotted since they diverge considerably from the statistical simulations.

For unequal reactant concentrations a divergence is observed to be small – Fig. 5.11. In physics of phase transition long-range interactions are known

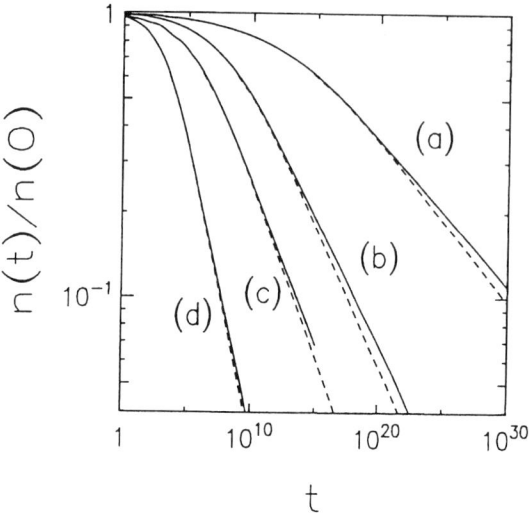

Fig. 5.10. The kinetics of multipole recombination $n(t)$ for $\bar{d}=1$ (curves a and b) and $\bar{d}=2$ (curves c and d). The powers of the multipole interaction: $m=10$ (curves a and c) and $m=6$ (curves b and d). Initial lattice densities $\rho_a(0) = N_A/N = \rho_b(0) = 0.01$ on the lattice of $N=10^6$ sites.

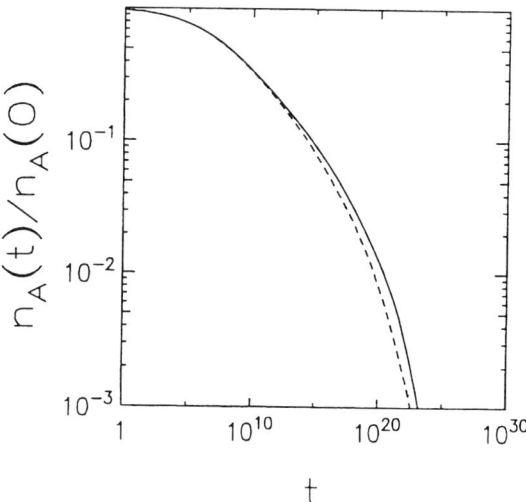

Fig. 5.11. The kinetics of the multipole recombination $n_A(t)$ for $\bar{d}=1$. The power of interaction $m=6$. Initial lattice densities $\rho_a(0) = N_A/N = 0.018$, $\rho_b(0) = N_B/N = 0.020$ on the lattice of $N=10^6$ sites.

to be well-described in terms of the *molecular (mean)*-field approximation [14]. Despite the fact that extrapolation of results obtained in physics of equilibrium critical phenomena into non-equilibrium bimolecular reactions should be carefully justified, results demonstrated in Figs 5.10 and 5.11 argue for such a similarity.

5.1.2.5 Diffusion-controlled reactions

Another important test of the accuracy of the superposition approximation is the diffusion-controlled A + B → 0 reaction. For the first time it was computer-simulated by Toussaint and Wilczek [27]. They confirmed existence of new asymptotic reaction laws but did not test different approximations used in the diffusion-controlled theories. Their findings were used in [28] to discuss divergence in the linear and the superposition approximations. Since analytical calculations [28] were performed for other sets of parameters as used in [27], their comparison was only qualitative. It was Schnörer et al. [29] who first performed detailed study of the applicability of the superposition approximation.

In computer simulations a discrete lattice of sites is considered, each site is occupied by not more than one particle. Particles A are localized in their sites for τ seconds and then possess hops. Thus, the mean diffusion coefficient $D_A = a^2/(2\bar{d}\tau)$ could be introduced. We assume that particles B are immobile, $D_B = 0$, since it permits to reduce greatly simulation time. Moreover, for $D_A = D_B$ and $\bar{d} = 1, 2$ the kinetics turn out to be quite similar. A hop of particle A into the site occupied by particle B results in their instant recombination.

To obtain reliable statistical data in the region of large reaction depths, $\Gamma \simeq 3 \div 4$, great *initial* lattice densities were used, $\rho_a, \rho_b \simeq 0.4$, so that most of lattice sites are initially occupied by particles. Since the given particle A cannot hop to a site occupied by another particle at great particle densities efficient diffusion coefficient is less than $D_A = a^2/(2\bar{d}\tau)$. In analytical calculations (in terms of the superposition and linear approximations) the black sphere model with radius r_0 equal to the lattice spacing was used. Certain difference in the lattice (discrete) and continuous statements of the problem is responsible for a small deviation of results at small reaction depths. In its turn, for greater Γ values the lattice densities are no longer large, $D_A = a^2/(2\bar{d}\tau)$ and the kinetics is governed not by the spatial scale-factor r_0 but by the diffusion length l_D. Therefore, the superposition approximation could be reliably tested for the *great* reaction rates – see Figs 5.12 and 5.13.

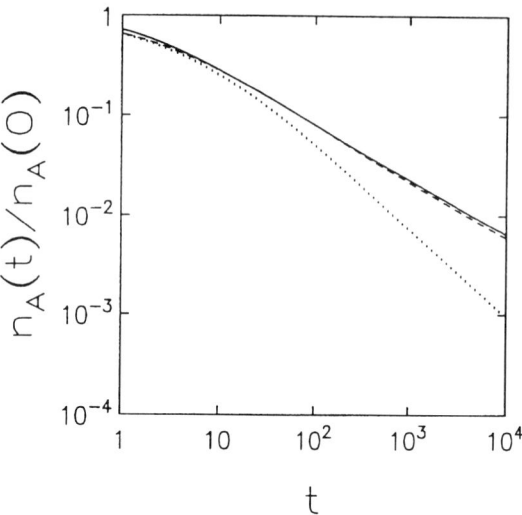

Fig. 5.12. The kinetics of diffusion-controlled recombination $n_A(t)$ for $\bar{d} = 2$. Initial lattice densities $\rho_a(0) = N_A/N = \rho_b(0) = 0.4$ on a lattice with $N = 1000 \times 1000$. Solid curve – computer simulations, broken line – the superposition approximation, dotted line – linear approximation. Time is in units τ (waiting time in a site).

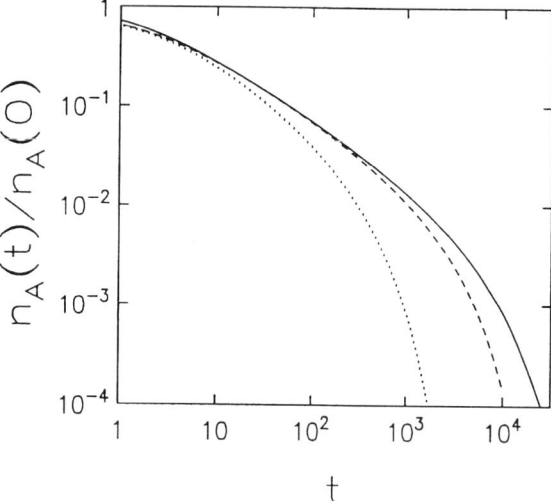

Fig. 5.13. The same as in Fig.5.12 with initial densities $\rho_a(0) = 0.40$, $\rho_b(0) = 0.41$.

In the small-Γ region the kinetic equations should be tested using *small lattice densities* which demonstrates their correctness.

The analysis of the diffusion-controlled computer simulations confirms once more conclusions drawn above for the static reactions of immobile particles. In particular, the superposition approximation gives the best *lower bound* estimate of the kinetics reaction, $n = n(t)$. Divergence of computer simulations and analytical theory being negligible for equal concentrations become essential for large depths and when one of reactants is in excess. The obtained results allow us to use the superposition approximation for testing the applicability of simple equations of the linear theory in those cases when computer simulations because of some reasons cannot be performed. Examples will be presented in Chapter 6.

Summing up, note that the direct statistical (computer) simulation does not demonstrate serious errors of the superposition approximation for equal reactant concentrations. Divergence begins first of all for unequal concentrations: for not very large reaction depths $\Gamma \leqslant 2$ it is almost negligible, at $\Gamma > 2$ and especially asymptotically (as $t \to \infty$) it becomes important, but the complete quantitative analysis cannot be done due to unreliable statistics of results. In this Section we have restricted ourselves to the A + B → 0 reaction *without particle generation*. Testing of the superposition approximation accuracy for the case of particle creation will be done in Chapter 7.

5.2 THE A + B → B REACTION

History supplies us mainly with bare facts rather than with understanding of their sense.

V. Klyuchevsky, old Russian historian

5.2.1 Correlation dynamics

5.2.1.1 Prehistory

For a long time the kinetics of the A+B → B reaction has been considered in the framework of a simple linear equation for the joint correlation function of dissimilar reactants defining the reaction rate [30]. The publication of the pioneering paper by Balagurov and Vaks [31], where the existence of the decay "tail" at long times for this reaction was shown for the first time, has stimulated a lot of similar investigations (e.g., [11, 32–45]). It was shown that the Balagurov–Vaks relation for the survival probability, $\omega(t)$, of donors A migrating with the diffusion coefficient D_A among randomly distributed non-saturable traps, B,

$$\lim_{t\to\infty} \frac{\ln \omega(t)}{(D_A t)^{\bar{d}/(\bar{d}+2)}} = -C_{\bar{d}} n_B^{2/(\bar{d}+2)} \qquad (5.2.1)$$

($C_{\bar{d}}$ is constant), is *both* the upper bound [32, 33, 39] and the lower [37] estimate of the exact solution. Equation (5.2.1) results from a strict mathematical theorem [32, 33] and is proved by its coincidence for $\bar{d} = 1$ with the exact solution for both continuous [31, 37] and lattice [43] statements of the problem.

Taking into account the exponential character of the solution, (5.2.1), let us introduce a new variable $F = -\ln \omega(t)$ and the corresponding critical exponent

$$\alpha = \lim_{t\to\infty} \frac{d \ln F}{d \ln t}. \qquad (5.2.2)$$

Now (5.2.1) is related to the magnitude $\alpha = \bar{d}/(\bar{d}+2)$ which is dependent on the space dimensionality, as in the case of the $A + B \to 0$ reaction discussed above.

All the above-mentioned approaches are essentially *asymptotic*, i.e., they do not allow the description of the $A + B \to B$ kinetics at *intermediate* times, and can estimate the exponent α only. In particular, the mathematical approach by Balagurov–Vaks and their successors is based on the idea that at long times the kinetics is determined exclusively by those reactants A which exist in large cavities free from reactants B (Chapter 2). (The made asymptotic analysis is very similar to that developed by Lifshitz [46] in the problem of density of the electronic states near the band edge of disordered semiconductors). For further estimates, it is assumed that inside these cavities the concentrations $C_A(\vec{r}, t)$ are governed by the diffusion equation with a zero boundary condition. The recombination law, $\sigma(r)$, is not detailed or discussed at all. Due to irregularity in the distribution of traps B, the boundaries of these cavities are also irregular. This point has been ignored in the approach [31–33] which aimed at an exponentially exact solution only. Assuming these cavities to be spherical, if $\bar{d} = 3$ (or circular, for $\bar{d} = 2$), and following Balagurov–Vaks, we obtain at long times equation (2.1.102). Averaging (2.1.102) over cavities of different sizes (with the Poisson distribution, equation (2.1.103)), we indeed arrive at (5.2.1). However, there remain serious doubts about the universality of the critical exponent $\alpha = \bar{d}/(\bar{d}+2)$, e.g., since the derivation of (5.2.1) just presented assumes the *immobility* of the reactants B ($D_B = 0$) and it *cannot* be generalised to the case of mobile B.

Moreover, for the immobile reactants A, the exact solution yields $\alpha = 1$ ($\bar{d} > 1$) rather than (5.2.1)! This argues for the relative diffusion coefficient $\kappa = D_A/(D_A + D_B)$ as one of the key parameters of this kinetics. However, at present there are no rigorous estimates in the general case $0 \leqslant \kappa \leqslant 1$ (see, e.g., [47] and discussion below).

Since the traditional approach to this reaction operates with the linearised equation for the joint densities, the parameter κ in which we are interested is absent here. In line with Section 5.1, we present here the novel approach valid for $0 \leqslant \kappa \leqslant 1$ and all reaction times [11, 48]. We pay more attention in this Section to mathematical formalism and discussion of the approximations involved.

If there is no interaction between similar reactants (traps) B, they are distributed according to the Poisson relation, $X_B(r,t) = 1$. Besides, since the reaction kinetics is linear in donor concentrations, the only quantity of interest is the survival probability of a *single* particle A migrating through traps B and therefore the correlation function $X_A(r,t)$ does not affect the kinetics under study. Hence the description of the fluctuation spectrum of a system through the joint densities $X_\nu(r,t)$, which was so important for understanding the A+B $\to$ 0 reaction kinetics, appears now to be incomplete. The fluctuation effects we are interested in are weaker here, thus affecting the critical exponent but not the exponential kinetics itself. It will be shown below that adequate treatment of these weak fluctuation effects requires a careful analysis of many-particle correlations.

5.2.1.2 The basic equations

To conform to Section 5.1 let us use again the approach of the many-point densities. The fact that the quantity of interest is the survival probability of a *single* particle A in terms of mathematics means that from the complete set of equations for many-point densities $\rho_{m',m}$, equation (2.3.38), we can restrict ourselves to those with only the first index equal to one: $m' = 1$; $m = 0, 1, \ldots, \infty$, that is

$$\frac{\partial \rho_{1,m}}{\partial t} = D_A \nabla_1'^2 \rho_{1,m} + D_B \sum_{j=1}^{m} \nabla_j^2 \rho_{1,m} - \sum_{j=1}^{m} \sigma(|\vec{r}_1' - \vec{r}_j|)\rho_{1,m} -$$

$$- \int \sigma(|\vec{r}_1' - \vec{r}_{m+1}|)\rho_{1,m+1} \, d\vec{r}_{m+1}. \quad (5.2.3)$$

Due to the homogeneity of the system, $\rho_{1,0}$ is independent of $\vec{r}_1'$ and is nothing but the concentration n_A, and the $\rho_{1,m}$ ($m \geq 1$) depend on the relative distances $r_j^* = \vec{r}_j - \vec{r}_1'$. Further, it is convenient to introduce the correlation functions $g_m = g_m(\{r^*\})$ through

$$\rho_{1,m} = n_A n_B^m g_m, \quad m \geq 1. \tag{5.2.4}$$

They describe the spatial correlation of a single A at the origin of coordinates with m B's at r_j^*. When $r_j^* \to \infty$, one gets $g_m \to 1$ (weakening of the correlations). Note also that $g_1 = g_1(r_1^*)$ rather than r_1^*.

After some mathematical manipulations with (5.2.3) we arrive at the equations of the concentration dynamics

$$\frac{d \ln n_A}{dt} = \frac{d \ln \omega(t)}{dt} = -n_B \int \sigma(r_1) g_1(r_1) \, d\vec{r}_1 \tag{5.2.5}$$

and its complementary equation for the correlation dynamics ($r_j^* \equiv r_j$)

$$\frac{\partial g_m}{\partial t} = D \left(\sum_{j=1}^{m} \nabla_j^2 + 2\kappa \sum_{i>j=1}^{m} \nabla^i \nabla_j \right) g_m - w_m g_m,$$
$$m = 1, 2, \ldots, \infty, \tag{5.2.6}$$

where

$$w_m = \sum_{j=1}^{m} \sigma(r_j) +$$
$$+ n_B \int \sigma(r_{m+1}) \left(\frac{g_{m+1}}{g_m} - g_1(r_{m+1}) \right) d\vec{r}_{m+1}. \tag{5.2.7}$$

To solve the problem, one has to consider the infinite (due to (5.2.7)) hierarchy of the non-linear equations (5.2.6). The set of correlation functions is complete and equation (5.2.6) is exact. The quantities w_m in (5.2.7) are no longer the effective trapping probabilities, since it is not self-evident that the integral term there is positively defined.

Due to the non-linearity of these equations and the complexity of the operator terms, the set (5.2.5) to (5.2.7) cannot be solved exactly and we have to cut off this hierarchy at some finite m.

5.2.1.3 The molecular field approximation

We assume hereafter the random initial distribution, $g_m = 1$ at $t = 0$. The important peculiarity of the A + B → B reaction is the existence for $\kappa = 0$ ($D_A = 0$) of the *exact* general *multiplicative* solution of (5.2.6) and (5.2.7)

$$g_m = \prod_{j=1}^{m} g_1(r_j) \tag{5.2.8}$$

which can be checked directly. For this reason we can restrict ourselves to the first equation only (since the spatial correlations of the particles B are absent):

$$\frac{\partial g_1}{\partial t} = D\nabla_1^2 g_1 - \sigma(r_1)g_1, \tag{5.2.9}$$

with $D = D_B$. As it is known [13], multiplicity of the correlation functions constitutes a foundation of the simplest *molecular field* approximation.

This is why the cut-off of set (5.2.6) at the first step $m = 1$, through substitution into (5.2.7), as $g_2 = g_1(r_1)g_1(r_2)$ yields the *exact* solution (for $\kappa = 0$).

In the case when $\kappa \neq 0$ the correlation functions g_m, $m \geqslant 2$ are no longer multiplicative but have angular dependence, e.g., g_2 depends on r_1, r_2 and the angle between them: $g_2 = g_2(r_1, r_2; \vartheta)$. However, the molecular field approximation neglects this fact and again yields (5.2.9) which is now only an approximate solution.

The conclusion can be drawn that the traditional approach [30] corresponds to a *zero-level* approximation, i.e., a cut-off of the hierarchy at $m = 1$ thus linearising (5.2.9) for g_1 and neglecting all integral terms in w_1 (equation (5.2.7)) which describe competition of several B's for some reactant A. It is not surprising that the "standard" critical exponent $\alpha = 1$ obtained from equations (5.2.5) and (5.2.9) is independent of the relative diffusion coefficient, since κ enters into equation (5.2.6) only for $m \geqslant 2$. Therefore, in an approximation at the next level we have to consider the equation for the correlation function g_2 at least. This level of approximation could be acceptable (at least to reproduce equation (5.2.1)) if the reaction for $\kappa \neq 0$ remains weakly non-ideal (the g_m is nearly multiplicative).

Equation (5.2.1) is derived for $\kappa \equiv 1$. There is a reason to think that this single *point* $\kappa = 1$ only is the range of the validity of the critical exponent

$\alpha = \bar{d}/(\bar{d}+2)$, since for this magnitude of κ the quadratic differential forms in equation (5.2.6) become parabolically degenerate: their diagonalisation is accompanied by the emergence of several zero coefficients. The peculiarity of the basic equations for $\kappa = 1$ is easily seen for g_2: without recombination ($w_2 = 0$), substitution of the solution in the form of plane waves $g_2 \propto \exp[-\varepsilon t + i(\vec{q}_1 \vec{r}_1 + \vec{q}_2 \vec{r}_2)]$ yields the dispersion law $\varepsilon = D(\vec{q}_1^2 + 2\kappa \vec{q}_1 \vec{q}_2 + \vec{q}_2^2)$. For $\kappa < 1$ the sole non-dissipative mode ($\varepsilon = 0$) with $\vec{q}_1 = \vec{q}_2 = 0$ exists, while for $\kappa = 1$ a set of coherent modes with $\vec{q}_1 = -\vec{q}_2$ arises.

Therefore, the weakness of the fluctuation effects in the A + B → B reaction as compared to the case of A + B → 0 (Section 5.1) does not permit us to restrict ourselves to the *radial* correlations between A's and B's; subtle effects leading to equation (5.2.1) arise due to the *angular* correlations of reactants B in the vicinity of a single A. The demonstrated peculiarity of the A + B → B reaction requires an adequate specified mathematical approach.

5.2.1.4 The black sphere model

The integral terms in equation (5.2.7) play a key role in obtaining relation (5.2.1). These terms, despite their limited or even small magnitudes as compared to $\sum_{j=1}^{m} \sigma(r_j)$, have a different action radius depending on time.

In the case of the *black sphere* model (equation (3.1.1)) describing instant recombination, using the dimensionless variables $r = r/r_0$, $t = Dt/r_0^2$, $n_B = \gamma_{\bar{d}} n_B r_0^{\bar{d}}$ (see Section 5.1) we can rewrite equation (5.2.5) in the form

$$\frac{d\ln \omega}{dt} = -K(t) n_B, \qquad K(t) = \left. \frac{\partial g_1}{\partial r_1} \right|_{r_1=1}. \qquad (5.2.10)$$

The boundary conditions for the correlation functions are $g_m = 0$ at $r_j \leqslant 1$ for each $j = 1, \ldots, m$. Thus we obtain instead of (5.2.6)

$$\frac{\partial g_m}{\partial t} = \left(\sum_{j=1}^{m} \nabla_j^2 + 2\kappa \sum_{i>j=1}^{m} \nabla_i \nabla_j \right) g_m - w_m g_m,$$

$$m = 1, 2, \ldots, \infty, \qquad (5.2.11)$$

$$w_m = n_B K(t) \left\{ \frac{1}{\gamma_{\bar{d}}} \int d\vec{S}_{m+1} \left(\frac{\nabla_{m+1} g_{m+1}}{K(t) g_m} - 1 \right) \right\}. \qquad (5.2.12)$$

The integration in (5.2.12) is carried out over the surface of a $\bar{d}$-dimensional unit radius sphere.

The reaction radius r_0 does not enter the asymptotic relation (5.2.1). Its absence at $t \to \infty$ in the spirit of Section 5.1 could be interpreted as emergence of a *new spatial scale* – the correlation length ξ. It arises in (5.2.11) through the terms $w_m g_m$ which play the role of the correlation sources.

5.2.1.5 The Kirkwood superposition approximation

The next approximation level requires the use of the three-point correlation functions g_3 expressed through the Kirkwood superposition approximation (Section 2.3.1)

$$g_3 = \frac{g_2(\vec{r}_1, \vec{r}_2) g_2(\vec{r}_2, \vec{r}_3) g_2(\vec{r}_3, \vec{r}_1)}{g_1(r_1) g_1(r_2) g_1(r_3)}. \tag{5.2.13}$$

Note that the introduction of the correlation functions g_m in (5.2.4) instead of $(m+1)$-point densities $\rho_{1,m}$ in fact enabled us to reduce the number of variables. For instance, the molecular field approximation, $g_2 = g_1(r_1) g_1(r_2)$, corresponds to that for superposition approximation (equation (2.3.55)) for $\rho_{1,2}$ whereas, in its turn, equation (5.2.13) for g_3 corresponds to the higher-order superposition approximation (equation (2.3.56)) for $\rho_{1,3}$. When substituting (5.2.13) into (5.2.12) with $m = 2$, we obtain an exact equation for g_1 with

$$w_1 = n_B K(t) \left\{ \frac{1}{\gamma_{\bar{d}}} \int d\Omega_3 \left(\frac{g'(r_1; \vartheta_1)}{K(t) g_1(r_1)} - 1 \right) \right\} \tag{5.2.14}$$

and the approximate (due to the use of equation (5.2.13)) equation for g_2 with

$$w_2 = n_B K(t) \left\{ \frac{1}{\gamma_{\bar{d}}} \int d\Omega_3 \left(\frac{g'(r_1; \vartheta_1) g'(r_2; \vartheta_2)}{K^2(t) g_1(r_1) g_1(r_2)} - 1 \right) \right\}. \tag{5.2.15}$$

In (5.2.14) and (5.2.15), the integration is carried out over the spatial angle and

$$g'(r_j; \vartheta_j) = \left. \frac{\partial g_2(r_j, r_3; \vartheta_j)}{\partial r_3} \right|_{r_3 = 1}, \tag{5.2.16}$$

where ϑ_j is an angle between the vectors $\vec{r}_j$ and $\vec{r}_3$.

5.2.1.6 Discussion

The equations derived above, describing the A + B → B reaction kinetics in terms of the correlation functions g_1 and g_2, have the form of the non-linear generalised multi-dimensional diffusion equation. Ignoring the multi-dimensionality of the operator terms in (5.2.11), these equations could be formally considered as similar to the basic non-linear equations for the A + B → 0 reaction (Section 5.1). Equations studied in both Sections 5.1 and 5.2 are derived with the help of the Kirkwood superposition approximation, the use of which leads to several equations for the correlation functions of similar and dissimilar reactants.

The main difficulty in solving the equation for g_2 arises from the *multi-dimensionality* of the problem; the operator terms contain derivatives with respect to r_1, r_2 and $\cos\vartheta$, including mixed derivatives, whereas in equations (5.2.14) and (5.2.15) the integration is done over the angular variable. Taking all this into account, the creation of an efficient and time-saving difference scheme for treating the sought solution in the long-time asymptote is not a trivial problem. On the other hand, in order to study correctly these subtle effects, alternative *analytical* methods for solving (5.2.11) and (5.2.12) have to be developed (as it has been done for the A + B → 0 reaction – see Section 5.1 and Chapter 6) and their results should be compared with the computer calculations.

We have shown that a formal analytical solution of the *steady-state* equations for the correlation functions g_1 and g_2 for $\bar{d} = 3$, using an expansion in a power series of the small parameters $\kappa \ll 1$, yields a correction to the steady-state reaction rate $K(t)$ of the order of κ^2, but the coefficient at κ^2 diverges. This indicates that the small parameter approach *fails* for the analysis of the set of equations describing the correlation dynamics, since even arbitrarily small κ violates the steady-state regime of $K(t)$ at $t \to \infty$, and thus the reaction rate becomes essentially time-dependent. The existence for $\kappa = 1$ and $\bar{d} = 1$ of an exact solution is associated with the degeneration of the correlation functions g_m, leading to the cut-off of the infinite hierarchy (5.2.11). For $\kappa = 1$ ($D_B = 0$) and the black sphere model, any reactant A can recombine with the traps B *nearest* to it on the left or right only since other traps are out of the game (the so-called *bus effect*). In terms of g_m, this screening effect results in the necessity to consider only the configurations BAB (two-coordinate functions) and AB (single-coordinate function). Due to the topological peculiarity of the case $\bar{d} = 1$, this is not the case for $\bar{d} = 2$

and 3. Numerical solution of equations (5.2.11) and (5.2.12) for $\bar{d} = 1$, taking into account the degeneracy of the solution mentioned, reproduces the result [31].

5.2.2 Other approaches

5.2.2.1 Density expansion

In this Section we consider several approaches which differ from the many-point density formalism discussed above. Szabo et al. [45] have introduced a novel method based on the *density expansion* for the survival probability, $\omega(t)$. Consider a system containing walkers (particles A) and N traps (quenchers B) in volume V in $\bar{d}$-dimensional space. We assume that the particles have a finite size but the traps can be idealized as points and hence are ignorant of each other. When the concentration of the walkers is sufficiently low so that excluded volume interactions between them are negligible, one might focus on a single walker.

Let $\vec{r}$ and $\vec{r}'_j$, $j = 1, \ldots, N$ be the coordinates of the walker and traps, respectively, in this $\bar{d}$-dimensional space at $t = 0$. The *survival probability* of the walker for the given initial positions of the traps, $W_{1,N} = W_{1,N}(\vec{r}; \vec{r}'_1, \ldots, \vec{r}'_N; t)$, satisfies the many-body diffusion equation (2.3.69):

$$\frac{\partial W_{1,N}}{\partial t} = \left(D_A \nabla^2 + D_B \sum_{j=1}^{N} \nabla'^2_j\right) W_{1,N}, \qquad (5.2.17)$$

which is to be solved subject to the initial condition that $W_{1,N} = 1$ ($t = 0$) and the boundary conditions that $W_{1,N} = 0$ whenever the particle comes in contact with any of the traps. Note that while the diffusion equation is separable in these coordinates, the boundary conditions are not. By introducing relative coordinates $\vec{r}_i = \vec{r}'_i - \vec{r}$, $W_{1,N} \equiv S_N(r_1, \ldots, r_N; t)$, equation (5.2.17) is transformed into equation (2.3.70),

$$\frac{\partial S_N}{\partial t} = D\left(\sum_{j=1}^{N} \nabla^2_j + 2\kappa \sum_{i>j=1}^{N} \nabla_i \nabla_j\right) S_N. \qquad (5.2.18)$$

Now the boundary conditions are separable, but if $\kappa \neq 0$, the diffusion equation is not separable. The problem is still hard.

Ultimately we wish to calculate the survival probability in the presence of uniformly distributed traps. This average is denoted by brackets and obtained by an integration over the volume accessible to the traps:

$$\langle S_n(t) \rangle = V^{-N} \int \cdots \int d\vec{r}_1 \ldots d\vec{r}_N S_N(\vec{r}_1, \ldots, \vec{r}_N; t). \tag{5.2.19}$$

The survival probability $\omega(t)$ for concentration n_B is a limit of the above $\bar{d} \times N$-dimensional integral as $N \to \infty$, and $V \to \infty$, taken in such a way that $n_B = N/V$ is constant.

Our starting point is a density analogous to that used in [49] in treating the migration of excitons between randomly distributed sites. This expansion is generalization of the cluster expansion in equilibrium statistical mechanics to dynamical processes. It is formally exact even when the traps interact, but its utility depends on whether the coefficients are well behaved as V and t approach infinity. For the present problem, the survival probability of equation (5.2.19) admits the expansion

$$\langle S_n(t) \rangle = 1 + \frac{N}{V} a_1(t) + \frac{N(N-1)}{V^2} a_2(t) +$$
$$+ \frac{N(N-1)(N-2)}{V^3} a_3(t) + \cdots. \tag{5.2.20}$$

The unknown coefficients $a_k(t)$ are determined by our requiring equation (5.2.20) to be exact for $1, 2, \ldots, N$ traps successively. In this way we find that

$$a_1(t) = \int d\vec{r}_1 \left(S_1(\vec{r}_1; t) - 1 \right),$$

$$a_2(t) = \iint d\vec{r}_1 \, d\vec{r}_2 \left(S_2(\vec{r}_1, \vec{r}_2; t) - 2S_1(\vec{r}_1; t) + 1 \right),$$

$$a_3(t) = \iiint d\vec{r}_1 \, d\vec{r}_2 \, d\vec{r}_3 \left(S_3(\vec{r}_1, \vec{r}_2, \vec{r}_3; t) - \right.$$
$$\left. -3S_2(\vec{r}_1, \vec{r}_2; t) + 3S_1(\vec{r}_1; t) - 1 \right), \tag{5.2.21}$$

and so on. To obtain $w(t)$ we take a limit of large N and V but finite $n_B = N/V$. To improve convergence we exponentiate the series to obtain

$$w(t) = \exp\left(n_B b_1(t) + \frac{1}{2!} n_B^2 b_2(t) + \frac{1}{3!} n_B^3 b_3(t) + \cdots\right). \quad (5.2.22)$$

This is cumulant-like expansion of the many-particle survival probability in terms of the many-body dynamics that determine the coefficients b_k. By demanding that the series expansions of equations (5.2.20) and (5.2.22) agree with the appropriate power of n_B, we get

$$b_1(t) = \int d\vec{r}_1 \left(S_1(\vec{r}_1;t) - 1\right) = a_1(t),$$

$$b_2(t) = \iint d\vec{r}_1 \, d\vec{r}_2 \left(S_2(\vec{r}_1, \vec{r}_2;t) - 2S_1(\vec{r}_1;t)S_1(\vec{r}_2;t)\right), \quad (5.2.23)$$

$$b_3(t) = \iiint d\vec{r}_1 \, d\vec{r}_2 \, d\vec{r}_3 \left(S_3(\vec{r}_1, \vec{r}_2, \vec{r}_3;t) - 3S_2(\vec{r}_1, \vec{r}_2;t)S_1(\vec{r}_3;t) \right.$$
$$\left. + 2S_1(\vec{r}_1;t)S_1(\vec{r}_2;t)S_1(\vec{r}_3;t)\right).$$

Expansion (5.2.23) is not necessarily valid for all times when truncated at an arbitrary level. The lowest-order truncation,

$$w(t) \approx \exp\left[n_B b_1(t)\right] = \exp\left\{-n_B \int \left(1 - S_1(\vec{r};t)\right) d\vec{r}\right\}, \quad (5.2.24)$$

is expected to be valid at short times, since initially reaction takes place only with (initially) nearby traps. To demonstrate the equivalence with the Smoluchowski approach, we differentiate equation (5.2.23) with respect to t. It follows that the above $w(t)$ solved equation (5.2.10) with

$$K(t) = -\frac{db_1(t)}{dt} = \int d\vec{r} \left(\frac{\partial S_1(\vec{r};t)}{\partial t}\right). \quad (5.2.25)$$

This is indeed the reactive flux for a trap–particle pair with a relative diffusion coefficient $D = D_A + D_B$.

In the limit of a static particle ($D_A = 0$) the cross terms in equation (5.2.18) vanish. In the absence of trap–trap excluded-volume interactions, the survival probability factorizes as

$$S_N(\vec{r}_1,\ldots,\vec{r}_N;t) = \prod_{j=1}^{N} S_1(\vec{r}_j;t). \tag{5.2.26}$$

From equation (5.2.23) it follows that $b_k = 0$ for all $k \geqslant 2$ and equation (5.2.22) reduces to equation (5.2.24). An alternative way, analogous to the treatment of static acceptors [50], is first to average the above expression to obtain

$$\langle S_N(t) \rangle = \langle S_1(t) \rangle^N. \tag{5.2.27}$$

Taking the limit $N, V \to \infty$, $N/V = n_B$, of the logarithm expanded to the first order gives equation (5.2.24).

We conclude that the first term in the expansion is exact for arbitrary diffusion coefficients at short times, for a static particle with noninteracting traps at all times, and that it precisely equals the Smoluchowski result in all dimensions. Equation (5.2.22) provides a systematic approach for improvement upon this approximation and can also be used to estimate the errors involved in such an approximation.

5.2.2.2 The $\bar{d} = 1$ case

One expects the Smoluchowski result for mobile traps and particles to improve as the dimensionality increases. Hence it is of interest to examine the worst case of diffusion in $\bar{d} = 1$. This may also be relevant to certain experimental systems [51]. To find the leading correction term in $\bar{d} = 1$ as a function of $\kappa = D_A/D$, we rewrite equation (5.2.22) as

$$\begin{aligned}\omega(t) &\approx \exp\left[n_B b_1(t)\right]\left(1 + \frac{1}{2} n_B^2 b_2(t)\right) \\ &= \exp\left[-f_1 \sqrt{\tau}\right](1 + f_2 \tau),\end{aligned} \tag{5.2.28}$$

where $\tau \equiv n_B^2 D t$ and the f_k's are related to the b_k's of equation (5.2.22). The distance between the j-th trap and the particle is denoted by y_j. The operator ∇_j in equation (5.2.18) becomes a partial derivative, $\partial/\partial y_j$.

The known result for an adsorbing boundary at the origin

$$S_1(y;t) = \text{erf}\left(-\frac{y}{\sqrt{4Dt}}\right), \quad y > 0 \qquad (5.2.29)$$

(erf is an error function) allows us to evaluate f_1 from (twice) the integral of $(1 - S_1)$ over y from 0 to ∞. This gives us $f_1 = -4/\sqrt{\pi}$ and well-known exact result for immobile particles (the so-called *target problem*) [4]

$$\omega(t) \propto \exp\left(-\frac{4}{\sqrt{\pi}} n_B \sqrt{D_B t}\right). \qquad (5.2.30)$$

We obtain f_2 by rewriting equation (5.2.18) for one particle, two traps, and positive y_j:

$$\frac{\partial S_2^\pm}{\partial t} = D\left(\frac{\partial^2}{\partial y_1^2} + \frac{\partial^2}{\partial y_2^2} \pm 2\kappa \frac{\partial^2}{\partial y_1 \partial y_2}\right) S_2^\pm(y_1, y_2; t). \qquad (5.2.31)$$

The cross term is positive for two traps on the same side of the particle (denoted by $+$) and negative for the two-sided $(-)$ configuration. $f_2^\pm$ is subsequently obtained from equation (5.2.23) (for $b_2(t)$) by use of S_1 of equation (5.2.29) and replacement of S_2 by $S_2^\pm$. Finally, $f_2 = f_2^+ + f_2^-$.

One may solve equation (5.2.31) for noninteracting point traps by transforming it into $2\bar{d}$ diffusion in a wedge of angle θ_0, $\cos\theta = \pm\kappa$, and absorbing sides [52, 53]. Integration of the Green's function for this problem in radial coordinates gives [45]

$$f_2^\pm = 8\sin\theta_0 \int_0^\infty r\,dr \int_0^{\theta_0} d\theta \left\{\sqrt{\frac{8}{\pi}} \sum_{n=0}^\infty \frac{\sin\left[(2n+1)\pi\frac{\theta}{\theta_0}\right]}{2n+1} \times\right.$$

$$\times r \exp(-r^2)\left[I_{\nu^+}(r^2) + I_{\nu^-}(r^2)\right] -$$

$$\left. - \text{erf}\left(\sqrt{2}r\sin\theta\right)\text{erf}\left[\sqrt{2}r\sin(\theta_0 - \theta)\right]\right\}, \qquad (5.2.32)$$

where $\nu^\pm \equiv [(2n+1)\pi/\theta_0 \pm 1]/2$, $I_\nu(r)$ are the modified Bessel functions of the first kind and order ν, and $\theta_0 = \arccos(\pm\kappa)$ is the wedge angle for the ($\pm$)-configuration.

The authors [45] were able to obtain f_2^- analytically for several special values of κ. For $\kappa = 1$ one gets

$$f_2 = 4\ln 2 - \frac{8}{\pi}, \qquad (5.2.33)$$

in agreement with the short-time expansion of the survival probability for static traps. To obtain the general solution for noninteracting traps equation (5.2.32) was integrated numerically and fitted to

$$f_2 \approx 0.197\kappa^2 + 0.0285\kappa^4, \qquad (5.2.34)$$

with an (absolute) error $\delta \leqslant 0.001$. This correction term is relatively small because the term $f_2^{\pm}$ have opposite signs.

The utility of equation (5.2.28) is illustrated in Fig. 5.14 for two κ values: $\kappa = 1/2$ ($D_A = D_B$) and $\kappa = 1$ ($D_B = 0$). $\kappa = 1$ is the worst case for the theory, yet the curve agrees remarkably well with the exact solution for static traps, except at very long times. The exact survival probability obtained from accurate simulations for $\kappa = 1/2$ is expected to be almost indistinguishable from the curve 2 calculated in the time range shown. The Smoluchowski

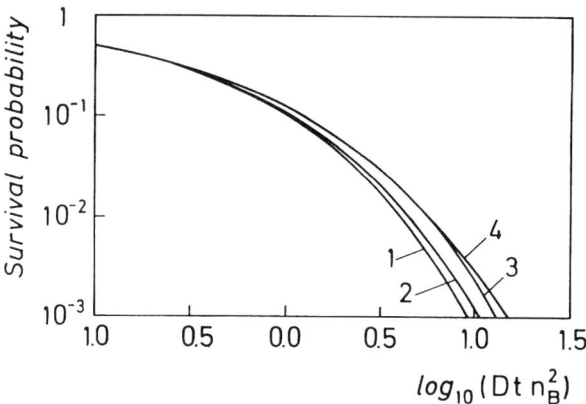

Fig. 5.14. The time dependence of the survival probability for one-dimensional diffusion with mobile, noninteracting traps for various values of κ. Curve 1 is exact for static particles ($\kappa = 0$) and identical to the Smoluchowski result; curve 4 is exact for static traps ($\kappa = 1$), curve 2 for equal diffusion coefficients ($\kappa = 1/2$), and curve 3 for $\kappa = 1$ are obtained from the density expansion [45].

result remains accurate for longer times as κ decreases. For example, for equal diffusion coefficients the error is $\leqslant 10\%$ when $\omega(t) \geqslant 0.5$. Since $\bar{d} = 1$ is most unfavorable for the mean-field-like theory, one expects the accuracy of the Smoluchowski theory to improve even more dramatically in higher dimensions. The density expansion introduced in this work can be used to verify this expectation.

5.2.2.3 The Wiener trajectories approach

An original formalism for the treatment of many-particle effects in the $A + B \to B$ reaction was developed in a series of papers by Berezhkovskii, Machnovskii and Suris [54–59]. It is based on the so-called *Wiener trajectories* and related the *Wiener sausages* concept (the spatial region visited by a spherical Brownian particle during its random walks) [55, 60, 61]. It was shown that the convential survival probability for a walker among traps, which could be presented in a form [47]

$$-\ln\omega(t) = \begin{cases} \dfrac{4}{\sqrt{\pi}} n_B \sqrt{Dt}, & \bar{d} = 1, \\ \dfrac{4\pi n_B Dt}{\ln(Dt/r_0^2)}, & \bar{d} = 2, \\ 4\pi n_B r_0 Dt + 8\sqrt{\pi}\, n_B r_0^2 \sqrt{Dt}, & \bar{d} = 3, \end{cases}$$

($D = D_A + D_B$, r_0 is a trapping radius and n_B is trap concentration) is valid only if one neglects *fluctuations* of the volume of the Wiener sausage. In the opposite case at long times the kinetics for mobile donors A becomes fluctuation-controlled and as $t \to \infty$ obeys finally equation (2.1.106). Of our special interest here are arguments and results for the case of *mobile* traps B and $\bar{d} \geqslant 3$ based on simple estimates similar to those which resulted in equation (2.1.106).

Since the particle will not die during the time t if it spends all this time inside a cavity free from traps, we introduce a $\bar{d}$-dimensional sphere of an arbitrary radius R, surrounding the starting point of a particle A and write down evident inequality

$$\omega(t) > Q_1(R)Q_2(R, t)Q_3(R, t), \tag{5.2.35}$$

where

$$Q_1(R) = \exp\left(-n_B V_{\bar{d}} R^{\bar{d}}\right) \tag{5.2.36}$$

is the probability of trap absence inside a sphere of the radius R at the initial instant of time $(V_{\bar{d}} = \pi^{\bar{d}/2}/\Gamma(1+\bar{d}/2))$; $Q_2(R,t)$, defined by

$$Q_2(R,t) = \exp\left(-n_{\rm B}\bar{d}(\bar{d}-2)V_{\bar{d}}R^{\bar{d}-2}D_{\rm B}t\right) \qquad (5.2.37)$$

is the probability of keeping a sphere of the radius R from the traps during the time t; $D_{\rm B}$ is a coefficient of trap diffusion; $Q_3(R,t)$, defined by

$$Q_3(R,t) \approx \exp\left(-\beta_{\bar{d}}\frac{D_{\rm A}t}{R^2}\right) \qquad (5.2.38)$$

is the probability of a particle staying during the time t inside a sphere of a radius R surrounding its starting point ($\beta_{\bar{d}}$ is the square of the first zero of the Bessel function of the first kind of the order $1/2\,[\bar{d}-2]$). The estimation (5.2.38) of the probability $Q_3(R,t)$ is true for long times which satisfy the inequality $D_{\rm A}t \gg R^2$.

Substitution of equations (5.2.36) to (5.2.38) into the inequality (5.2.35) gives

$$w(t) > \exp\left\{-\left[n_{\rm B}V_{\bar{d}}R^{\bar{d}} + n_{\rm B}\bar{d}(\bar{d}-2)V_{\bar{d}}R^{\bar{d}-2}D_{\rm B}t + \right.\right.$$
$$\left.\left. + \beta_{\bar{d}}\frac{D_{\rm A}t}{R^2}\right]\right\}. \qquad (5.2.39)$$

Up to now, no restriction have been imposed on the radius of the sphere R. Now we shall optimise our estimation choosing $R = R_t$ so that the right-hand side of the inequality (5.2.39) is maximal at a certain instant time t. Substituting the value R_t obtained in this way, into equation (5.2.39) one obtains the probability of particle survival:

$$w(t) > \begin{cases} \exp\left[-\dfrac{\bar{d}+2}{2}\left(\dfrac{2\beta_{\bar{d}}}{\bar{d}}(n_{\rm B}V_{\bar{d}})^{2/\bar{d}}D_{\rm A}t\right)^{\bar{d}/(\bar{d}+2)}\right], & t < t_{\rm c} \\ \exp\left[-\dfrac{\bar{d}\beta_{\bar{d}}}{\bar{d}-2}\left(\dfrac{\bar{d}(\bar{d}-2)^2}{2\beta_{\bar{d}}}\right)^{2/\bar{d}} \times \right. & \\ \left. \times (n_{\rm B}V_{\bar{d}})^{2/\bar{d}}D_{\rm A}^{1-2/\bar{d}}D_{\rm B}^{2/\bar{d}}t\right], & t > t_{\rm c} \end{cases} \qquad (5.2.40)$$

where

$$t_c = \frac{\bar{d}+2}{\bar{d}(\bar{d}-2)}\left(1 - \frac{4}{\bar{d}^2}\right)^{\bar{d}/2}\left(\frac{2\beta_{\bar{d}}}{\bar{d}(\bar{d}-2)^2}\right)^{2/\bar{d}} \times$$

$$\times \frac{D_A^{2/\bar{d}}}{D_B^{1+2/\bar{d}}(n_B V_{\bar{d}})^{2/\bar{d}}}. \qquad (5.2.41)$$

This estimate is of a special interest since it predicts a higher probability of particle survival than the conventional expression having the form ($\bar{d} \geqslant 3$)

$$\omega(t) = \exp\left\{-n_B V_{\bar{d}} \bar{d}(\bar{d}-2) r_0^{\bar{d}-2}(D_A + D_B)t\right\}.$$

A comparison of this equation with equations (5.2.40) shows that the fluctuational decrease of particle disappearance rate takes place *both* for $D_B = 0$ and $D_B \neq 0$, if the trap diffusion takes place slowly enough,

$$\frac{D_B}{D_A} < \frac{2}{\bar{d}}\left(\frac{(\bar{d}-2)^2}{\beta_{\bar{d}}}\rho\right)^{(\bar{d}-2)/2}, \qquad (5.2.42)$$

where $\rho = n_B V_{\bar{d}} r_0^{\bar{d}}$ is the volume fraction of traps, which we assume to be small, $\rho \ll 1$. For long times which satisfy the condition

$$\frac{\bar{d}+2}{2}\left(\frac{\beta_{\bar{d}}(\bar{d}+2)}{\bar{d}^2(\bar{d}-2)}\right)^{\bar{d}/2}\frac{1}{\rho^{(\bar{d}-2)/2}}$$
$$< \tau = n_B V_{\bar{d}} \bar{d}(\bar{d}-2) r_0^{\bar{d}-2}(D_A + D_B)t$$
$$< \frac{\bar{d}}{(\bar{d}-2)}\left(\frac{2\beta_{\bar{d}}}{\bar{d}(\bar{d}-2)^2}\right)^{2/\bar{d}}\left(\frac{D_A}{D_B}\right)^{(\bar{d}+2)/\bar{d}}\rho^{(\bar{d}-2)/\bar{d}}, \qquad (5.2.43)$$

the probability of particle survival is described by (2.1.106). Here τ is the dimensionless time resulting from the reference of time t to the characteristic time $1/n_B V_{\bar{d}} \bar{d}(\bar{d}-2) r_0^{\bar{d}-2}(D_A + D_B)$, during which the probability of particle survival decrease by e times, according to the conventional equation (5.2.41).

For the stationary traps, $D_B = 0$, equation (2.1.106) is an asymptotic limit of the survival probability $w(t)$ as $t \to \infty$, but if $D_B \neq 0$ it is only an intermediate asymptotic limit. With the time determined by the inequality

$$\tau > \frac{\bar{d}}{(\bar{d}-2)} \left(\frac{2\beta_{\bar{d}}}{\bar{d}(\bar{d}-2)^2} \right)^{2/\bar{d}} \left(\frac{D_A}{D_B} \right)^{(\bar{d}+2)/\bar{d}} \rho^{(\bar{d}-2)/\bar{d}} \qquad (5.2.44)$$

it passes into another expression having the form

$$w(t) \propto \exp\left[-\frac{2\beta_{\bar{d}}}{(\bar{d}-2)} \left(\frac{\bar{d}(\bar{d}-2)^2}{2\beta_{\bar{d}}} \right)^{2/\bar{d}} \times \right.$$

$$\left. \times (n_B V_{\bar{d}})^{2/\bar{d}} D_A^{1-2/\bar{d}} D_B^{2/\bar{d}} t \right]. \qquad (5.2.45)$$

This equation describes the fluctuational decrease of particle decay rate when $t \to \infty$ in the case of mobile traps. Although in equation (5.2.45) the exponent is the linear function of time, as in the conventional expression (5.2.41) and (5.2.45) however, describes a slower particle decay rate than that predicted by the conventional equation due to the small value of the parameter

$$\left[\frac{D_B}{D_A} \frac{\bar{d}}{2} \left(\frac{(\bar{d}-2)^2}{\beta_{\bar{d}}} \rho \right)^{1-\bar{d}/2} \right]^{2/\bar{d}} \qquad (5.2.46)$$

(see the inequality (5.2.42)). This conclusion supports what was said above in Section 5.2.1.6 about a role of the *ratio* of particle mobilities D_A and D_B.

Note that similar estimates of the trap diffusion influence on the fluctuational reduction of particle decay rate, as $t \to \infty$ were made also in [62, 63]. However the latter estimates for the particle survival probability with $D_B \neq 0$ in the case $\bar{d} \geqslant 3$ have quite different form

$$w(t) > \exp(-\lambda n_B t), \qquad (5.2.47)$$

where λ is a constant. Note the different dependence of the coefficient with t in the exponent on the concentration of traps in this expression: according to equation (5.2.47), it is proportional to n whereas according to (5.2.40) it is proportional to $n_B^{2/\bar{d}}$.

According to the conventional expression, the survival probability depends on the mobility of particles and traps only via the sum of their diffusion coefficients, $D = D_A + D_B$, but not their individual values. Equation (5.2.45) shows that this is no longer so. According to inequality (5.2.42), the larger the space dimensionality, the narrower is the interval of values of the relationship D_B/D within which the fluctuational decrease of the decay rate occurs. It will be recalled that with $\bar{d} = 1$ and 2 the fluctuational decay rate reduction takes place for an arbitrary value of the trap mobility, $D_B/D < 1$ [54].

Furthermore, an increase in the space dimensionality results in restricting the time interval where an intermediate asymptotic limit (2.1.106) is true. This happens at the expense of a shift in its lower boundary towards larger values of τ, while the upper limit of the interval with $\bar{d} \gg 1$ does not depend on the space dimensionality. Indeed, with $\bar{d} \gg 1$, $\beta_{\bar{d}} \approx \bar{d}^2/4$ and the inequality (5.2.43) it takes the form

$$\frac{\bar{d}}{2} \frac{1}{(4\rho)^{\bar{d}/2}} < \tau < \frac{D_A}{D_B} \rho. \tag{5.2.48}$$

This shows that with an increase in the space dimensionality, there is an extension of the time interval where the conventional expression of particle survival probability (5.2.41) is valid.

The calculations of a mean square particle displacement, $\langle r^2(t) \rangle$, have demonstrated that at the initial stage of reaction (small t) the conventional expression is valid,

$$\langle r^2(t) \rangle = 2\bar{d}Dt, \tag{5.2.49}$$

whereas at asymptotically large times it is lower and transforms into the expression characteristic for the anomalous diffusion with $\gamma = 2/(\bar{d}+2)$, i.e.,

$$\langle r^2(t) \rangle \propto \left(\frac{Dt}{n_B}\right)^\gamma, \tag{5.2.50}$$

n_B is trap concentration [64].

Lastly, studies of the many-particle effects for reaction with *correlated distributions of traps* [65, 66] have shown that a trapping process could be accelerated or become slower depending on whether traps attract or repulse each other.

5.2.2.4 The probability density function

The non-uniform distribution of particles arising in the course of bimolecular reactions and characterizing self-organization effects could be studied in terms of the *probability density functions* for the nearest-neighbour distance between reactants. These functions could be calculated analytically and found also from computer simulations. In the particular case of the A + B → B reaction, which is under study in this Chapter, it was shown that in the two limiting cases – of either mobile particles A or mobile traps B – the mean nearest-neighbour AB distance $\langle L(t) \rangle$ increases in time for $\bar{d} = 1$ as [67, 68]

$$\lim_{t \to \infty} \langle L(t) \rangle \propto t^{-\alpha} = \begin{cases} t^{1/2}, & D_A = 0, \ D_B > 0, \\ t^{1/4}, & D_A > 0, \ D_B = 0. \end{cases} \quad (5.2.51)$$

Computer simulations in the intermediate case when *both* A's and B's are mobile, have led to the critical exponent α of the form [69]

$$\alpha = \frac{1}{\pi} \arctan \sqrt{1 + 2\frac{D_A}{D_B}}. \quad (5.2.52)$$

This equation reproduces correctly the limiting cases mentioned above and has been derived in [53] for the particular case of a single B particle surrounded by two A particles. Careful checking has shown that this exponent α depends indeed only on the ratio, but not separately on the individual values of the diffusion coefficients, D_A and D_B.

5.2.2.5 Reversible diffusion-controlled reactions

Up to now, we have been discussing mainly either contact reactions (black sphere or clear-cut recombination radius approximation) or irreversible short-range (tunnelling) and long-range (multipole) reactions. Quite general theory of *reversible* diffusion-controlled reactions has been developed recently in [70, 71]. The model accepted has both phenomenological and microscopic features and is specified as follows: all particles A and B are spherical and obey the diffusion equation. The B's are assumed to be point particles; when A and B come in contact (at $r = r_0$), they may react to form AB. This reaction is described in terms of the so-called *radiation boundary* condition [72–74], equation (4.2.8), which involves an intrinsic *association rate* parameter k entering the total diffusional flux through a recombination sphere.

A partially absorbing boundary is equivalent to a finite strength, *delta-function sink* located at the boundary [75]; the sink does not need to coincide with the boundary. Theory [70, 71] was successfully applied to reversible reactions of isolated (geminate) pairs but its generalizations to pseudo-first order bimolecular reactions (const $= n_B \gg n_A$), even at the level of the Smoluchowski approximation for noninteracting B's and independent AB pairs turned out to be non-trivial problem (see also [76]). Even when the Smoluchowski theory for irreversible pseudo-first-order reactions is exact, no rigorous theory that is valid for an arbitrary set of kinetic parameters was developed; in [70] the additional assumption was made that every time a bound AB pair dissociates forming an unbound pair at contact, this pair behaves as if it was surrounded by an *equilibrium* distribution of B's independent of the history of previous associations and dissociations (see also [77]).

Approximate treatment of the many-particle effects in *reversible* bimolecular reactions has been undertaken in several papers (see for a review [78]); we would like also to note here pioneering studies of Ovchinnikov's group [79–82] and Kang and Redner's paper [83]. The former approach was discussed above in Section 2.1.2.3 where the kinetics of the approach to equilibrium for the simple reaction $A \leftrightarrows B + B$ (dissociation and association of molecules A) was shown to approach the equilibrium as $t^{-3/2}$. Note also that in the paper [84] a new elegant quantum-field formalism has been developed for the first time and applied to the diffusion-controlled reactions in the fluctuation regime; its results agree completely with the phenomenological estimate (2.1.61).

It should be stressed that despite the same critical exponent, $\alpha = 3/2$, was obtained in studies of reversible reaction $A \leftrightarrows B + B$ discussed above

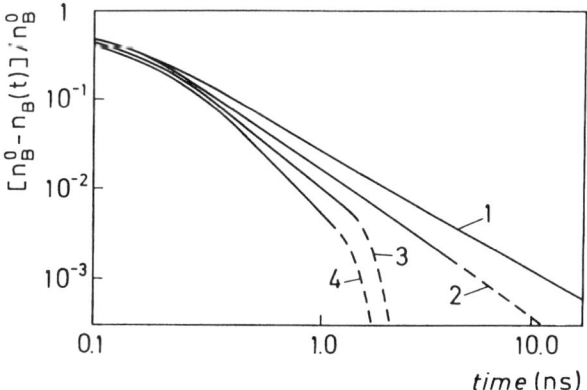

Fig. 5.15. Power-law approach to the equilibrium concentration in excited dye molecules for different proton concentrations in solution (mM) [85]: 1 – 0.01, 2 – 4; 3 – 1.5 and 4 – 15. Full lines show smoothed experimental curves, dashed lines correspond to large statistical noise.

and for pseudo-monomolecular one, $A + B \to AB$, $n_A \gg n_B$, considered by Agmon et al., their background physical mechanisms could be different and related to the fluctuations in the initial random distribution of particles and the probability of a single walker return to the origin respectively. (We are greatly indebted to N. Agmon for this comment.)

It should be stressed that the reversible chemical reactions give us better chance to observe many-particle effects since there is no need here to monitor vanishing particle concentrations over many orders of magnitude. Indeed, the fluctuation-controlled law of the approach to the reaction equilibrium similar to (2.1.61) was observed recently experimentally [85] for the pseudo-first-order reaction $A + B \leftrightarrows AB$ of laser-excited ROH dye molecules which dissociate in the excited state to create a geminate proton-excited anion pair. The solvated proton is attracted to the anion and recombines with it reversibly. After several dissociation–association cycles it finally diffuses to long distances and further recombination becomes unobservable.

Figure 5.15 shows that the asymptotic approach of the concentration in the wide time interval indeed obeys the power-law

$$\left(n_B(0) - n_B(t)\right) \propto t^{-\alpha}. \tag{5.2.53}$$

However, the relevant *critical exponent* surprisingly increases with proton concentration in solution (Fig. 5.16) which is not expected by present level of theory (see also discussion in [86, 87]).

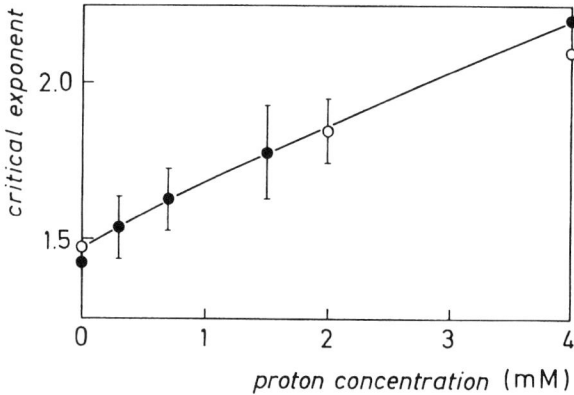

Fig. 5.16. The dependence of the critical exponent α on proton concentration in solution [85] derived from Fig. 5.15. Open and black cycles denote two sets of independent experiments.

5.3 THE A + A → 0 AND A + A → A REACTIONS

The simplest class of bimolecular reactions involves only *one* type of mobile particles A and could result either in particle *coagulation* (coalescence, fusion) A + A → A, or *annihilation*, A + A → 0 (inert product). Their simplicity in conjunction with the simple topology of $\bar{d} = 1$ allows us to solve the problem exactly, which makes it very attractive for testing different approximations and computer simulations. In the standard chemical kinetics (i.e., mean-field theory, Section 2.1.1) we expect in $\bar{d} = 2$ and 3 for both reactions mentioned trivial behaviour quite similar to the A+B → 0 reaction, i.e., $n_A(t) \propto t^{-1}$, as $t \to \infty$. For $\bar{d} = 1$ in terms of the Smoluchowski theory the joint density obeys respectively the equation (4.1.56) with $\nabla^2 \equiv \frac{\partial^2}{\partial r^2}$ and $D = 2D_A$.

With reaction rate given by equation (4.1.69) the $\bar{d} = 1$ concentration decay obeys

$$\frac{dn_A(t)}{dt} = -\sqrt{\frac{8D_A}{\pi t}}\, n_A^2(t), \quad (5.3.1)$$

that is

$$n_A(t) = \frac{n_A(0)}{1 + 4n_A(0)\sqrt{\frac{2D_A t}{\pi}}}. \quad (5.3.2)$$

As $t \to \infty$, $n_A(t) \propto t^{-1/2}$.

Exact solution of this problem obtained in [88, 89] is, however

$$n_A(t) = n_A(0) \exp\left(8D_A t n_A^2(0)\right) \mathrm{erfc}\left(\sqrt{8D_A t}\, n_A(0)\right), \quad (5.3.3)$$

where erfc is a standard error function.

Despite the fact that (5.3.3) reveals the same asymptotic behaviour ($n_A(t) \propto t^{-1/2}$, as $t \to \infty$), the relative concentration is always *smaller* than predicted by the Smoluchowski theory; for $n_A = 1/2 n_A(0)$, the discrepancy is 9%. In other words, the Smoluchowski approach slightly underestimates a real reaction rate due to its neglect of reactant density fluctuations stimulating. (Note that in the case of second reaction, A + A → A, the reaction rate $K(t)$ in the Smoluchowski approach has to be corrected by a factor 1/2 meaning that only one particle disappears in each reaction event.) It should be stressed that peculiarity of one-dimensional topology leads to the absence of the reaction radius r_0 in the reaction rate defining time development of concentrations, $n_A(t)$.

The particular striking discrepancy between the exact and the Smoluchowski solution is observed, however, for the non-uniform, *clustered* distribution of particles A (Fig. 5.17) [89] – in a line with what was said above (Section 3.2) about shortcoming of the use of Kirkwood's superposition approximation for initially strongly correlated particle distributions.

Studies of the fluctuation-controlled asymptotics of the A + A-type reactions have led to the universal decay law [63, 90]

$$n_A(t) \propto \left(m_{\bar{d}} g_{\bar{d}}(t)\right)^{-1}, \qquad (5.3.4)$$

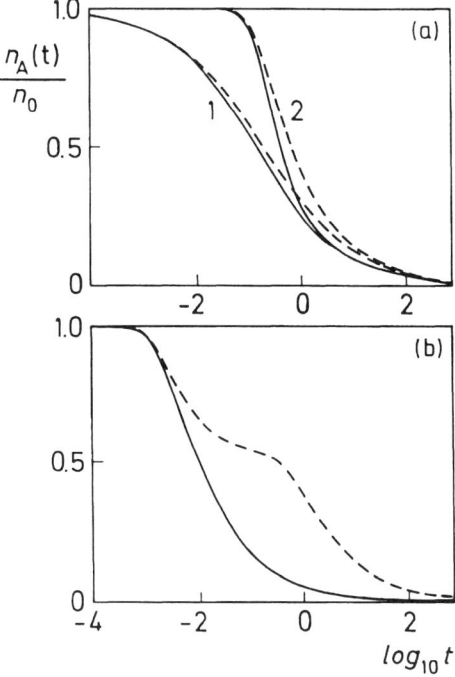

Fig. 5.17. Exact results (full curves) and Smoluchowski theory (dashed curves) for the A + A → 0 reaction in one-dimension for: (a) random particle distribution (curves 1), equidistant particle distribution (curves 2); and clustered distribution (b) [89].

where $m_{\bar{d}}$ are constants and

$$g_{\bar{d}}(t) = \begin{cases} \sqrt{t}, & \bar{d} = 1, \\ t/\ln t, & \bar{d} = 2, \\ t, & \bar{d} \geqslant 3. \end{cases} \quad (5.3.5)$$

Unlike the $A + B \to 0$ reaction with marginal dimension $\bar{d}_0 = 4$ (see [63, 90] for more detail) here $\bar{d}_0 = 2$ only. The decay law for $\bar{d} = 1$, $n_A(t) \propto t^{-1/2}$, was established in a number of theoretical papers [27, 63, 83, 90–93], including field-theoretical formalism [94, 95], and confirmed in several computer simulations [27, 93, 96, 97]. The $t^{-\bar{d}/2}$ law could be extended for fractal spaces replacing $\bar{d}$ by $\tilde{d}$ (spectral dimension) – see Section 6.3 and [97–99] for more details.

Little attention is paid usually to the manifestation of many-particle effects not at asymptotically long, but relatively *short* times, e.g., during the transient process when a random particle distribution changes for the quasi-steady state and the reaction rate, in terms of standard chemical kinetics [100], equation (4.1.61), has to obey the following equation (for $\bar{d} = 3$, in dimensionless units)

$$K(t) = 1 + \frac{1}{\sqrt{\pi t}}. \quad (5.3.6)$$

Figure 5.18 demonstrates the *non-stationary part* of the reaction rate [second term in r.h.s. of equation (5.3.6)] taking into account the many-particle effects as they are incorporated in Kirkwood's superposition approximation for the three-dimensional $A+A \to 0$ reaction (curves 1 to 3) and $A+B \to 0$ reaction (curve 4) [101]. The reference dashed line corresponds to equation (5.3.6). Unlike the reaction between dissimilar particles, where aggregation of similar non-reacting species means simultaneous segregation of dissimilar and thus lowering of the reaction rate (curve 4), for the reaction between similar particles aggregation leads to reaction *acceleration;* which is the larger, the higher initial reactant concentration. Curves 1 to 3 could be well-approximated in a universal form (as $t > 10^2 r_0^2/D$):

$$K(t) = 1 + \frac{\gamma}{\sqrt{\pi t}}, \quad \gamma \approx 2.1. \quad (5.3.7)$$

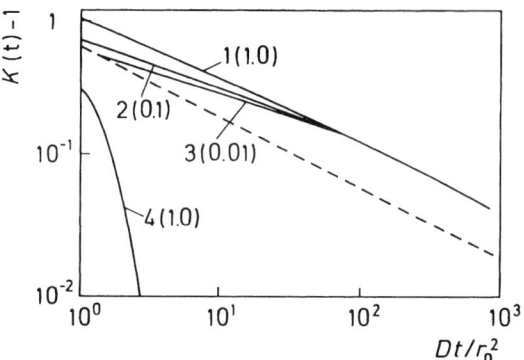

Fig. 5.18. The non-stationary part of the reaction rate $K(t)$ for the transient kinetics of the $A + A \to 0$ reaction (curves 1 to 3) and $A + B \to 0$ reaction (curve 4), $\bar{d} = 3$ (random initial distribution) [101]. The broken line shows prediction of the standard chemical kinetics, equation (5.3.10). The initial concentrations are given.

Note that these calculations complement well computer simulations done by Toussaint and Wilczek [27] clearly revealing the same transient process before reaching the asymptotical regime $n_A \propto t^{-1/2}$.

This asymptotic decay law means that at long time reaction is described formally by the *third*-order kinetics [68, 102, 103] which is very unusual for the standard chemical kinetics:

$$\frac{dn_A(t)}{dt} = -\text{const} \cdot n_A^3(t). \tag{5.3.8}$$

It was also shown [104] that (5.3.8) is valid at *all* times if (and only if!) the initial interparticle distribution function obeys the scaling distribution

$$\phi(r,t) = \frac{\pi}{2} n_A^2(t) r \exp\left(-\frac{\pi}{2} n_A^2(t) \frac{r^2}{2}\right). \tag{5.3.9}$$

Computer simulations [104, 105] of the evolution of interparticle distribution function $\phi(r,t)$ confirm validity of equation (5.3.9) derived for the $A + A \to A$ reaction and demonstrate its discrepancy from the asymptotic distribution function for the similar (but not identical) annihilation reaction $A + A \to 0$. The latter distribution function has its maximum shifted to the shorter distances and a longer tail at large distances. This distinction between two distributions is not very surprising since in the annihilation reaction we have the stronger effect of simultaneous disappearance of *two* particles instead of one particle in the fusion reaction. Thus, the generally-accepted statement that the reactions $A + A \to A$ and $A + A \to 0$ fall into the same *universality class* does not extended to the relevant spatial distributions, i.e., there is many-to-one correspondence between distribution function and the asymptotic decay law.

The effect of *grey boundary condition* (partial reflection caused by particle contact) in the $A + A \to 0$ $1\bar{d}$-lattice reaction was simulated in [106] whereas more refined problem for the *combined reaction*

$$A + A \to \begin{cases} A & \text{with probability } p, \\ 0 & \text{with probability } 1 - p, \end{cases} \qquad (5.3.10)$$

solved in [107]. Explicit results in the latter paper were derived for the cases where the initial distribution of particles is the Poissonian and a deterministic distribution with each site occupied on the integer lattice.

Exact solution of one-dimensional *reversible* coagulation reaction $A + A \to A$ was presented in [108, 109] (see also Section 6.5). In these studies a *dynamical phase transition* of the second order was discovered, using both continuum and discrete formalisms. This shows that the relaxation time of particle concentrations on the equilibrium level depends on the initial concentration, if the system starts from the concentration smaller than some critical value, and is independent of the $n_A(0)$ otherwise.

These results are complemented by theoretical calculations and computer simulations [110, 111] for $\bar{d} = 1, 2$ and 3 of bimolecular trapping/annihilation reaction $A + A \to 0$, $A + T \to A_T$ and $A + A_T \to T$ (T is an immobile trap making A particle to become immobile too) and unimolecular trapping/annihilation, $A + A \to 0$, $A \to A_T$, $A + A_T \to 0$. It was found that the kinetics of trapped particles can be described by the mean-field theory for *bimolecular* but not for unimolecular reactions. The kinetics of free A's is described by mean-field theory at short times, but at long times and low trap concentrations the concentration of free A's decays as (2.1.106).

Lastly, in conclusion of the Chapter 5 a study of the effect of the shortened and complete Kirkwood superposition approximations on asymptotic kinetics of a number of single-species reaction should be noted [112, 113]. The general conclusion was drawn that for irreversible reactions the complete superposition approximation, (2.3.62), gives better results than its shortened version, (2.3.64). However, the opposite situation occurs for *reversible* reactions, probably, due to the boundary condition imposed on the two-point density.

References

[1] V.N. Kuzovkov and E.A. Kotomin, Chem. Phys. 81 (1983) 335.
[2] W. Ebeling, Structurbildung bei irrevesiblen Prozessen (Teubner, Leipzig, 1976).
[3] G. Nicolis and I. Prigogine, Self-Organization in Non-Equilibrium Systems. From Dissipative Structures to Order through Fluctuations (Wiley, New York, 1977).
[4] R. Balescu, Equilibrium and Non-Equilibrium Statistical Mechanics (Wiley, New York, 1975).
[5] J.M. Ziman, Models of Disorder (Cambridge Univ. Press, London, 1979).
[6] H. Schnörer, V. Kuzovkov and A. Blumen, J. Chem. Phys. 92 (1990) 2310.
[7] H. Schnörer, V. Kuzovkov and A. Blumen, J. Chem. Phys. 93 (1990) 7148.
[8] S. Luding, H. Schnörer, V. Kuzovkov and A. Blumen, J. Stat. Phys. 65 (1991) 1261.
[9] V.N. Kuzovkov and E.A. Kotomin, Chem. Phys. 98 (1985) 351.
[10] V.N. Kuzovkov and E.A. Kotomin, Chech. J. Phys. B 35 (1985) 541.
[11] V.N. Kuzovkov and E.A. Kotomin, Rep. Prog. Phys. 51 (1988) 1479.
[12] V.N. Kuzovkov, Teor. Eksp. Khim. 21 (1985) 33 (Sov. Theor. Exp. Chem., in Russian).
[13] R. Brout, Phase Transitions (North-Holland, Amsterdam, 1965).
[14] H. Stanley, Introduction to Phase Transitions and Critical Phenomena (Clarendon Press, Oxford, 1971).
[15] R. Baxter, Exactly Solved Models in Statistical Mechanics (Academic Press, London, 1982).
[16] R. Kikuchi and S.G. Brush, J. Chem. Phys. 47 (1967) 195.
[17] A.A. Ovchinnikov, S.F. Timashev and A.A. Belyi, Kinetics of Diffusion-Controlled Chemical Processes (Nova Science, New York, 1989).
[18] G. Lück and R. Sizmann, Phys. Status Solidi 17 (1966) K61.
[19] G. Lück and R. Sizmann, Nucleonik 8 (1966) 256.
[20] I.A. Tale, D.K. Millers and E.A. Kotomin, J. Phys. C: 8 (1975) 2366.
[21] E.A. Kotomin, PhD Thesis (University of Latvia, Riga, 1975).
[22] V.L. Vinetsky, Yu.H. Kalnin, E.A. Kotomin and A.A. Ovchinnikov, Sov. Phys. Usp. 33 (1990) 793.
[23] V.P. Zhdanov, Sov. Phys. Solid State 27 (1985) 733.
[24] V.P. Zhdanov, Khim. Fiz. 6 (1987) 278 (Sov. Chem. Phys., in Russian).
[25] V.N. Kuzovkov, Khim. Fiz. 6 (1987) 1146 (Sov. Chem. Phys., in Russian).
[26] H. Schnörer, V. Kuzovkov and A. Blumen, Phys. Rev. Lett. 63 (1989) 805.
[27] D. Toussaint and F. Wilczek, J. Chem. Phys. 78 (1983) 2642.
[28] V.N. Kuzovkov, Khim. Fiz. 6 (1987) 831 (Sov. Chem. Phys., in Russian).
[29] H. Schnörer, PhD thesis (Universität Bayreuth, 1991).

References

[30] V.M. Agranovich and M.D. Galanin, Modern Problems in Condensed Matter Science, Vol. 3: Electron Excitation Energy Transfer in Condensed Matter (North-Holland, Amsterdam, 1982).
[31] B.Ya. Balagurov and V.G. Vaks, Sov. Phys. JETP 38 (1974) 968.
[32] M.D. Donsker and S.R.S. Varadhan, Commun. Pure Appl. Math. 28 (1975) 279.
[33] M.D. Donsker and S.R.S. Varadhan, Commun. Pure Appl. Math. 28 (1975) 525.
[34] M.D. Donsker and S.R.S. Varadhan, Commun. Pure Appl. Math. 32 (1979) 721.
[35] U. Gösele, J. Nucl. Mater. 78 (1978) 83.
[36] R. Kapral, J. Chem. Phys. 68 (1978) 1903.
[37] P. Grassberger and I. Procaccia, Phys. Rev. A: 77 (1982) 6291.
[38] A. Onipko, in: Physics of Many-Particles Systems, Vol. 2 (Naukova Dumka, Kiev, 1982) p. 60.
[39] R.F. Kayser and J.B. Hubbard, Phys. Rev. Lett. 51 (1983) 79.
[40] Yu. Kalnin, Paper N37 at the Int. Conf. on Defects in Insulating Crystals (Inst. of Physics, Salaspils, 1981).
[41] F. Delyon and B. Souillard, Phys. Rev. Lett. 51 (1983) 1720.
[42] S. Havlin, M. Dishon, J.E. Kiefer and G.H. Weiss, Phys. Rev. Lett. 53 (1984) 407.
[43] J.K. Anlauf, Phys. Rev. Lett. 52 (1984) 1845.
[44] J. Klafter, G. Zumofen and A. Blumen, J. Phys. (Paris) Lett. 45 (1984) 49.
[45] A. Szabo, R. Zwanzig and N. Agmon, Phys. Rev. Lett. 61 (1988) 2498.
[46] I.M. Lifshitz, Sov. Phys. Usp. 83 (1964) 617.
[47] A. Blumen, J. Klafter and G. Zumofen, in: Optical Spectroscopy of Glasses, ed. I. Zchokke (Reidel, Dordrecht, 1986) p. 199.
[48] V.N. Kuzovkov, Latv. PSR Zinat. Akad. Vestis Fiz. Teh. Zinat. Ser. 5 (1986) 46 (Proc. Latv. Acad. Sci. Phys. Technol. Ser., in Russian).
[49] S.W. Haan and R. Zwanzig, J. Chem. Phys. 68 (1978) 1879.
[50] M. Inokuti and F. Hirayama, J. Chem. Phys. 43 (1965) 1978.
[51] I.G. Hunt, D. Bloor and B. Movaghar, J. Phys. C: 18 (1985) 3497.
[52] D. Ben-Avraham, Philos. Mag. A 56 (1987) 1015.
[53] D. Ben-Avraham, J. Chem. Phys. 88 (1988) 941.
[54] A.M. Berezhkovskii, Yu.A. Makhnovskii and R.A. Suris, Sov. Phys. JETP 64 (1986) 1301.
[55] A.M. Berezhkovskii, Yu.A. Makhnovskii and R.A. Suris, J. Stat. Phys. 57 (1989) 333.
[56] A.M. Berezhkovskii, Yu.A. Makhnovskii and R.A. Suris, Chem. Phys. 137 (1989) 41.
[57] A.M. Berezhkovskii, Yu.A. Makhnovskii and R.A. Suris, Chem. Phys. Lett. 175 (1989) 499.
[58] A.M. Berezhkovskii, Yu.A. Makhnovskii and R.A. Suris, J. Stat. Phys. 65 (1991) 1025.
[59] A.M. Berezhkovskii and Yu.A. Makhnovskii and R.A. Suris, J. Phys. A: 22 (1989) L615.
[60] F. Spitzer, Z. Wahrscheinlichkeitstheor. Verwandte Gebiete 4 (1965) 248.
[61] J.-F. Le Gall, Ann. Probability, 16 (1988) 991.
[62] S. Redner and K. Kang, J. Phys. A: 17 (1984) 451.
[63] M. Bramson and J.L. Lebowitz, Phys. Rev. Lett. 61 (1988) 2397.

[64] A.M. Berezhkovskii, Yu.A. Makhnovskii and R.A. Suris, Phys. Lett. 150 (1990) 296; 157 (1991) 146.
[65] A.M. Berezhkovskii, Yu.A. Makhnovskii, L.V. Bogachev and S.A. Molchanov, Phys. Rev. E: 47 (1993) 4564.
[66] A.M. Berezhkovskii, Yu.A. Makhnovskii, R.A. Suris, L.V. Bogachev and S.A. Molchanov, Phys. Rev. A: 45 (1992) 6119.
[67] D. Ben-Avraham and G.H. Weiss, Phys. Rev. A: 39 (1989) 6436.
[68] G.H. Weiss, R. Kopelman and S. Havlin, Phys. Rev. A: 39 (1989) 466.
[69] R. Schoonover, D. Ben-Avraham, S. Havlin, R. Kopelman and G.H. Weiss, Physica A 171 (1991) 232.
[70] N. Agmon and A. Szabo, J. Chem. Phys. 92 (1990) 5270.
[71] N. Agmon, Phys. Rev. E: 47 (1993) 2415.
[72] F.C. Collins and G.E. Kimball, J. Colloid. Sci. 4 (1949) 425.
[73] D. Peak and J.W. Corbett, Phys. Rev. B: 5 (1972) 1226.
[74] E.A. Kotomin and A.B. Doktorov, Phys. Status Solidi B: 74 114 (1982) 287.
[75] A. Szabo, G. Lamm and G.H. Weiss, J. Stat. Phys. 34 (1984) 225.
[76] S. Lee and M. Karplus, J. Chem. Phys. 86 (1987) 1883.
[77] N. Agmon, H. Schnörer and A. Blumen, J. Phys. Chem. 95 (1991) 7326.
[78] A. Szabo, J. Chem. Phys. 95 (1991) 2481.
[79] Ya.B. Zeldovich and A.A. Ovchinnikov, Sov. Phys. JETP 47 (1978) 829.
[80] G.S. Oshanin and S.F. Burlatsky, J. Phys. A: 22 (1989) L973; 22 (1989) L977.
[81] S.F. Burlatsky, A.A. Ovchinnikov and G.S. Oshanin, Sov. Phys. JETP 68 (1989) 1153.
[82] S.F. Burlatsky, G.S. Oshanin and A.A. Ovchinnikov, Chem. Phys. 152 (1991) 13.
[83] K. Kang and S. Redner, Phys. Rev. A: 32 (1985) 435.
[84] Ya.B. Zeldovich and A.A. Ovchinnikov, Sov. Phys. JETP Lett. 26 (1977) 440.
[85] D. Huppert, S.Y. Goldberg, A. Masad and N. Agmon, Phys. Rev. Lett. 68 (1992) 3932.
[86] A. Edelstein and N. Agmon, J. Chem. Phys. 99 (1993) 5396.
[87] N. Agmon and A. Edelstein, J. Chem. Phys. 100 (1994) 4181.
[88] D.C. Torney and H.M. McConnell, J. Phys. Chem. 87 (1983) 1941.
[89] D.J. Balding and N.J.B. Green, Phys. Rev. A: 40 (1989) 4585.
[90] M. Bramson and J.L. Lebowitz, J. Stat. Phys. 65 (1991) 941.
[91] A.A. Ovchinnikov and Ya.B. Zeldovich, Chem. Phys. 28 (1978) 215.
[92] K. Kang and S. Redner, Phys. Rev. Lett. 52 (1984) 955.
[93] P. Meakin and H.E. Stanley, J. Phys. A: 17 (1984) L173.
[94] A.A. Lushnikov, Phys. Lett. A 120 (1987) 135.
[95] T. Ohtsuki, Phys. Rev. A: 43 (1991) 6917.
[96] Z. Jiang and C. Ebner, Phys. Rev. A: 41 (1990) 5333.
[97] P. Argyrakis and R. Kopelman, Phys. Rev. A: 45 (1992) 5814.
[98] K. Lindenberg, W.S. Sheu and R. Kopelman, Phys. Rev. A: 43 (1991) 7070.
[99] G. Zumofen, J. Klafter and A. Blumen, Phys. Rev. A: 43 (1991) 7068.
[100] T.R. Waite, Phys. Rev. 107 (1957) 463.
[101] E.A. Kotomin and V.N. Kuzovkov, Chem. Phys. Lett. 117 (1985) 226.
[102] R. Kopelman, Science 241 (1988) 1620.
[103] R. Kopelman, S.J. Parus and J. Prasad, Chem. Phys. 128 (1988) 209.

References

[104] C.R. Doering and D. Ben-Avraham, Phys. Rev. A: 38 (1988) 3035.
[105] P. Argyrakis and R. Kopelman, Phys. Rev. A: 41 (1990) 2114.
[106] L. Braunstein, H.O. Martin, M.D. Grynberg and H.E. Roman, J. Phys. A: 25 (1992) L255.
[107] J. Zhuo, Phys. Rev. A: 43 (1991) 5689.
[108] M.A. Burschka, C.R. Doering and D. Ben-Avraham, Phys. Rev. Lett. 63 (1989) 700.
[109] J.-Ch. Lin, Phys. Rev. A: 45 (1992) 3892.
[110] J.C. Rasaiah, J.B. Hubbard, R.J. Rubin and S.H. Lee, J. Phys. Chem. 94 (1990) 652.
[111] J.C. Rasaiah, J. Zhu, J.B. Hubbard and R.J. Rubin, J. Chem. Phys. 93 (1990) 5768.
[112] J.-Ch. Lin, C.R. Doering and D. Ben-Avraham, Chem. Phys. 146 (1990) 355.
[113] J.-Ch. Lin, Phys. Rev. A: 44 (1991) 6706.

Chapter 6

The Many-Particle Effects in Irreversible A + B → 0 Reaction

6.1 LONG-RANGE RECOMBINATION OF IMMOBILE PARTICLES

> He kills the matter and then says that it is dead.
> (Er tötet die Materie und sagt hernach, daß sie tot sei)
>   Lichtenberg

6.1.1 Tunnelling recombination

Let us illustrate an analysis of the accuracy of the superposition approximation presented in Section 5.1 by both numerical calculations and analytical estimates. Following [1], in Figs 6.1 and 6.2 a solution of a set of equations (4.1.19), (4.1.28), (6.1.14) to (6.1.16) is presented for different choices of parameters assuming that particle concentrations are equal: $n_A(t) = n_B(t) = n(t)$.

It is well seen from Fig. 6.1 that the reaction rate in systems with greater space dimension $\bar{d}$ is also accelerated. It results from the fact that if $\bar{d}$ increases, a number of AB pairs separated by a given distance also does so. The asymptotic behaviour of $n(t)$ as $t \to \infty$ depends also on $\bar{d}$ and needs a special analysis.

In its turn Fig. 6.2 illustrates the effect of the initial concentration on the static tunnelling recombination kinetics. The latter is defined by a competition of *three* distinctive scales – the tunnelling recombination radius r_0, mean distance between particles $l_0 = n(0)^{-1/\bar{d}}$ and lastly, the time-dependent correlation radius ξ. At long time curves corresponding to different initial concentrations could be coincided by their displacements along ordinate axis, which confirms existence of the *universal asymptotic decay law*.

The correlation functions plotted in Figs 5.5 and 5.6 present the statistical information about relative spatial distribution of reacting particles. In particular, the distinctive scale ξ emerges here, thus indicating that all dissimilar

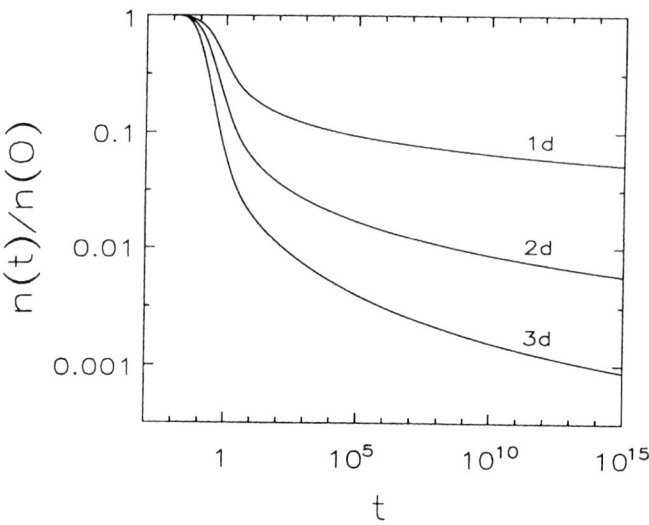

Fig. 6.1. Decay of particle concentrations $n(t)$ for $\bar{d} = 1, 2, 3$ due to the tunnelling recombination. Initial concentration $n(0) = 1$ (in units $r_0^{-\bar{d}}$), whereas time is given in units σ_0^{-1}.

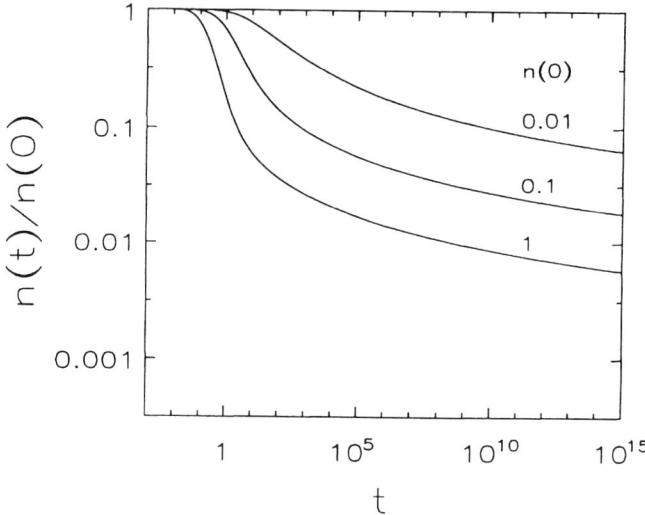

Fig. 6.2. Concentration decay $n(t)$ for $\bar{d} = 2$ and different initial concentrations: $n(0) = 0.01$, 0.1 and 1.0.

AB pairs separated by the distance $r < \xi$ have recombined: $Y(r < \xi, t) \approx 0$. In its turn, on the same distance interval a number of similar AA and BB pairs is in excess (compared to the Poisson distribution), $X_\nu(r,t) \gg 1$. Therefore, a system structure could be symbolically presented as a pattern in Fig. 2.8 where, however, the diffusion length l_D should be replaced now by the correlation length ξ.

Quantitative information available from the correlation functions could be completed by the *qualitative* one, i.e., a visualized distribution of particles obtained by the direct statistical simulations [1–3]. Figures 1.20 and 1.21 demonstrate time development of spatial structures for $\bar{d} = 1$ and 2. Note that computer simulations are restricted in time t since the scale ξ should be considerably less that the system size.

Figures 1.20 and 1.21 show that at long time $t \geqslant 10^5$ aggregates of similar particles are well pronounced. Their formation is associated with emergence of the *non-Poisson* fluctuation spectrum in a system. Similar particle aggregation leads to the effective dissimilar particle separation (as compared to their random distribution) and thus – to a *reduced* reaction rate which is essentially less than expected by the linear approximation.

6.1.2 Correlation length ξ and critical exponents

As it was mentioned in Section 5.1, computer simulations demonstrate existence of the correlation length ξ whose time development is, however, difficult to investigate in detail. At any rate, it corresponds approximately to the length scale $\zeta_0 \propto \ln t$ introduced earlier in the linear approximation. We can introduce the asymptotic ($t \to \infty$) exponent α for the *static* tunnelling recombination similarly to (4.1.68) used for the diffusion-controlled problem:

$$n(t) \propto \xi_0^{-\alpha}. \tag{6.1.1}$$

The *intermediate* asymptotic exponent could be also defined as

$$\alpha(t) = -\frac{d \ln n(t)}{d \ln \xi_0}. \tag{6.1.2}$$

Obviously,

$$\alpha = \lim_{t \to \infty} \alpha(t). \tag{6.1.3}$$

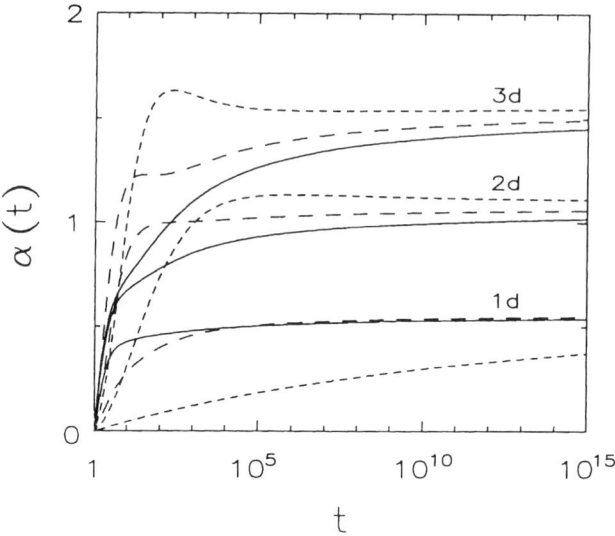

Fig. 6.3. Exponent $\alpha(t)$ for the three different initial concentrations $n(0)$ in $\bar{d} = 1, 2, 3$; $n(0)$: (full curves) 1.0, (broken) 0.1, (dashed) 0.01.

This intermediate asymptotic exponent is useful to demonstrate the formation in time of a new asymptotic law shown in Fig. 6.3. In a given temporal interval $\alpha(t)$ approaches its asymptotic value close to $\alpha = \bar{d}/2$.

This estimate should be made more precise. To do it, let us use some results of the numerical solution of a set of the kinetic equations derived in the superposition approximation. The definition of the correlation length ξ_0 in the linear approximation was based on an analysis of the time development of the correlation function $Y(r,t)$ as it is noted in Section 5.1. Its solution is obtained neglecting the *indirect* mechanism of spatial correlation formation in a system of interacting particles, i.e., omitting integral terms in equations (5.1.14) to (5.1.16). Taking now into account such indirect interaction mechanism, the dissimilar correlation function, obtained as a solution of the *complete* set of equations in the superposition approximation

$$Y(r,t) = Y_0(r,t)z(r,t), \qquad (6.1.4)$$

where $Y_0(r,t) = \exp[-\sigma(r)t]$ is a solution of linear approximation (4.1.40), whereas a co-factor $z(r,t)$ arises due to the indirect interactions and equals just the unity in the linear approximation. As it is noted in Section 5.1, the correlation lengths in the linear (ξ_0) and superposition (ξ) approximations in principle can diverge. Tunnelling recombination can serve as an example of the *short-range* recombination mechanism (according to the definition (5.1.65)), for which $\xi \equiv \xi_0$. This statement is clarified below in a course of the analysis of the reaction asymptotics for *long-range* multipole interaction.

Assuming that the reaction kinetics in the region where (6.1.5) holds is defined entirely by the single scale $\xi_0 \gg (r_0, l_0)$, we arrive at the asymptotic solution for the correlation functions (for equal initial concentrations) in a form $X_\nu(r,t) = X(\eta)$, $Y(r,t) = Y(\eta)$, $z(r,t) = z(\eta)$, where the auto-model variable $\eta = r/\xi_0$ is introduced. Analysis of the numerical solution of equations at long times, $\xi_0 \gg r_0, l_0$, shows that $Y(r,t) \approx z(r,t)$ for all r. In other words, the function $z(r,t)$ is qualitatively close to the solution of (4.1.40): $z(r,t) \approx 0$, as $r < \xi_0$, and $z(r,t) \approx 1$ as $r \gg \xi_0$. Since $\xi_0 \to \infty$, equation (4.1.48) behaves like a step function: $Y(\eta) = Y_0(\eta) = 0$ as $\eta < 1$, $z(\eta) = 0$ does so. As $\eta > 1$, $Y_0(\eta) = 1$ for $\xi_0 \to \infty$, and respectively $Y(\eta) = z(\eta)$.

Solution of the linear approximation (4.1.48) reveals the transient region $\Delta r \simeq 1$, in which the correlation function $Y_0(r,t)$ increases from zero to unity. For the auto-model variable $\eta = r/\xi_0$ in equation (4.1.48) the transient region width $\Delta \eta = 1/\xi_0 \to 0$, which corresponds to the step-like function. The function $Y(r,t)$ in the superposition approximation reveals the transient region Δr *increasing* in the time which is confirmed by computer simulations shown in Figs 5.5 and 5.6.

The performed analysis permits to assume that an indirect channel of the correlation emergence results in the correlation function $Y(\eta) - z(\eta)$ whose shape is a *smoothed step*, besides near $\eta = 1$ because of the continuity reasons one gets $z(\eta) \equiv 0$ as $\eta \leqslant 1$

$$z(\eta) \approx b(\eta - 1)^\lambda, \quad (\eta - 1) \ll 1, \tag{6.1.5}$$

and $z(\eta) \equiv 0$ as $\eta \leqslant 1$, where $\lambda > 0, b$ is a constant. Let us consider now what follows from the assumption (6.1.5), making use of explicit equations (4.1.19) and (4.1.28). (To simplify equations, we use distances in units of r_0 and time in σ_0^{-1}. Thus the reaction rate is $\sigma(r) \equiv \exp(-r)$ and $\xi_0 = \ln t$ respectively.) Rewrite (4.1.28) in a form

$$\frac{\mathrm{d} \ln n(t)}{\mathrm{d} \ln t} = \frac{\mathrm{d} \ln n(t)}{\mathrm{d} \xi_0} = -tK(t)n(t). \tag{6.1.6}$$

Calculate now the product $tK(t)$. By definition (4.1.19),

$$tK(t) = \int t\sigma(r)Y_0(r,t)z(r,t)\,\mathrm{d}\vec{r}. \tag{6.1.7}$$

Write an expression $t\sigma(r)Y_0(r,t)$ in a form $\mathrm{e}^{-\Psi(\eta)}$. Making use of the definitions ξ_0 and η, expend function $\Psi(\eta)$ into the Taylor series near $\eta = 1$,

$$\Psi(\eta) = \mathrm{e}^{(\xi_0-r)} - (\xi_0 - r) \approx 1 + \frac{1}{2}\xi_0^2(\eta-1)^2 + \cdots. \tag{6.1.8}$$

therefore, as $\xi_0 \to \infty$, when $z(r,t) \to z(\eta)$, integrating over η, we get

$$tK(t) \approx \gamma_{\bar{d}}\xi_0^{\bar{d}}\int_1^\infty \mathrm{e}^{-\Psi(\eta)}z(\eta)\eta^{\bar{d}-1}\,\mathrm{d}\eta. \tag{6.1.9}$$

Taking into account the behaviour of $\Psi(\eta)$ near $\eta = 1$, in (6.1.9) instead of the function $z(\eta)$ its expansion (6.1.5) could be used, whereas co-factor $\eta^{\bar{d}-1}$, limited at $\eta = 1$, replaces for its value at $\eta = 1$. In doing so, one gets, at last, an asymptotic estimate as $\xi_0 \to \infty$:

$$tK(t) \approx \gamma_{\bar{d}}b\Gamma\left(\frac{\lambda+1}{2}\right)\mathrm{e}^{-1}2^{(\lambda-1)/2}\xi_0^{(\bar{d}-\lambda-1)} = c\xi_0^{(\bar{d}-\lambda-1)}, \tag{6.1.10}$$

c is constant. Similar estimate neglecting an indirect mechanism of correlations, $z(r,t) = 1$, leads to equation (6.1.10) with $\lambda = 0$ and another constant c.

Using (6.1.1) to (6.1.3), let us rewrite (6.1.6) in a form

$$\alpha = -\lim_{\xi_0 \to \infty}\frac{\mathrm{d}\ln n(t)}{\mathrm{d}\ln \xi_0} = tK(t)n(t)\xi_0. \tag{6.1.11}$$

Since α is a constant, powers of ξ_0 in r.h.s. of (6.1.11) have to cancel each other. Taking into account (6.1.1) and (6.1.10), we arrive at the following relation between two exponents:

$$\alpha = \bar{d} - \lambda. \tag{6.1.12}$$

To obtain λ or α, equations for the correlation functions have to be treated. In equations (5.1.14) or (5.1.15) for equal concentrations, $n_A(t) = n_B(t) =$

$n(t)$, $X_\nu(r,t) \to X(\eta)$) the derivative in t could be transformed into that in η:

$$t\frac{\mathrm{d}\ln X(\eta)}{\mathrm{d}t} = \frac{\mathrm{d}\ln X(\eta)}{\mathrm{d}\xi_0} = -\xi_0^{-1}\frac{\mathrm{d}\ln X(\eta)}{\mathrm{d}\ln \eta}, \qquad (6.1.13)$$

using $\eta = r/\xi_0$. Taking into account the functional $I[Z]$ definition (5.1.7) the equation for the correlation functions $X(\eta)$ and $z(\eta)$ could be rewritten:

$$\frac{\mathrm{d}\ln X(\eta)}{\mathrm{d}\ln \eta} = 2n(t)\xi_0^{\bar{d}+1}S[z], \qquad (6.1.14)$$

$$\frac{\mathrm{d}\ln z(\eta)}{\mathrm{d}\ln \eta} = 2n(t)\xi_0^{\bar{d}+1}S[X], \qquad (6.1.15)$$

where the functional

$$S[Z] = \gamma_{\bar{d}} \int e^{-\Psi(\eta')} z(\eta') \eta'^{\bar{d}-1} \left(\widetilde{Z}(\eta^*) - 1 \right) \mathrm{d}\eta', \qquad (6.1.16)$$

is defined and $\widetilde{Z}(\eta^*)$ is similar to equation (5.1.8):

$$\widetilde{Z}(\eta^*) = \frac{\int Z(\eta^*)\,\mathrm{d}\Omega_{\bar{d}}}{\int \mathrm{d}\Omega_{\bar{d}}}, \qquad \eta^* = (\eta^2 + \eta'^2 - 2\eta\eta'\cos\vartheta)^{1/2}. \qquad (6.1.17)$$

Equation (6.1.15) should be solved at $\eta > 1$, since $\eta < 1$, $z(\eta) = 0$. Keeping in mind properties of $\Psi(\eta')$, an asymptotic estimate of the integral over η' in (6.1.16) could be obtained analogously to the integration (6.1.9). Thus we get simple equations

$$\frac{\mathrm{d}\ln X(\eta)}{\mathrm{d}\ln \eta} = 2\alpha\big(\tilde{z}(\eta^*) - 1\big), \qquad (6.1.18)$$

$$\frac{\mathrm{d}\ln z(\eta)}{\mathrm{d}\ln \eta} = 2\alpha\big(\widetilde{X}(\eta^*) - 1\big), \qquad (6.1.19)$$

where

$$\widetilde{Z}(\eta^*) = \frac{\int Z(\eta^*)\,\mathrm{d}\Omega_{\bar{d}}}{\int \mathrm{d}\Omega_{\bar{d}}}, \qquad \eta^* = \left(\eta^2 + 1 - 2\eta\cos\vartheta\right)^{1/2}. \qquad (6.1.20)$$

When deriving (6.1.18) and (6.1.19) relations (6.1.10), (6.1.11) were used. As one can see, r.h.s.'s of equations do not depend on ξ_0. It justifies the search of solutions in a form containing auto-model variable $\eta = r/\xi_0$ only.

To define the asymptotic exponent α, there is no need to solve numerically a set of non-linear equations (6.1.18), (6.1.19). Assuming as $\eta \to 0$ that the correlation function of similar particles has power singularity, $X(\eta) \propto \eta^{-\beta}$, we find value of β with the help of (6.1.18).

$$\beta = -\lim_{\eta \to 0} \frac{\mathrm{d} \ln X(\eta)}{\mathrm{d} \ln \eta} = 2\alpha, \tag{6.1.21}$$

since

$$\lim_{\eta \to 0} \tilde{z}(\eta^*) = z(1) = 0. \tag{6.1.22}$$

On the other hand, the exponent λ, according to (6.1.5) characterizing power behaviour of $z(\eta)$ in the vicinity of the point $\eta = 1$, could be found using (6.1.19):

$$\lambda = \lim_{\eta \to 1+0} (\eta - 1) \frac{\mathrm{d} \ln y(\eta)}{\mathrm{d} \ln \eta} = \lim_{\eta \to 1+0} (\eta - 1)\left(\widetilde{X}(\eta^*) - 1\right). \tag{6.1.23}$$

If $X(\eta) \propto \eta^{-\beta}$ is singular at zero, function $\widetilde{X}(\eta^*)$ has a singularity at $\eta = 1$; $\widetilde{X}(\eta^*) \propto |\eta - 1|^{-\mu}$. It could be shown by an analysis of the definition (6.1.20). Note that there is a simple relation between exponents μ and β

$$\mu = \beta - \bar{d} + 1. \tag{6.1.24}$$

Since l.h.s. of equation (6.1.23) is constant, the powers of $|\eta - 1|$ have to cancel each other; $\mu - 1 = 0$. Taking into account (6.1.21) and (6.1.24), one gets

$$\alpha = \bar{d}/2. \tag{6.1.25}$$

Therefore, analytical estimates of the solution of a set of non-linear kinetic equations based on the superposition approximation confirm reduction of the asymptotic exponent.

The law $n(t) \propto \xi_0^{-\bar{d}/2}$ thus found in accompanied by the estimates of the correlation function behaviour. A singularity at zero coordinate of the similar particle function (Figs 5.5 and 5.6) reveals the power character: $X(\eta) \propto \eta^{-\bar{d}}$. The function $z(\eta)$ tends to zero, as $\eta \to 1$, in a line with the law (6.1.5) and with $\lambda = \bar{d}/2$ coming from (6.1.12). Note that a new asymptotic law of tunnelling recombination with the exponent (6.1.25) has been derived in [4] making use of another technique which, however, does not permit to calculate the reaction kinetics at *all* times.

6.1.3 Reactions on fractals

In this Section, we will discuss briefly results of the reaction on idealized disordered systems called *fractals* [5]. The concept of fractals, which was introduced by Mandelbrot [6], makes it possible to study also reactions on irregular structures with non-integer dimensionality. Such regular geometries find, for instance, application in modeling the structure of complex systems like polymeric materials. Properties of fractal objects related to physical quantities (such as mass distribution, density of vibrational states, etc.) are describable through several, not necessarily integer parameters, which play a role similar to that of the spatial dimension. Thus the *fractal (Hausdorff) dimension* $\dot{d}$ characterizes the distribution of mass. Denoting by N the number of lattice points inside a sphere of radius R, one has

$$N \propto R^{\dot{d}}, \tag{6.1.26}$$

where $\dot{d}$ is defined by

$$\dot{d} = \lim_{R \to \infty} \frac{\ln N}{\ln R}. \tag{6.1.27}$$

Another important parameter which appears in connection with dynamical properties of fractals (such as diffusion) is the spectral (fracton) dimension $\tilde{d}$. Thus, in the diffusion-limited reactions, one has to replace $\bar{d}$ in (2.1.78) by $\tilde{d}$, i.e.,

$$n(t) \propto t^{-\tilde{d}/4}, \tag{6.1.28}$$

as it was shown previously [7, 8]. For the Euclidean spaces, $\bar{d}$, $\dot{d}$ and $\tilde{d}$ coincide. Thus, one can interpolate between the Euclidean dimensions by using suitably tailored fractal structures.

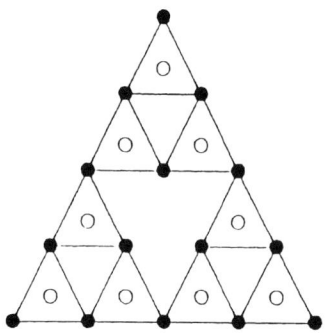

Fig. 6.4. The Sierpinski gasket on the 3rd stage. The open and filled circles give the positions of sites for type 'a' and type 'b' gaskets, respectively.

In this Section following [9], we analyse the A + B → 0 reaction with immobile reactants on the so-called *Sierpinski gasket* described below. We will proceed to show that in this case equation (6.1.1) with $\alpha = \bar{d}/2$ transforms into

$$n(t) \propto \xi_0^{-\tilde{d}/2}, \qquad (6.1.29)$$

i.e., the fractal dimension $\tilde{d}$ arises instead of $\bar{d}$.

The starting point for the simulations are random distributions of A and B particles (N_0 each) on the (finite) Sierpinski gasket. Starting with an equilateral triangle, the Sierpinski gasket is created by repeatedly adding two copies of the given structure to the left and right lower corner of the structure. As demonstrated in Fig. 6.4, there are two possibilities to position the particles on the gasket. One can either choose the centres of the smallest triangles (type 'a', open circles) or the corners of these (type 'b', filled circles) as possible positions for the reactants. Thus the number of sites for the Sierpinski gasket on the nth stage is 3^{n-1} for the first case and $(3^n + 3)/2$ for the second case. (The gasket in Fig. 6.4 corresponds to the 3rd stage.) Both structures, however, have the same fractal dimension $\tilde{d} = \ln 3/\ln 2 = 1.5849\ldots$

In simulations [9] Sierpinski gaskets on the 12th stage, containing 177147 or 265722 sites, were used respectively. The number N_0 of randomly distributed A or B particles was 10 percent of the total number of sites. The random mutual annihilation of dissimilar particles was simulated through a minimal process method [10]: from all AB pairs at each reaction step one pair was selected randomly, according to its reaction rate (3.1.2); the time

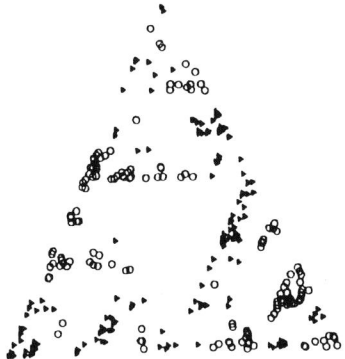

Fig. 6.5. Distribution of A and B particles on a type a gasket on the 9th stage at $t = 100$.

increment τ for this step was computed as $\tau = -\ln T/R$, with R being a sum of the rates of all AB pairs present and T – a random number from the homogeneous distribution in the unit interval. The time was measured in units of σ_0^{-1} and all distances are given in units of the nearest-neighbor distance a_0 of the underlying structure.

Figure 6.5 displays a typical distribution of A and B particles at $t = 100$ for a realization of the annihilation process on the Sierpinski gasket of the first kind at the 9th stage. The segregation of dissimilar particles, resulting from initial concentration fluctuations, is clearly visible at this reaction stage.

In order to extend the analytical equations to a fractal lattice, we will need the radial distribution function rdf(r) of the Sierpinski gasket, rdf(r) dr being the average number of sites with distance between r and $r + dr$ from a given site. For fractal lattices one has

$$\text{rdf}(r) = \gamma r^{\tilde{d}-1}, \tag{6.1.30}$$

where γ is a constant that depends on the structure of the particular fractal object. In regular dimensions $\tilde{d} = \bar{d}$, whereas γ is equal to the surface of the unit sphere, $\gamma = \gamma_{\bar{d}}$. Deviations from equation (6.1.30) may be obtained, when working with a discrete lattice.

In order to determine the constant γ, we computed the radial distribution functions for the two types of the Sierpinski gaskets under consideration. In Fig. 6.6 these functions are plotted, as averaged over all sites of the finite gaskets at the 11th stage. Due to the finite size of the structures, deviations

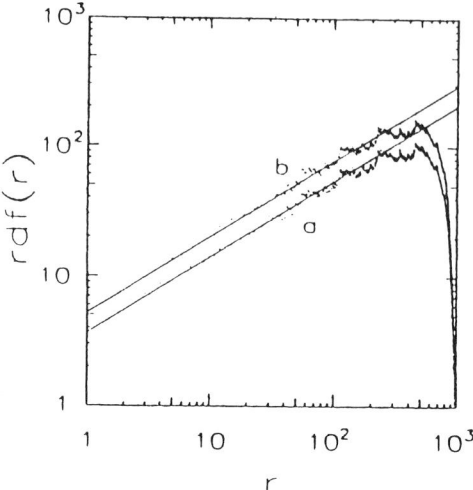

Fig. 6.6. Radial distribution functions rdf(r) for the two types of the Sierpinski gaskets 'a' and 'b' (dots) and ideal rdfs (solid curves) with $\bar{d} = 1.58$ and $\gamma = 3.65$ (a) or $\gamma = 5.2$ (b) (see equation (6.1.30)). Note, that both axes are logarithmic.

from the expected law in equation (6.1.1), $\alpha = \bar{d}/2$, must occur for radii comparable to the size of the whole gasket (which is $1024 a_0$ in Fig. 6.6). Fitting equation (6.1.1) to the initial, linear part of the data in Fig. 6.6, one obtains in both cases $\bar{d} = 1.58$ and $\gamma = 3.65 \pm 0.2$ for the gasket of type 'a' and $\gamma = 5.2 \pm 0.2$ for the type 'b' gasket. For radii comparable to the size of the system the real rdf deviates strongly from the ideal rdf in equation (6.1.30). Thus in the simulations one has to see to it that the mean cluster size does not become comparable to the size of the structure, if one wants to exclude finite size effects.

The starting point for both the simulation and also for the analytical development according to equations (5.1.14) to (5.1.16) is an uncorrelated distribution of A and B particles with initial densities $n(0) = n_0$ and $X(r,t) = Y(r,t) = 1$, $X_\nu(r,t) = X(r,t)$.

Now we want to apply the above described formalism, Section 5.1, to fractal lattices. Here one has to be careful when performing ensemble averages. Due to the lack of translational invariance, the ensemble averaged concentration $\langle \hat{n}(\vec{r},t) \rangle$ on a fractal is no longer independent of the position vector $\vec{r}$, as it is in normal dimensions. In order to avoid a further complication of the formalism, however, we neglect in the following analytical approach this position dependence of the concentration and also that of the corresponding correlation functions.

The main problem in the numerical evaluation of equations (5.1.14) to (5.1.16) is the evaluation of the integrals for fractal lattices. Integrals (4.1.19) of the form $\int f(r)\,d\vec{r}$, where the integrand $f(r)$ depends only on $r = |\vec{r}|$, may be expressed in polar coordinates, equation (5.1.6), as $\gamma_{\bar{d}} \int f(r) r^{\bar{d}-1}\,dr$, where $\gamma_{\bar{d}}$ for regular dimensions is identical to the surface of the unit sphere. Integrals of this form can be evaluated for fractal lattices, if one replaces $\bar{d}$ by $\tilde{d}$, and $\gamma_{\bar{d}}$ by γ.

The situation is more complicated for integrals of the form like equation (5.1.5). Introducing again polar coordinates with $\bar{d}$-dimensional angular element $d\Omega_{\bar{d}}$, equation (5.1.6), one obtains from equations (5.1.5) and (5.1.7). Thus, in equation (5.1.8), for every r' one has to perform additional integration over all angular orientations of the vector $\vec{r}'$. In order to calculate integrals of this type for the Sierpinski gasket, we used the following approximation: as the Sierpinski gasket is embedded in two-dimensional space, we perform the angular integration for the two-dimensional case ($d\Omega_2 = d\varphi$) and again replace $\bar{d}$ by $\tilde{d}$ for the radial integration. Thus, equation (5.1.7) results in

$$I[Z] = c\gamma \int \sigma(r')Y(r',t)\big(\widetilde{Z}(r^*,t) - 1\big) r'^{\tilde{d}-1}\,dr', \qquad (6.1.31)$$

where the co-factor c was introduced to correct possible errors of the approximation. In the particular case of the integrals in equations (5.1.14) to (5.1.16), which we want to evaluate for the Sierpinski gasket, the choice $c = 1$ yields reasonable results.

We performed numerical simulations of the annihilation process for the Sierpinski gaskets up to the 12th stage, starting with 10 percent of all sites randomly filled with A particles and another 10 percent with B particles. We chose the interaction range $r_0 = 4.27$ (in units of the nearest-neighbor distance a_0, which leads to the dimensionless initial reactant concentration $n(0) = 1$ (in units of $r_0^{-\tilde{d}}$, $\tilde{d} = 1.585$). Note, that time is still measured in units of σ_0^{-1}.

For the same parameter values we also evaluated numerically equations (5.1.14) to (5.1.16). Figure 6.7 shows a comparison of the results. The curves a and b show the decay of reactant concentration $n(t)$ for the direct annihilation process on gaskets of type 'a' or 'b', respectively, each averaged over 6 realizations of the process (the dotted curves indicate the scatter

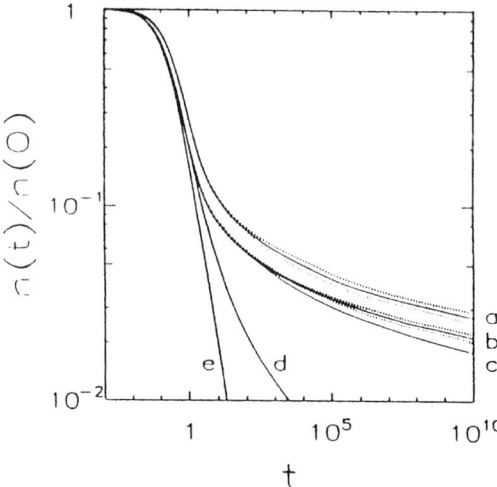

Fig. 6.7. Decay of reactant concentration obtained from direct computer simulations of the reaction on the Sierpinski gaskets of type 'a' and 'b' (curve a and b, respectively), from the numerical evaluation of equations (5.1.14) to (5.1.16) (curve c) and from the lower-level approximations, neglecting correlations between similar particles ($X(r,t) = 1$, curve d) or neglecting *all* spatial correlations ($X(r,t) = Y(r,t) = 1$, curve e).

of the different realizations). Curve c is the solution of equations (5.1.14)–(5.1.16), whereas curve d resulted from the lower level approximation, equation (4.1.38), where we set $X(r,t) \equiv 1$ (linear approximation). Curve e finally is the solution of the standard chemical kinetics, equations (5.1.63) and (5.1.64), which neglects all spatial correlations, $X(r,t) = Y(r,t) = 1$. Curves c–e are all calculated for the type 'b' gasket, i.e., $\gamma = 5.2$.

As we already found for one and two dimensions, a proper use of Kirkwood's superposition approximation, which led to equations (5.1.14) to (5.1.16) (curve c), gives a very good description for the strictly bimolecular annihilation process $A + B \to 0$. Lower-level approximations (curves d and e), however, differ strongly from the behaviour observed in direct simulations of the reaction process.

In order to demonstrate the predicted asymptotic behaviour $n(t) \propto (\ln t)^{-\tilde{d}/2}$, equation (6.1.29), we replotted the curves a–c in Fig. 6.8, displaying $\log[n(t)]$ versus $\log[\ln t]$. In this presentation a curve following equation (6.1.29) would appear as a straight line with the slope $-\tilde{d}/2$. Indeed, the three curves a–c come quite close to the expected slope $-\tilde{d}/2 \approx -0.79$

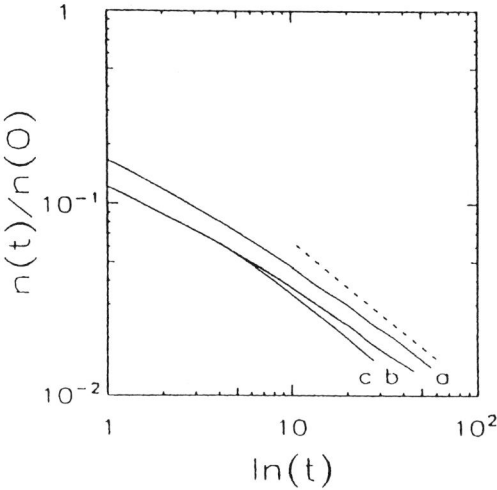

Fig. 6.8. Curves a–c of Fig. 6.7 replotted as $\log[n(t)/n(0)]$ versus $\log[\ln t]$. The dashed line shows the expected asymptotic slope $-\bar{d}/2 \approx -0.79$.

(which is indicated as a broken line) in the long-time range. Slight deviations may be due to statistical errors or to the finiteness of the gaskets.

In summary, we have shown that the kinetics of the bimolecular reaction $A + B \to 0$ with immobile reactants follows equation (6.1.1), even on a fractal lattice, if $\bar{d}$ is replaced by $\tilde{d}$, equation (6.1.29). Moreover, the analytical approach based on Kirkwood's superposition approximation [11, 12] may also be applied to fractal lattices and provides the correct asymptotic behaviour of the reactant concentration. Furthermore, an approximative method has been proposed, how to evaluate integrals on fractal lattices, using the polar coordinates of the embedding Euclidean space.

6.1.4 Multipole interaction

For the multipole interaction (4.1.44) the dissimilar correlation function could be also presented in a form of product (6.1.4), where $Y_0(r, t) = \exp[-\sigma(r)t]$. Neglecting indirect correlation mechanism, the dissimilar particle function $Y_0(r, t) = \exp[-(r/\xi_0)^{-m}]$, with ξ_0 defined by (4.1.45), is stationary in term of variable $\eta_0 = r/\xi_0$. Indirect mechanism of the correlation formation, as follows from a solution of equations derived in the superposition approximation, results at long times in $Y(r, t) \approx z(r, t)$, thus a direct mechanism of correlations becomes unimportant. The function $z(r,t)$ is characterized by a new scale ξ, respectively such that $\xi/\xi_0 \to \infty$ as $\xi_0 \to \infty$. Introducing new auto-model variable $\eta = r/\xi$, we have at long times $z(r,t) \to z(\eta)$. In its turn, as $\xi/\xi_0 \to \infty$, the function $Y_0(r,t) = \exp[-(\xi_0/\xi)^m \eta^{-m}]$ strives for unity. This point permits us searching of the asymptotic solution of a set of equations for the correlation functions (5.1.14) to (5.1.16) to omit in (5.1.16) the term $-\sigma(r)$, as responsible for "*short*-range" interaction characterized by the scale ξ_0 and thus to take into account the integral terms only producing "*long*-range" effects with the new scale ξ. Simultaneously we put into the kinetic equations $Y(r,t) = z(r,t)$.

In analogy with (6.1.1), let us define the asymptotic exponent α of the multipole interaction as

$$n(t) \propto \xi^{-\alpha}, \qquad (6.1.32)$$

where a new correlation length ξ must be defined, and estimate the reaction rate of the tunnelling recombination $K(t)$, equation (4.1.19), as $\xi \to \infty$:

$$K(t) = \int \sigma(r) Y(r,t)\, d\vec{r} \approx \xi^{-(m-\bar{d})} \int \sigma(\eta) z(\eta)\, d\vec{\eta}$$

$$= \xi^{-(m-\bar{d})} k_0, \qquad (6.1.33)$$

where

$$k_0 = \gamma_{\bar{d}} \int_0^\infty \sigma(\eta) z(\eta) \eta^{\bar{d}-1}\, d\eta. \qquad (6.1.34)$$

Equation (4.1.28) could be rewritten as

$$\frac{d \ln n(t)}{dt} = \frac{d \ln n(t)}{d \ln \xi} \frac{d \ln \xi}{dt} = -K(t) n(t). \qquad (6.1.35)$$

Making use of definition (6.1.33), we obtain from (6.1.35) the asymptotic relation

$$\frac{d \ln \xi}{dt} = \frac{K(t) n(t)}{\alpha} = c \xi^{-(m+\alpha-\bar{d})}, \qquad (6.1.36)$$

c is constant. Relation (6.1.36) defines the time-dependence of the correlation length ξ assuming the exponent α is known. Thus we obtain $\xi \propto t^{1/(m+\alpha-\bar{d})}$. In the linear approximation $\alpha = \bar{d}$ and $\xi = \xi_0 \propto t^{1/m}$.

Let us transform a set (5.1.14) to (5.1.16). At long times the solution has auto-model form when passing to the variable $\eta = r/\xi$, i.e., $X_\nu(r,t) = X(r,t) \to X(\eta)$, $Y(r,t) \to z(\eta)$. Time derivative is

$$\frac{\mathrm{d}\ln X(\eta)}{\mathrm{d}t} = -\frac{\mathrm{d}\ln X(\eta)}{\mathrm{d}\ln \eta}\frac{\mathrm{d}\ln \xi}{\mathrm{d}t}. \qquad (6.1.37)$$

Derivative $\frac{\mathrm{d}\ln \xi}{\mathrm{d}t}$ is defined via equation (6.1.36). Recalling now definition of the functionals (5.1.5) entering equations (5.1.14) to (5.1.16), we obtain a set of equations

$$\frac{\mathrm{d}\ln X(\eta)}{\mathrm{d}\ln \eta} = \frac{2\alpha I[z]}{k_0}, \qquad (6.1.38)$$

$$\frac{\mathrm{d}\ln z(\eta)}{\mathrm{d}\ln \eta} = \frac{2\alpha I[X]}{k_0}. \qquad (6.1.39)$$

The functional

$$I[Z] = \int \sigma(\eta')z(\eta')\bigl(Z(\eta^*) - 1\bigr)\,\mathrm{d}\vec{\eta}',$$

$$\eta^* = |\vec{\eta} - \vec{\eta}'|, \quad Z = X, Y \qquad (6.1.40)$$

has here a form similar to equation (5.1.5). The functional $I[Z]$, and the integral

$$k_0 = \int \sigma(\eta)z(\eta)\,\mathrm{d}\vec{\eta}$$

are defined in a similar way; in particular, $k_0 = -I[0]$. Ratio of functionals like $I[Z]/k_0$ entering r.h.s. (6.1.38) and (6.1.39) is analogous to the earlier derived expressions of the $(Z(\eta^*) - 1)$-type in r.h.s. of equations (6.1.18), (6.1.19): they arise due to $(Z(\eta^*) - 1)$ averaging. In the case of tunnelling recombination the angle average is calculated (see definition (6.1.20)), whereas for the multipole recombination it is performed using weighting function $\sigma(\eta)z(\eta)$.

In a set of equations (6.1.38) and (6.1.39), as well as in a set (6.1.18) and (6.1.19) the sought for exponent α plays a role of the effective eigenvalue of the problem when searching solution with desired asymptotics, $X(\infty) =$

$z(\infty) = 1$. Solutions $X(\eta)$ and $z(\eta)$ depend on a particular recombination mechanism, but the eigenvalue α is universal. A numerical solution of the asymptotic equations for the multipole reactions for all space dimensions, $\bar{d} = 1, 2, 3$, and power exponents $m = 6, 8, 10$ performed in [13] has demonstrated that relation (6.1.25) for α holds very well (errors of the difference scheme were $< 1\%$). In other words, the concentration decay law $n(t) \propto \xi^{-\bar{d}/2}$ is *universal* for immobile particles in equal concentrations.

When taking into account (6.1.36), it follows from $\alpha = \bar{d}/2$ that the correlation length of the multipole recombination

$$\xi \propto t^{2/(2m-\bar{d})} \tag{6.1.41}$$

differs from that known in the linear approximation $\xi_0 \propto t^{1/m}$; thus $\xi/\xi_0 \propto t^{\bar{d}/m(2m-\bar{d})} \to \infty$ as $t \to \infty$. A novel concentration decay law

$$n(t) \propto \xi^{-\bar{d}/2} \propto t^{-\bar{d}/(2m-\bar{d})} \tag{6.1.42}$$

is *slower* than the linear approximation estimate

$$n(t) \propto \xi_0^{-\bar{d}} \propto t^{-\bar{d}/m}. \tag{6.1.43}$$

Equation (6.1.42) was also derived in [14] by means of another method based on the variational principle.

The difference in the exponents (6.1.42) and (6.1.43) is large, which was indeed confirmed by computer simulations [13]. However, change of the correlation length in computer simulations is a more complicated problem. Indeed, assuming $\alpha = \bar{d}/2$, and $\xi = \xi_0$, the exponent in the expression

$$n(t) \propto \xi_0^{-\bar{d}/2} \propto t^{-\bar{d}/2m} \tag{6.1.44}$$

differs insignificantly from that in (6.1.42). Nevertheless, their divergence could be established – see Figs 6.9 and 6.10. For the multipole interaction kinetic curves $n(t)$ are universal: if kinetics is calculated once for given initial concentration, all necessary kinetics corresponding to other concentrations can be easily obtained just by changing time scale. The conclusion can be drawn from figures presented above that a new asymptotic reaction law is reached even at the reaction depths $\Gamma \approx 1$, i.e., range of the applicability of the linear approximation is small.

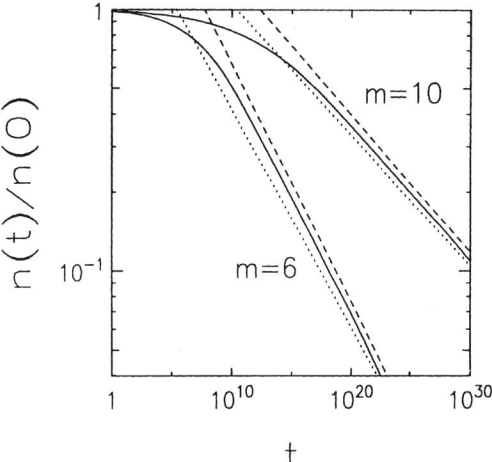

Fig. 6.9. Decay of the reactant concentrations $n(t)$ for $\bar{d} = 1$ obtained in the computer simulations. Realizations of the decay process were considered with 10^4 particles of each kind, initially distributed on a chain of 10^6 lattice sites, periodic boundary conditions are imposed. Full curves give the simulation data for $m = 6$ and $m = 10$, as indicated; the dashed line gives the slope $-\bar{d}/(2m - \bar{d})$, the dotted line gives the slope $-\bar{d}/2m$, to be compared with equations (6.1.41) and (6.1.43).

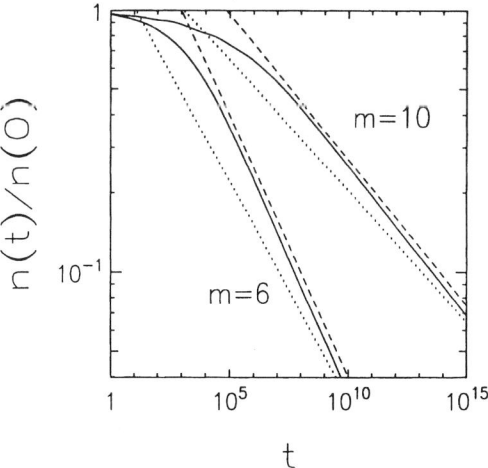

Fig. 6.10. Same as Fig. 6.9 for $\bar{d} = 2$ and a square lattice with 1000×1000 sites, periodic boundary conditions are imposed.

The ratio of the correlation lengths $\xi/\xi_0 \propto t^{\bar{d}/m(2m-\bar{d})}$ depends on the power exponent m of the multipole interaction. As m increases, the interaction becomes more and more short-range and as $m \to \infty$ it becomes in fact short range; characterised by $\xi/\xi_0 \to t^0 = \text{const}$. It clarified existence of the correlation length ξ_0 for the short-range tunnelling interaction.

6.1.5 The nearest-available-neighbour approximation

The accuracy of the Kirkwood superposition approximation was questioned recently [15] in terms of the new reaction model called NAN (*nearest available neighbour reaction*) [16–20]. Unlike previous reaction models, in the NAN scheme AB pairs recombine in a *strict* order of separation; the closest pair in an initially random distribution is removed first, then the next one and so on. Thus for NAN, the *recombination distance* R, e.g., the separation of the closest pair of dissimilar particles at any stage of the recombination, replaces real time as the ordering variable; time does not enter at all the NAN scheme. R is conveniently measured in units of the initial pair separation. At large R in $\bar{d}$-dimensions, NAN scaling arguments [16] lead rapidly to the result that the pair population decreases asymptotically as $cR^{-d/2}$ (c is constant) for $R \to \infty$, c could be calculated in the Monte Carlo simulations.

The kinetic equations [1–3, 12] were rewritten in [15] for a special choice of the recombination law $\sigma(r)$, adequate to the NAN model, and solved numerically for $\bar{d} = 1$. The general conclusion was drawn that the Kirkwood approximation is quite correct but leads to the error of the order of 10% for the critical exponent α in the asymptotic decay law $n(R) \propto R^{-\alpha}$. This quantity (10%) was suggested to be used as a measure of the accuracy of the Kirkwood approximation in the kinetics of the bimolecular reaction $A + B \to 0$.

Before analyzing critically the formalism [15], let us rewrite the basic equations of the NAN model in the superposition approximation. Equations for concentration $n(R)$ come from (4.1.28) using (4.1.19) with a formal substitution of time t for R:

$$\frac{dn(R)}{dR} = -K(R)n^2(R), \tag{6.1.45}$$

$$K(R) = \int \sigma(r) Y(r, R) \, d\vec{r}. \tag{6.1.46}$$

The same should be done in the joint correlation functions introducing $Y(r, R)$ and $X(r, R)$. The relevant equations come from a set (5.1.14) to (5.1.16):

$$\frac{\mathrm{d} \ln X(r, R)}{\mathrm{d}R} = -2n(R)I[Y], \qquad (6.1.47)$$

$$\frac{\mathrm{d} \ln Y(r, R)}{\mathrm{d}R} = -\sigma(r) - 2n(R)I[X], \qquad (6.1.48)$$

where the functional $I[Z]$ is defined by equation (5.1.5). We use below dimensionless units (length scale is the mean distance between particles, $l_0 = n(0)^{-1/\bar{d}}$) for R and $n(R)$; in particular, $n(0) \equiv 1$. Following [15], we study the case $\bar{d} = 1$ when equation (6.1.46) transforms into

$$K(R) = 2 \int_0^\infty \sigma(r) Y(r, R) \, \mathrm{d}r. \qquad (6.1.49)$$

The reaction rate $\sigma(r)$ satisfying the NAN model should be substituted into equations (6.1.45) to (6.1.48) and these equations then simplified for the analytical solution. It was assumed in [15] that the NAN model corresponds to the choice of the function $\sigma(r) \equiv \delta(r - R)$, where $\delta(x)$ is the Dirac delta function, which allowed the authors to simplify kinetic equations. Indeed, formal use of the main property of δ-function leads to $K(R) \equiv 2Y(R, R)$ and simplifies integral terms in equations (6.1.47) and (6.1.48). However, it leads to a number of errors not noticed in [15]. In fact, direct use of the property of the function $\sigma(r) \equiv \delta(r - R)$ in (6.1.49) is justified only if the function $Y(r,t)$ is an *analytical* one near the point $r = R$. This is not true in our case, since $Y(r, R)$ is a solution of the singular equation (6.1.48) with $\sigma(r) \equiv \delta(r-R)$ and consequently is also a generalized function (distribution). As it is well known from the theory, properties of products of such distributions $\sigma(r)Y(r, R)$ are not well defined. This is why we will introduce here less singular function $\sigma(r) = w_\varepsilon(r - R)$ where $w_\varepsilon(x)$ is an analytical function of the argument x and parameter $\varepsilon \to +0$. The choice of $\sigma(r) = \delta(r - R)$ in [15] suits in particular the well-known representation

$$w_\varepsilon(x) = \frac{1}{\sqrt{2\pi}\,\varepsilon} \exp\left(-\frac{x^2}{2\varepsilon^2}\right). \qquad (6.1.50)$$

Similarly to (6.1.4), let us present $Y(r, R)$ in the form

$$Y(r, R) = Y_0(r, R) z(r, R), \qquad (6.1.51)$$

where the function $Y_0(r, R)$ is a solution of (6.1.48) in the linear approximation, i.e., neglecting the term containing the functional $I[X]$ on its r.h.s.:

$$\frac{d \ln Y_0(r, R)}{dR} = -\sigma(r), \qquad (6.1.52)$$

whereas $z(r, R)$ gives the correction caused by many-particle effects. Defining

$$\Psi_\varepsilon(x) = -\int_x^\infty w_\varepsilon(x') \, dx', \qquad (6.1.53)$$

we arrive at the solution of a linear problem

$$Y_0(r, R) = \exp\big[-\Psi_\varepsilon(r - R)\big]. \qquad (6.1.54)$$

It could be easily shown that the (6.1.50)-type function is too weak for the NAN model. Indeed, in the limit $\varepsilon \to +0$ one gets from equation (6.1.54) a desired result: as $r > R$, $Y_0(r, R) = 1$ (all AB pairs with $r > R$ do not recombine) but if $r < R$ the result is not correct: $Y(r, R) = e^{-1}$ – AB pairs with $r < R$ still exist, what contradict the idea of the NAN model. As a result, the choice of equation (6.1.50) (or $\sigma(r) = \delta(r - R)$) does not allow to reproduce the well-known result of the linear approximation:

$$n(R) = \frac{1}{1 + 2R}, \qquad (6.1.55)$$

as it was assumed in [15]. As a more suitable choice of the function $w_\varepsilon(x)$ in the NAN model, the following "stronger" (in the sense $\lim_{\varepsilon \to 0} \int w_\varepsilon(x) \, dx = \infty$) function could be used:

$$w_\varepsilon(x) = \frac{1}{\sqrt{2\pi \varepsilon^2}} \exp\left(-\frac{x^2}{2\varepsilon^2}\right). \qquad (6.1.56)$$

In the limit $\varepsilon \to 0$ one gets

$$Y_0(r, R) = \begin{cases} 1, & r > R, \\ 0, & r \leqslant R. \end{cases} \qquad (6.1.57)$$

Obviously, solution (6.1.57) has a step-like form, besides

$$\frac{dY_0(r, R)}{dr} = \delta(r - R).$$

Substitution of (6.1.51) into (6.1.49) and use of $\sigma(r) = w_\varepsilon(r - R)$ allow us to separate finally a product which in the limit $\varepsilon \to 0$ gives δ-function

$$w_\varepsilon(r - R)Y_0(r, R) \equiv \frac{d}{dr} Y_0(r, R) = \delta(r - R). \qquad (6.1.58)$$

Only *now* one can use in equation (6.1.49) the main property of the Dirac δ-function and get a simple relation

$$K(R) = 2z(R, R), \qquad (6.1.59)$$

where $z(r, R)$ is an analytical function of r for finite R. Use of equations (6.1.51) and (6.1.58) in a set of equations (6.1.47) and (6.1.48) results in a much simpler set of the kinetic equations:

$$\frac{d \ln X(r, R)}{dR} = -n(R)K(R) \times$$
$$\times \big(Y(|r - R|, R) + Y(r + R, R) - 2\big), \qquad (6.1.60)$$

$$\frac{d \ln z(r, R)}{dR} = -n(R)K(R) \times$$
$$\times \big(X(|r - R|, R) + X(r + R, R) - 2\big). \qquad (6.1.61)$$

Keeping in mind the property of $Y_0(r, R)$, equation (6.1.57), we can assume reasonably that $Y(r, R) \equiv z(r, R)$ at $r > R$ but $K(R) \equiv 2Y(R + 0, R)$. Therefore in equation (6.1.61) from the function $z(r, R)$ one can again return to the correlation function $Y(r, R)$

$$\frac{d \ln Y(r, R)}{dR} = -n(R)K(R) \times$$
$$\times \big(X(|r - R|, R) + X(r + R, R) - 2\big), \qquad (6.1.62)$$

but now equation (6.1.62) should be solved in the region $r > R$ only since if $r < R$ $Y(r,R) \equiv 0$. Note that only after these corrections a set (6.1.45), (6.1.60), (6.1.62) accompanied with the definition $K(R) \equiv 2Y(R+0,R)$ corresponds to that derived in [15], except the additional singular term $\delta(r-R)$, present in [15] in the equation analogous to equation (6.1.62).

Let us consider now the asymptotic solution of the derived system in the limit $R \to \infty$. Analogously to equation (6.1.2), let us define the sought for exponent as

$$\alpha(R) = -\frac{d \ln n(R)}{d \ln R} \equiv n(R)K(R)R \qquad (6.1.63)$$

and look for its asymptotic solution, $\alpha = \alpha(\infty)$ as $R \to \infty$. Proceeding in a usual way, widely used in this Section, we introduce auto-model variables $\eta = r/R$, $\tau = \ln R$. In these new variables equations for the functions $X = X(r,R) \to X(\eta,\tau)$, $Y = Y(r,R) \to Y(\eta,\tau)$ are transformed into

$$\frac{\partial \ln X}{\partial \tau} = \frac{\partial \ln X}{\partial (\ln \eta)} - \alpha(R)\big(Y(|\eta-1|,\tau) + Y(\eta+1,\tau) - 2\big), \qquad (6.1.64)$$

$$\frac{\partial \ln Y}{\partial \tau} = \frac{\partial \ln Y}{\partial (\ln \eta)} - \alpha(R)\big(X(|\eta-1|,\tau) + X(\eta+1,\tau) - 2\big), \qquad (6.1.65)$$

provided that

$$K(R) = 2Y(1,\tau). \qquad (6.1.66)$$

As $R \to \infty$, we expect $\alpha < 1$. Since the l.h.s. of (6.1.63) strives to $\alpha = $ const, a constant value of its r.h.s. is achieved by a cancellation of powers of R in a product $n(R)K(R)R \propto K(R)R^{1-\alpha} \propto R^0$, from where one gets $K(R) \propto R^{-(1-\alpha)} \to 0$ as $R \to \infty$. Therefore, in the limit $\tau \to \infty$ according to (6.1.66) $Y(1,\tau) \to 0$ has to be fulfilled. Results of the numerical calculations presented below confirm this analysis. We will restrict ourselves now to the analysis of the limiting case of $\tau = \infty$ corresponding to the formation of stationary solutions of equations $X(\eta, \infty) \equiv X(\eta)$, $Y(\eta, \infty) \equiv Y(\eta)$, $\alpha = \alpha(\infty)$ with the auto-model variable η:

$$\frac{\partial \ln X}{\partial (\ln \eta)} = \alpha \big(Y(|\eta - 1|) + Y(\eta + 1) - 2 \big), \tag{6.1.67}$$

$$\frac{\partial \ln Y}{\partial (\ln \eta)} = \alpha \big(X(|\eta - 1|) + X(\eta + 1) - 2 \big), \tag{6.1.68}$$

subject to the boundary conditions $X(\infty) = Y(\infty) = 1$ (correlation weakening at infinity) and

$$Y(1) = 0. \tag{6.1.69}$$

It comes immediately from equation (6.1.67) that the correlation function $X(\eta)$ at the coordinate origin behaves as

$$\left. \frac{\partial \ln X}{\partial (\ln \eta)} \right|_{\eta=0} = \alpha \big(2Y(1) - 2 \big) \equiv -2\alpha, \tag{6.1.70}$$

i.e., $X(\eta) \propto \eta^{-2\alpha}$ as $\eta \to 0$ (there is a singularity!).

Let us assume (and confirm it in the self-consistent way) that due to its continuity the function $Y(\eta)$ strives to its limiting value, equation (6.1.69), in a power form, $Y(\eta) \propto (\eta - 1)^\beta$ (the actual value of β is not important for the further analysis and could be calculated numerically). Now let us consider equation (6.1.68) in a small vicinity of the point $\eta = 1$ ($\eta \to 1 + o$). The left-side derivative of equation (6.1.68) with $Y(\eta) \propto (\eta - 1)^\beta$ yields the term of the order of $(\eta - 1)^{-1}$ (singularity). In its turn, the only singularity source in r.h.s. of (6.1.68) is the function $X(\eta - 1) \propto (\eta - 1)^{-2\alpha}$ as $(\eta - 1) \to 0$. The equality of these powers gives us the desired exponent $\alpha = 1/2$, that is $n(R) \propto R^{-1/2}$, which is in a complete agreement with the NAN model [15–20]. This result could be easily generalized for the arbitrary space dimensionality $\bar{d}$: $\alpha = \bar{d}/2$.

Numerical solution of a set of the kinetic equations (6.1.45) and (6.1.63) to (6.1.66) for the joint correlation functions is presented in Figs 6.11 and 6.12. (To make them clear, double logarithmic scale is used.) The auto-model variable η is plotted along abscissa axis in Fig. 6.11 showing the correlation function $X(r, R)$.

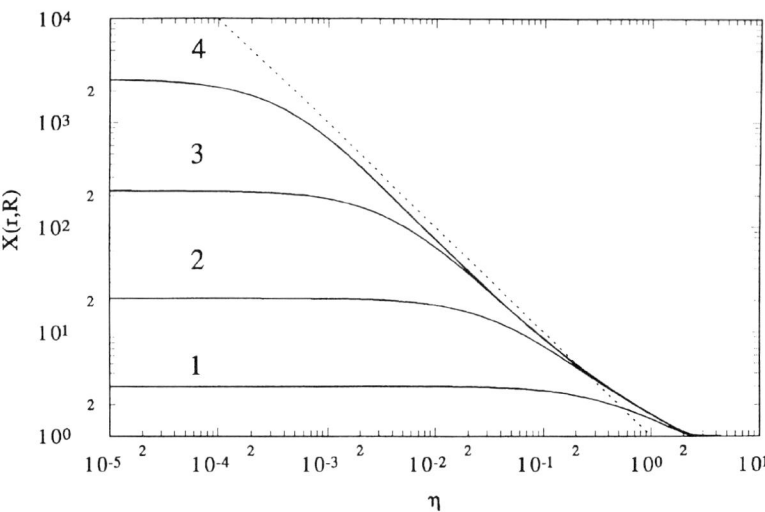

Fig. 6.11. The joint correlation functions of similar particles, $X(r,R)$ as a function of separation R and distance $\eta = r/R$. Recombination has proceeded to R: 1 (1), 10^1 (2), 10^2 (3), 10^3 (4). Dotted line is for η^{-1}.

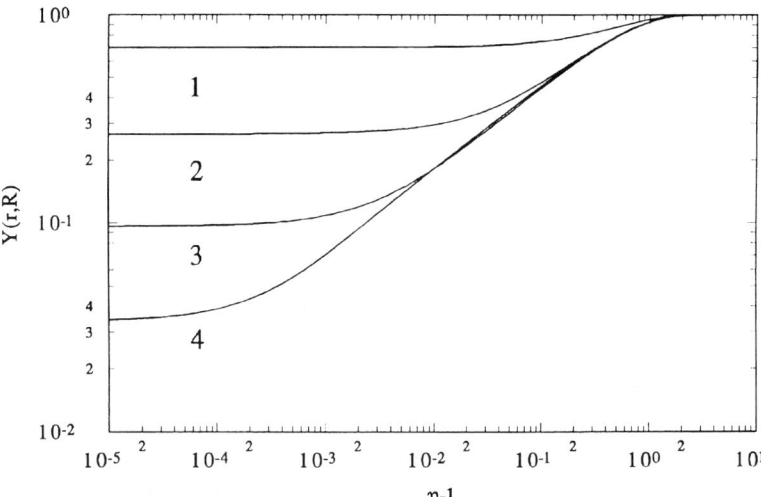

Fig. 6.12. The correlation functions of dissimilar particles, $Y(r,R)$ as a function of separation R. Recombination has proceeded to R: 1 (1), 10^1 (2), 10^2 (3), 10^3 (4).

Keeping in mind the expected limiting behaviour $X(\eta) \propto \eta^{-2\alpha} = \eta^{-1}$, the calculated curves are compared with this function given by the dotted line. For any value of parameter R deviation from the limiting law, $X(\eta) \propto \eta^{-1}$, as it is seen in Fig. 6.11, occurs on abscissa scale $\eta_0 \propto R^{-1}$ (the exponent, strictly speaking, is only close to -1 but this deviation is not important for our conclusions). In the graphical representation of the function $Y(r, R)$ (Fig. 6.12) the variable $\eta - 1$ is used which is convenient for the observation how the power law $Y(\eta) \propto (\eta-1)^\beta$ is formed. As in Fig. 6.11, the distinctive deviations from the power law are observed for $x = (\eta_0 - 1) \propto R^{-1}$. The length scale $\delta\eta_0 \propto R^{-1}$ could serve as a measure of the deviation of the correlation functions $X(\eta)$ and $Y(\eta)$ from their limiting values. At $R = 1$ the value of $Y(R, R) \approx 0.7$ seems to be close to the initial value of 1 (at $R = 0$); but on the other hand, the deviation from the limiting value, $Y(R, R) = 0$ as $R \to \infty$ – see equation (6.1.69), is very large. The limiting value of $Y(R, R) = 0$ is achieved, as easy follows from equation (6.1.63), according to the law $Y(R, R) \propto R^{-1/2}$. It is quite natural to assume that at large R the power law takes place also when approaching to the asymptotic value of $\alpha = 1/2$:

$$\alpha(R) = \alpha + O(R^{-\zeta}), \tag{6.1.71}$$

where the term $O(R^{-\zeta})$ could be neglected only as $R \gg 1$. This additional term cannot be found by means of simple analytical methods but much easier in numerical calculations. For a comparison, let us consider the result of the linear approximation (6.1.55) leading to equation (6.1.71) with $\zeta = 1$ but now with a classical value of the exponent, $\alpha(\infty) = \alpha_0 = 1$. In (6.1.55) the correction term $O(R^{-\zeta})$ could be neglected as $R > 10^2$.

In this connection the asymptotic equation suggested by Eggert [17], for $\alpha = 1/2$,

$$\alpha(R) = \alpha - \exp[-4R]/2 = \alpha + O\big(\exp[-4R]\big), \tag{6.1.72}$$

is derived using the scaling arguments and creates some doubts. Indeed, according to equation (6.1.72), an approach to the limiting value of α is *exponentially* fast, so $\alpha(R) \approx \alpha$ takes place even at $R \sim 1$! Besides, the function $\alpha(R)$ increases monotonously. Respectively, at $R > 1$ the concentration decay should be [17]

$$n(R) \approx 0.3746 R^{-1/2}. \tag{6.1.73}$$

The value of $R = 1$ [by the definition of R and the dimensionless concentration $n(R)$, such that $n(0) = 1$] corresponds to the case when the reaction has destroyed pairs AB separated by the relative distances r less or equal to the mean distance between particles. In other words, according to equation (6.1.73), a new asymptotic law with $\alpha = 1/2$ occurs already at *very small reaction depths*, $\Gamma \approx 0.5$!

However, the investigation of numerous processes done in this Chapter, allows us to establish a general law for the reaction asymptotics, $\alpha(R)$ (or $\alpha(\xi)$): in all cases studied the intermediate exponent $\alpha(\xi)$ is *non*-monotonous function reaching a maximum at some time moment (reaction rate), which is better pronounced for large spatial dimensions $\bar{d}$. This maximum in $\alpha(\xi)$ occurs at that reaction stage where we have crossover from the reaction regime where the fluctuation effects are not important (linear approximation is correct and the critical exponent α monotonously approaches the "classical" value of $\alpha_0 = \bar{d}$) to the regime characterized by the smaller exponent $\alpha = \bar{d}/2$. This transition happens typically at the reaction depth $\Gamma > 1$ when the curve $\alpha(\zeta)$ corresponding to the linear approximation exceeds its fluctuation-defined limiting value of α. As a result, the fluctuation-controlled curve $\alpha(\xi)$ approaches from *above* to its asymptotic value. Testing of the Kirkwood approximation accuracy presented in Section 5.1 has demonstrated that it gives always the best *lower bound estimate* for the concentration decay underestimating usually many-particle effects. Keeping in mind the monotonous increase of the exponent $\alpha(R)$ in equation (6.1.72), one can reasonably assume that the scaling approximation gives the *upper* estimate for concentration development thus *overestimating* the many-particle effects under study. As an illustration, let us consider now Figs 6.13 and 6.14. In the former picture several concentration dependencies $n(R)$ are compared being based on different approximations. In particular, the approximation (6.1.73) and numerical solution of a set (6.1.45) and (6.1.63) to (6.1.66) are quite close to each other but differ strongly from the linear approximation. A certain difference in slopes of curves $n(R)$ is observed which was used in [15] as an argument for incorrectness of the superposition approximation (in an original paper [15] the kinetic equations were solved for a smaller range $R \leqslant 10$ using (6.1.73) as a standard [17]).

Figure 6.14 presents the deviation in slopes just mentioned in a more obvious way. In the region $1 < R < 10^3$ numerical solution of equations (6.1.45) and (6.1.63) to (6.1.66) yields $\alpha(R) \approx 0.56 > \alpha = 0.50$ which has led to

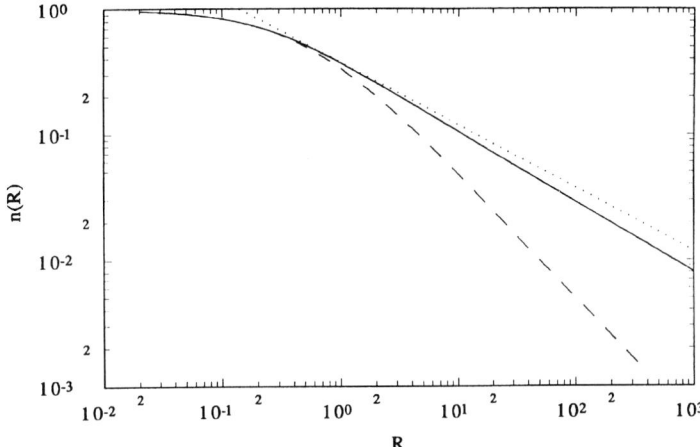

Fig. 6.13. The pair population $n(R)$ remaining after the recombination proceeds to a distance R. Full curve – equations (6.1.45), (6.1.63) to (6.1.66). Dashed curve – equation (6.1.55) – linear approximation. Dotted curve – equation (6.1.73).

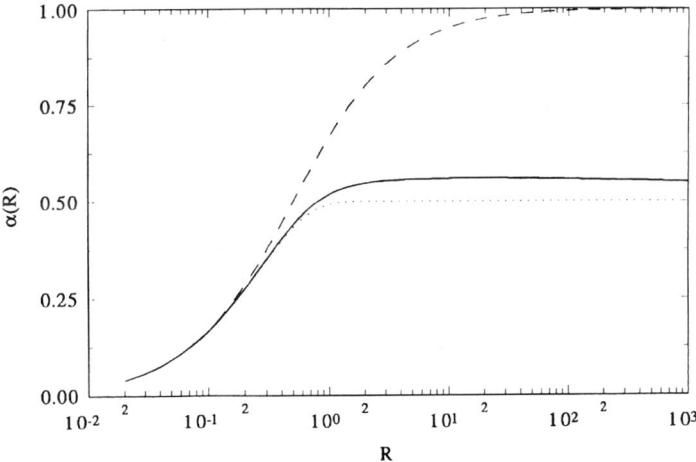

Fig. 6.14. The critical exponent $\alpha(R)$: full curve – equations (6.1.45), (6.1.63)–(6.1.66); dashed curve – equation (6.1.55) – linear approximation; dotted curve – equation (6.1.72).

the conclusion about the 10% error of the Kirkwood superposition approximation. Calculations show that $\alpha(R)$ approaches to its asymptotic value *very slowly*, in agreement with equation (6.1.71). The realistic power-law approach to the asymptotics makes a fast observation of the limiting value of α to be quite complicated (it is difficult to establish the relevant exponent ζ). As it was mentioned, equation (6.1.72) gives the limiting value of α already at $R \sim 1$. Therefore, as follows from this study, there is no chance of deriving *exact* approach to the asymptotic value of $\alpha = 1/2$ in terms of the NAN model. On the other hand, a problem of the accuracy of asymptotic equations (6.1.71) and (6.1.22) cannot be solved staying in the framework of the approximate methods [11, 12, 17].

6.2 KINETICS OF DIFFUSION-CONTROLLED REACTIONS OF MOBILE NON-INTERACTING PARTICLES

> There are two kinds of fools: some do not understand what everybody knows, others understand things which nobody is able to understand.
>
> V. Klyuchevsky, old Russian historian

6.2.1 Spatial structure of a system of mobile non-interacting particles

In the case of diffusion-controlled A + B → 0 reaction distinctive spatial distributions of reactants observed in computer simulations (e.g., [21]) are qualitatively the same as were presented earlier in Figs 1.20 and 1.21. Quite similar aggregation of similar particles into loose clusters occurs in agreement with a distinctive block-structure characterized by the diffusion length $l_D = \sqrt{Dt}$ shown in Fig. 2.8. When the reaction is controlled by the particle diffusion, these clusters (domains) are less pronounced since diffusion is known to *smooth* nonuniform particle distribution created in a course of reaction.

Behaviour of the joint correlation functions (see Figs 6.15 to 6.17 as typical examples of the black sphere model) resembles strongly those demonstrated above for immobile particles: at the scale $r < \xi = l_D$ the similar particle function exceeds its asymptotic value: $X_\nu(r,t) \gg 1$. As $r \gg l_D$, both the correlation functions strive for their asymptotics: $Y(r,t)$, $X_\nu(r,t) \approx 1$. The only peculiarity is that for mobile particles the boundary condition (5.1.40) tends to smooth similar particle correlation near the point $r = 0$. On the other hand, if one of diffusion coefficients, say D_A, is zero, the corresponding

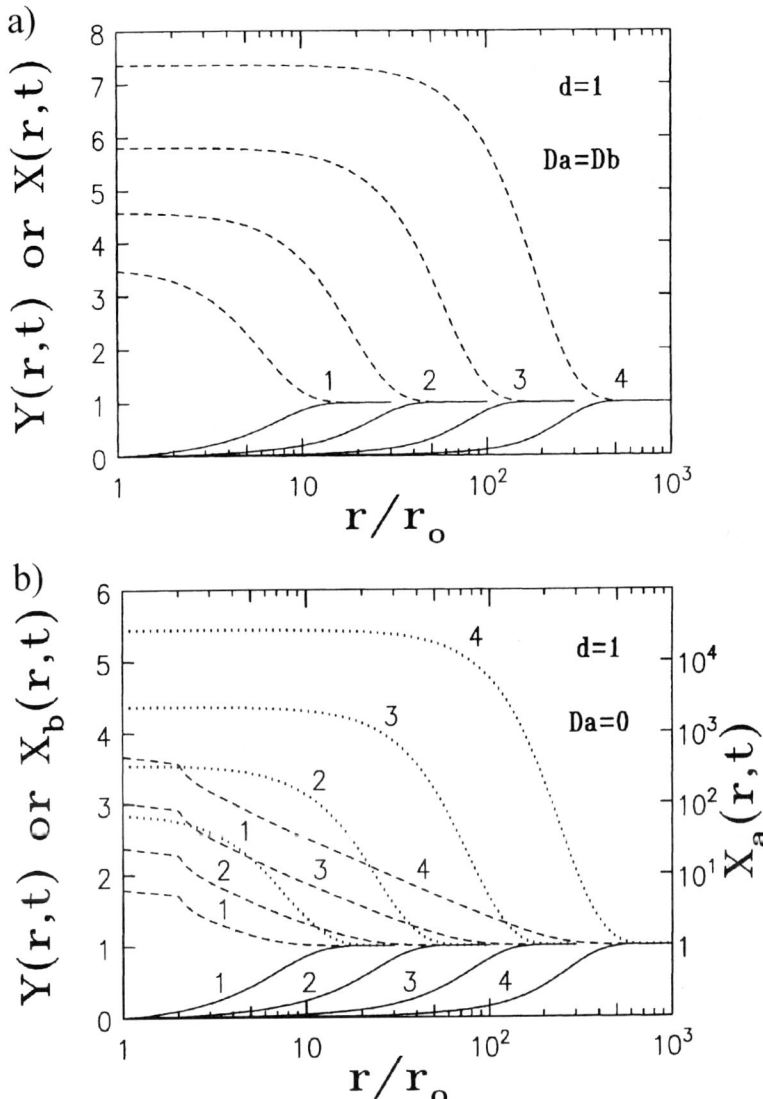

Fig. 6.15. A logarithmic plot of the joint correlation functions for similar, $X_\nu(r,t)$, and dissimilar, $Y(r,t)$, particles for $\bar{d}=1$ and symmetric (a), $D_A = D_B$, and asymmetric (b), $D_A = 0$, cases, respectively. Full curves are $Y(r,t)$, broken and dotted lines $X_A(r,t)$ and $X_B(r,t)$. The initial concentration $n(t) = 1.0$. The dimensionless time Dt/r_0^2 is: 10^1 (curve 1); 10^2 (2); 10^3 (3); 10^4 (4).

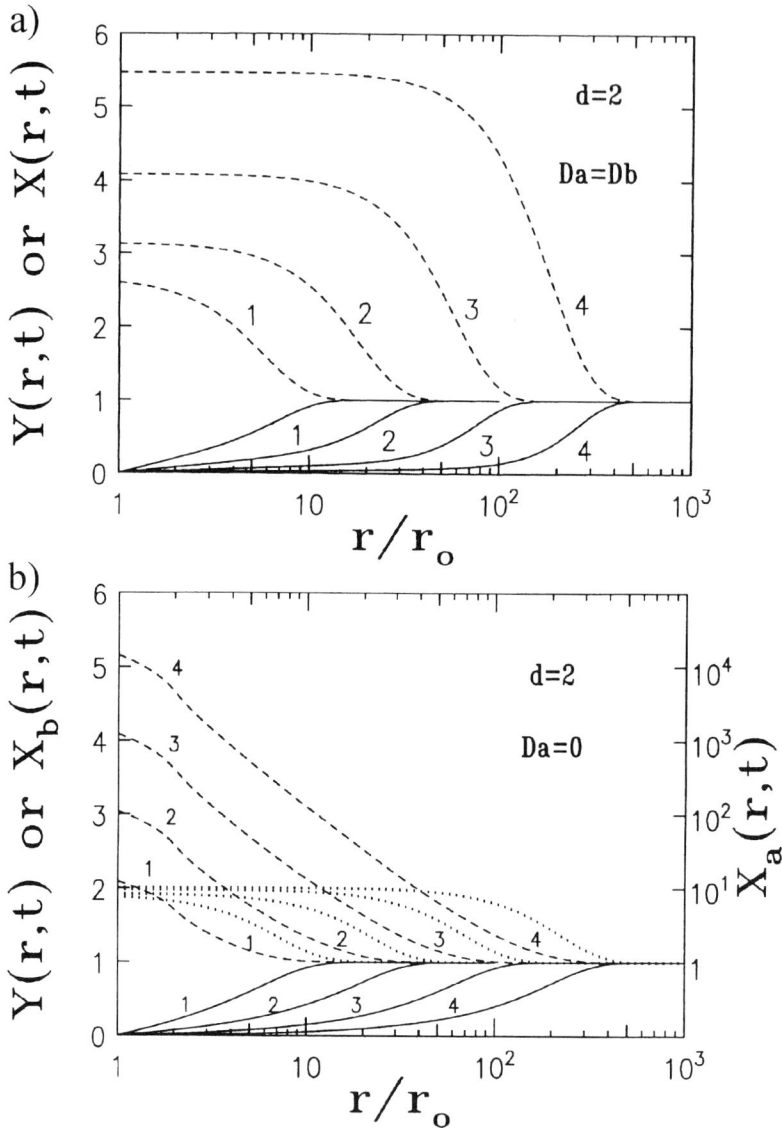

Fig. 6.16. A logarithmic plot of the joint correlation functions for similar, $X_\nu(r,t)$, and dissimilar, $Y(r,t)$, particles for $\bar{d}=2$ and symmetric (a), $D_A = D_B$, and asymmetric (b), $D_A = 0$, cases, respectively. Full curves are $Y(r,t)$, broken and dotted lines $X_A(r,t)$ and $X_B(r,t)$. The initial concentration $n(t) = 1.0$. The dimensionless time Dt/r_0^2 is: 10^1 (curve 1); 10^2 (2); 10^3 (3); 10^4 (4).

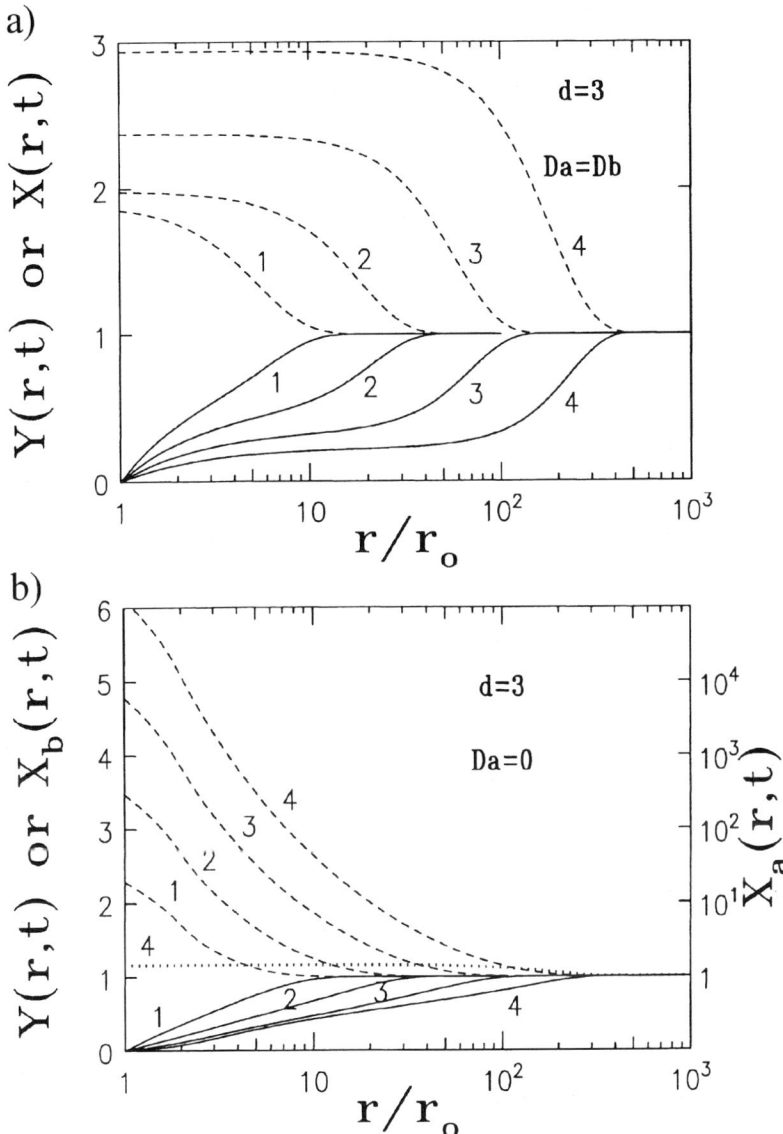

Fig. 6.17. A logarithmic plot of the joint correlation functions for similar, $X_\nu(r,t)$, and dissimilar, $Y(r,t)$, particles for $\bar{d}=3$ and symmetric (a), $D_A = D_B$, and asymmetric (b), $D_A = 0$, cases, respectively. Full curves are $Y(r,t)$, broken and dotted lines $X_A(r,t)$ and $X_B(r,t)$. The initial concentration $n(t) = 1.0$. The dimensionless time Dt/r_0^2 is: 10^1 (curve 1); 10^2 (2); 10^3 (3); 10^4 (4).

correlation function $X_A(r,t)$, remains non-smoothed, as it is shown in Figs 6.15 to 6.17.

The temporal evolution of spatial correlations of both similar and dissimilar particles for $\bar{d} = 1$ is shown in Fig. 6.15 (a) and (b) for both the symmetric, $D_A = D_B$, and asymmetric, $D_A = 0$ cases. What is striking, first of all, is rapid growth of the non-Poisson density fluctuations of similar particles; e.g., for $Dt/r_0^2 = 10^4$ the probability density to find a pair of close $(r \geqslant r_0)$ A (or B) particles, $X_A(r_0, t)$, by a factor of 7 exceeds that for a random distribution. This property could be used as a good *aggregation criterion* in the study of reactions between actual defects in solids, e.g., in ionic crystals, where concentrations of monomer, dimer and tetramer F centres (1 to 3 electrons trapped by anion vacancies which are 1 to 3nn, respectively) could be easily measured by means of the optical absorption [22]. Namely in this manner non-Poissonian clustering of F centres was observed in KCl crystals X-irradiated for a very long time at 4 K [23].

A *shape* of such clusters or domains containing alternatively A or B particles transforms rapidly to the trapezoidal profile, in a complete agreement with both computer simulations and theory presented in [27]. Note also that in the asymmetric case (broken lines and r.h.s. scale in Fig. 6.15(b)) immobile particles A form much more dense and compact clusters than mobile B's [26].

The analysis of the spatial distribution of particles for $\bar{d} = 2$ and 3 (Figs 6.16 and 6.17) shows that: (i) the deviation from the Poisson fluctuation spectrum is much smaller than for $\bar{d} = 1$. Comparison of the values of $X(0,t)$ for $Dt/r_0^2 = 10^4$ in the symmetric case yields $X(0,t) \approx 7, 5, 3$ for $\bar{d} = 1, 2$ and 3, respectively; (ii) the diffusion length $\xi = \sqrt{Dt}$ remains to be the characteristic length of the A- or B-rich domain size; and (iii) the gap between alternating domains with different-kind particles is less pronounced. For $\bar{d} = 1$ the joint correlation function for dissimilar particles is practically equal to zero up to a certain distance ("gap") after which it abruptly increases. (Its time behaviour was studied in detail in [27].) In contrast, for $\bar{d} = 3$ shown here $Y(r,t)$ has a long tail to the short distances and thus a gap becomes quite smoothened out.

6.2.2 Reaction asymptotic behaviour at long times

As it was shown above, in Section 6.1.1 using the reaction of immobile particles as example, the kinetic equations (5.1.2) to (5.1.4), being derived under the superposition approximation, indeed are able to reproduce reaction asymptotics at long times. Good agreement between theory and computer simulations, presented in Fig. 5.12, gives a support to an idea that an accuracy of our formalism would not be changed after a passage to the diffusion-controlled kinetics. In a line with the asymptotic law (2.1.78) discussed in Section 2.1.2, of great interest is its independent confirmation in the framework of the method pretending on the complete description of reaction kinetics at *any* times.

Indeed, the law (2.1.78) has been confirmed in [28–30] for different space dimensions. However, since instead of general equations (5.1.2) to (5.1.4), their black sphere analog (5.1.33) to (5.1.35) was analyzed, a set of non-linear equations lost its transparency. This is why let us start now with a complete set of equations for the superposition approximation, $n_A(t) = n_B(t) = n(t)$ assuming that the recombination $\sigma(r)$ is short-range and could be characterized by a finite radius r_0 so that the integral

$$k_0 = \int \sigma(r)\,d\vec{r} \tag{6.2.1}$$

is finite. Tunnelling recombination (3.1.2), or grey sphere model, corresponding to (3.1.16) with a finite σ_0 value can serve as good examples.

New reaction asymptotic law (2.1.78) emerges due to formation during the reaction course of a new spatial scale – the correlation length $\xi = l_D$. Similar to the case of immobile particles, we can expect here that at long times the coordinate r enters into the correlation function in a scaling form $\eta = r/l_D$, so that $Y(r,t) \to Y(\eta,\tau)$, $X_\nu(r,t) \to X_\nu(\eta,\tau)$, where the second variable is $\tau = \ln t$. Let us consider consequences of the new scale formation. The reaction rate $K(t)$ (4.1.19) at long times can be estimated as

$$K(t) = \int \sigma(r)Y(r,t)\,d\vec{r}$$

$$= \int \sigma(r)Y(\eta,\tau)\,d\vec{r} \approx k_0 Y(0,\tau). \tag{6.2.2}$$

It was used in deriving (6.2.2), that since the function $Y(\eta,\tau)$ changes considerably on the distances of the order of $l_D \gg r_0$ only, it can be taken out of the integral in (6.2.2) with the maximal contribution on the scale r_0, along with the definition (6.2.1). In doing so, the value of $Y(\eta,\tau)$ is taken at $\eta = 0$.

In the same way the functionals (5.1.5) entering (5.1.2) to (5.1.4) could be estimated as

$$I[Z] \approx k_0 Y(0, \tau)\bigl(Z(\eta, \tau) - 1\bigr). \tag{6.2.3}$$

Let us use now the definition of the critical exponent α in the reaction asymptotics (4.1.68). As it follows from (4.1.28) and (4.1.68) for equal concentrations

$$\alpha = -\lim_{t \to \infty} \frac{\mathrm{d}\ln n(t)}{\mathrm{d}\ln t} = \lim_{t \to \infty} tK(t)n(t). \tag{6.2.4}$$

This exponent α defines the reaction asymptotics at long times: $n(t) \propto t^{-\alpha}$ as $t \to \infty$. Since l.h.s. (6.2.4) is constant, exponents in its r.h.s. should be canceled, thus for the time development of $K(t)$ we get

$$K(t) \propto t^{-(1-\alpha)}. \tag{6.2.5}$$

If the exponent α turns out to be less than its "classical" value found in the linear approximation, $\alpha_0 = 1$ ($\bar{d} \geqslant 2$), the relevant reaction rate $K(t)$ is *reducing* at long times so that in the long-time limit $K(\infty) = 0$ (the reaction rate's zerofication). On the other hand, when searching the asymptotic solution of non-linear equations, the asymptotic relation (6.2.4) permits to replace $tK(t)n(t)$ for α.

To simplify mathematical manipulations, let us consider now the case of equal diffusion coefficients, $D_A = D_B$, in which case the similar correlation functions just coincide, $X_\nu(\eta, \tau) = X(\eta, \tau)$. Taking into account the definition of correlation length $l_D = \sqrt{Dt}$, where $D = D_A + D_B = 2D_A$, as well as time-dependence of new variables η and τ, one gets from (5.1.2) to (5.1.4) a set of equations

$$\frac{\partial X(\eta, \tau)}{\partial \tau} = \widehat{L}X(\eta, \tau) - 2\alpha X(\eta, \tau)\bigl(Y(\eta, \tau) - 1\bigr), \tag{6.2.6}$$

$$\frac{\partial Y(\eta, \tau)}{\partial \tau} = \widehat{L}Y(\eta, \tau) - 2\alpha Y(\eta, \tau)\bigl(X(\eta, \tau) - 1\bigr). \tag{6.2.7}$$

In these equations the motion operator is

$$\widehat{L} = \frac{\partial^2}{\partial \eta^2} + \frac{\bar{d}-1}{\eta}\frac{\partial}{\partial \eta} + \frac{\eta}{2}\frac{\partial}{\partial \eta}. \tag{6.2.8}$$

The first and second terms in (6.2.8) come from the Laplace operator (3.2.8), whereas the third one arises due to passage from the variable r to η. In deriving (6.2.6) and (6.2.7), the relations (6.2.2)–(6.2.4) were used. The boundary conditions imposed on these equations are $Y(\infty, \tau) = X(\infty, \tau) = 1$ (correlation weakening at the infinity). The condition (5.1.40) corresponds to

$$\lim_{\eta \to 0} \eta^{\bar{d}-1} \frac{\partial X(\eta, \tau)}{\partial \eta} = 0. \tag{6.2.9}$$

The function $Y(\eta, \tau)$ is expected to be limited at $\eta = 0$.

The "eigenvalue" α of the non-linear problem enters a set of equations (6.2.6), and (6.2.7) as it did also in the asymptotic equations for immobile particles considered above – equations (6.1.18), (6.1.19) or (6.1.31), (6.1.32) – due to which the boundary conditions are met. Let us introduce now the quantity $z(\eta, \tau) = X(\eta, \tau) - Y(\eta, \tau)$, whose equation follows from subtracting (6.2.7) from (6.2.6):

$$\frac{\partial z(\eta, \tau)}{\partial \tau} = \widehat{L} z(\eta, \tau) + 2\alpha z(\eta, \tau). \tag{6.2.10}$$

The boundary conditions for $z(\eta, \tau)$ are

$$\lim_{\eta \to \infty} \eta^{\bar{d}-1} \frac{\partial z(\eta, \tau)}{\partial \eta} = 0, \quad z(\infty, \tau) = 0, \tag{6.2.11}$$

arising naturally from the boundary conditions imposed on the two joint correlation functions.

Introduction of $z(\eta, \tau)$ is argued by the following reasons. The correlation function of similar particles enters (6.2.7) in a form of the difference $[X(\eta, \tau) - 1]$ which can be approximated by the function $z(\eta, \tau)$. To visualize this point, let us look at Figs 6.15(a) to 6.17(a) and analyze the behaviour of the correlation functions $Y(r, t)$ and $X_\nu(r, t)$. As $r \gg l_D$, the function $Y(r, t) \approx 1$, and therefore $[X(r, t) - 1] \approx [X(r, t) - Y(r, t)] = z(\eta, \tau)$. A deviation is observed at relative distances $r \leqslant l_D$. Here $z(\eta, \tau) = [X(r, t) - 1] + O(1)$, where absolute error $O(1)$ does not exceed unity, but a *relative* error for the strong *positive* spatial correlations observed $X(r, t) \gg 1$ as $r < l_D$, is negligibly small. Therefore, the function $z(\eta, \tau)$ is able not only to reproduce the *asymptotic* behaviour of the unknown function $[X(\eta, \tau) - 1]$, but also its behaviour at *any* η with an error unimportant for the asymptotic analysis.

An asymptotic solution of (6.2.10) could be found when separating variables η and τ in a form

$$z(\eta, \tau) = \sum_j \varphi_j(\tau) u_j(\eta). \tag{6.2.12}$$

Note peculiarity of (6.2.10) – this asymptotic equation is valid in the limit $\tau \to \infty$ (or $t \to \infty$), and thus *no* initial conditions should be imposed. The expansion (6.2.12) is also asymptotic one ($\tau \to \infty$). Defining the increase of the exponents of $\varphi_j(\tau)$ as

$$\varepsilon_j = \lim_{\tau \to \infty} \frac{d \ln \varphi_j(\tau)}{d\tau}, \tag{6.2.13}$$

one arrives at equations for the "eigenfunctions" $u_j(\eta)$:

$$\widehat{L} u_j(\eta) + (2\alpha - \varepsilon_j) u_j(\eta) = 0. \tag{6.2.14}$$

Taking now into account the boundary conditions (6.2.11), the maximal exponent

$$\varepsilon_0 = 2\alpha - \bar{d}/2 \tag{6.2.15}$$

is associated with the eigenvalue

$$u_0(\eta) = \exp(-\eta^2/4). \tag{6.2.16}$$

Therefore, a leading term of the asymptotic ($\tau \to \infty$) expansion (6.2.12), which we use below, reads as

$$z(\eta, \tau) \approx \left[X(\eta, \tau) - 1 \right] \approx \varphi_0(\tau) u_0(\eta). \tag{6.2.17}$$

Formation of the non-Poisson fluctuation spectrum corresponds to the $\varepsilon_0 \geqslant 0$, otherwise the asymptotic solution of (6.2.10) is trivial: $z(\eta, \tau) = 0$. Note that a definition of the exponent (6.2.13) of the $\varphi_j(\tau)$ increase does not mean that $\varphi_j(\tau) = c_j \exp(\varepsilon_j \tau)$, $c_j = $ const only, c_j can be also a *power* function of τ. Since for the power behaviour $c_j(\tau)$,

$$\lim_{\tau \to \infty} \frac{d \ln c_j}{d\tau} = 0$$

is valid, the power corrections to the exponent $\exp(\varepsilon_j \tau)$ are not essential for obtaining eigenfunctions, despite the fact that in principle these correlations will be important for a further consideration. Let us consider now equation (6.2.7). Substituting there $[X(\eta, \tau) - 1]$ by $z(\eta, \tau)$, we get

$$\frac{\partial Y(\eta, \tau)}{\partial \tau} = \widehat{L}Y(\eta, \tau) - h^2(\tau)u_0(\eta)Y(\eta, \tau). \tag{6.2.18}$$

Here time-dependent function $h^2(\tau) = 2\alpha\varphi_0(\tau)$ is introduced. Since $\varepsilon_0 \geqslant 0$, $h(\tau)$ increases in time τ. As $h(\tau) \to \infty$, the asymptotic solution of (6.2.18) coincides with the solution of *degenerate* differential equation

$$\widehat{L}Y(\eta, \tau) - h^2(\tau)u_0(\eta)Y(\eta, \tau) = 0, \tag{6.2.19}$$

corresponding to the *quasi-steady-state* ($\frac{\partial Y(\eta, \tau)}{\partial \tau} = 0$).
Substituting

$$Y(\eta, \tau) = g(\eta, \tau) \frac{\exp[-\eta^2/8]}{\eta^{(\bar{d}-1)/2}} \tag{6.2.20}$$

we arrive at an equation for the function $g(\eta, \tau)$

$$\frac{d^2 g(\eta, \tau)}{d\eta^2} = \Psi^2(\eta, \tau) g(\eta, \tau), \tag{6.2.21}$$

where

$$\Psi^2(\eta, \tau) = h^2(\tau)u_0(\eta) + \frac{(\bar{d}-1)(\bar{d}-3)}{4\eta^2} + \frac{\bar{d}}{4} + \frac{\eta^2}{16}. \tag{6.2.22}$$

Equation (6.2.21) is an ordinary differential equation having large parameter $h(\tau)$. Its solution under certain standard conditions could be obtained by means of the steepest descent (called also VKB) method.

This problem is simplified for $\bar{d} = 1, 3$ since in these two particular cases the singular term $\frac{(\bar{d}-1)(\bar{d}-3)}{4\eta^2}$ becomes zero and thus there is *no* limitations for use of the VKB method and we can easily find the solution satisfying both asymptotics at infinity, $Y(\infty, \tau) = 1$, as well as its limited value $Y(\eta, \tau)$ at the origin. For instance, for $\bar{d} = 3$ the solution of (6.2.19) is

$$Y(\eta, \tau) = c \exp\left[-\frac{\eta^2}{8}\right] \frac{\sinh\left(\int_0^\eta \Psi(\eta', \tau) \, d\eta'\right)}{\eta \sqrt{\Psi(\eta, \tau)}}. \tag{6.2.23}$$

The constant c can be found from the asymptotic condition $Y(\eta, \tau) = 1$. The main result here is asymptotic behaviour of $Y(\eta, \tau)$ at $\eta = 0$ [28–30]:

$$Y(0, \tau) \propto \exp\left[-ah(\tau)\right], \tag{6.2.24}$$

where a is a constant.

In its turn, for $\bar{d} = 2$ the singular term $(\bar{d}-1)(\bar{d}-3)/(4\eta^2)$ prevents use of the VKB in a whole interval of η values. However, a special analysis carried out in [30] has demonstrated that this singularity does not affect the final result (6.2.24). So far, the relation (6.2.24) holds for the space dimensions $\bar{d} \leqslant 3$.

For $\bar{d} \geqslant 4$ singular term $(\bar{d}-1)(\bar{d}-3)/(4\eta^2)$ does not allow to find the solution *finite* at $\eta = 0$. It has simple interpretation: in systems with so large space dimensionalities *no* variable $\eta = r/l_D$ exists there. Similar to $\bar{d} = 3$ in the linear approximation, for $\bar{d} \geqslant 4$ we can find the stationary solutions, $Y(r, \infty) = Y_0(r)$. For them the reaction rate $K(\infty) = K_0 = $ const and the "classical" asymptotics $n(t) \propto t^{-\alpha_0}$, $\alpha_0 = 1$ hold. Therefore, for a set of kinetic equations derived in the superposition approximation the *critical* space dimension could be established for the diffusion-controlled reactions.

Equation (6.2.15) with the exponent $\alpha < \alpha_0 = 1$ after transition from variable t to $\tau = \ln t$ corresponds to the reaction rate

$$K(t) \propto \exp\left[-(1-\alpha)\tau\right]. \tag{6.2.25}$$

By a comparison of (6.2.11), (6.2.24) and (6.2.25), we arrive at the condition imposed on the function $h(\tau)$: $h(\tau) \propto \tau$. Keeping in mind definition of $h(\tau)$, we conclude that $\varphi_0(\tau)$ is the power of function: $\varphi_0(\tau) \propto \tau^2$. By definition (6.2.13), its exponent $\varepsilon_0 \equiv 0$. It follows from (6.2.15) that $\alpha = \bar{d}/4$ thus finally confirming the asymptotic reaction law (2.1.78).

6.2.3 Correlation length and critical exponent

Let us discuss now results [26] of the numerical solution of equations (5.1.33) to (5.1.35) shown in Figs 6.18 to 6.20. First of all, Fig. 6.18 shows the decay of particle concentrations for $\bar{d} = 1$ and different initial

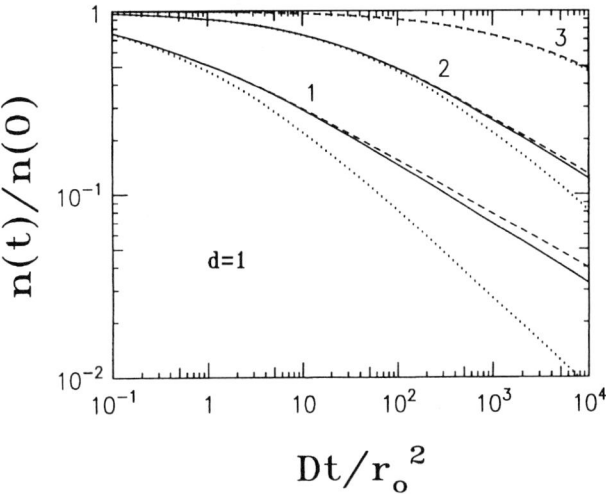

Fig. 6.18. The concentration decay as a function of dimensionless time for $\bar{d} = 1$. Curves 1 to 3 correspond to the initial concentrations $n(0) = 1; 0.1; 0.01$, respectively. Full curves show the asymmetric case, $D_A = 0$; broken curves are symmetric case, $D_A = D_B$. Dotted lines show neglect of the many-particle effects.

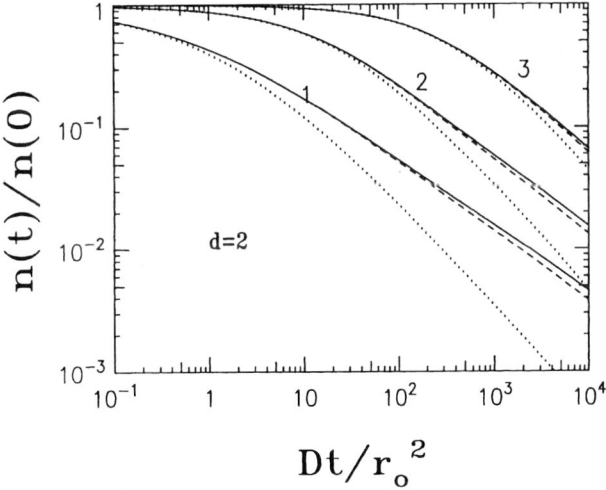

Fig. 6.19. The concentration decay as a function of dimensionless time for $\bar{d} = 2$. Curves 1 to 3 correspond to the initial concentrations $n(0) = 1; 0.1; 0.01$, respectively. Full curves show the asymmetric case, $D_A = 0$; broken curves are symmetric case, $D_A = D_B$. Dotted lines show neglect of the many-particle effects.

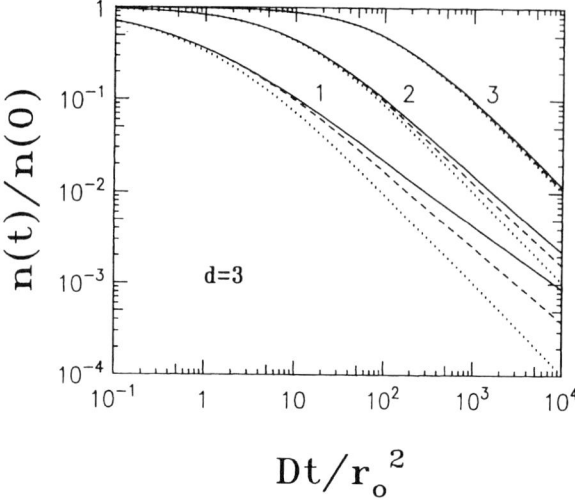

Fig. 6.20. The concentration decay as a function of dimensionless time for $\bar{d} = 3$. Curves 1 to 3 correspond to the initial concentrations $n(0) = 1; 0.1; 0.01$, respectively. Full curves show the asymmetric case, $D_A = 0$; broken curves are symmetric case, $D_A = D_B$. Dotted lines show neglect of the many-particle effects.

concentrations, $n(0)$. As it is well seen, the higher $n(0)$, the faster the kinetics deviates from the standard approach (Chapter 4) given by a dotted line. For the symmetric mobilities, $D_A = D_B$, decay is slightly slower than for asymmetric, $D_A = 0$. These theoretical results agree well with the recent independent computer simulations [31–33].

Transformation from t to the diffusion length l_D permits us to present (2.1.78) in the form $n(t) \propto l_D^{-\bar{d}/2}$, completely corresponding to the asymptotic reaction law for immobile particles $n(t) \propto \xi^{-\bar{d}/2}$, with the correlation length $\xi = l_D$. Therefore $n(t) \propto \xi^{-\bar{d}/2}$, is a *universal* law of the bimolecular A + B → 0 reaction (provided equal reactant concentrations). Peculiarity of the diffusion-controlled regime of reaction is the existence of the *marginal* space dimensionality $\bar{d}_0$.

In the time region where asymptotic law $n(t) \propto \xi^{-\bar{d}/2}$ takes place, the *spatial* region of the essential change of correlation functions is defined by the scale ξ. However, this statement should be specified for unequal diffusion coefficients. Strictly speaking, there are *several* correlation (diffusion) lengths for the diffusion-controlled A + B → 0 reaction: $\xi_A = \sqrt{2D_A t}$, $\xi_B = \sqrt{2D_B t}$

and $\xi = \sqrt{Dt}$. If both diffusion coefficients D_A and D_B are nonzero, ξ variable could be taken as an independent one, since the correlation lengths ξ_A, $\xi_B \propto \xi$ are defined by ξ value. For such a choice the reaction asymptotics being defined by ξ does not depend on the factor $\kappa = D_A/D$, which characterizes the *difference* of the diffusion coefficients. Indeed, studies [26, 29, 34, 35] have demonstrated that change of κ affects the correlation functions only, the latter still could be expressed through the correlation lengths ξ_A or ξ_B. For example, the equation $[X(\eta, \tau) - 1] \approx \varphi_0(\tau) u_0(\eta)$ could be generalized for the case of unequal diffusion coefficients as

$$\left(X_\nu(r,t) - 1\right) \propto \tau^2 \exp[-\eta_\nu^2/4]$$

$$= (\ln t)^2 \exp\left(-\frac{r^2}{8D_\nu t}\right), \quad \nu = A, B, \tag{6.2.26}$$

where $\eta_\nu = r/\xi_\nu$. The functions (6.2.26) characterize statistically the dynamics of similar particle aggregation. Their size, being defined by ξ_ν, is controlled entirely by particle *diffusion*. It was shown also in [29] that universality of the $n(t) \propto l_D^{-\bar{d}/2}$ law can be violated if one kind of reactants is immobile, say $D_A \equiv 0$, i.e., $\kappa = 0$. In such a situation the relevant diffusion length ξ_A disappears; i.e., aggregates of A particles are "frozen" (Figs 6.15(b) to 6.17(b)) and their size is changed only due to diffusion of B particles to their boundary. In other words, there is *no* mechanism of the diffusion smoothening of inhomogeneities in particles A *inside* their aggregates.

The analysis of the relevant equations of the superposition approximation for the asymmetric case done in [26, 29] shows that the *universality* $n(t) \propto l_D^{-\bar{d}/2}$ is violated for $\bar{d} = 3$; in this case instead of $n(t) \propto t^{-3/4}$ one gets $n(t) \propto t^{-1/2}$, i.e., an additional great reduction of the reaction rate is expected.

Direct establishment of the asymptotic reaction law (2.1.78) requires performance of computer simulations up to certain reaction depths Γ, equation (5.1.60). In general, it depends on the initial concentrations of reactants. Since both computer simulations and real experiments are limited in time, it is important to clarify which values of the *intermediate* asymptotic exponents $\alpha(t)$, equation (4.1.68), could indeed be observed for, say, $\Gamma \leqslant 3$. The relevant results for the black sphere model (3.2.16) obtained in [25, 26] are plotted in Figs 6.21 to 6.23. The illustrative results for the linear approximation are also presented there.

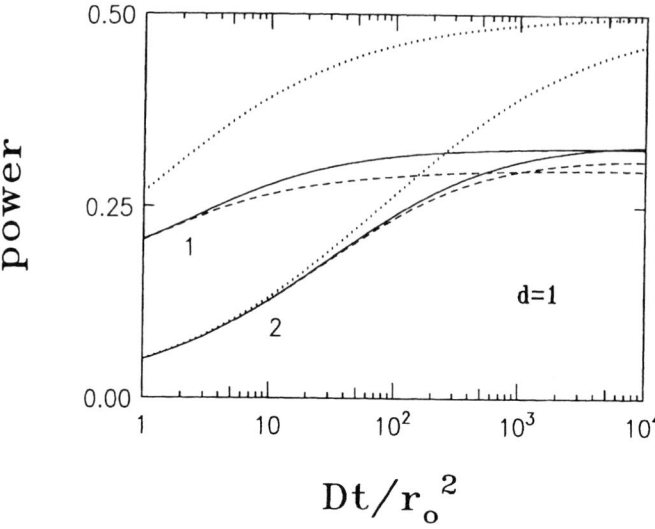

Fig. 6.21. The critical exponent $\alpha(t)$ as a functions of time for $\bar{d} = 1$. Curves 1 and 2 correspond to the initial concentrations $n(0) = 1; 0.1; 0.01$, respectively. Full curves show the asymmetric case, $D_A = 0$; broken curves are symmetric case, $D_A = D_B$. Dotted lines show neglect of the many particle effects.

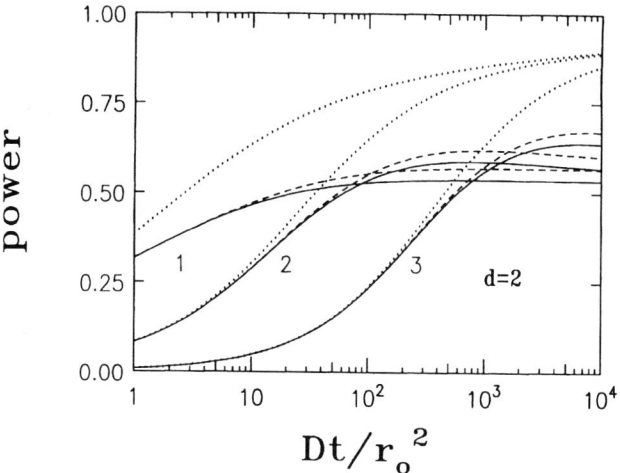

Fig. 6.22. The critical exponent $\alpha(t)$ as a function of time for $\bar{d} = 2$. Curves 1 and 2 correspond to the initial concentrations $n(0)=1; 0.1; 0.01$, respectively. Full curves show the asymmetric case, $D_A = 0$; broken curves are symmetric case, $D_A = D_B$. Dotted lines show neglect of the many particle effects.

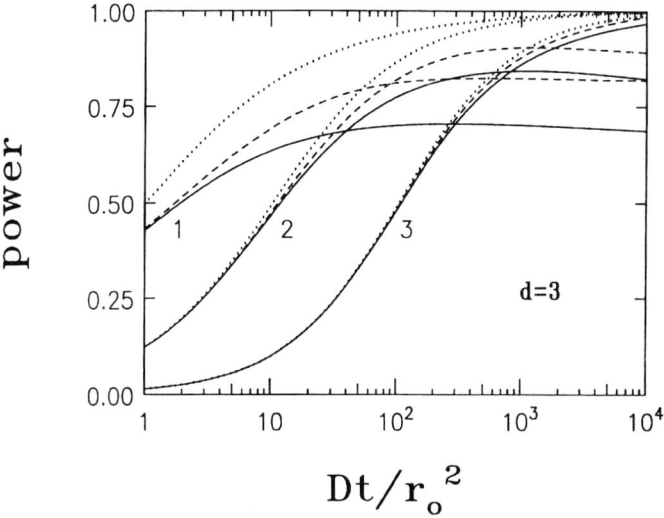

Fig. 6.23. The critical exponent $\alpha(t)$ as a function of time for $\bar{d} = 3$. Curves 1 and 2 correspond to the initial concentrations $n(0) = 1$; 0.1; 0.01, respectively. Full curves show the asymmetric case, $D_A = 0$; broken curves are symmetric case, $D_A = D_B$. Dotted lines show neglect of the many particle effects.

As it is seen in these figures, the higher $n(0)$, the faster the asymptotics is achieved. For the immobile reactant A and $\bar{d} = 1$, $\alpha(t)$ systematically exceeds that for the equal mobilities which leads to faster concentration decay in time. The results for $\bar{d} = 2$ and 3 are qualitatively similar. Their comparison with the one-dimensional case demonstrates that the concentration decay is now much faster since the critical exponents strive for $\alpha = 3/4$ and $\alpha = 1/2$ for the symmetric and asymmetric cases, respectively, which differ greatly from the "classical" value of $\alpha = 1$. Respectively, the gap between symmetric and asymmetric decay kinetics grows much faster than in the $\bar{d} = 1$ case. Therefore, the conclusion could be drawn that the effect of the relative particle mobility is pronounced better and thus could be observed easier in *three*-dimensional computer simulations rather than in one-dimensional ones, in contrast to what was intuitively expected in [33].

In the linear approximation the critical exponent $\alpha(t)$ for $\bar{d} = 2$ and 3 approaches monotonously its asymptotic value $\alpha = \alpha_0 = 1$. In contrast, incorporation of the fluctuation effects leads to the emergence of the *maximum* $\alpha(t)$, better pronounced for $\bar{d} = 2$; the position of a maximum $\alpha(t)$

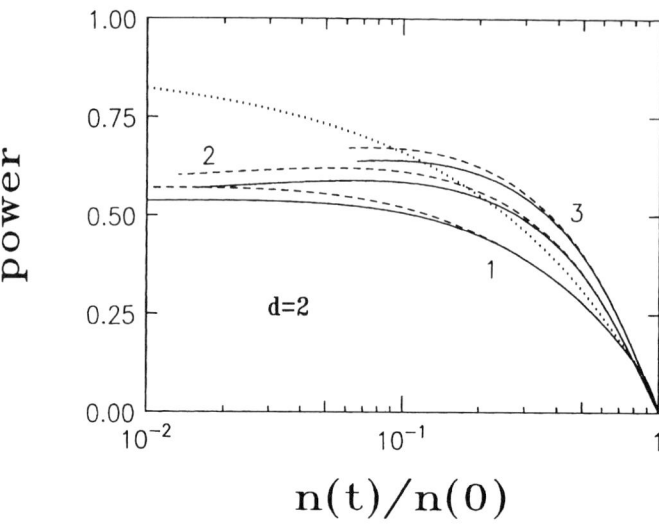

Fig. 6.24. The critical exponent $\alpha(t)$ as a function of $n(t)/n(0)$ for $\bar{d} = 2$. Notation as in Fig. 6.22.

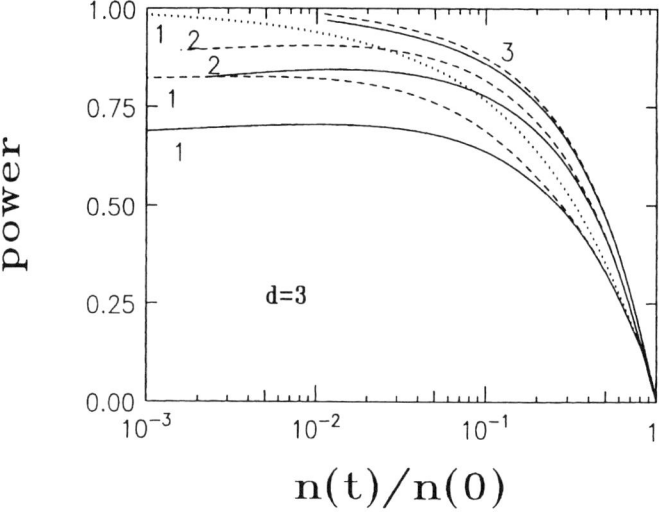

Fig. 6.25. The critical exponent $\alpha(t)$ as a function of $n(t)/n(0)$ for $\bar{d} = 3$. Notation as in Fig. 6.23.

TABLE 6.1
The critical exponents obtained in computer simulations for $D_A = D_B$ using the minimal process method [38].

d	α_{theor}	α_{simul}
1	0.25	0.249 ± 0.001
2	0.50	0.512 ± 0.001
3	0.75	0.759 ± 0.003
4	1.0	1.037 ± 0.003

practically remains the same for a wide range of initial concentrations and corresponds to the reaction depth of $\Gamma \simeq 1$ – Fig. 6.24. It means that for $\bar{d} = 2$ existence of a new asymptotic law (2.1.78) could be established already at $\Gamma > 1$ since the linear approximation *deviates* greatly from the results of computer simulations. Thus, linear approximation *fails* to describe kinetics in this region. The rate of the approach to the asymptotics $\alpha = 1/2$ depends on the initial concentration. Note that for $\bar{d} = 2$ under consideration the factor $\kappa = D_A/D$ affects the kinetic curve but not the critical exponent.

An observation of the fluctuation effects for $\bar{d} = 3$ is more problematic since the $\alpha(t)$ reveals only slight maximum and the rate of the approaching to the asymptotic α value is slow – Fig. 6.25. For the reaction depth $\Gamma = 3$ $\alpha(t)$ still differs from its asymptotic limit for *any* initial concentration. For example, computer simulations [36] were carried out up to $\Gamma_0 = 1.5$ with very high initial concentration $n_0 \simeq 1$. For equal diffusion coefficients *only qualitative reduction* of $\alpha(t)$ was observed – in a line with its behaviour in Fig. 6.25. Near the point $\Gamma = \Gamma_0$ computer simulation revealed $\alpha(t) \approx 0.79$, being again in good agreement with Fig. 6.25 and close to the limit $\alpha = 0.75$. A novel *minimal process method* (known also as the Gillespie algorithm [37]) allowed very recently to reach real asymptotics for the $A + B \to 0$ reactions with equal reactant mobilities [38]. Results for different space dimensions are summarized in Table 6.1.

6.2.4 Asymmetric particle mobilities

More favourable conditions of the fluctuation effect manifestation are met if one of reactants is immobile, e.g., $D_A = 0$. In this case *considerable* deviation of kinetic curves as compared to those just presented for *equal* diffusion coefficients is observed in computer simulations [36]: $\alpha(t) = 0.69$ at $\Gamma \simeq \Gamma_0$, which is already *less* than asymptotic limit $\alpha = 0.75$ expected for the $D_A = D_B$ case, but still greater than the limiting theoretical exponent $\alpha = 0.50$. Note that this computer-calculated value of $\alpha(t) = 0.69$ agrees also with Fig. 6.25 (see also Section 6.3).

If for equal particle concentrations the reaction depth $\Gamma \leqslant 3$ is *enough* to confirm the asymptotic reaction law (it is well demonstrated in [21] for $\bar{d} \leqslant 2$), it is no longer true for unequal concentrations, $n_A(t) < n_B(t)$. In this situation the region $0 < \Gamma \leqslant 3$ turns out to be too short for checking up different hypotheses of the reaction asymptotics (see, e.g., [32]). Say, it is predicted on the basis of the superposition approximation that $n_A(t) \propto t^{-\bar{d}/2} \propto l_D^{-\bar{d}}$ [29] but the range of its applicability estimated via the time development of the exponent $\alpha(t)$, in the simplest for the observation case $\bar{d} = 1$ lies at very large values of $\Gamma > 5$.

The role of asymmetry in particle mobilities and initial concentrations was studied in several papers (see, e.g., [32, 33, 39] and references therein). For the latter case an analytical asymptotic $(t \to \infty)$ estimate was obtained for $\bar{d} = 1$ and $D_A = D_B$ in a form:

$$n_A(t) \approx (Dt)^{-1/4} \exp\left\{ -\frac{\delta n^2}{[n_A(0) + n_B(0)]} \sqrt{2\pi Dt} \right\},$$

$$\delta n = n_B(0) - n_A(0). \tag{6.2.27}$$

For equal particle concentrations, $n_A = n_B$, equation (6.2.27) transforms into the standard result $\alpha = \bar{d}/4$, but demonstrates additional reaction *acceleration* if one of reactants is in minority, $n_A < n_B$.

In the more difficult case of asymmetry in *both* mobilities ($D_A = 0$, $D_B > 0$) and initial concentrations, $\delta n > 0$, a kind of master equation was derived and solved numerically, along with the Monte Carlo simulations [32]. As it is seen in Fig. 6.26, for equal particle concentrations the decay asymptotics, shown by dashed lines, is reached very fast for both symmetrical and asymmetrical reactant mobilities (curves 1). It is no longer true, however, for asymmetrical concentrations shown in curves 2. These latter demonstrate clearly an existence of the *two* different classes of universality for reactions with unequal concentrations; one class corresponds to the case when both (A and B) reactants are mobile or those which are in majority whereas

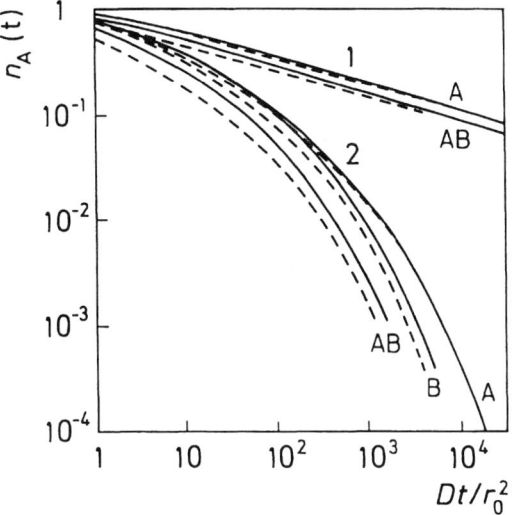

Fig. 6.26. Time decay of the particle concentration, $n_A(t)$, for cases of equal (curves 1) and unequal (curves 2) concentrations involved into the $A+B \to 0$ reaction for $d=1$ [32]. Curves 1: $n_A(0) = n_B(0) = 0.1$; curves 2: $n_A = 0.1$, $n_B(0) = 0.2$. Symbols A, B, AB correspond to types of mobile species (one type or both types respectively). Full lines show results of the Monte Carlo simulations whereas dashed lines – solution of a master equation.

another class corresponds to mobility of *minority* species (A). This situation reminds the difference observed for the asymptotic kinetics for discussed in Section 5.2 target and trapping problems in the $A + B \to B$ reaction when either $D_A = 0$ or $D_B = 0$ holds respectively.

This conclusion has not been supported however by rigorous mathematical bounds derived for the long-range asymptotics [4, 39] of the $A + B \to 0$ reaction with $n_A < n_B$ presented in a form

$$n_A(t) \propto \exp\left(-c_{\bar{d}} \delta n \, g_{\bar{d}}(t)\right), \qquad (6.2.28)$$

where $c_{\bar{d}}$ is constant and $g_{\bar{d}}$ is defined in equation (5.3.5).

That is, in the above-discussed case of $\bar{d} = 1$ we arrive at

$$n_A(t) \propto \exp\left(-\text{const} \cdot \delta n \sqrt{t}\right). \qquad (6.2.29)$$

Keeping in mind mentioned in Section 5.1 errors of the superposition approximation for unequal concentrations, this problem has no large outlook.

6.2.5 Reaction rate $K(t)$

Equation (6.2.15) with $\alpha = \bar{d}/4$ corresponds to the reaction rate whose time decay is characterized by the reaction rate $K(t) \propto t^{-(1-\bar{d}/4)}$, $\bar{d} \leqslant 3$. The analysis of its behaviour permits to find simple relation between the kinetic descriptions done in terms of the superposition and linear approximations.

As it is noted in Section 5.1, a distinctive feature of the linear approximation is the absence of back-coupling between the concentration $n(t)$ and the correlation function $Y(r,t)$ which is also independent of the *initial* reactant concentrations. Moreover, in the linear approximation the parameter $\kappa = D_A/D$ does not play any role at all. So, in the black-sphere model for the standard random distribution, $Y(r > r_0, t) = 1$, one gets universal relations (4.1.69), (4.1.65) and (4.1.61) for $\bar{d} = 1, 2$ and 3 respectively. In contrast, in the superposition approximation the law $K = K(t)$ loses its universality, since along with space dimension $\bar{d}$ it depends also on *both* the parameter κ and the initial reactant concentrations.

Figures 6.27 to 6.29 demonstrate results of the reaction rate calculations and its time-development: $K = K(t)$ [11, 12, 25, 26]. The relation between

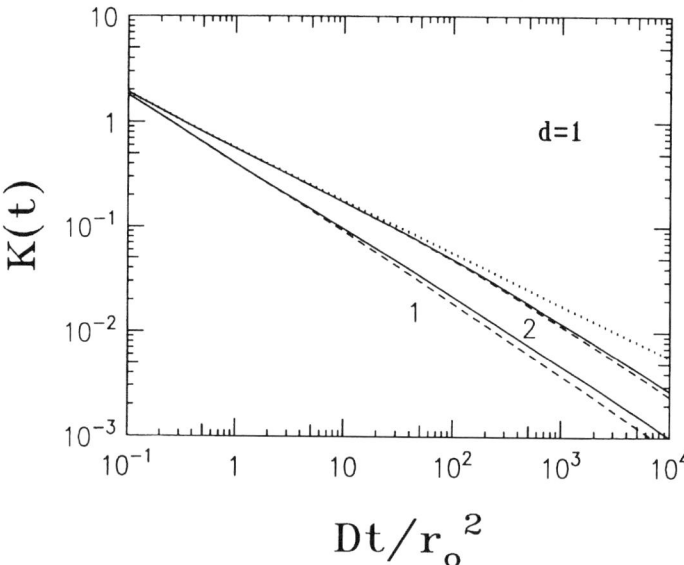

Fig. 6.27. Dependence of the reaction rate $K(t)$ upon time for $\bar{d} = 1$. Notations as in Fig. 6.21.

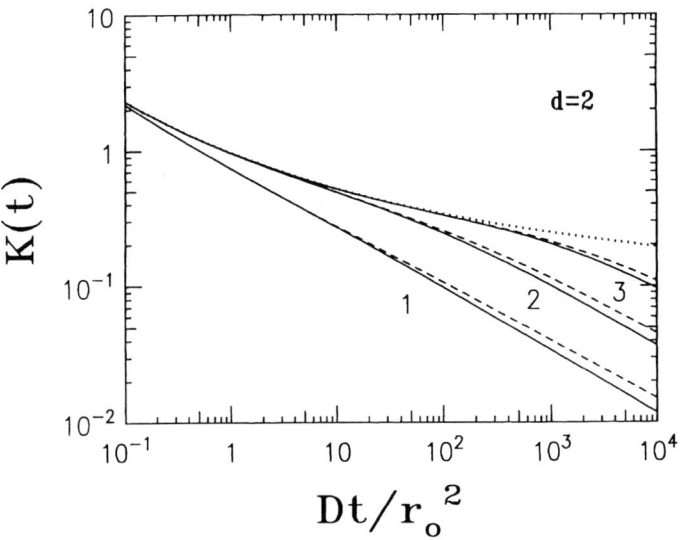

Fig. 6.28. Dependence of the reaction rate $K(t)$ upon time for $\bar{d} = 2$. Notations as in Fig. 6.22.

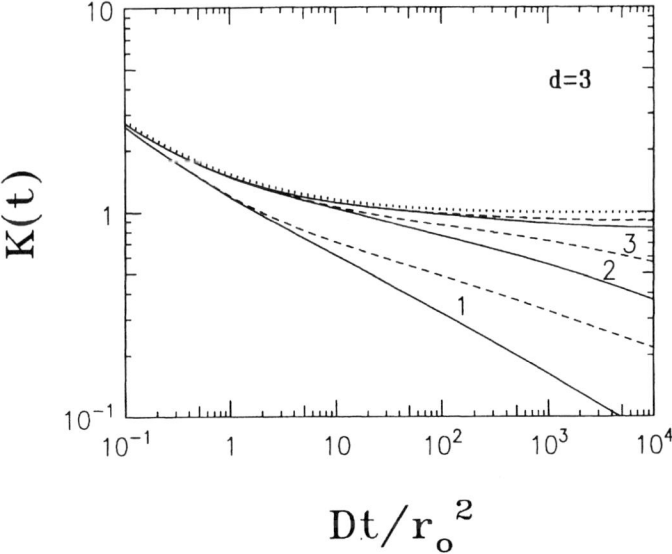

Fig. 6.29. Dependence of the reaction rate $K(t)$ upon time for $\bar{d} = 3$. Notations as in Fig. 6.23.

different approximations is very well seen in Figs 6.27 and 6.28: the function $K(t)$ in the linear approximation is a *round* curve (upper bound) of a set of curves obtained for the different initial concentrations and parameters κ.

When the initial concentrations are small, fluctuation effects are not pronounced and the curve of the superposition approximation degenerates to those of the linear approximation. However, for *any* initial concentration n_0 there exists distinctive time $t_0 = t_0(n_0)$, at which these approximations start to diverge from each other. For $\bar{d} = 2$ parameter κ slightly affects behaviour of the $K = K(t)$ – full and broken curves 1 at long times t reveal the same slopes with the only difference in co-factors. At last, for $\bar{d} = 3$ curves at $\kappa = 0$ and $\kappa = 0.5$ differ considerably and have different slopes at long t (this peculiarity of the kinetics when one of the reactants is immobile has been already stressed above).

6.2.6 *Other formalisms and approaches*

Lattice models for four basic kinds of bimolecular reactions, A + B → 0 and A + A → 0, (annihilation) A + A → A (coagulation) and A + B → B (energy transfer) were studied in [40] exploring the utility of using *time-power series* for determining the critical exponents of the kinetics. The particle concentration is presented in a form

$$C(t) = \sum_{n=0}^{\infty} C^{(n)} \frac{t^n}{n!} \qquad (6.2.30)$$

with coefficients

$$C^{(n)} = \left. \frac{d^n C(t)}{dt^n} \right|_{t=0}. \qquad (6.2.31)$$

A finite number of the coefficients $C^{(n)}$ could be found from a hierarchy of differential equations for single densities, joint densities etc. These series were found well-behaved with the coefficients all positive and monotonously increasing. Even moderate length series of 5–6 terms gave exponents quite close to those analytically estimated for above-mentioned basic bimolecular reactions (see Section 5.3).

The use of an elegant field-theoretical formalism has been already discussed in Section 2.3.2 (see also review articles and a book [11, 41, 42] and references therein). It is complemented in a series of recent papers [43–45] dealing with the second quantization version of the *cluster expansion method*, called also *the coupled cluster approach* [46]. The advantage of its use is: (i) automatic accounting the indistinguishability of classical particles when the Bose statistics for creation and annihilation operators is chosen and (ii) option of selective sum lader, ring and other diagrams without requiring perturbation analysis. It is important since ring diagrams, i.e., terms responsible for random phase approximation are known to allow better understanding of the appearance of collective modes in other many-body systems. (iii) Lastly, it treats on an equal footing both closed and open systems with varying number of particles and due to its non-variational character allows to study such processes governed by the non-Hermitian Liouvillean, as diffusion-controlled reactions under study.

An extension of the coupled-cluster approximation to the non-equilibrium classical systems [43–45] has allowed to study asymptotics of bimolecular reactions. It resulted in a rather unexpected conclusion that now the generally-accepted time dependence of the $A+B \to 0$ reaction for $\bar{d} = 3$, $n(t) \propto t^{-3/4}$, is only the pre-asymptotic stage, with the true asymptotics $n(t) \propto t^{-1}$! Similar technique was used also for the study of diffusion-limited aggregation and structure formation processes [47].

Lastly, we would like to mention here results of the two kinds of large-scale computer simulations of diffusion-controlled bimolecular reactions [33, 48]. In the former paper [48] reactions were simulated using random walks on a $\bar{d}$-dimensional (1 to 4) hypercubic lattice with the imposed periodic boundary conditions. In the particular case of the $A + B \to 0$ reaction, $D_A = D_B$ and $n_A(0) = n_B(0)$, the critical exponents 0.26 ± 0.01; 0.50 ± 0.02 and 0.89 ± 0.02 were obtained for $\bar{d} = 1$ to 3 respectively. The theoretical value of $\alpha = 0.75$ expected for $\bar{d} = 3$ was not achieved due to cluster size effects. The result for $\bar{d} = 4$, $\alpha = 1.02 \pm 0.02$, confirms that this is a marginal dimension. However, in the case of the $A + B \to B$ reaction with $D_B = 0$, the asymptotic long-time behaviour, equation (2.1.106), was not achieved at all – even at very long reaction times of $\approx 10^5$ Monte Carlo steps, which were sufficient for all other kinds of bimolecular reactions simulated. It was concluded that in practice this theoretically derived asymptotics is hardly accessible.

Other careful computer simulations [33] focused more attention on the two diffusion-controlled reactions: $A + B \to 0$ and $A + A \to 0$, on both fractal and $1\bar{d}$ lattices. In the case of the Euclidean space it was well demonstrated that achievement of the theoretical limit of $\alpha = 0.25$ is a quite long-time

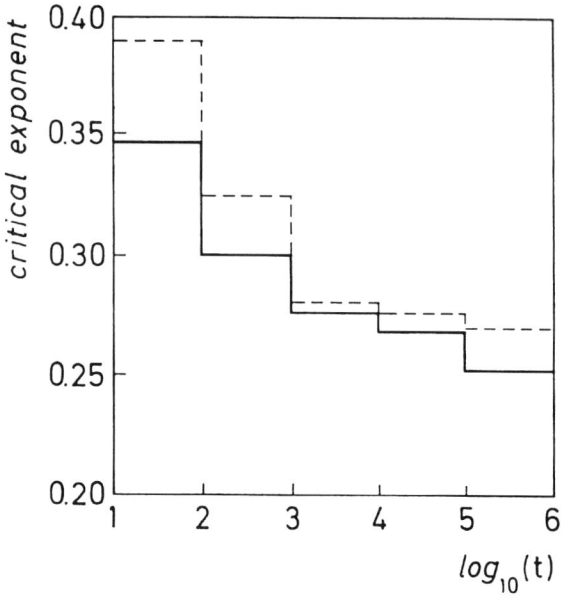

Fig. 6.30. Results of computer simulations for the time-development of critical exponent α (average for each time decade) observed in the reaction $A + B \to 0$, $\bar{d} = 1$ and $n_A(0) = n_B(0) = 0.4$. Lattice contains 10^5 sites, results are for cyclic boundary conditions and 10 runs averaged. Full lines are for the excluded volume case, whereas in dashed line any number of particles could occupy a given site.

process (Fig. 6.30), which supports our arguments against the NAN model (Section 6.1). This study focused also on some important methodological points, e.g., a comparison of discrete-lattice simulations and continuum-space analytical theory; the effect of the excluded volume (no multiple occupancy of any lattice site); which quantities should be averaged in computer simulations (say $n(t)$, $n^{-1}(t)$ or $\alpha(t)$); what is a role of finite-cluster size effects. The excluded-volume effect was found to be not too important – as it is seen from full and dashed lines in Fig. 6.30.

Lastly, Argyrakis and Kopelman [33] have simulated $A + B \to 0$ and $A + A \to 0$ reactions on two- and three-dimensional critical *percolation clusters* which serve as representative random fractal lattices. (The critical thresholds are known to be $p_c = 0.5931$ and 0.3117 for two and three dimensions respectively.). The expected important feature of these reactions is superuniversality of the kinetics *independent* on the spatial dimension and

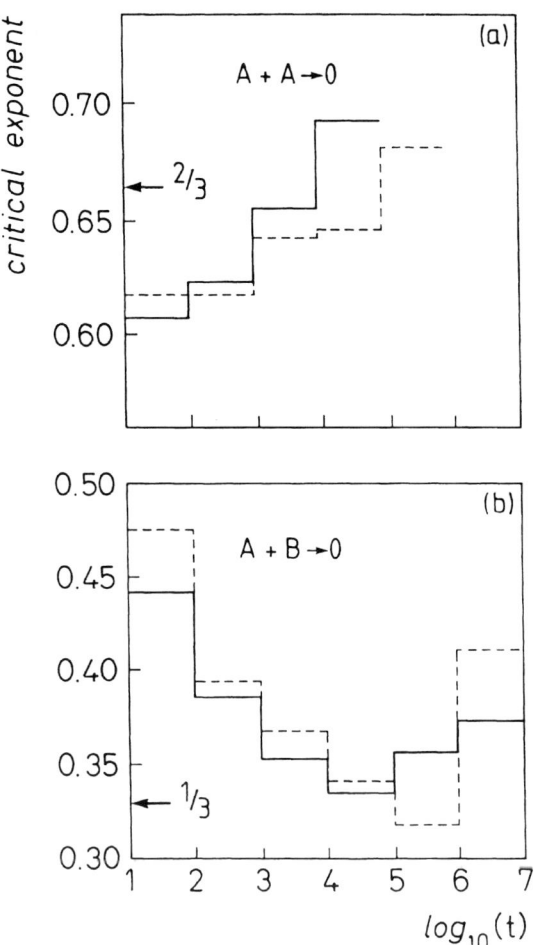

Fig. 6.31. As Fig. 6.30 for the A + A → 0 (a) and A + B → 0 (b) reactions on a critical percolation clusters (curves 1 – two-dimensional, curves 2 – three-dimensional clusters).

governed by a *fracton* dimension $\tilde{d} = 4/3$ [5, 7, 49]. Figure 6.31 confirms this prediction for both kinds of reactions: the limiting values of $\alpha = \tilde{d}/2$ are 2/3 and 1/3 for the A + A → 0 and A + B → 0 respectively, it demonstrates as well the final-size effects at extremely long times (α shows an increase in time after reaching theoretical limit when particles are very likely to "hit" the boundaries, periodical in these simulations).

6.3 DYNAMIC PARTICLE AGGREGATION INDUCED BY ELASTIC INTERACTIONS

Up to now we have been discussing in this Chapter many-particle effects in bimolecular reactions between non-interacting particles. However, it is well known that point defects in solids interact with each other even if they are not charged with respect to the crystalline lattice, as it was discussed in Section 3.1. It should be reminded here that this *elastic* interaction arises due to overlap of displacement fields of the two close defects and falls off with a distance r between them as $U(r) = -\lambda r^{-6}$ for two symmetric (isotropic) defects in an isotropic crystal or as $U(r) = -\lambda[\alpha_4]r^{-3}$, if the crystal is weakly anisotropic [50, 51] ($[\alpha_4]$ is an angular dependent cubic harmonic with $l = 4$). In the latter case, due to the presence of the cubic harmonic α_4 an interaction is attractive in some directions but turns out to be repulsive in other directions. Finally, if one or both defects are anisotropic, the angular dependence of $U(\vec{r})$ cannot be presented in an analytic form [52]. The role of the elastic interaction within pairs of the complementary radiation the Frenkel defects in metals (vacancy-interstitial atom) was studied in [53–55]; it was shown to have considerable impact on the kinetics of their recombination, $A + B \to 0$.

The elastic interaction between defects was studied also in alkali halide crystals; in particular, for the primary Frenkel defects – the so-called F and H centres (anion vacancy trapped an electron and interstitial halogen atom forming a quasi-molecule X_2^- with a regular anion X^- nearby). Since the latter are anisotropic, the interaction decays as r^{-3}; the elastic constant was calculated to be $\lambda \approx 3$ eV Å^{-3} in KBr [56]. Taking into account that H centres become mobile above 30 K [57], we obtain that the distinctive dimensionless parameter $r_\mathrm{e} = \sqrt[3]{\lambda/k_\mathrm{B}T} = \sqrt[3]{\beta\lambda} \gg r_0$ (the annihilation radius, which is close to the interionic spacing) and therefore the elastic interaction is expected to play an essential role in alkali halides too. (F centres in alkali halides are immobile below room temperature.) The role of elastic defect interaction in alkali halides was studied in [58–63]. Since an H centre starts to rotate on a lattice site already at 10 K, i.e., much before it starts to migrate, one can reasonably assume that when approaching F centre, it takes energetically the most favourable orientation corresponding to the attraction and thus one may consider hereafter F, H interaction as *isotropic* attraction. Note also that careful variational calculations [54] have also demonstrated that mobile interstitials in metals avoid the repulsive directions around vacancies and recombine approaching them from another side, which effectively results in an isotropic attraction with the recombination radius corrected by a factor close to 0.5.

Mobile H centres in alkali halides are known to aggregate in a form of complex hole centres [64]; this process is stimulated by elastic attraction. It was estimated [65, 66] that for such *similar* defect attraction the elastic constant λ is larger for a factor of 5 than that for dissimilar defects – F, H centres. Therefore, elastic interaction has to play a considerable role in the colloid formation in alkali halides observed at high temperatures [67]. In this Section following [68] we study effects of the elastic interaction in the kinetics of concentration decay whereas in Chapter 7 the concentration *accumulation* kinetics under permanent particle source will be discussed in detail.

To our knowledge, until now an interaction between *similar* defects (particles) AA or BB during their bimolecular A + B → 0 reactions (the Frenkel defect recombination) were practically neglected in theoretical literature, unlike interaction between dissimilar defects AB. (One of rare exceptions is the paper [69] dealing with the Coulomb interaction of similar particles.) The traditional approach for describing this kinetics [70–72] is based on a simple mathematical scheme described in Chapter 4. It self-evidently neglects any interaction of *similar* particles; the partial diffusion coefficients enter here only as their sum $D = D_A + D_B$, and potentials $U_{AA}(r), U_{BB}(r)$ are also absent. As it was noted for the first time in [73], the very fact of the incorporation of the relative distribution of similar particles in the kinetics of bimolecular reactions gives rise to a *non-linear many-particle effect* which requires a new more complete level of the correlation treatment including, along with the joint functions $Y(r,t)$ mentioned above, the analogous joint functions $X_A(r,t)$, $X_B(r,t)$ for *similar* particles. Such an approach has been described in previous Sections of this Chapter but neglecting particle dynamic interactions and including only those interactions which lead to their recombination. The many-particle effects here reveal themselves as a statistical particle aggregation.

In this Section we consider the following problem. Defects B are mobile ($D_B > 0$) and interact with each other elastically as $U_{BB}(r) = -\lambda r^{-3}$; we call hereafter this interaction *dynamical*. Their counter-partners A involved into the bimolecular recombination, A + B → 0, could be both immobile, $D_A = 0$, and mobile, $D_A = D_B$. Obviously to calculate the kinetics of this reaction, we have to go beyond the framework of the traditional approach, Section 4.1, which neglects the interaction of similar particles.

In the dimensionless units: $r' = r/r_0$, $t' = Dt/r_0^2$, $K'(t') = K(t)/(4\pi Dr_0)$, $n'(t') = 4\pi r_0^3 n(t)$ and using relative diffusion coefficients $\mathfrak{D}_A = 2\kappa$, $\mathfrak{D}_B = 2(1-\kappa)$ where $\kappa = D_A/D$ (primes are omitted below) the joint correlation functions describing spatial correlations of dissimilar, $Y(r,t)$, and similar, $X(r,t)$, particles obey the following non-linear kinetic equations

$$\frac{\partial Y(r,t)}{\partial t} = \nabla\big(\nabla Y(r,t) + Y(r,t)\nabla\mathfrak{U}_{AB}(r,t)\big) -$$
$$- n(t)K(t)Y(r,t)\sum_{\nu=A,B} J[X_\nu], \qquad (6.3.1)$$

$$\frac{\partial X_A(r,t)}{\partial t} = \mathfrak{D}_A\nabla\big(\nabla X_A(r,t) + X_A(r,t)\nabla\mathfrak{U}_{AA}(r,t)\big) -$$
$$- 2n(t)K(t)X_A(r,t)J[Y], \qquad (6.3.2)$$

$$\frac{\partial X_B(r,t)}{\partial t} = \mathfrak{D}_B\nabla\big(\nabla X_B(r,t) + X_B(r,t)\nabla\mathfrak{U}_{BB}(r,t)\big) -$$
$$- 2n(t)K(t)X_B(r,t)J[Y]. \qquad (6.3.3)$$

In these equations the dimensionless potentials $\mathfrak{U}_{AA}$, $\mathfrak{U}_{BB}$ and $\mathfrak{U}_{AB}$ are defined by β times the potential of mean force given by equation (2.3.50).

In the formal limit of the infinitely diluted system ($n(t) \to 0$) the dimensionless interaction potential $\mathfrak{U}(r,t) = \beta U(r)$. (Note that $\mathfrak{U}(r,t)$ could give rise to another source of the non-linearity in the kinetic equations, additional to that caused by terms J.) In this limiting case equation (6.3.1) transforms into well-known in the standard kinetics equation (4.1.57) modified for the black-sphere model of recombination, as it was described above. This demonstrates the relation between traditional and our many-particle approaches. In turn, the equations for the similar correlation functions X_ν ($\nu = A, B$) become decoupled off the complete set of kinetic equations and thus no longer affect the reaction kinetics. Their steady-state solution ($t \to \infty$) is the Boltzmann distribution

$$X_\nu^0(r) = X_\nu(r,\infty) = \exp\big[-\beta U_{\nu\nu}(r)\big]. \qquad (6.3.4)$$

Below we take into account the non-linear terms in the kinetic equations containing functionals J (coupling spatial correlations of similar and dissimilar particles) but neglect the perturbation of the pair potentials assuming that $\mathfrak{U}(r,t) = \beta U(r)$. This is justified in the diluted systems and for the moderate particle interaction which holds for low reactant densities and loose aggregates of similar particles. However, potentials of mean force have to be taken into account for strongly interacting particles (defects) and under particle accumulation when colloid formation often takes place [67].

Following the arguments given above, we restrict ourselves to the case of the *isotropic* attraction, $U(r) = -\lambda r^{-3}$. In terms of mechanics this long-range potential is characterised by an infinite action radius, but in the kinetics it results in the *effective recombination radius*. In the traditional approach [53, 54] it is $R_{\text{eff}} = 3r_e/\Gamma(1/3) = 1.12\sqrt[3]{\lambda\beta}$ (Γ is gamma function), i.e., decays with the temperature as $T^{-1/3}$. (To distinguish the elastic interactions within different pairs of defects, say AA, BB and AB, *several* radii r_e could be used.) The Boltzmann distribution, equation (6.3.4), could be rewritten as $X^0(r) = \exp[(r_e/r)^3]$; it shows how the dynamic interaction leads to the increased concentration of close similar particles, at $r \leqslant r_e$, as compared to their Poisson (random) distribution, $X \equiv 1$. Keeping in mind that the law $U(r) = -\lambda r^{-3}$ holds only asymptotically and fails for nearest neighbours, $r \leqslant a_0$ (a_0 is interionic spacing), let us introduce the cut-off potential: $U(r) = -\lambda r_0^{-3}$ as $r \leqslant r_0$. It means that we use in equations (6.3.1) to (6.3.3) the dimensionless potentials

$$\mathfrak{U} = \begin{cases} -(r_e/r_0)^3, & r < r_0, \\ -(r_e/r_0)^3 \dfrac{1}{r^3}, & r \geqslant r_0. \end{cases} \quad (6.3.5)$$

The ratio r_e/r_0 entering this equation is a distinctive parameter, defining the effect of dynamic particle aggregation. Remember that the statistical aggregation, arising due to many-particle effects, is characterized by another spatial scale – the so-called *diffusion length* $\xi = \sqrt{Dt}$ (see Chapter 5). Therefore, the recombination kinetics under study is governed by an *interplay* of these two parameters; depending on the reaction time t, both cases – $\xi < r_e$ and $\xi > r_e$ – are possible. In *both* kinds of aggregation the joint correlation function $X(r,t)$ at small relative distances r exceeds considerably the asymptotic value of the unity and thus it is not a trivial task to distinguish these two competing effects.

6.3.1 Asymmetric particle mobilities

In calculations presented below we assume first one kind of defects to be immobile ($D_A = 0$, $\kappa = 0$) and their dimensionless initial concentration $n(0) = 0.1$ is not too high; it is less than 10 per cent of the defect saturation level accumulated after prolonged irradiation [41]. Its increase (decrease) does not affect the results qualitatively but shorten (lengthen, respectively) the distinctive times when the effects under study are observed. To stress the effects of defect mobility, we present in parallel in Sections 6.3.1 and 6.3.2 results obtained for immobile particles A ($D_A = 0$, *asymmetric case*) and equal mobility of particles A and B ($D_A = D_B$, *symmetric* case). In both cases only pairs of similar particles BB interact via elastic forces, (6.3.5), but not AA or AB. The initial distribution ($t = 0$) of all defects is assumed to be random, $Y(r > 1, 0) = X_\nu(r, 0) = 1$; $\nu = A, B$.

In Fig. 6.32(a) the decay of the defect concentration vs time is shown. (Note that all parameters given hereafter are dimensionless.) In the standard chemical kinetics, the reaction rate is time-dependent only at short times, equation (4.2.21). The dotted curve gives results for the standard chemical kinetics, whereas the full curves 1 to 3 are calculated for different values of the parameter r_e/r_0. Curve 1 coincides practically with that for $r_e = 0$, i.e., for the case of non-interacting particles. In other words, deviation of the full curve at long times from that for the standard kinetics demonstrates the above-discussed reduction of the reaction rate caused by the *statistical* aggregation of similar particles. In turn, curve 2 for $r_e/r_0 = 2$ considerably deviates from curve 1 at the *intermediate* times $10^1 < t < 10^3$ due to the contribution of the *dynamic* aggregation. At longer times these two curves tend to coincide again thus demonstrating unimportance of the dynamic effects. Finally, curve 3 calculated for a strong mutual attraction of B particles, $r_e/r_0 = 3$, shows a considerable reduction of the reaction rate (in a given time interval); it will approach the curve 1 only at very long times.

These results could be complemented well with the curve slopes in the double logarithmic coordinates as plotted in Fig. 6.33(a) using idea of the intermediate *critical exponent* $\alpha(t)$, equation (4.1.68). In the traditional chemical kinetics its asymptotic limit $\alpha_0 = \alpha(\infty) = 1$ is achieved already during the presented dimensionless time interval, $t \approx 10^4$. For non-interacting particles and if one of two kinds is immobile, $D_A = 0$, it was earlier calculated analytically [11] that the critical exponent is additionally reduced down to $\alpha_0 = 0.5$. However, for a weak interaction (curve 1) it is observed that in the time interval $t \leqslant 10^4$ $\alpha_{\max} \approx 0.8$ is achieved only for a given $n(0) = 0.1$, i.e., the

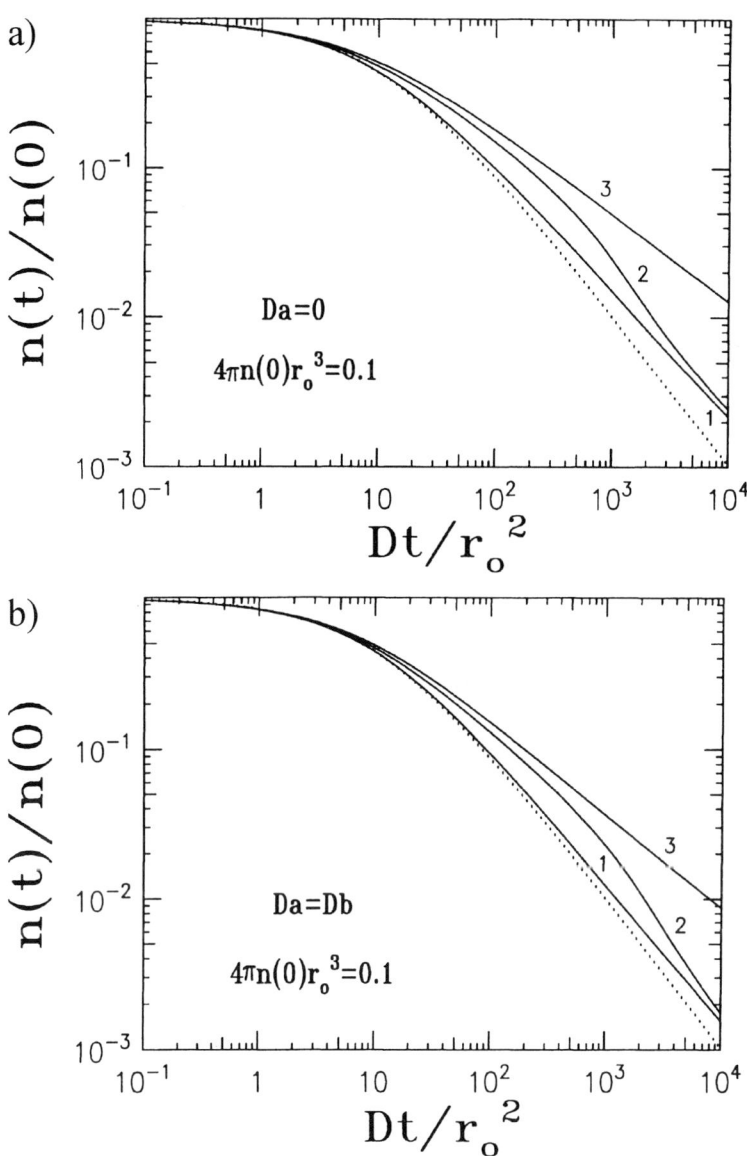

Fig. 6.32. The decay in time of the macroscopic particle concentration for asymmetric (a) and symmetric (b) cases corresponding to $D_A = 0$ and $D_A = D_B$ respectively. The initial concentration $n(0) = 0.1$. Parameter r_c/r_0: 1(1); 2(2); 3(3). Full curves show many-particle effects, whereas dotted line shows results of their neglect.

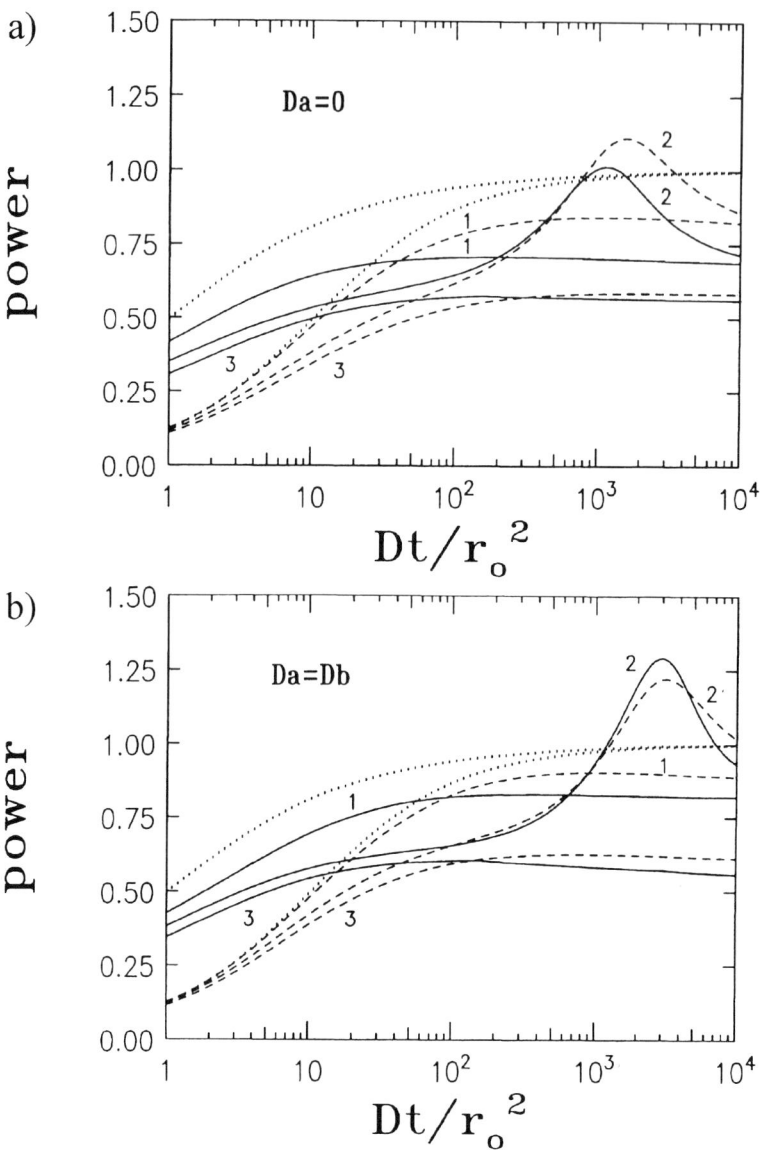

Fig. 6.33. The critical exponent, equation (4.1.68), as a function of time for asymmetric (a) and symmetric (b) cases. The initial particle concentrations $n(0)$: 1.0 (full curves), 0.1 (dashed curves). Dotted lines are obtained neglecting the many-particle effects. Parameter r_c/r_0: 1(1); 2(2); 3(3).

reaction rate is *reduced* as compared to the standard chemical kinetics without particle interaction (the dotted curve). In contrast, curve 3 (strong interaction) shows a rapid transition to the analytically expected value of $\alpha_0 = 0.5$. In turn, the intermediate value of $r_e/r_0 = 2$ (curve 2) demonstrates non-trivial *non-monotonous* transition at about $t \approx 10^3$ between the two intermediate asymptotic values of power-law with $\alpha = 0.50$ and $\alpha_0 = 0.75$. Increase in initial concentration leads to a faster transition to the critical exponent limit, $t \to \infty$.

The origin of this unusual behaviour is partly clarified from Fig. 6.34(a) where the relevant curves 2 demonstrate the same kind of the non-monotonous behaviour as the critical exponents above. Since, according to its definition, equation (4.1.19), the reaction rate is a functional of the joint correlation function, this non-monotonicity of curve 2 arises due to the spatial re-arrangements in defect structure. It is confirmed by the correlation functions shown in Fig. 6.34(a). The distribution of BB pairs is quasi-stationary, $X_B(r,t) \approx X^0(r) = \exp[(r_e/r)^3]$, which describes their dynamic aggregation. (The only curve is plotted for X_B in Fig. 6.35(a) for $t = 10^2$ (the dotted line) since for other time values X_B changes not more than by 10 per cent.) This quasi-steady spatial particle distribution is formed quite rapidly; already at $t \approx 10^0$ it reaches the maximum value of $X_B(r,t) \approx 10^3$. The effect of the statistical aggregation practically is not observed here, probably, due to the diffusion separation of mobile B particles.

The behaviour of the correlation functions of immobile defects X_A is less obvious. The monotonous increase of its maximum with time shown in Fig. 6.35(a) means a strong A particle aggregation, but now it has the *statistical* nature. Maximum of the critical exponent in curve 2 of Fig. 6.33(a) corresponds to the time interval $t \approx 10^3$–10^4 when the values of both X_A and X_B become of the same order of magnitude. At longer times the reaction kinetics is defined by the relative distribution of A particles only; if the dynamic aggregation of particles B is characterized by a *single* parameter r_e, the time development of $X_A(r,t)$ shows a formation of the spatial scale (where X_A deviates from its asymptotic value of unity) which could be associated with the diffusion length $\xi = \sqrt{Dt}$. The same scale is also clearly seen for the joint density of *dissimilar* defects $Y(r,t)$. However, from these results we cannot make a final conclusion which of the two factors (i.e., an increase of the $X_A(r,t)$ maximum or formation of the diffusion length ξ) is responsible for the transition from the kinetic stage (dynamic particle aggregation) to the stage of the statistical A aggregation. So strong increase

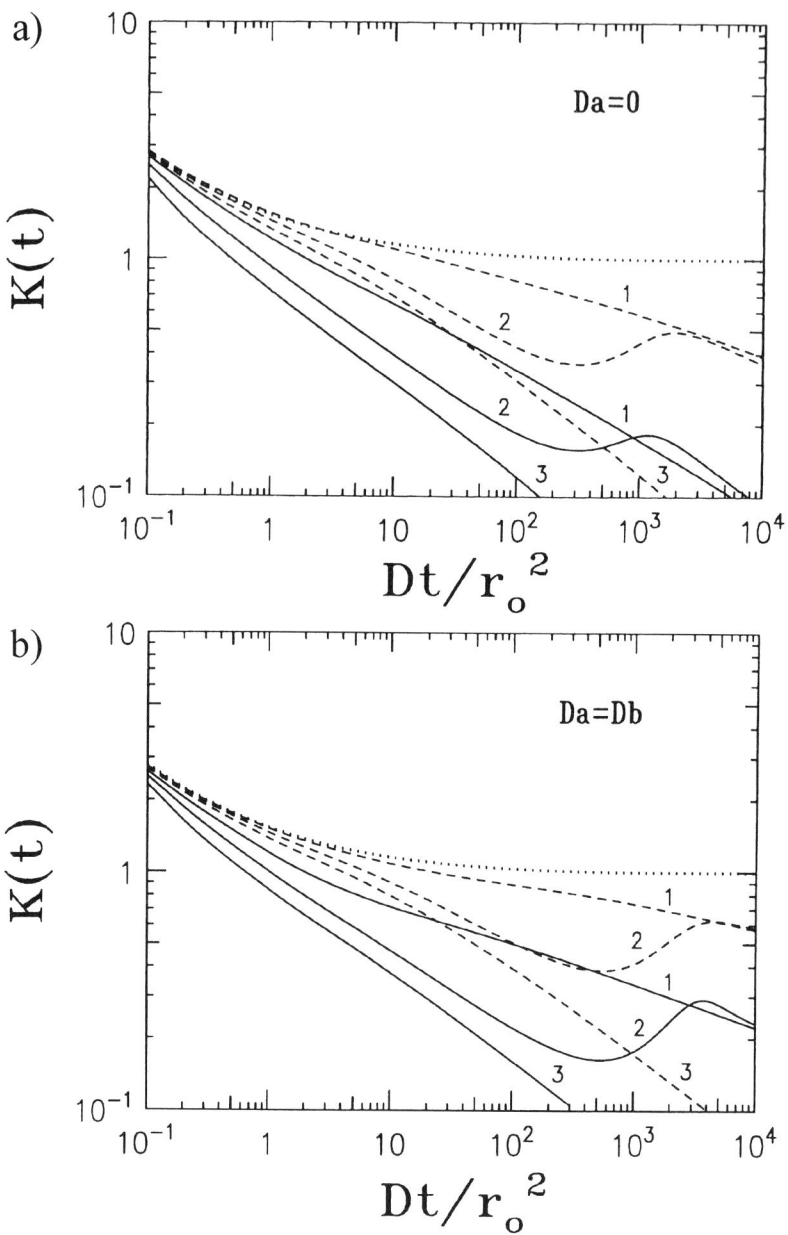

Fig. 6.34. The dimensionless reaction rate as a function of time for asymmetric (a) and symmetric (b) cases. Notations and parameters as in Fig. 6.32.

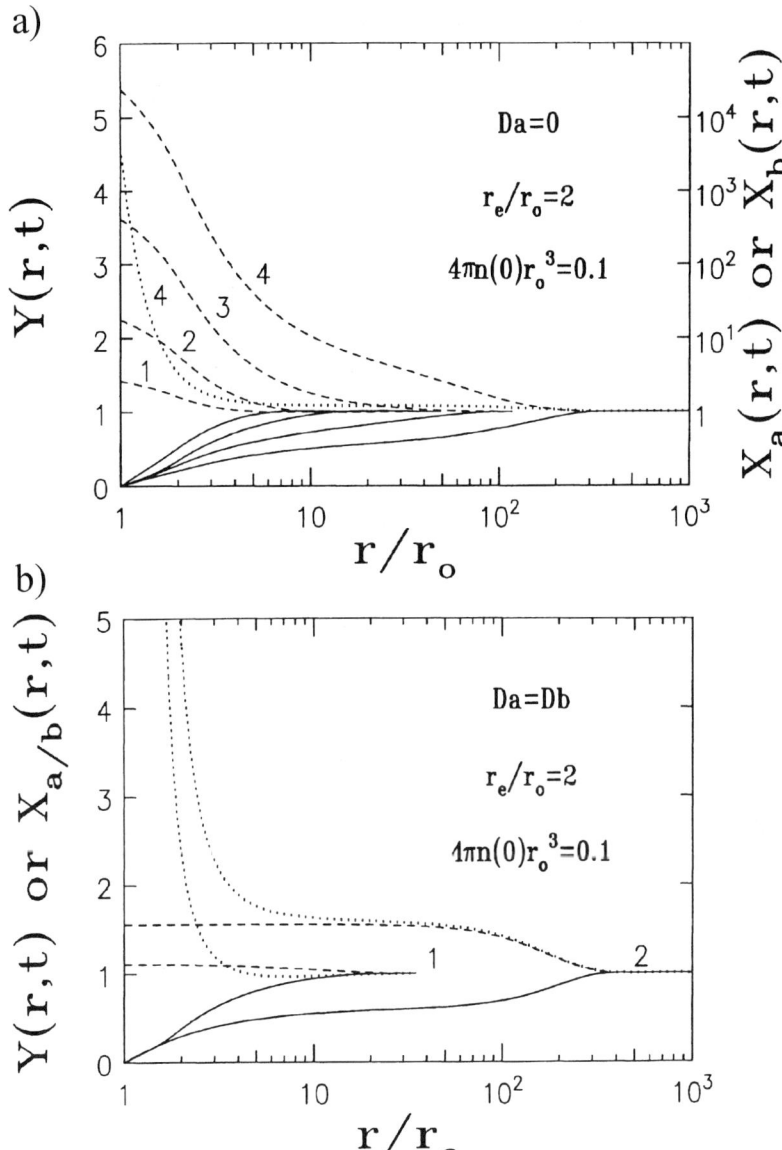

Fig. 6.35. The joint correlation functions for dissimilar particles $Y(r,t)$ (full curves), immobile similar particles $X_A(r,t)$ (broken curve) and mobile particles $X_B(r,t)$ (dotted curve). Parameters used are given. The dimensionless time (in units r_0^2/D) is: (a) $t = 10^1$ (1); 10^2 (2); 10^3 (3); 10^4 (4); (b) $D_A = D_B$, $t = 10^2$ (1); 10^4 (curve 2).

in the maximum of $X_A(r,t)$ is defined entirely by great asymmetry in the diffusion coefficients of particles D_A and D_B. Therefore, to answer this question, a more general case when both kinds of defects are mobile should be analyzed, which is done in the next Section 6.3.2.

6.3.2 Equally mobile A and B particles

From general consideration one can expect that A-rich aggregates will be essentially destroyed by the motion of A particles which should be also accompanied by the reduced $X_A(r,t)$ maximum at short distances; this effect is indeed observed in Fig. 6.35(b). To check these ideas numerically, the calculations presented in a) were repeated here for the case $D_A = D_B$ and the results obtained are plotted in Figs 6.32(b) to 6.35(b) respectively. (It should be remembered that particles A do not interact dynamically.) The conclusion suggests itself that the symmetric case does not quantitatively differ from the asymmetric one, when one kind of defects was immobile. The main difference is that for weak dynamic interactions, $r_e/r_0 = 1$ (curves 1), the asymptotic magnitude of the critical exponent shown in Fig. 6.33(b) tends to change from the value of $\alpha = 0.5$ to $\alpha = 0.75$ analytically calculated in [11]. This effect is well seen also in Fig. 6.32(b) where particle concentrations decay faster than in the asymmetric case, Fig. 6.32(a). The reaction rates shown in Fig. 6.34(b) exceed always the same value in Fig. 6.34(a) and for the intermediate values $r_e/r_0 = 2$ reveal quite similar non-monotonic character. However, for a given time interval and the initial concentration $n(0) = 0.1$ the critical exponent does not reach this limiting value and one observes just some acceleration of the reaction as compared to the previous case of $D_A = 0$.

Of greater interest is the behaviour of the joint correlation functions presented in Fig. 6.35(b). At any reaction time $X_B(r,t) > X_A(r,t)$ holds; now an increase of the maximum of $X_A(r,t)$ in time is very slow. According to the above-given estimates for neutral non-interacting particles, it has a logarithmic character:

$$X_A(r,t) = 1 + \text{const} \cdot (\ln t)^2 \exp\left[-\frac{r^2}{4D_A t}\right]. \tag{6.3.6}$$

The deviation of the joint correlation function X_B from X_A arises due to the additional effect of the *dynamic* aggregation, which is observed mainly at the relative distances $r \leqslant r_e$. It follows from equation (6.3.6), that the joint density of similar interacting particles A exceeds its asymptotic value at distances less than the correlation length $r \leqslant \xi = \sqrt{Dt}$. It is clearly seen from Fig. 6.35(b), that ξ is the *common* length scale for *all* correlation functions, both X and Y.

The whole reaction volume at long times may be qualitatively considered as consisting of domains with the linear size ξ, each domain has particles of one kind only, A or B. Particles A are distributed inside such domains randomly (which follows from the fact that $X_A(r,t) \approx$ const, as $r < \xi$). In contrast, the spatial distribution of particles B reveals an additional structure formation at $r < r_e$ caused by the dynamic interaction: large B-domains contain inside themselves small dynamic aggregates with the distinctive sizes r_e. Since the reaction rate is defined by the recombination of two kinds of particles on the common boundaries of the nearest domains (which reduces the reaction rate), such an *internal* structure of domains cannot affect the kinetics under study.

Therefore, in the symmetric situation, $D_A = D_B$, the recombination kinetics may be also separated into two subsequent stages of dynamic and statistical aggregation. At long times the particle density in these aggregates (domains), characterized by the maximum values of the correlation functions $X_\nu(r \to 0, t)$, is not very high. The reaction rate is governed by the ratio of two distinctive spatial scales $-\xi$ and r_e.

6.3.3 An elastic attraction of dissimilar particles

As it was mentioned above, up to now only the dynamic interaction of dissimilar particles was treated regularly in terms of the standard approach of the chemical kinetics. However, our generalized approach discussed above allow us for the first time *to compare effects* of dynamic interactions between similar and dissimilar particles. Let us assume that particles A and B attract each other according to the law $U_{AB}(r) = -\lambda r^{-3}$, which is characterized by the elastic reaction radius $r_e = (\beta\lambda)^{1/3}$. The attraction potential for BB pairs is the same at $r > r_0$ but as earlier it is cut-off, as $r \leqslant r_0$. Finally, pairs AA do not interact dynamically. Let us consider now again the symmetric and asymmetric cases. In the standard approach the relative diffusion coefficient D_A/D and the potential $U_{BB}(r)$ do not affect the reaction kinetics; besides at long times the reaction rate tends to the steady-state value of $K(\infty) \propto r_e$.

The results of the incorporation of AB attraction into the kinetic equations are plotted in Figs 6.36 to 6.38. The decay of the defect concentration shown

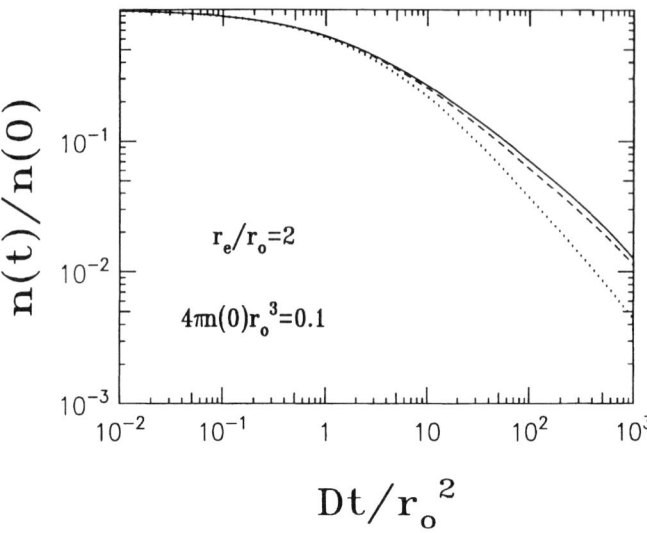

Fig. 6.36. The defect concentration vs time. The dotted line shows neglect of the similar particle correlation; in the full line it is incorporated for the case $D_A = 0$ and in the broken line – for the $D_A = D_B$ case. The elastic interaction constant λ of similar and dissimilar particles is the same; $U_{AB} = U_{BB} = -\lambda r^{-3}$, whereas $U_{AA} = 0$.

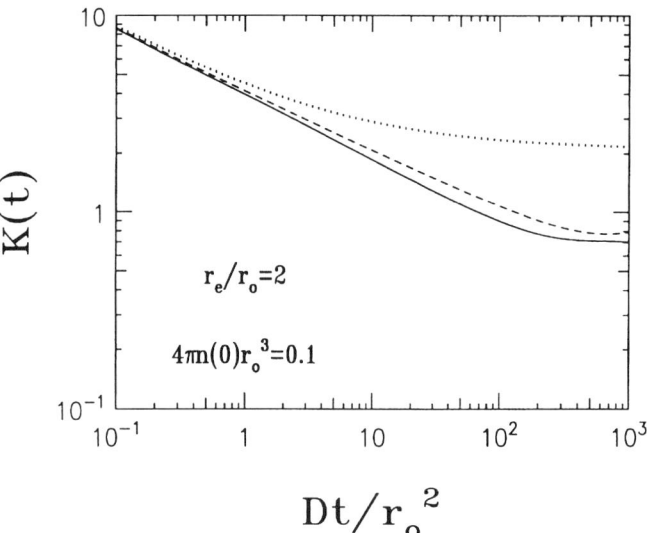

Fig. 6.37. The dimensionless reaction rate vs time. Notations as in Fig. 6.36.

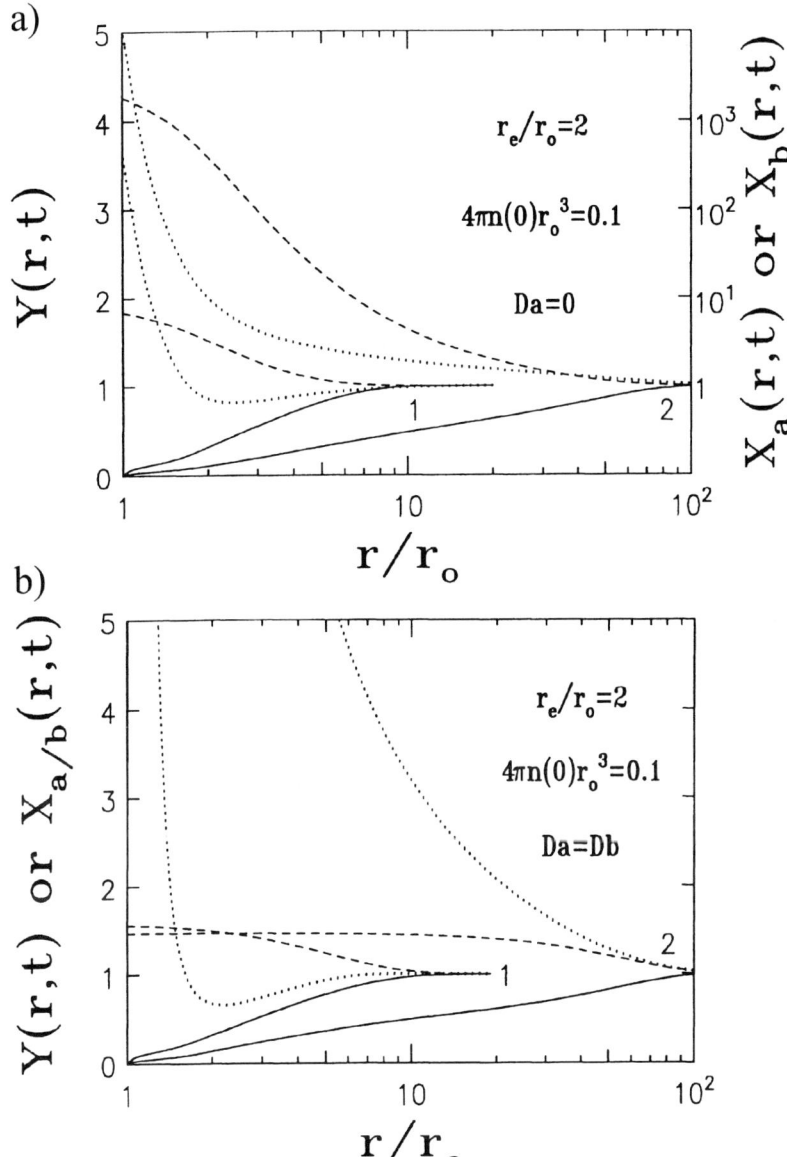

Fig. 6.38. The joint correlation function for the asymmetric (a) and symmetric (b) cases. Full curves are $Y(r,t)$; broken curves $X_A(r,t)$; dotted curves $X_B(r,t)$. Dimensionless time is: 10^1 (1); 10^3 (2). Note that $U_{AA} = 0$.

in Fig. 6.36 demonstrates competition of the effects of the elastic interaction of similar and dissimilar particles. A comparison of the dotted lines in Fig. 6.36 and Fig. 6.32 (a) and (b) shows clearly reaction acceleration due to mutual attraction of dissimilar particles. The correlation of the spatial distribution of similar particles is neglected in these dotted lines. The incorporation of the attraction of similar particles (full and broken curves in Fig. 6.36) demonstrates quite opposite effect of the reaction reduction caused by their aggregation. It leads to the *lowered* reaction rate (Fig. 6.37); this effect is greater for the asymmetric case. Since the dynamic and statistical particle aggregations change the reaction rate, the use here of traditional equations of the chemical kinetics for interpretation of the experimental kinetics (e.g., for obtaining the r_e value) may lead to the considerable systematic errors.

As earlier for the case $U_{AB} = 0$ (Fig. 6.35), the correlation functions $X_A(r,t)$ and $Y(r,t)$ shown in Fig. 6.38 (a) and (b) demonstrate appearance of the *domain structure* in a reaction volume with interacting particles, having the distinctive size $\xi = \sqrt{Dt}$. Interaction within AB pairs holds at the relative distances $r \leqslant r_e$ (at long times $r_e < \xi$ takes place) and only slightly modifies the AB pair distribution on the domain boundaries, where the reaction takes place, but do not influence essentially the entire mechanism of the domain formation (the effect of *statistical aggregation*).

In conclusion of Section 6.3 we wish to stress that the elastic attraction of similar defects (reactants) leads to their *dynamic* aggregation which, in turn, reduces considerably the reaction rate. This effect is mostly pronounced for the *intermediate* times (dependent on the initial defect concentration and spatial distribution), when the effective radius of the interaction $r_e = \sqrt[3]{\beta\lambda}$ exceeds greatly the diffusion length $\xi = \sqrt{Dt}$. In this case the reaction kinetics is governed by the elastic interaction of *both* similar and dissimilar particles. A comparative study shows that for equal elastic constants λ the elastic attraction of *similar* particles has greater impact on the kinetics than interaction of dissimilar particles.

At longer times, when $r_e \ll \xi$, the effect of the *statistical* aggregation of similar particles begins to dominate, which takes also place for neutral non-interacting particles as it was discussed in Chapter 5. At this stage the reaction leads to the formation of A- or B-rich domains with the linear size ξ; in turn, these domains are structured *inside* themselves into smaller blocks having the typical size of r_e, which however no longer affects the kinetics.

6.4 EFFECT OF NON-EQUILIBRIUM CHARGE SCREENING (COULOMB INTERACTION)

In this Section we continue studies of particle dynamical interactions. For this purpose the formalism of many-particle densities is applied to the study of the cooperative effects in the kinetics of bimolecular $A + B \to 0$ reaction between oppositely charged particles (reactants) interacting via the Coulomb forces. We show that unlike the Debye–Hückel theory in statistical physics, here charge screening has essentially a *non-equilibrium* character. For the asymmetric mobility of reactants ($D_A = 0$, $D_B \neq 0$) the joint spatial distribution of similar immobile reactants A reveals at short distances a singular character associated with their aggregation. The relevant reaction rate does not strive for a steady-state (as it does in the symmetric case, $D_A = D_B$) but infinitely increases in time, thus leading to the concentration decay which is *quicker* than the algebraic law generally-accepted in chemical kinetics, $n \propto t^{-1}(\bar{d} = 3)$.

The kinetics of the $A + B \to 0$ bimolecular reaction between charged particles (reactants) is treated traditionally in terms of the *law of mass action*, Section 2.2. In the transient period the reaction rate $K(t)$ depends on the initial particle distribution, but as $t \to \infty$, it reaches the steady-state limit $K(\infty) = K_0 = 4\pi D R_{\text{eff}}$, where $D = D_A + D_B$ is a sum of diffusion coefficients, and R_{eff} is an effective reaction radius. In terms of the *black sphere* approximation (when AB pairs approaching to within certain critical distance r_0 instantly recombine) this radius is [74]

$$R_{\text{eff}} = r_0 \frac{L}{1 - \exp(-L)}, \qquad (6.4.1)$$

where the dimensionless parameter $L = R/r_0$ contains the so-called *Onsager* radius discussed in Section 3.2 which for equal charges of particles reads:

$$R = \frac{e^2}{\varepsilon k_B T}. \qquad (6.4.2)$$

The value of the parameter L entering equation (6.4.1) defines whether the Coulomb attraction or recombination is predominant: as $L \ll 1$, $R_{\text{eff}} \approx r_0$, whereas in the opposite case $L \gg 1$, $R_{\text{eff}} \approx R$ (the effective recombination sphere equals the Onsager radius).

Despite the fact that formalism of the standard chemical kinetics (Chapter 2) was widely and successfully used in interpreting actual experimental data [70], it is not well justified theoretically: in fact, in its derivation the solution of a *pair* problem *with non-screened* potential $U_{AB}(r) = -e^2/(\varepsilon r)$ is used. However, in the statistical physics of a system of charged particles the so-called *Coulomb catastrophes* [75] have been known for a long time and they have arisen just because of the neglect of the essentially *many-particle* charge screening effects. An attempt [76] to use the screened Coulomb interaction characterized by the phenomenological parameter – the Debye radius R_D [75] does not solve the problem since $K(\infty)$ has been still traditionally calculated in the same pair approximation.

As it was noted in [77], reduction of the reaction rate with time observed for non-interacting particles at high concentrations/long reaction times, Section 6.2, is unlikely to occur for charged particles since spatial fluctuations in particle densities are now governed not by $\xi(t)$ but the *screening radius;* in other words, the Coulomb repulsion of similar particles prevents their aggregation.

As it was demonstrated for the first time by us [69], charge screening in the bimolecular reaction has a non-equilibrium character and is not determined uniquely by particle densities in an *equilibrium* system of charged particles. Indeed, diffusion (spatial particle motion) determining the reaction kinetics is quite slow. On the other hand, unlike equilibrium systems now oppositely charged, particles recombine. Let us represent the effective force of the Coulomb attraction by $F = -e^2 S/\varepsilon r^2$, where $S = S(r,t)$ is a *screening factor*. In the thermodynamic equilibrium state $S = S(r)$ only and is characterized by a single parameter R_D, besides $\lim_{r\to\infty} S(r) = 0$. For non-equilibrium screening $S_0(t) = \lim_{r\to\infty} S(r,t)$ which generally speaking is non-zero. It is shown [69] that in the case of equal diffusion coefficients, $D_A = D_B$, $\lim_{t\to\infty} S_0(t) = 0$, i.e., the *quasi*-equilibrium charge distribution is formed at long reaction times. In other words, equations (6.4.1), (6.4.2) remain valid but in the framework of the many-particle formalism [69] it is *dynamic interplay* between the reaction rate acceleration due to the Coulomb repulsion of similar particles and the cut-off of their interaction caused by a screening.

In this Section we consider the case when charge screening could be *principally non-equilibrium* and thus equations of the standard chemical kinetics are no longer valid [24, 78]. To demonstrate it, let us assume that particles of one kind, say A, are immobile ($D_A = 0$). This situation was considered in Section 6.2 for *neutral* particles and a strong statistical aggregation of immobile particles was found to occur. This aggregation is not destroyed by the diffusive motion of reactants and leads to an additional reaction rate reduction, which was confirmed by computer simulations [36]. We show below that for the *charged particles* in question an aggregation also takes place but has other consequences.

At long distances from the A-rich aggregate existing on the background of the uniform B distribution it resembles a super-particle with the effective charge $e_\text{eff} \approx N_A e_A$, N_A is a number of similar particles in the aggregate. As $L > 1$, the effective recombination radius is proportional to the product of charges (equation (6.4.2)), i.e., $R_\text{eff} \approx N_A R$. For the qualitative estimate we take into account that diffusive motion of B's destroys A aggregates, so that at the moment t the reaction kinetics is determined by the interaction between the aggregate A occupying the distinctive volume $V \approx \xi^3(t)$ and the background of B. As the upper limit estimate of N_A let us take a mean number of particles A in the volume V; $N_A = n(t)\xi^3(t)$. Substituting now the reaction rate $K(t) \approx 4\pi D R N_A$ into equation (4.1.28), one arrives at the new asymptotic law $n(t) \propto t^{-5/4}$, indicating that the reaction now is *accelerated*. This problem applies to several fields of the chemical physics of condensed matter. In particular, it is true for radiation physics of ionic crystals [57, 64] where the primary Frenkel defects – vacancies are immobile below room temperature, whereas interstitial ions become mobile above typically 30 K; these defects are oppositely charged and interact via the Coulomb law.

Before discussing mathematical formalism we should stress here that the Kirkwood approximation cannot be used for the modification of the drift terms in the kinetics equations, like it was done in Section 6.3 for elastic interaction of particles, since it is too rough for the Coulomb systems to allow us the correct treatment of the charge screening [75]. Therefore, the cut-off of the hierarchy of equations in these terms requires the use of some *principally new approach*, keeping also in mind that it should be consistent with the level at which the fluctuation spectrum is treated. In the case of joint correlation functions we use here it means that the only acceptable for us is the Debye–Hückel approximation [75], equations (5.1.54), (5.1.55), (5.1.57).

Therefore, the approximate treatment of the A+B → 0 reaction for charged particles inavoidably requires a combination of several approximations: the Kirkwood superposition approximation for the reaction terms and the Debye–Hückel approximation for modification of the drift terms with self-consistent potentials. Not discussing here the accuracy of the latter approximation, note only that we found it to be correct enough for both qualitative and semi-quantitative description of the charge screening.

The equation for the time development of macroscopic concentrations formally coincides with the law of mass action but with dimensionless reaction rate $\mathcal{K}(t) \equiv K(t)/(4\pi D r_0)$ which is, generally speaking, *time-dependent* and defined by the flux of the dissimilar particles via the recombination sphere of the radius r_0, equation (5.1.51). Using dimensionless units $n(t) \equiv 4\pi r_0^3 n(t)$, $r \equiv r/r_0$, $t \equiv Dt/r_0^2$, and the condition of the reflection of similar particles upon collisions, equation (5.1.40) (zero flux through origin), we obtain for the joint correlation functions the equations (6.3.2), (6.3.3). Note that we use the dimensionless diffusion coefficients, $\mathcal{D}_A = 2\kappa$, $\mathcal{D}_B = 2(1 - \kappa)$, $\kappa = D_A/(D_A + D_B)$ entering equation (6.3.2).

To demonstrate the charge screening importance in a many-particle system, it should be mentioned that the substitution of potentials $\mathfrak{U}_\nu(r, t)$ entering equations (6.3.2), (6.3.3) by non-screened potentials $U_\nu(r) = L/r$ leads to the *Coulomb catastrophe* manifesting itself by the unlimited increase of the reaction rate $\mathcal{K}(t)$ [24, 78].

If particles have different diffusion coefficients, $D_A \neq D_B$, their correlation functions no longer coincide, $X_A \neq X_B$, which results also in *different* screening effects for these two kinds of particles.

Unlike the case of the neutral reactants, where analytical solution reveals the auto-model behaviour in coordinated r/ξ, in our case of charged particles the singular solutions arise on the spatial scale of the order of the recombination radius r_0 thus preventing us from such a simplified analytical analysis. Therefore, we will compare semi-qualitative arguments for the new law, $n(t) \propto t^{-5/4}$, with numerical calculations of our kinetic equations.

A solution of the kinetic equations with singular (e.g., Coulomb) potentials is a non-trivial problem. The efficient calculation of equations (6.3.2) and (6.3.3) needs to use difference schemes whose coefficients depend on potentials. Since these potentials $\mathfrak{U}_\nu(r, t)$ in their turn depend on the correlation functions, we have to handle the non-linear equations, for which the iterative procedure earlier developed in [24] was used.

The asymptotic $(t \to \infty)$ treatment of the set of non-linear integro-differential equations is quite difficult even as a computational problem since solution stability requires the use of small time increments in a mesh and thus we could reach $t = 10^5$ only (in the dimensionless units). Besides, in the particular case of the asymmetric mobility, $D_A = 0$, we observe the spatial particle distribution revealing strongly developed singular properties, which requires the additional reduction of the coordinate increment.

Effect of non-equilibrium charge screening

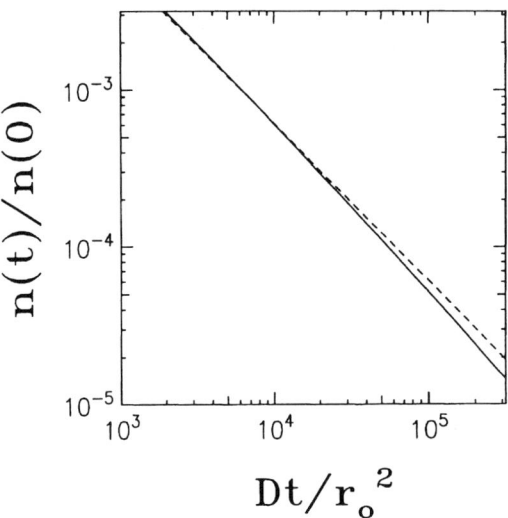

Fig. 6.39. Particle concentration $n(t)$ as a function of time. Full curve – $D_A = 0$, broken curve – $D_A = D_B$. Parameters are: $L = 1$, $n(0) = 0.1$.

Figure 6.39 shows the time development of particle concentrations. At long times the kinetics for a symmetric ($D_A = D_B$) and asymmetric ($D_A = 0$) cases differ significantly: in the latter case reaction proceeds more quickly. Note that the choice of the parameter $L = 1$ corresponds to the weak electrostatic field; the Onsager radius R is small and coincides with the recombination sphere radius r_0. The initial dimensionless concentration $n(0) = 0.1$ is not also too large: it is only 10 percent of the maximum concentration which could be achieved under irradiation [12]. The magnitudes of these two parameters were chosen to make our computations more time-saving.

The difference in the kinetics for two limiting cases $D_A = 0$ and $D_A = D_B$ becomes more obvious in terms of the *current critical exponents* defined earlier, equation (4.1.68). It yields the *slope* of decay curves shown in Fig. 6.39. The conclusion can be drawn from Fig. 6.40 that in the symmetric case we indeed observe well-known algebraic decay kinetics with $\alpha(\infty) = 1$ corresponding to time-independent reaction rate. However, in the asymmetric case the critical exponent increases in time thus indicating the peculiarity of the kinetics as we qualitatively estimated in the beginning of this Section 6.4.

The dimensionless reaction rate $K(t)$ as a function of time for different values of $\kappa = D_A/D$ is plotted in Fig. 6.41. One can see again that as $t \to \infty$,

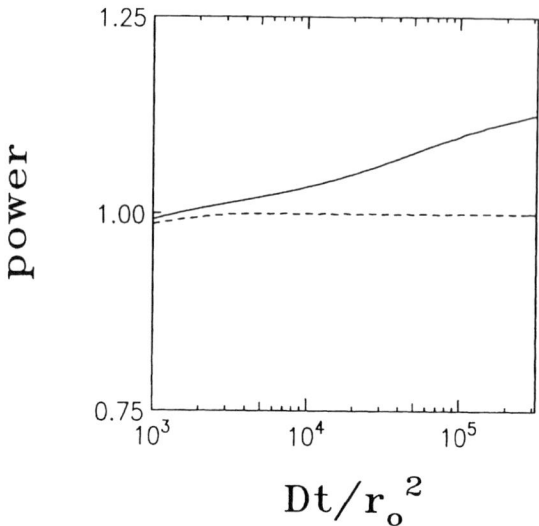

Fig. 6.40. The critical exponent as a function of time. Parameters and notations as in Fig. 6.39.

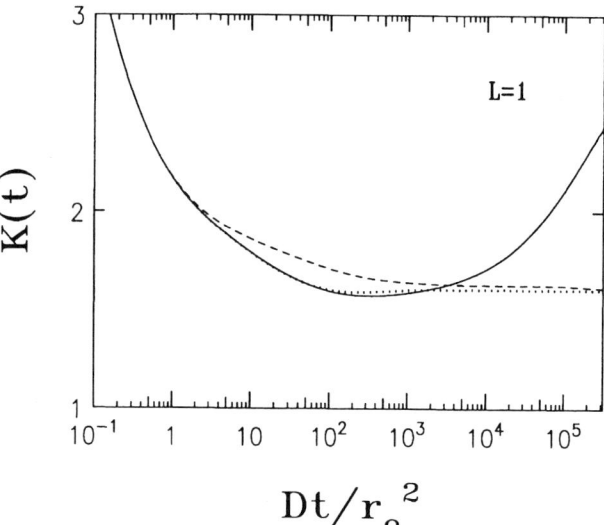

Fig. 6.41. The time-development of the reaction rate $K(t)$. Full curve – symmetric case ($D_A = D_B$), broken curve – asymmetric case ($D_A = 0$), dotted curve – $D_A/D = 0.01$.

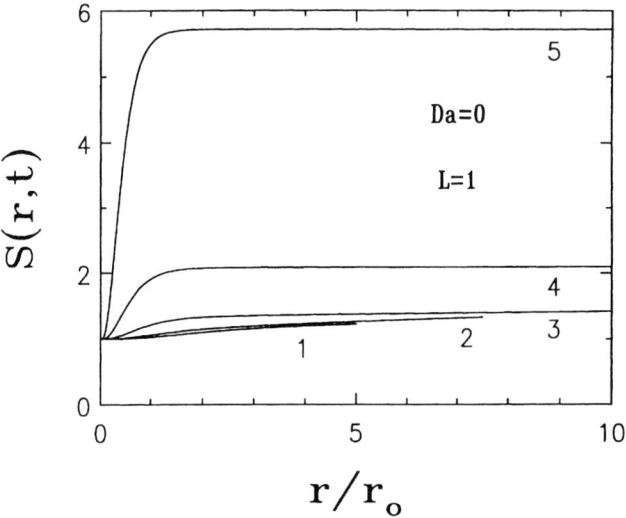

Fig. 6.42. The screening factor in the asymmetric case. Parameters as in Fig. 6.39. Curves 1 to 5 correspond to the dimensionless time Dt/r_0^2: 10^1, 10^2, 10^3, 10^4 and 10^5, respectively.

in the symmetric case it indeed reaches the steady-state value described by equations (6.4.1), (6.4.2). However, in the asymmetric case there is *no* steady-state: it is seen very well from the additional curve for $D_A/D = 0.01$ – here the kinetics is the same as for the asymmetric case up to $Dt/r_0^2 \leqslant 10^3$ but at greater times it follows the kinetics known for the symmetric situation.

Unusual behaviour of the reaction rate is clarified in Fig. 6.42: the screening factor $S(r,t)$ shown here is obviously *non-equilibrium*; its asymptotic value $S_0(t)$ increases in time. (As it was said above, at low concentrations the values of $S_0(t)$ correspond to the mean number of particles A in their aggregates existing on the background of uniform B distribution.) Note that this aggregation has purely statistical character.

For the asymmetric case the spatial distribution of A particles reveals quite singular behaviour (*raisins in dough*) (Fig. 6.43). The joint correlation function for similar particles, $X_A(r,t)$, has a sharp maximum near the coordinate origin: its amplitude increases monotonously with time, but it decreases by several orders of magnitude as r increases from zero up to several times r_0. Correspondingly, the screening factor shown in Fig. 6.42 approaches at the same distance its asymptotic value. The power-law increase in $X_A(r,t)$ max-

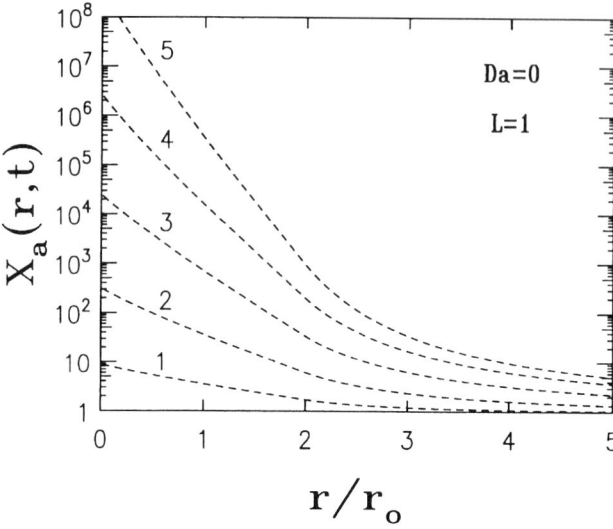

Fig. 6.43. The joint correlation function X_A of similar particles A in the asymmetric case. Notations as in Fig. 6.42.

imum with time seen in Fig. 6.43 indicates clearly *strong aggregation* of immobile A particles.

Joint distribution of BB and AB pairs is shown in Fig. 6.44. The distribution of similar mobile particles B at long times in the asymmetric case practically is the same as in the symmetric case (when $X_A = X_B$). The behaviour of $X_B(r, t)$ is determined by the Coulomb repulsion of B's for which the non-equilibrium screening effect does not take place. In its turn, some deviation for the joint dissimilar functions $Y(r, t)$ seen in Fig. 6.44 for the symmetric and asymmetric cases is a direct consequence of different screening effects: in the latter case the effective recombination radius increases in time which results in an increase of the $Y(r, t)$ gradient at $r = r_0$; at long times this correlation function itself strives for the Heaviside step-like form.

Lastly, Fig. 6.45 demonstrates the *intermediate case* of slow-mobile A particles. It is seen very well that now aggregates of A are dissolved by diffusive motion and their distribution is no longer singular. At long times the difference between correlation function X_A and X_B becomes negligible.

The screening parameter $S(r, t)$ for the symmetric case, $D_A = D_B$ is shown in Fig. 6.46. It demonstrates clearly the formation of a quasi-

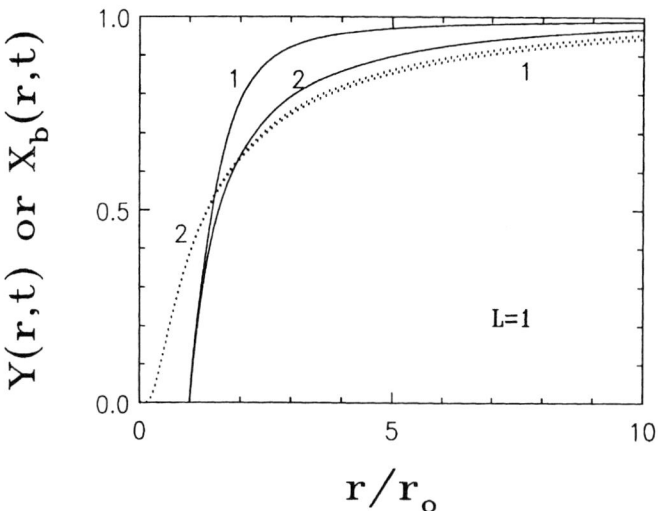

Fig. 6.44. The joint correlation function of dissimilar particles, $Y(r,t)$ – full curve, and that of similar particles $X_B(r,t)$ – dotted curve. Parameters are: $L = 1$, $n(0) = 0.1$. Curves 1 – asymmetric case, curves 2 – symmetric case.

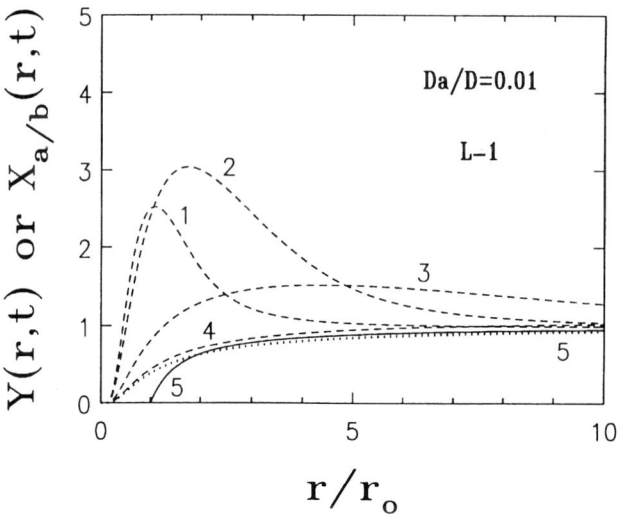

Fig. 6.45. The joint correlation functions in the intermediate case $D_A/D = 0.01$. Full curve – $Y(r,t)$, broken curve – $X_A(r,t)$, dotted curve – $X_B(r,t)$.

380   The many-particle effects in irreversible A + B → 0 reaction   Ch. 6

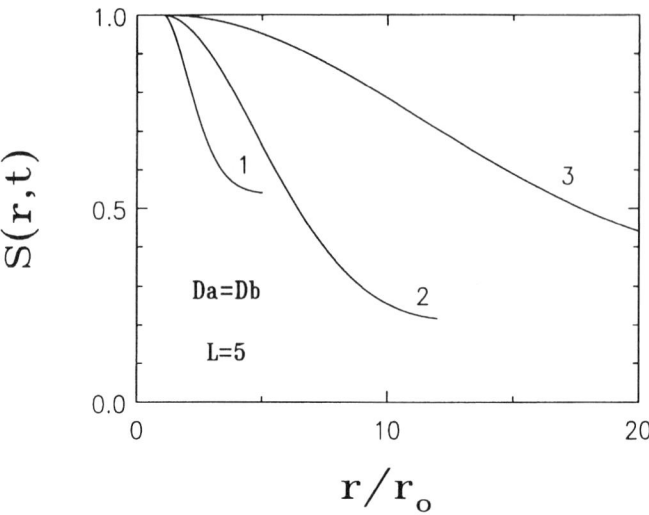

Fig. 6.46. The screening parameter in the symmetric case, $D_A = D_B$. Parameters are $n(0) = 1$, $L = 5$. Curves 1 to 3 correspond to the dimensionless time $Dt/r_0^2 = 1, 10^1, 10^2$ respectively.

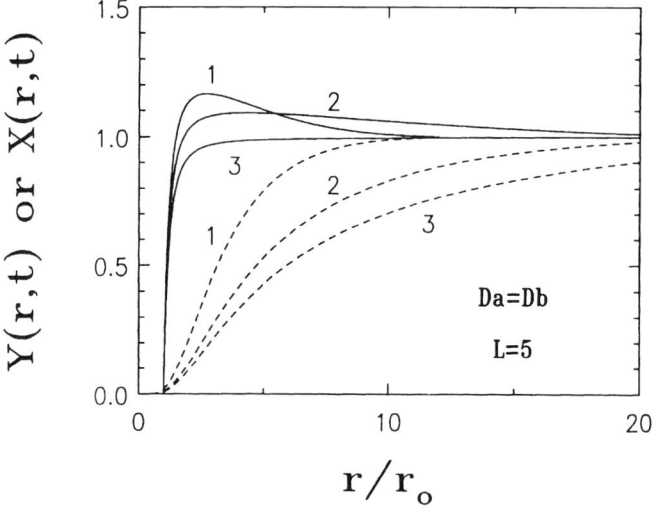

Fig. 6.47. The joint correlation functions for the case $D_A = D_B$. Full curves – dissimilar defects, $Y(r, t)$, broken curves – similar defects, $X_A(r, t) = X_B(r, t)$. Parameters $n(0) = 1$, $L = 5$. Curves 1 to 3 correspond to the dimensionless times equal to $10^1, 10^2, 10^5$, respectively.

equilibrium charge screening; at any time $S(r,t) \leqslant 0$ and its asymptotic value $S_0(t)$ decays monotonically in time.

The non-equilibrium particle distribution is clearly observed through the joint correlation functions plotted in Fig. 6.47. Note that under the linear approximation [74] the correlation function for the dissimilar defects $Y(r,t)$ increases *monotonically* with r from zero to the asymptotic value of unity; $Y(r \to \infty, t) = 1$. In contrast, curve 1 in Fig. 6.47 ($t = 10^1$) demonstrates *a maximum* which could be interpreted as an enriched concentration of dissimilar pairs, AB, near the boundary of the recombination sphere, $r \geqslant r_0$. With increasing time this maximum disappears and $Y(r,t)$ assumes the usual smoothed-step form. The calculations show that such a maximum in $Y(r,t)$ takes place within a wide range of the initial defect concentrations and for a random initial distribution of both similar and dissimilar particles used in our calculations: $X_\nu(r,0) = Y(r > 1,0) = 1$. The mutual Coulomb repulsion of similar particles results in a rapid disappearance of close AA (BB) pairs separated by a distance $r < L$ (seen in Fig. 6.47 as a decay of $X_\nu(r,t)$ at short r with time). On the other hand, it stimulates strongly the mutual approach (aggregation) of *dissimilar* particles leading to the maximum for $Y(r,t)$ at intermediate distances observed in Fig. 6.47.

In its turn, the non-monotonous behaviour of $Y(r,t)$ results in a similar behaviour of the reaction rate $K(t)$ in time (Fig. 6.48). The local maximum of $K(t)$ observed at $t = 10^1$ (for a given L) for different initial concentrations likely arises due to the initial conditions used, which do not take into account peculiarities of the spatial distribution of charged particles: a more adequate one would be a quasi-equilibrium pair distribution with incorporated potential screening.

The role of the non-equilibrium charge screening is emphasized by calculations neglecting such screening, i.e., when equations (5.1.54) are omitted and $U_\nu(r) = L/r$ is postulated. In this case mutual repulsion of similar particles accompanied by the attraction between dissimilar particles are characterized by the *infinite* interaction radius between particles which leads immediately to the *Coulomb catastrophe* – an infinite increase in $K(t)$ in time shown in Fig. 6.47. This effect is independent of the choice of the initial defect distributions for both similar and dissimilar particles. On the other hand, incorporation of the Coulomb screening makes equations (6.4.1), (6.4.2) asymptotically valid for any initial distribution of particles.

Summing up this Section, we note that study of charge screening in the kinetics of bimolecular reaction $A + B \to 0$ with the Coulomb interaction

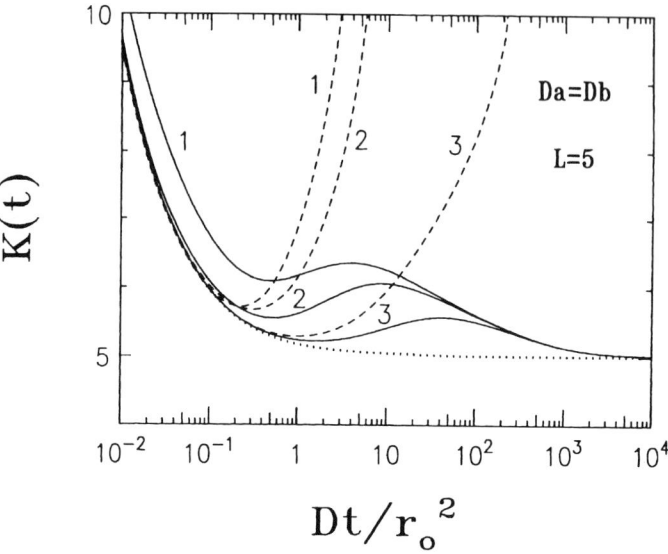

Fig. 6.48. The role of non-equilibrium charge screening in eliminating the Coulomb catastrophe: the dimensionless reaction rate vs time. Dotted curve – the Debye theory (no screening and similar particle correlation); broken curves – the solution of kinetic equations incorporating these correlations but neglecting screening; full curves, screening is taken into account. Parameters: $L = 5$, $D_A = D_B$. Curves 1 to 3 correspond to dimensionless concentrations: 10^0, 10^{-1} and 10^{-2} respectively.

between reactants has demonstrated that this screening – unlike the standard Debye–Hükel theory – has essentially *non-equilibrium* character. For equal reactant mobilities (the so-called symmetric case) neglect of this fact leads to the Coulomb catastrophe (an infinite increase of the reaction rate in time) whereas its proper incorporation into kinetic equations results in both the moderate increase of the reaction rate at intermediate times and non-monotonic behaviour of the joint correlation functions of dissimilar reactants.

In the asymmetric case ($D_A = 0$) similar immobile particles A become aggregated in the course of reaction and, as $t \to \infty$, the relevant reaction rate no longer has steady-state but increases in time leading to the *accelerated* particle recombination (see also [79]).

References

[1] H. Schnörer, V. Kuzovkov and A. Blumen, J. Chem. Phys. 92 (1990) 2310.
[2] H. Schnörer, V. Kuzovkov and A. Blumen, Phys. Rev. Lett. 63 (1989) 805.
[3] H. Schnörer, V. Kuzovkov and A. Blumen, J. Chem. Phys. 93 (1990) 7148.
[4] S.F. Burlatsky and A.A. Ovchinnikov, Sov. Phys. JETP 65 (1987) 908.
[5] S. Havlin and D. Ben-Avraham, Adv. Phys. 36 (1987) 695.
[6] B.B. Mandelbrot, The Fractal Geometry of Nature (Freeman, San Francisco, 1982).
[7] K. Kang and S. Redner, Phys. Rev. Lett. 52 (1984) 955.
[8] A. Blumen, J. Klafter and G. Zumofen, in: Optical Spectroscopy of Glasses, ed. I. Zchokke (Reidel, Dordrecht, 1986) p. 199.
[9] H. Schnörer, S. Luding and A. Blumen, in: Fractals in the Fundamental and Applied Sciences, eds H.-O. Peitgen, J.M. Henriques and L.F. Penedo (North-Holland, Amsterdam, 1991) p. 395.
[10] P. Hanusse and A. Blanche, J. Chem. Phys. 74 (1981) 6148.
[11] V.N. Kuzovkov and E.A. Kotomin, Rep. Prog. Phys. 51 (1988) 1479.
[12] E.A. Kotomin and V.N. Kuzovkov, Rep. Prog. Phys. 55 (1992) 2079.
[13] S. Luding, H. Schnörer, V. Kuzovkov and A. Blumen, J. Stat. Phys. 65 (1991) 1261.
[14] S.F. Burlatsky and A.I. Chernoustan, Phys. Lett. A 145 (1990) 56.
[15] T.M. Searle, Phys. Rev. Lett. 69 (1992) 3256.
[16] J.R. Eggert, Phys. Rev. B: 29 (1984) 6664.
[17] J.R. Eggert, Phys. Rev. B: 37 (1988) 4051.
[18] T.M. Searle and J.E.L. Bishop, Philos. Mag. B 53 (1986) L9.
[19] J.E.L. Bishop and T.M. Searle, Philos. Mag. B 57 (1988) 329.
[20] A.S. Andrews, J.E.L. Bishop, I.H. Clayton, J.S. Mitchell and T.M. Searle, Phys. Rev. B: 41 (1990) 7224.
[21] D. Toussaint and F. Wilczek, J. Chem. Phys. 78 (1983) 2642.
[22] G. Zumofen, A. Blumen and J. Klafter, J. Chem. Phys. 82 (1985) 3198.
[23] V.N. Kuzovkov and E.A. Kotomin, J. Phys. C: 17 (1984) 2283.
[24] V.N. Kuzovkov and E.A. Kotomin, J. Stat. Phys. 72 (1993) 127.
[25] V.N. Kuzovkov, Khim. Phys. 6 (1987) 831 (Sov. Chem. Phys., in Russian).
[26] E. Kotomin, V. Kuzovkov, A. Seeger and W. Frank, J. Phys. A 27 (1994) 1453.
[27] F. Leyvraz and S. Redner, Phys. Rev. Lett. 66 (1991) 2168.
[28] V.N. Kuzovkov and E.A. Kotomin, Chem. Phys. Lett. 87 (1982) 575.
[29] V.N. Kuzovkov and E.A. Kotomin, Chem. Phys. 81 (1983) 335.
[30] V.N. Kuzovkov, Teor. Eksp. Khim. 19 (1983) 528 (Sov. Theor. Exp. Chem., in Russian).
[31] H. Schnörer, PhD thesis (Universität Bayreuth, 1991).
[32] I.M. Sokolov, H. Schnörer and A. Blumen, Phys. Rev. A: 44 (1991) 2388.
[33] P. Argyrakis and R. Kopelman, Phys. Rev. A: 45 (1992) 5814.
[34] E.A. Kotomin and V.N. Kuzovkov, Chem. Phys. Lett. 117 (1985) 266.
[35] V.N. Kuzovkov and E.A. Kotomin, Czech. J. Phys. B 35 (1985) 541.
[36] A.G. Vitukhnovsky, B.L. Pyttel and I.M. Sokolov, Phys. Lett. A 128 (1988) 161.
[37] T.D. Gillespie, Markov Processes: An Introduction for Physical Scientists (Academic Press, New York, 1992).

[38] D. Wendt, T. Fricke, and J. Schnakenberg, Z. Physik B 96 (1995) 541.
[39] M. Bramson and J.L. Lebowitz, Phys. Rev. Lett. 61 (1988) 2397; J. Stat. Phys. 63 (1991) 297; J. Stat. Phys. 65 (1991) 941.
[40] S. Song and D. Poland, J. Phys. A: 25 (1992) 3913.
[41] V.L. Vinetsky, Yu.H. Kalnin, E.A. Kotomin and A.A. Ovchinnikov, Sov. Phys. Usp. 33 (1990) 793.
[42] A.A. Ovchinnikov, S.F. Timashev and A.A. Belyi, Kinetics of Diffusion-Controlled Chemical Processes (Nova Science, New York, 1989).
[43] M.G. Rudavets, J. Phys. A 25 (1992) 5283.
[44] M.G. Rudavets, J. Phys. A: 26 (1993) 5313.
[45] M.G. Rudavets, Phys. Lett. A 176 (1993) 62.
[46] R.F. Bishop, Theor. Chim. Acta, 80 (1991) 95.
[47] M.G. Rudavets, J. Phys. C: 5 (1993) 1039.
[48] Z. Jiang and C. Ebner, Phys. Rev. A: 41 (1990) 5333.
[49] P. Meakin and H.E. Stanley, J. Phys. A: 17 (1984) L173.
[50] I.D. Eshelby, Solid State Phys. 3 (1956) 79.
[51] H.E. Schaefer and H. Kronmüller, Phys. Status Solidi B: 67 (1975) 63.
[52] W. Scheu, W. Frank and H. Kronmüller, Phys. Status Solidi B: 82 (1977) 523.
[53] K. Schröder, Diffusion Reactions of Point Defects, Report 1083-FF, Jülich, 1974.
[54] K. Schröder and K. Dettmann, Z. Phys. B 22 (1975) 343.
[55] M.H. Yoo and W.H. Butler, Phys. Status Solidi B: 77 (1977) 181.
[56] K. Bachmann and H. Peisl, J. Phys. Chem. Solids 31 (1970) 1525.
[57] M.N. Kabler, Point Defects in Solids, ed. J. Crawford (Plenum Press, New York, 1972) p. 327.
[58] E.A. Kotomin and I.I. Fabrikant, Radiat. Eff. 46 (1980) 85; 46 (1980) 91.
[59] E.A. Kotomin and A.B. Doktorov, Phys. Status Solidi B: 114 (1982) 287.
[60] E.A. Kotomin, Latv. PSR Zinat. Akad. Vestis Fiz. Zinat. Teh. Ser. 5 (1985) 122 (Proc. Latv. Acad. Sci. Phys. Technol. Ser., in Russian).
[61] E.A. Kotomin and A.S. Chernov, Sov. Phys. Solid State, 22 (1980) 1575.
[62] E.A. Kotomin, DSc Thesis (Inst. of Physics Latv. Acad. Sci., Riga, 1988).
[63] P.V. Bochkanov, V.I. Korepanov and V.M. Lisitsin, Izv. Vyssh. Ucheb. Zaved. Fiz. 3 (1989) 16 (in Russian).
[64] N. Itoh, Adv. Phys. 31 (1972) 491.
[65] M. Saidoh and N. Itoh, J. Phys. Chem. Solids 34 (1973) 1165.
[66] N. Itoh and M. Saidoh, Phys. Status Solidi 33 (1969) 649.
[67] K.K. Schwartz and Yu.A. Ekmanis, Dielectric Materials: Radiation-Induced Processes and Radiation Stability (Zinatne, Riga, 1989).
[68] V.N. Kuzovkov and E.A. Kotomin, J. Chem. Phys. 98 (1993) 9107.
[69] V. N. Kuzovkov and E. A. Kotomin, Chem. Phys. 98 (1985) 357.
[70] H. Eyring, S.H. Lin and S.M. Lin, Basic Chemical Kinetics (Wiley, New York, 1980).
[71] A.I. Onipko, in: Physics of Many-Particle Systems (Naukova Dumka, Kiev, 1982) Ch. 2, p. 60.
[72] T.R. Waite, Phys. Rev. 107 (1957) 463.
[73] V.N. Kuzovkov, Teor. Eksp. Khim. 21 (1985) 33 (Sov. Theor. Exp. Chem., in Russian).

References

[74] P. Debye, J. Electrochem. Soc. 32 (1942) 265.
[75] R. Balescu, Equilibrium and Non-Equilibrium Statistical Mechanics (Wiley, New York, 1975).
[76] A.B. Doktorov, A.A. Kipriyanov and A.I. Burstein, Sov. Opt. Spectr. 45 (1978) 497.
[77] A.A. Ovchinnikov and Ya.B. Zeldovich, Chem. Phys. 28 (1978) 215.
[78] V.N. Kuzovkov and E.A. Kotomin, Physica A 191 (1992) 172.
[79] V.N. Kuzovkov, E.A. Kotomin and W. von Niessen, J. Chem. Phys. (1996) (submitted).

Chapter 7

The Many-Particle Effects in A + B → 0 Reaction with Particle Generation

7.1 THE KINETICS OF DEFECT ACCUMULATION UNDER IRRADIATION

> Great fleas have lesser fleas upon their backs to bite them. And lesser fleas have lesser still. And so *ad infinitum*.
>
> Johnathan Swift

7.1.1 Introduction

In this Chapter the kinetics of the Frenkel defect accumulation under permanent particle source (irradiation) is discussed with special emphasis on many-particle effects. Defect accumulation is restricted by their diffusion and annihilation, A + B → 0, if the relative distance between dissimilar particles is less than some critical distance r_0. The formalism of many-point particle densities based on Kirkwood's superposition approximation, other analytical approaches and finally, computer simulations are analyzed in detail. Pattern formation and particle self-organization, as well as the dependence of the saturation concentration after a prolonged irradiation upon spatial dimension ($\bar{d} = 1, 2, 3$), defect mobility and the initial correlation within geminate pairs are analyzed. Special attention is paid to the conditions of aggregate formation caused by the elastic attraction of particles (defects).

The irradiation of all kinds of solids produces pairs of the point *Frenkel defects* hereafter denoted just AB-vacancies, v, and interstitial atoms, i, which usually are well correlated spatially [1–10]. In many ionic crystals these Frenkel defects are the so-called F and H *centres* discussed in Chapter 3. The function of the initial distribution of complementary defects – v, i pairs (called also *geminate*) over relative distances depends strongly not only on the particular mechanism of defect creation but also on the particular irradiation kind (e.g., X-rays or photons) [9]. Under creation of v, i pair an interstitial atom has an excessive kinetic energy due to which it can be displaced, e.g., by a chain of *focusing collisions* (usually along axis of compactly packed rows of atoms). This displacement ranges from $10\text{--}20a_0$ (a_0 is lattice spacing) in metals [11] down to second nearest neighbours in alkali halide crystals [1, 2], i.e., complementary radiation-induced defects in alkali halides are spatially well correlated.

An initial distribution function within geminate pairs defines directly their stability. In terms of *black sphere* model pairs of (v, i) dissimilar defects disappear instantly, $A + B \rightarrow 0$, when i approach v to within, or is just created by an irradiation, at the critical relative distance r_0 (the so-called *clear-cut* annihilation radius). Its typical value varies between $3\text{--}5a_0$ for metals [3, 4] down to nearest neighbours for alkali halides. At temperatures when interstitial atoms and ions become *mobile* (typically, 20–30 K in both metals and insulators), they perform thermoactived incoherent hops with the frequencies (diffusion coefficients) which are exponentially dependent on the temperature. The basic mobile defects participating in such *diffusion-controlled recombination* processes in alkali halides are hole H and V_k centres (X_2^--quasimolecule occupying one or two lattice sites respectively, X stands for halogen ion) [12]. For example, their thermoactivated motion in KBr begins at $T \geqslant 35$ K and 150 K with the relevant activation energies entering the diffusion coefficients of 0.09 eV and 0.47 eV respectively.

In the last several decades, both experimental data and theoretical studies [5, 9, 13–15] have revealed the effect of *similar defect aggregation* in the course of the bimolecular $A+B \rightarrow 0$ reaction under permanent particle source (irradiation) – the phenomenon similar to that discussed in previous Chapters for the diffusion-controlled concentration decay. Radiation-induced aggregation of similar defects being observed experimentally at 4 K after prolonged X-ray irradiation [16] via both anomalously high for random distribution concentration of dimer F_2 centres (two nearest F centres) and directly in the electronic microscope [17], permits to accumulate defect concentrations whose saturation value exceed by several times that of the Poisson distribution.

At low temperatures dissimilar-defect segregation arises under permanent particle source due to *local fluctuations of particle (defect) densities*. Radiation-induced production by a change of two or more similar defects (say, A) nearby creates a *germ* of their aggregate which is more stable and has a greater chance to survive rather than two isolated defects A since the probability that *two* or more defects B will be created statistically in the same region to destroy this A-germ is very small. In other words, similar defect aggregate is an example of a system which has a larger chance to survive as compared to the equivalent number of isolated and randomly distributed defects. Since creation of some defect B nearby the A-rich aggregate results in disappearance of a single peripheral defect A but doesn't affect defects in a centre of the aggregate, the effect of the *statistical defect screening* arises: *effective* volume v_0 per a vacancy in an aggregate is much smaller than for an isolated vacancy, $v_0 = \frac{4}{3}\pi r_0^3$ [18–23].

In order to study theoretically defect aggregation, several methods of physical and chemical kinetics were developed in recent years. Irrespective of the particular method used, the two basic approaches – a continuous and discrete-lattice ones – are used. In the former model intrinsic defect volume is ignored and thus a number of similar defects in any volume element is unlimited. In its turn, in the latter model any lattice site could be occupied by no more than a single particle (v or i) [15].

To compare computer simulations with an analytical theory, it is convenient to introduce a distinctive parameter – dimensionless saturation concentration $U_0 = n(\infty)v_0$, where $n(\infty)$ is stationary concentration of accumulated defects at their saturation $(t \to \infty)$. (It is assumed that $n(t) = n_A(t) = n_B(t)$, v_0 is volume of the $\bar{d}$-dimensional sphere having the recombination radius r_0; $v_0 = \gamma_{\bar{d}} r_0^{\bar{d}}/\bar{d}$.) In the continuous model it is clear that the quantity U_0, if it exists, is a *universal* parameter dependent on $\bar{d}$ only but not on v_0. Indeed, most of previous theoretical studies were aimed mainly to obtain U_0.

First attempts to find closed equation for concentration accumulation, $n = n(t)$, neglected any analysis of the particle spatial correlations, even on the level of the joint correlation functions. Since macroscopic particle concentration is a characteristic of the system's structure for the random (Poisson) distribution only and neglects the fluctuation effects, more adequate correlation function formalism is replaced inavoidably by *a priori* assumptions using additional structure parameters and based on the simplest probabilistic models. Since the results thus obtained depend considerably on assumptions made, even agreement between computer simulations and theory cannot serve as a proof of the model. Distinctive example is the paper [24] where inconsistent and contradictory treatment of spatial correlations has led for $\bar{d} = 3$ to the physically wrong conclusion about *infinite* concentration increase at high irradiation doses. *Assuming* existence of the stationary concentration for $\bar{d} = 1$, Antonov–Romanovskii [25–28] considered a model of periodically repeated A and B aggregates of equal size (to be discussed in detail below). Using simple probabilistic arguments, he estimated the dimensionless parameter $U_0 = 3.43$. On the other hand, under certain conditions (system's size L does not greatly exceeds the recombination radius r_0) *complete segregation* of the reaction volume has been also observed in computer simulations into two large domains containing either vacancies or interstitials respectively [15, 29]. These *macroscopic aggregates* move in time and sometimes decay into several smaller aggregates. An increase in a size of $1\bar{d}$-system, but keeping v_0 fixed, demonstrates existence of similar defect aggregates but without breaking systems into two regions, the parameter U_0 increases with the L increase. Therefore, strong segregation effect for $\bar{d} = 1$ prevents reaching the *thermodynamical limit* in which results are no longer dependent on the system's size L. In more detail computer simulations of particle accumulation are discussed in Section 7.4.

Let us start with first-principles approach in the particular case of *immobile* particles (low temperatures). An infinite set of equations for *many-point densities* $\rho_{m,m'} = \rho_{m,m'}(\{\vec{r}\}_m; \{\vec{r}'\}_{m'}; t)$, where $\{\vec{r}\}_m = \vec{r}_1, \ldots, \vec{r}_m$ (below a symbol $\{\vec{r}\}_m^i$ denotes the vector $\vec{r}_i$ omitted in a set $\{\vec{r}\}_m$) could be derived by summing recombination and generation contributions which yields:

$$\frac{\partial \rho_{m,m'}}{\partial t} = p \sum_{i=1}^{m} \rho_{m-1,m'}(\{\vec{r}\}_m^i; \{\vec{r}'\}_{m'}; t) +$$

$$+ p \sum_{j=1}^{m'} \rho_{m,m'-1}(\{\vec{r}\}_m; \{\vec{r}'\}_{m'}^j; t) +$$

$$+ p \sum_{i=1}^{m} \sum_{j=1}^{m'} f(|\vec{r}_i - \vec{r}'_j|) \rho_{m-1,m'-1}(\{\vec{r}\}_m^i; \{\vec{r}'\}_{m'}^j; t) -$$

$$- \sum_{i=1}^{m} \sum_{j=1}^{m'} \sigma(|\vec{r}_i - \vec{r}'_j|) \rho_{m,m'} -$$

$$- \sum_{i=1}^{m} \int \sigma(|\vec{r}_i - \vec{r}'_{m'+1}|) \rho_{m,m'+1} d\vec{r}'_{m'+1} -$$

$$- \sum_{j=1}^{m'} \int \sigma(|\vec{r}'_j - \vec{r}_{m+1}|) \rho_{m+1,m'} d\vec{r}_{m+1}. \qquad (7.1.1)$$

Here p is a defect creation rate per unit time and volume, called also *dose rate*, $f(r)$ is their initial distribution function over relative distances, normalised according to $\int f(r)\,d\vec{r} = 1$, $\sigma(r)$ the AB pair recombination rate. For the annihilation mechanism $\sigma(r) = \sigma_0 \theta(r_0 - r)$, θ is the Heaviside step-function (Section 3.1).

As in previous Chapters, for practical use this infinite set (7.1.1) has to be decoupled by the Kirkwood – or any other superposition approximation, which permits to reduce a problem to the study of *closed* set of densities $\rho_{m,m'}$ with indices $(m+m') \leqslant 2$. As earlier, this results in several equations: for macroscopic concentrations and three joint correlation functions, for similar, $X_A(r,t), X_B(r,t)$, and dissimilar defects $Y(r,t)$. However, unlike the kinetics of the concentration decay discussed in previous Chapters, for processes with particle sources direct use of Kirkwood's superposition approximation gives good results for small dimesionless concentration parameters $U_\nu(t) = n_\nu(t)v_0 \ll 1$ only (v_0 is $\bar{d}$-dimensional sphere's volume, r_0 is its radius). The accumulation kinetics predicted has a very simple form [30, 31]

$$\frac{dn_A(t)}{dt} = 2pv_0\big(n(\infty) - n_A(t)\big), \qquad (7.1.2a)$$

or

$$n_A(t) = n(\infty)\big(1 - \exp(-2pv_0 t)\big), \qquad (7.1.2b)$$

where $n(\infty) = 1/(2v_0)$ is *saturation concentration*. Note that the kinetics (7.1.2) is *not* coupled with the correlation functions of similar and dissimilar reactants, as it was observed before in previous Chapters, and the saturation concentration is predicted to be very low, even lower than that for the *random* particle distribution [32].

It was demonstrated in [31] that it is namely the black sphere model which introduces large errors into this accumulation kinetics treated in terms of the superposition approximation. The way of avoiding the superposition approximation's shortcomings was developed in [33] and discussed below.

7.1.2 Virtual particle configurations

The distinctive feature of equation (7.1.1) is that all many-point densities $\rho_{m,m'}$ become zero in the limiting case of instant annihilation, $\sigma_0 \to \infty$, provided for any pair of indices i, j $|\vec{r}_i - \vec{r}'_j| \leqslant r_0$ holds. (Dissimilar AB pairs separated by a distance $r \leqslant r_0$ do not exist.) On the other hand, in the equation (7.1.1) the integral terms contain many-point densities which tend to zero at coordinates in the integration region, but they are multiplied by a large parameter $\sigma_0 \to \infty$. Finally, it results in their *finite* contribution into the kinetic equations.

Therefore, for correct use of the black sphere model as $\sigma_0 \to \infty$ it is important to know the behaviour of the many-point densities $\rho_{m,m'}$ corresponding to the *"virtual"* defect configurations (when some defect falls into recombination region around defects of another kind). The treatment of the Frenkel defect accumulation is simplified by the fact that many-point densities for these virtual configurations differ by their orders of magnitudes in σ_0 parameter [31, 34]. It is sufficient to study either equations for the densities $\rho_{m,1}$ with arguments obeying $|\vec{r}_i - \vec{r}'_1| \leqslant r_0$, $(i = 1, \ldots, m)$ or taking into account problem's symmetry with respect to defect kinds A and B, the densities $\rho_{1,m'}$ with $|\vec{r}'_j - \vec{r}_1| \leqslant r_0$ $(j = 1, \ldots, m')$. These many-point densities are of the same order of magnitude $\rho_{1,m'} \propto \sigma_0^{-1}$ and correspond to virtual configurations where just created defect A has in his recombination sphere of the radius r_0 m defects of another kind B (and vice versa). Other virtual configurations have smaller order of magnitude in parameter σ_0^{-1} and thus could be omitted in the relevant equations.

The direct consequence of this statement for Kirkwood's superposition approximation is as follows. Substitution of equation (2.3.62) into $\rho_{2,1}$ yields correct order of its magnitude, σ_0^{-1}, provided $|\vec{r}_1 - \vec{r}'_1| < r_0$, $|\vec{r}_2 - \vec{r}'_1| > r_0$ (i.e., there is a single A in the recombination sphere around B), since two-point density $\rho_{1,1}(\vec{r}_1; \vec{r}'_1; t) \propto \sigma_0^{-1}$ and $\rho_{1,1}(\vec{r}_2; \vec{r}'_2; t) \propto (\sigma_0)^0$ (i.e., is *limited* as well as another density $\rho_{2,0}$, which does not fall into category of virtual configurations). On the other hand, for coordinates satisfying $|\vec{r}_1 - \vec{r}'_1| < r_0$, $|\vec{r}_2 - \vec{r}'_1| < r_0$ (i.e., defect B has in its recombination sphere two defects A) substitution of equation (2.3.62) results in $\rho_{1,2} \propto \sigma_0^{-2}$ instead of the correct σ_0^{-1}! Due to this the superposition approximation neglects in the limit $\sigma_0 \to \infty$ a number of terms in equations which finally leads to a considerable error in the accumulation kinetics.

In order to improve the bottleneck of the standard superposition approximation, let us consider this point more carefully. Let us introduce a function

$$\Psi(r) = \frac{1}{v_0}\theta(r_0 - r), \qquad \int \Psi(r)\,d\vec{r} = 1.$$

Taking into account preceding discussion of the virtual configurations for many-point densities $\rho_{m,1}$ or $\rho_{1,m'}$ let us define new functions

$$g_{m,1} = g_{m,1}(\{\vec{r}\}_m; \vec{r}_1'; t) = \sigma_0 v_0^m \rho_{m,1} \prod_{i=1}^{m} \Psi(|\vec{r}_i - \vec{r}_1'|). \qquad (7.1.3)$$

The function $g_{m,1}$ is nonzero if for all $i = 1, \ldots, m$ $|\vec{r}_i - \vec{r}_1'| < r_0$. It also remains constant as $\sigma_0 \to \infty$ (we use hereafter the same notation for the limiting value too). In the same manner we define functions $g_{1,m'}$.

7.1.3 The asymptotically exact equations

The first equations for macroscopic densities $n_\nu(t)$ come from a set (7.1.1) provided $(m + m') = 1$. Consider, for instance, an equation for the density $n_A(t)$:

$$\frac{\partial \rho_{1,0}}{\partial t} \equiv \frac{dn_A(t)}{dt} = p - \int \sigma(|\vec{r}_1 - \vec{r}_1'|) \rho_{1,1} \, d\vec{r}_1'$$

$$= p - \int g_{1,1} \, d\vec{r}_1. \qquad (7.1.4)$$

Making transformations here we used equation (7.1.3) and the fact that $g_{1,1}$ as well as $\rho_{1,1}$ depend on $r = |\vec{r}_1 - \vec{r}_1'|$ only. (It permits to change the integration variable in equation (7.1.4).)

Equation for the $g_{1,1}$ function comes for the joint density $\rho_{1,1}$ multiplied by $v_0 \Psi(|\vec{r}_1 - \vec{r}_1'|)$ and then making use of the definition (7.1.3):

$$v_0 \Psi(|\vec{r}_1 - \vec{r}_1'|) \frac{\partial \rho_{1,1}}{\partial t} = \sigma_0^{-1} \frac{\partial g_{1,1}}{\partial t} \to 0$$

$$= p v_0 \Psi(|\vec{r}_1 - \vec{r}_1'|) \left(\rho_{1,0}(\vec{r}_1; t) + \rho_{0,1}(\vec{r}_1'; t) + f(|\vec{r}_1 - \vec{r}_1'|) \right) -$$

$$- g_{1,1} - \int g_{2,1} \, d\vec{r}_2 - \int g_{1,2} \, d\vec{r}_2'. \qquad (7.1.5)$$

Note that after making limiting transition $\sigma_0 \to \infty$ the equation for $g_{1,1}$ is steady-state. Equation (7.1.5) contains integrals of functions $g_{2,1}$ and $g_{1,2}$. A study of the equations defining these functions leads to relations with functions like $g_{m,1}$ and $g_{1,m'}$. Let us consider for definiteness equation for $g_{2,1}$

following from that for $\rho_{2,1}$ after it was multiplied by $v_0^2 \prod_{i=1}^2 \Psi(|\vec{r}_i - \vec{r}_1'|)$,

$$\sigma_0^{-1} \frac{\partial g_{2,1}}{\partial t} \to 0$$

$$= p\sigma_0^{-1}\left(g_{1,1}(\vec{r}_1; \vec{r}_1'; t) + g_{1,1}(\vec{r}_2; \vec{r}_1'; t)\right) +$$

$$+ pv_0^2 \prod_{i=1}^2 \Psi(|\vec{r}_i - \vec{r}_1'|)\left(\rho_{1,0}(\vec{r}_1; t) f(|\vec{r}_2 - \vec{r}_1'|) + \right.$$

$$\left. + \rho_{1,0}(\vec{r}_2; t) f(|\vec{r}_1 - \vec{r}_1'|)\right) +$$

$$+ pv_0^2 \prod_{i=1}^2 \Psi(|\vec{r}_1 - \vec{r}_1'|)\rho_{2,0} - 2g_{2,1} - \int g_{3,1}\,\mathrm{d}\vec{r}_3 -$$

$$- \left(\int \sigma(|\vec{r}_1 - \vec{r}_2'|)\rho_{2,2}\,\mathrm{d}\vec{r}_2' + \int \sigma(|\vec{r}_2 - \vec{r}_2'|)\rho_{2,2}\,\mathrm{d}\vec{r}_2'\right) \times$$

$$\times v_0^m \prod_{i=1}^2 \Psi(|\vec{r}_i - \vec{r}_1'|). \tag{7.1.6}$$

In equation (7.1.6) the first and the last term in the r.h.s. turns out to zero as $\sigma_0 \to \infty$. The integral terms contain functions $\rho_{2,2}$ in which there exist several defects A in the recombination sphere around B which, in their turn, have partners to recombine at $r \leqslant r_0$. In the limit of instant annihilation such functions have another order of magnitude in σ_0^{-1}.

Analogously general relation ($m > 2$) could be obtained

$$0 = pv_0^m \prod_{i=1}^m \Psi(|\vec{r}_i - \vec{r}_1'|) \sum_{k=1}^m f(|\vec{r}_k - \vec{r}_1'|) \rho_{m-1,0}(\{\vec{r}\}_m^k; t) +$$

$$+ pv_0^m \prod_{i=1}^m \Psi(|\vec{r}_i - \vec{r}_1'|) \rho_{m,0} - mg_{m,1} - \int g_{m+1,1}\,\mathrm{d}\vec{r}_m. \tag{7.1.7}$$

A set (7.1.7) could be solved simply with respect to functions $g_{m,1}$. To shorten these equations let us introduce more compact expressions $f_i = f(|\vec{r}_i - \vec{r}_1'|)$, $\Psi_i = \Psi(|\vec{r}_i - \vec{r}_1'|)$, $\rho_{m-1,0}(\{\vec{r}\}_m^i; t) = \rho_{m-1,0}^{(i)}$. Finally, we arrive at

$$g_{2,1} = \frac{pv_0^2}{2}\Psi_1\Psi_2\left(\rho_{2,0} + \rho_{1,0}^{(1)}f_1 + \rho_{1,0}^{(2)}f_2\right) +$$

$$+ p\Psi_1\Psi_2 \sum_{m=3}^{\infty} \frac{(-v_0)^m}{m!} \int \ldots \int d\vec{r}_3 \ldots d\vec{r}_m \times$$

$$\times \prod_{i=3}^{m} \Psi_i\left(\rho_{m,0} + \rho_{m-1,0}^{(1)}f_1 + \rho_{m-1,0}^{(2)}f_2 + \right.$$

$$\left. + (m-2)\rho_{m-1,0}^{(3)}f_3\right). \tag{7.1.8}$$

The function $g_{1,1}$ could be presented, respectively in a form

$$g_{1,1} = pv_0\Psi_1\left(\rho_{1,0}(\vec{r}_1; t) + \rho_{0,1}(\vec{r}_1'; t) + f_1\right) - G_A - G_B, \tag{7.1.9}$$

with

$$G_A = \int g_{2,1}\,d\vec{r}_2, \qquad G_B = \int g_{1,2}\,d\vec{r}_2'. \tag{7.1.10}$$

Substitution of (7.1.8) into (7.1.10) leads to the following expression

$$G_A = p\Psi_1 \sum_{m=2}^{\infty} \frac{(-v_0)^m}{m!} \int \ldots \int d\vec{r}_2 \ldots d\vec{r}_m \times$$

$$\times \prod_{i=2}^{m} \Psi_i\left(\rho_{m,0} + \rho_{m-1,0}^{(1)}f_1 + (m-1)\rho_{m-1,0}^{(2)}f_2\right). \tag{7.1.11}$$

(The G_B function hereafter is calculated quite similarly.)

Equation (7.1.4) for macroscopic density contains the integral

$$\int g_{1,1}\,d\vec{r}_1 = -pv_0 \int \Psi_1 f_1\,d\vec{r}_1 + H_A + H_B, \tag{7.1.12}$$

where

$$H_A = -p \sum_{m=1}^{\infty} \frac{(-v_0)^m}{m!} \int \ldots \int d\vec{r}_1 \ldots d\vec{r}_2 \times$$

$$\times \prod_{i=1}^{m} \Psi_i\left(\rho_{m,0} + m\rho_{m-1,0}^{(1)}f_1\right). \tag{7.1.13}$$

Let us define now the averaging procedure over coordinates with the help of the normalised distribution $\prod_{i=1}^{m} \Psi_i$

$$\langle F(\{\vec{r}\}_m)\rangle = \int \ldots \int d\vec{r}_1 \ldots d\vec{r}_m \prod_{i=1}^{m} \Psi_i F(\{\vec{r}\}_m). \tag{7.1.14}$$

Let us introduce the quantity

$$\omega = v_0 \langle f_1 \rangle = \int f(r)\theta(r_0 - r)\, d\vec{r}, \tag{7.1.15}$$

which gives a proportion of genimate AB pairs created at $r < r_0$ and thus inavoidably disappearing. The accumulation kinetics described by (7.1.4) now reads

$$\frac{dU_A(t)}{dt} = v_0 \frac{dn_A(t)}{dt}$$
$$= pv_0(1-\omega)(1-\delta_A-\delta_B), \tag{7.1.16}$$

where

$$\delta_A = -\sum_{i=1}^{\infty} \frac{(-v_0)^m}{m!} \langle \rho_{m,0} \rangle. \tag{7.1.17}$$

The $A + B \to 0$ reaction with correlated defect creation is characterized by the time-independent quantity $\delta n = n_A(t) - n_B(t) = \text{const}$. It is self-evident since the equation for macroscopic density $n_B(t)$ or $U_B(t)$ follows from equation (7.1.16) replacing an index A for B (and vice versa).

Equation (7.1.16) is asymptotically ($\sigma_0 \to \infty$) exact. It shows that the accumulation kinetics is defined by: (i) a fraction of AB pairs, $1 - \omega$, created at relative distances $r > r_0$, (ii) recombination of defects created inside the recombination volume of another-kind defects. The co-factor $(1-\delta_A-\delta_B)$ in equation (7.1.16) gives just a fraction of free folume available for new defect creation. Two quantities δ_A and δ_B characterizing, in their turn, the whole volume fraction forbidden for creation of another kind defects are defined *entirely* by quite specific many-point densities $\rho_{m,0}$ and $\rho_{0,m'}$, i.e., by the relative distribution of *similar* defects only (see equation (7.1.17)).

7.1.4 The superposition approximation

If we succeeded in calculating the series in equation (7.1.17), the accumulation kinetics problem under question would be solved. However, an infinite set of coupled equations for $\rho_{m,0}$ turns out to be too complicated and thus we restrict ourselves to its cut-off by means of Kirkwood's superposition approximation, in order to get a closed set of non-linear equations for macriscopic densities $n_A(t)$ and $n_B(t)$, as well as for the three joint correlation functions $X_A(r,t)$, $X_B(r,t)$ and $Y(r,t)$.

Let us return now to equation (2.3.62), which is equivalent to

$$\rho_{2,1} \Rightarrow n_A^2(t) n_B(t) X_A(|\vec{r}_1 - \vec{r}_2|, t) Y(|\vec{r}_1 - \vec{r}_1'|, t) \times$$
$$\times Y(|\vec{r}_2 - \vec{r}_1'|, t). \tag{7.1.18}$$

Its structure is self-evident: each coordinate entering the many-point density $\rho_{m,m'}$ is associated with the relevant concentration co-factor $n_A(t)$ (if the coordinate is from a set $\{\vec{r}\}_m$ or $n_B(t)$ if it is from a set $\{\vec{r}'\}_{m'}$). Each pair of coordinates corresponds to the particular joint correlation function, all these functions are multiplied. An analog of the substitution (7.1.18) in the case of many-point densities is an expression

$$\rho_{m,0} \Rightarrow n_A(t)^m \prod_{i>}^{m} \prod_{j=1}^{m'} X_A(|\vec{r}_i - \vec{r}_j|, t). \tag{7.1.19}$$

Substitution of equation (7.1.19) into equation (7.1.17) leads to the problem equivalent to the calculation of statistical integrals in a system of m particles placed in volume v_0 and interacting by the pair-wise forces. It is clear that such a problem could be solved only approximately. The main approximation we use hereafter is the replacement of the mean value of a product of functions for a product of two mean values. Thus

$$\langle \rho_{m,0} \rangle = n_A(t)^m \langle X_A \rangle^{m(m-1)/2}. \tag{7.1.20}$$

Let us define

$$S_A = \ln\langle X_A \rangle. \tag{7.1.21}$$

To calculate (7.1.17) employing equation (7.1.20), one can use the integral presentation (depending on a sign of S_A)

$$\exp\left(\frac{S_A}{2}m^2\right) = -\frac{1}{\sqrt{\pi}}\int_{-\infty}^{\infty}\exp\left(-z^2 + zm\sqrt{2S_A}\right)dz,$$
$$S_A > 0, \qquad (7.1.22)$$

$$\exp\left(\frac{S_A}{2}m^2\right) = -\frac{1}{\sqrt{\pi}}\int_{-\infty}^{\infty}\exp\left(-z^2 + izm\sqrt{-2S_A}\right)dz,$$
$$S_A < 0. \qquad (7.1.23)$$

Calculation of infinite sums therein leads to an expression $\delta_A = \Xi_A(1)$, where

$$\Xi_A(x) = \frac{1}{x^2\sqrt{\pi}}\int_{-\infty}^{\infty} e^{-z^2}(1 - \exp(-x\varphi_A))dz, \qquad (7.1.24)$$

and

$$\varphi_A = U_A(t)\exp\left(-\frac{S_A}{2} + z\sqrt{2S_A}\right), \quad S_A > 0,$$

$$\varphi_A = U_A(t)\exp\left(-\frac{S_A}{2} + iz\sqrt{-2S_A}\right), \quad S_A < 0. \qquad (7.1.25)$$

Therefore, the r.h.s. of equation (7.1.16) could be presented via the dimensionless concentrations $U_\nu(t)$, $\nu = A, B$ and mean values like $\langle X_\nu \rangle$, which characterize short-range order in similar defect distribution. A set of equations for the correlation functions is added to equation (7.1.16).

An equation for the joint correlation function of dissimilar defects $Y(r,t)$ is simpler. Since as $\sigma_0 \to \infty$, the joint density $\rho_{1,1}$ having arguments $|\vec{r}_1 - \vec{r}_1'| < r_0$ becomes zero,

$$Y(r < r_0, t) = 0. \qquad (7.1.26)$$

As $|\vec{r}_1 - \vec{r}_1'| > r_0$, the definition of $\rho_{m,m'}$ gives us

$$\frac{\partial \rho_{1,1}}{\partial t} = p(\rho_{1,0} + \rho_{0,1} + f_1) - \int \sigma(|\vec{r}_2 - \vec{r}_1'|)\rho_{2,1}\,d\vec{r}_2 -$$
$$- \int \sigma(|\vec{r}_2' - \vec{r}_1|)\rho_{1,2}\,d\vec{r}_2'. \qquad (7.1.27)$$

There is no limitation for the use of Kirkwood's superposition approximation (2.3.62) in the integral terms of equation (7.1.27) since there is a *single* defect in the recombination sphere of dissimilar defect. For example, a product $\sigma(|\vec{r}_2 - \vec{r}_1'|)\rho_{1,1}(\vec{r}_2;\vec{r}_1';t)$ in a line with definition (7.1.3) could be replaced by $g_{1,1}(\vec{r}_2;\vec{r}_1';t)$. Therefore, we have

$$\sigma(|\vec{r}_2 - \vec{r}_1'|)\rho_{2,1} \Rightarrow g_{1,1}(\vec{r}_2;\vec{r}_1';t)\frac{\rho_{1,1}(\vec{r}_1;\vec{r}_1';t)\rho_{2,0}(\vec{r}_1,\vec{r}_2;t)}{n_A(t)^2 n_B(t)}$$

$$= g_{1,1}(\vec{r}_2;\vec{r}_1';t)n_A(t)Y(|\vec{r}_1 - \vec{r}_1'|,t)X_A(|\vec{r}_1 - \vec{r}_2|,t). \qquad (7.1.28)$$

The function $g_{1,1}$, as well as the correlation functions depend also on the-modulus of the coordinate difference which permits to present integrals in a form of the convolution of two functions

$$(\eta * \zeta) = \int \eta(r')\zeta(|\vec{r} - \vec{r}'|)d\vec{r}'. \qquad (7.1.29)$$

In particular, using the relative coordinate $r = |\vec{r}_1 - \vec{r}_1'|$, we arrive at

$$\int \sigma(|\vec{r}_2 - \vec{r}_1'|)\rho_{2,1}\,d\vec{r}_2 = n_A(t)Y(r,t)(g_{1,1} * X_A). \qquad (7.1.30)$$

The function $g_{1,1}$ could be calculated by means of the method discussed above. In doing so, and using equation (7.1.30), r.h.s. of (7.1.27) could be finally found.

An alternative approach we shall use to analyze the further equation of a set results in more simple relations. Its idea is to rewrite the convolution $(g_{1,1} * X_A)$ in equation (7.1.30) as an average over normalized distributions $\prod_{i=1}^{m} \Psi_i$ of infinite series of products of the correlation functions and f_i. (As earlier, a mean value of the product is replaced by a product of mean values.) Finally, one gets

$$(g_{1,1} * X_A) = \chi_A(r)p(1 - (1-\omega)(1 - \delta_A - \delta_B)), \qquad (7.1.31)$$

where

$$\chi_A(r) = (\Psi * X_A), \qquad (7.1.32)$$

and

$$\langle X_A \rangle = (\Psi \chi_A * 1). \qquad (7.1.33)$$

The function $\chi_A(r)$ is numerically close to $X_A(r,t)$, it results as an average of the $X_A(r,t)$ taken with some weighting function in the coordinate interval $(|r-r_0|, r+r_0)$.

Time derivative in equation (7.1.27) could be found in the following way (keeping also in mind equation (7.1.16)):

$$\frac{\partial \rho_{1,1}}{\partial t} = \frac{\partial}{\partial t}\left(n_A(t)n_B(t)Y(r,t)\right)$$
$$= n_A(t)n_B(t)\frac{\partial Y(r,t)}{\partial t} +$$
$$+ \left(n_A(t)+n_B(t)\right)Y(r,t)p(1-w)(1-\delta_A - \delta_B). \qquad (7.1.34)$$

Denoting $\mu = (1-w)(1-\delta_A - \delta_B)$, we arrive at

$$U_A(t)U_B(t)\frac{\partial Y(r,t)}{\partial t}$$
$$= pv_0\Big\{\left(U_A(t)+U_B(t)+v_0 f(r)\right) -$$
$$- \mu Y(r,t)\left(U_A(t)+U_B(t)\right) -$$
$$- (1-\mu)Y(r,t)\left(U_A(t)\chi_A + U_B(t)\chi_B\right)\Big\}. \qquad (7.1.35)$$

The correlation function $X_A(r,t)$ is obtained using equation for the joint density $\rho_{2,0}$:

$$\frac{\partial \rho_{2,0}}{\partial t} = p\left(\rho_{1,0}(\vec{r}_1;t) + \rho_{1,0}(\vec{r}_2;t)\right) - \int \sigma\left(|\vec{r}_1 - \vec{r}_1'|\right)\rho_{2,1}\mathrm{d}\vec{r}_1' -$$
$$- \int \sigma\left(|\vec{r}_2 - \vec{r}_1'|\right)\rho_{2,1}\mathrm{d}\vec{r}_1'. \qquad (7.1.36)$$

Due to equivalence of points $\vec{r}_1$ and $\vec{r}_2'$, both integrals in this equation are equal. Let us consider the first one. The integration domain here consists of two intervals. For the first one $|\vec{r}_2 - \vec{r}_1'| \leqslant r_0$ and consequently

$$\sigma\left(|\vec{r}_2 - \vec{r}_1'|\right)\rho_{2,1} \equiv g_{2,1}.$$

For the second interval $|\vec{r}_2 - \vec{r}_1'| \geqslant r_0$ and, as one can see in equation (7.1.27), equation (7.1.28) holds. Keeping in mind equation (7.1.26) one gets

$$\sigma(|\vec{r}_2 - \vec{r}_1'|)\rho_{2,1}$$
$$\Rightarrow g_{2,1} + g_{1,1}(\vec{r}_2; \vec{r}_1'; t) n_A(t) \times$$
$$\times Y(|\vec{r}_1 - \vec{r}_1'|, t) X_A(|\vec{r}_1 - \vec{r}_2|, t). \qquad (7.1.37)$$

Here each of terms is nonzero in its own region of definition. An integral of the second term in equation (7.1.37) could be found by the method just described, which gives

$$n_A(t) X_A(r, t)(g_{1,1} * Y) = n_A(t) X_A(r, t) p(1 - \mu) \vartheta(r), \qquad (7.1.38)$$

where

$$\vartheta(r) = (\Psi * Y). \qquad (7.1.39)$$

The function $\vartheta(r)$ is numerically close to $Y(r, t)$ and arises as its average in the interval $(|r - r_0|, r + r_0)$.

The integral $\int g_{2,1} d\vec{r}_1'$ is more difficult. As we once did in equation (7.1.14), let us define new averages through

$$\langle\langle F(\{\vec{r}\}_m)\rangle\rangle$$
$$= \frac{v_0}{\Omega(r)} \int \ldots \int d\vec{r}_1' d\vec{r}_3 \ldots d\vec{r}_m \prod_{i=1}^{m} \Psi_i F(\{\vec{r}\}_m), \qquad (7.1.40)$$

where co-factors in front of the integral are used for distribution normalisation by unity. The function

$$\Omega(r) = v_0(\Psi * \Psi) \qquad (7.1.41)$$

is a share of overlaped volumes of two $\bar{d}$-dimensional spheres of radius r whose centres are separated by r, the function $\Omega(r) \equiv 0$ as $r \geqslant 2r_0$. Function $\Omega(r)$ depends on the space dimension $\bar{d}$:

$$\Omega(r) = 1 - \frac{r}{2r_0}, \quad \bar{d} = 1,$$

$$\Omega(r) = \frac{2}{\pi} \left\{ \arcsin\left(1 - \frac{r^2}{4r_0^2}\right)^{1/2} - \frac{r}{2r_0}\left(1 - \frac{r^2}{4r_0^2}\right)^{1/2} \right\}, \quad \bar{d} = 2,$$

$$\Omega(r) = 1 - \frac{3r}{4r_0} + \frac{r^3}{16r_0^3}, \quad \bar{d} = 3, \tag{7.1.42}$$

respectively. If the function under averaging does not depend on coordinates $\vec{r}_1, \vec{r}_2$, by definitions (7.1.14) and (7.1.40) the mean values coincide. Taking into account equation (7.1.8) one gets

$$\int g_{2,1} d\vec{r}_1' = p \frac{\Omega(r)}{v_0} \sum_{m=2}^{\infty} \frac{(-v_0)^m}{m!} \left\langle \left\langle \left(\rho_{m,0} + \sum_{k=1}^{m} \rho_{m-1,0}^{(k)} f_k \right) \right\rangle \right\rangle. \tag{7.1.43}$$

The use to the described scheme to calculate a mean value in equation (7.1.40) leads to the expresion

$$\int g_{2,1} d\vec{r}_1' = p \frac{\Omega(r)}{v_0} \sum_{m=2}^{\infty} \frac{(-U_A)^m}{m!} \times$$

$$\times \left\{ X_A(r,t) \left(\tilde{\chi}_A^{2(m-2)} \langle X_A \rangle^{(m-2)(m-3)/2} + \right. \right.$$

$$+ (m-2) \frac{\tilde{\omega}}{U_A} \tilde{\chi}_A^{2(m-3)} \langle X_A \rangle^{(m-3)(m-4)/2} \right) +$$

$$\left. + 2 \frac{\tilde{\omega}}{U_A} \tilde{\chi}^{(m-2)} \langle X_A \rangle^{(m-2)(m-3)/2} \right\}$$

$$= p \frac{\Omega(r)}{v_0} Q_A. \tag{7.1.44}$$

New functions are introduced here

$$\tilde{\chi}_A(r) = \frac{v_0}{\Omega(r)} (\Psi * \Psi \chi_A) \tag{7.1.45}$$

and $\tilde{\omega} = v_0 \tilde{f}$, where

$$\tilde{f}(r) = \frac{v_0}{\Omega(r)} (\Psi * \Psi f). \tag{7.1.46}$$

Functions $\tilde{\chi}_A$ and $\tilde{f}$ are numerically close to χ_A and f respectively and arise due to their averaging with the normalised distribution in the interval $(|r - r_0|, r + r_0)$. Assiming that defects are created *uniformly* at up to certain generation radius R_g, i.e.,

$$f(r) = V_g^{-1}\theta(R_g - r), \qquad (7.1.47)$$

the problem is simplified since $\tilde{\omega} = \omega = v_0/V_g = (r_0/R_g)^{\bar{d}}$. Calculation of the infinite sums in equation (7.1.44) using equations (7.1.22), (7.1.23) yields

$$Q_A = U_A \frac{\langle X_A \rangle}{\tilde{\chi}_A^2} X_A(r,t) - 2\omega \frac{\langle X_A \rangle^3}{\tilde{\chi}_A^4} X_A(r,t) + 2\tilde{\omega} \frac{\langle X_A \rangle}{\tilde{\chi}_A} -$$

$$- (1-\omega) X_A(r,t) \langle X_A \rangle^{-1} \Xi_A \left(\frac{\tilde{\chi}_A^2}{\langle X_A \rangle^2} \right) +$$

$$+ \frac{2\omega}{U_A} X_A(r,t) \tilde{\chi}_A^{-2} \Xi_A \left(\frac{\tilde{\chi}_A^2}{\langle X_A \rangle^3} \right) -$$

$$- \frac{2\tilde{\omega}}{U_A} \langle X_A \rangle^{-1} \Xi_A \left(\frac{\tilde{\chi}_A}{\langle X_A \rangle^2} \right). \qquad (7.1.48)$$

It results in equation for the correlation function of similar defects $X_A(r,t)$

$$X_A(t) \frac{\partial X_A(r,t)}{\partial t}$$

$$= 2pv_0 \left\{ \left(1 - \Omega(r) \frac{Q_A}{U_A(t)} \right) - X_A(r,t)(\mu + (1-\mu)\vartheta(r)) \right\}. \qquad (7.1.49)$$

7.1.5 The basic kinetic equations

The set of equations (7.1.16), (7.1.35) and (7.1.49) is derived in a quite general form. In the particular case of the accumulation of the *immobile* Frenkel defects (whose concentrations are zero at $t = 0$) these equations are considerably simplified. Note the symmetry properties: $U_\nu(t) = U(t)$, $X_\nu(r,t) = X(r,t)$, $\delta_\nu = \delta$, $\chi_\nu(r) = \chi(r)$, $\nu = A, B$. Let us write down the kinetic equations provided $U_\nu(0) = 0$ or $U_\nu(t) = U(t)$. Keeping in mind further generalization of results for defect diffusion, the condition $X_\nu(r,t) = X(r,t)$ is not used. Introducing dimensionless time, $t \equiv pv_0 t$, the kinetic equations read

$$\left.\frac{dU(t)}{dt}\right|_{cr+rec} = \mu, \quad \mu = (1-\omega)(1-2\delta), \tag{7.1.50}$$

$$\left.\frac{\partial Y(r,t)}{\partial t}\right|_{cr+rec} = \frac{1}{U(t)}\left\{\left(2 + \frac{v_0 f(r)}{U(t)}\right) - Y(r,t)\big(2\mu + (1-\mu)(\chi_A + \chi_B)\big)\right\}, \tag{7.1.51}$$

$$\left.\frac{\partial X(r,t)}{\partial t}\right|_{cr+rec} = \frac{2}{U(t)}\left\{\left(1 - \Omega(r)\frac{Q}{U(t)}\right) - X(r,t)\big(\mu + (1-\mu)\vartheta\big)\right\}. \tag{7.1.52}$$

As it was first demonstrated in [30], in Kirkwood's superposition approximation (2.3.62) the series for $g_{m,1}$ are cut-off; in particular $g_{2,1} \equiv 0$, whereas in the $g_{1,1}$ only the first term is retained; $g_{1,1} = pv_0\Psi_1(\rho_{1,0} + \rho_{0,1} + f_1)$. Neglecting the geminate correlation, $f(r) = 0$, we arrive at equation

$$\frac{dU(t)}{dt} = 2\big(U_0 - U(t)\big) \quad \text{or} \quad U(t) = U_0\big(1 - \exp[-2t]\big) \tag{7.1.53}$$

with $t \equiv pv_0 t$, $U_0 = U(\infty) = n(\infty)v_0 = 0.5$ which is nothing but a dimensionless form of equation (7.1.2b). Note that irrespective on the spatial dimension $\bar{d}$ the saturation concentration obtained here, $U_0 = 0.5$, is an essentially underestimate as compared to that found in computer simulations ([15, 35] and Section 7.4). It should be stressed once again that concentration dynamics (7.1.53) *is not* coupled with correlation dynamics.

Another, more exact estimate presented first in [34] also does not come out from any analysis of the spatial defect correlations. It is based on *a priori* assumption of the *Poisson (random) distribution* of similar defects, $\rho_{m,0} = n_A(t)^m$, corresponding to the $X_A(r,t) \equiv 1$ in substitution (7.1.19). Under this assumption a series defining function $\delta(t)$ could be elementary summed up:

$$\delta(t) = 1 - \exp[-U(t)]. \qquad (7.1.54)$$

The relevant kinetic equation reads

$$\frac{d(t)}{dt} = 2\bigl(\exp[-U(t)] - \exp[-U_0]\bigr). \qquad (7.1.55)$$

It gives slightly higher saturation concentration $U_0 = \ln 2 \approx 0.69$ (again the same for all dimensions $\bar{d}$) than the superposition approximation does, but it is still essentially underestimated. For the first time the function $\delta(t)$ was successfully calculated in [31], as defined by the correlation function of similar defects, $X(r,t)$. However, the only *linear* corrections in the correlation functions were taken into account. The saturation concentration $U_0 = 1.08$ for $\bar{d} = 3$ agrees with computer simulations $U_0 = 1.01 \pm 0.10$ [36]. However, the saturation predicted for the *low* dimensions, e.g., $\bar{d} = 1$, $U_0 = 1.36$ is much lower than computer simulations (see, e.g., [15, 35]).

7.1.6 Results for immobile particles

Let us consider now results of the numerical solution of much more refined accumulation kinetics – equations (7.1.50) to (7.1.52). For the initially uncorrelated pairs, $f(r) \equiv 0$, $\omega = 0$, different asymptotic laws are observed for different spatial dimensions $\bar{d}$ (Fig. 7.1). Thus, the steady-state found for $\bar{d} = 3$, $U_0 = 1.02$, is in good agreement with the mentioned computer simulation [36]. In contrast, at low dimensions ($\bar{d} = 1$ and 2) these kinetic equations reveal *non*-steady-state solutions only. For $\bar{d} = 1$ and long t the magnitude of $U(t)$ increases as a power of time, $U(t) \propto t^\alpha$, $\alpha \approx 0.20$, whereas for $\bar{d} = 2$ $U(t)$ nearly roughly logarithmically. Irrespective of $\bar{d}$, at long t the function δ strives from below to the limiting value of $\delta = 0.5$, so that $U(t)$ is the quasy-steady-state solution of equation (7.1.50). Equation $\delta = \delta_A = \Xi_A(1) = 0.5$ and the definition (7.1.24) relates the steady-state value of U_0 (or current value of $U(t)$ for the quasi-steady-state) to the averaged correlation function $\langle X_A \rangle$ (see equations (7.1.21) to (7.1.24)). For high space dimensions $\bar{d}$ possible peculiarities of the joint correlation function of similar defects (maximum at the coordinate origin) are weakened after averaging due to the r-dependence of the volume unit, $dV \propto r^{\bar{d}-1}dr$, which strongly affects the asymptotic reaction rate.

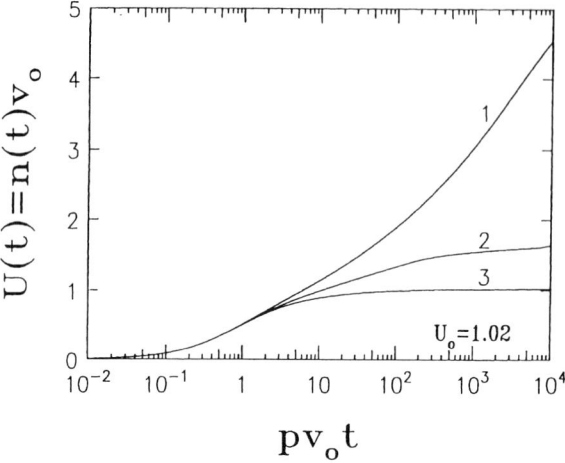

Fig. 7.1. Accumulation kinetics of immobile defects for different spatial dimensions: $\bar{d} = 1$ (curve 1), $\bar{d} = 2$ (curve 2), $\bar{d} = 3$ (curve 3). Concentration and irradiation time are in dimensionless units, time scale logarithmic.

Figures 7.2 and 7.3 show the relevant correlation functions. In three dimensions at the dimensionless time $pv_0t = 10^3$ the steady-state is already nearly achieved (the deviation from the unity is seen only at $r < 10r_0$). Since the correlation length ξ at large t is *finite*, microscopic defect segregation takes place for $\bar{d} = 3$. Quite contrary, for low ($\bar{d} \leqslant 2$) dimensions the correlation functions *are no longer stationary*. Similarly to the recombination *decay* kinetics treated in [14], the accumulation kinetics demonstrates also an *infinite* increase in time of the correlation length ξ (defined by a coordinate where $X(r,t) \gg 1$ or $Y(r,t) \ll 1$ holds). In other words, reaction volume is divided into blocks (domains) of the distinctive size ξ, each block contains mainly similar defects, either A or B. For a *finite* system with a linear size L condition $L \approx \xi$ means in fact nothing but *macroscopic* defect segregation: reaction volume is divided into several domains of similar defects. This effect was indeed observed in computer simulations for low dimensional systems [15, 35]. For instance, for $\bar{d} = 1$ defects are grouped into *two* large clusters of only A's and B's slowly walking with time in space.

A comparison of the calculated correlation length ξ for the infinite system with distinctive L used in computer simulations permits us to understand a nature of the instability of results observed in many statistical simulations.

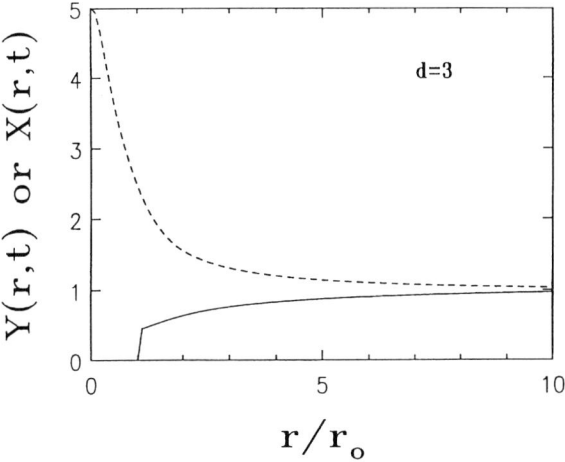

Fig. 7.2. The stationary $(t \to \infty)$ joint correlation functions for $\bar{d} = 3$. Full curve is for dissimilar defects, $Y(r, \infty)$, broken line is for similar defects, $X(r, \infty)$.

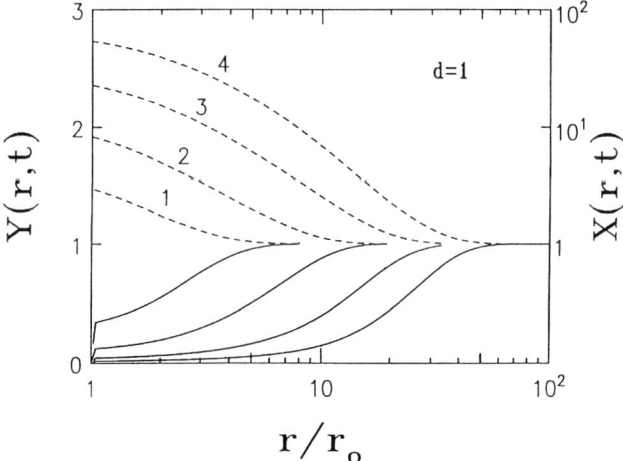

Fig. 7.3. The time development of the joint correlation functions for $\bar{d} = 1$. Curves 1 to 4: $pv_0 t = 10, 10^2, 10^3$ and 10^4 respectively. Full curves are $Y(r,t)$, broken curves are $X(r,t)$ (in the logarithmic inits). Note that the correlation length $\xi(t)$ infinitely increases in time thus indicating *macroscopic* defect segregation.

Say, Pirogov and Palagashvili [36] in their $3\bar{d}$ study used the system's size L close to the correlation length ξ, which resulted in large concentration dispersion ($\approx 10\%$) well seen in their steady-state.

For $\bar{d} = 3$ deviations of the correlation functions from their asymptotics in the limited the coordinate interval ($\approx 2r_0$) are considerable. A fraction of very close similar defects, $X(r \to 0, \infty)$, exceeds 3–4 *times its Poisson value*. It gives an explanation and is in excellent agreement with the experimental data for unexpectedly high concentration of dimer F_2 centres in KCl (two nearest F centres) produced at 4 K after prolonged X-ray irradiation [13].

If radiation defects are *charged*, their aggregation is accompanied by the internal electric fields (which are the greater, the less particle mobility and larger irradiation intensity p). Quantitative analysis of this problem has been done in [37]. (Probably, this effects has been observed experimentally in [38].)

It should be stressed once more that the accumulation curve $n(t)$ (or $U(t)$), especially at high doses, *cannot* be described by a simple equation (7.1.53) which is often used for interpreting the real experimental data (e.g., [19, 20]). Despite there is the *only* recombination mechanism, the $A + B \to 0$ accumulation kinetics at long t due to many-particle effects is no longer exponential function of time (dose). Therefore, successful expansion of the experimental accumulation curve $U = U(t)$ in several exponentials (stages) does not mean that several *different mechanisms* of defect creation are necessarily involved (as sometimes they suggest, e.g., [39, 40]).

7.1.7 Correlated particle creation

The accumulation kinetics of *correlated* (geminate) pairs is less studied. In [30, 41] it was demonstrated that in this case the aggregation effect is *weakened* and the saturation concentration is reduced essentially. This makes use of Kirkwood's superposition approximation more reliable. Employing the latter, the following relation was derived [30]

$$U_0 = \frac{1}{2}\left(1 - \int_{v_0} f(r)\,d\vec{r}\right) = \frac{1}{2}(1 - \omega). \tag{7.1.56}$$

Integration here is over the recombination sphere. The accumulation kinetics remains to be equation (7.1.53), but with U_0 from equation (7.1.56). It follows from (7.1.56) that creation of the Frenkel partners separated by the distinctive

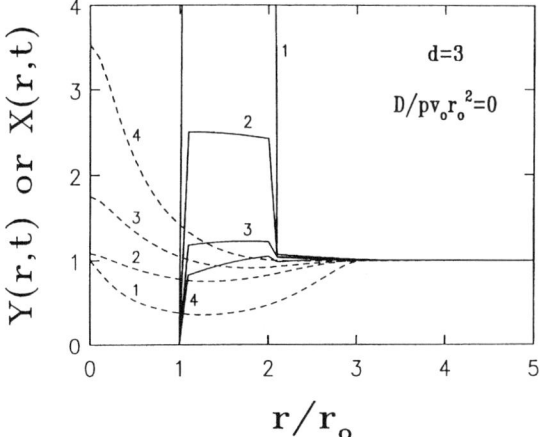

Fig. 7.4. The joint correlation functions for $\bar{d} = 3$ and the initially correlated immobile geminate pairs (random AB distribution within the narrow interval of relative distances $[r_0, 2r_0]$). Curves 1 to 4 correspond to the dimensionless times $10^{-2}, 10^{-1}, 10^0, 10^1$ respectively.

distance $r_0 \leqslant r \leqslant R_g$ is associated with a very small saturation concentration U_0.

In Fig. 7.4 the joint correlation functions are plotted for distribution of geminate partners created randomly within narrow interval $r_0 \leqslant r \leqslant R_g$. Two important conclusions suggest themselves from this figure: (i) due to *similar and dissimilar reactant correlation back-coupling* the narrow peak of Y at short distances is accompanied by the *decay* in X, (ii) for great doses, $n \lesssim n(\infty)$, the joint correlation functions are quite similar to those observed for uncorrelated distribution, i.e., an aggregation manifests itself mainly at high defect concentrations.

In the review article [15] numerous analytical expressions have been presented to describe experimental accumulation curves $n(t)$ – see Section 7.4. Note here that such kinetics are studied widely in *all* kinds of solids, from alkali halides [5, 9, 10, 42] to metals [43–46].

7.1.8 Incorporation of diffusion

When temperature increases, defects become mobile and the pattern becomes more complicated due to additional reactions. On the other hand, if we neglect dynamic particle interaction (to be discussed in Section 7.2), the aggregation effect is expected to be *reduced* since diffusion acts to smooth inhomogeneities in particle densities and thus for a large enough diffusion coefficients Kirkwood's superposition approximation could be safely used. Several techniques were used to solve this problem. Simple and approximate results were obtained by means of the *shortened* superposition approximation, $X_\nu(r,t) \equiv 1$, equation (2.3.64) [15, 29, 47–49]. In this theory the recombination rate and the steady-state concentration are defined in the self-consistent way. The accumulation equations thus derived were checked by means of the direct computer simulations for $\bar{d} = 2$ [50]. It has been demonstrated that for a small dimensionless parameter $2pv_0/Dn(\infty)$ (i.e., for large D) the following relation defining the saturation concentration $n(\infty)$

$$p = 2\pi x \frac{K_1(x)}{K_0(x)}, \qquad (7.1.57)$$

turns out to agree well with the statistical simulations. Here $x = [2pv_0/Dn(\infty)]^{1/2}$, and K_0 and K_1 are the modified Bessel's functions. For $\bar{d} = 3$ the shortened superposition approximation gives

$$p = 4\pi D r_0 \left(1 + \sqrt{3U_0}\right) n^2(\infty), \qquad (7.1.58)$$

where as usual, $U_0 = n(\infty)v_0$.

Note that equations (7.1.57) and (7.1.58) are of rather limited use since they are derived for large diffusion coefficients D when defect aggregation is not well pronounced. Moreover, equation (7.1.57) assumes existence of the steady-state for $\bar{d} = 2$ whereas other methods discussed in [15] argue for the *macroscopic* defect segregation occuring here even for mobile defects. In this respect of great interest is the generalization of the more correct accumulation equations (7.1.50) to (7.1.52) presented below for the case of mobile defects.

A specific feature of the black sphere model is trivial functional discrimination of terms entering the kinetic equations depending if they are related on the defect production or on the spatial correlations. The r.h.s. of equations (7.1.50) to (7.1.52) describe decay of newly-created defect if they find themselves in the recombination volumes of dissimilar defects. If this is not the case, newly created defects can further disappear during diffusive migration. The latter problem was already considered in [14] (see equations (2.1) to (2.3) therein).

Making use of the dimensionless variables $r' = r/r_0$, $t' = pv_0 t$, $D'_A = 2\kappa$, $D'_B = 2(1-\kappa)$, $\kappa = D_A/(D_A + D_B)$ (primes are omitted below) and introducing the new dimensionless parameter characterizing defect mobility

$$\lambda = \frac{D}{pv_0 r_0^2}, \qquad (7.1.59)$$

we can rewrite the diffusion–recombination contributions into our previous kinetic equations as

$$\left.\frac{dU(t)}{dt}\right|_{\text{diff+rec}} = -\lambda \bar{d} K(t) U^2(t), \quad K(t) = \left.\frac{\partial Y(r,t)}{\partial r}\right|_{r=1}, \qquad (7.1.60)$$

$$\left.\frac{\partial Y(r,t)}{\partial t}\right|_{\text{diff+rec}} = \lambda \Big\{ \Delta Y(r,t) - \bar{d} K(t) U(t) Y(r,t) \sum_{\nu=A,B} J[X_\nu] \Big\}, \qquad (7.1.61)$$

$$\left.\frac{\partial X_A(r,t)}{\partial t}\right|_{\text{diff+rec}} = \lambda \Big\{ D_A \Delta X_A(r,t) - \bar{d} K(t) U(t) X_A(r,t) J[Y] \Big\}. \qquad (7.1.62)$$

The functionals $J[Z]$ for $\bar{d} = 1, 2$ and 3 were presented earlier, equations (5.1.36) to (5.1.38).

The parameter λ could be interpreted as follows. The ratio $r_0^2/D = \tau_D$ is a distinctive time necessary for a particle's passage over the distance r_0. If the production rate is p, $\tau_p = 1/pv_0$ gives a mean time between two defect births in given volume v_0. Thus, the quantity $\lambda = \tau_p/\tau_D$ is a ratio of these two distinctive times demonstrating which of two effects – defect migration or its production – is predominant.

The concentration time-development is a sum of terms (7.1.50) and (7.1.60), thus giving us the kinetic equation sought for (a set of equations for the correlation functions is derived quite similarly)

$$\frac{dU(t)}{dt} = (1-\omega)(1-\delta_A-\delta_B) - \lambda \bar{d} K(t) U^2(t). \qquad (7.1.63)$$

The linear approximation leading to equations (7.1.57) and (7.1.58) neglects the aggregation effect (which manifests itself in an increase in time

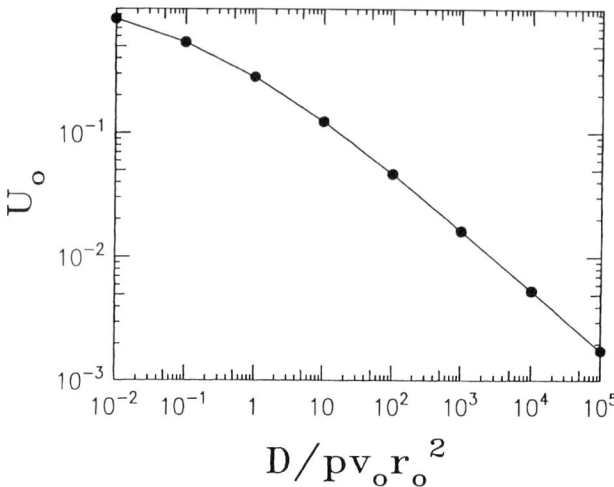

Fig. 7.5. Dependence of the saturation concentration ($\bar{d} = 3$, $D_A = D_B$) upon the dimensionless generation-recombination parameter $\lambda = D/(pv_0 r_0^2)$ (in double logarithmic units). As $\lambda \to \infty$, $U_0 \propto \lambda^{-1/2}$.

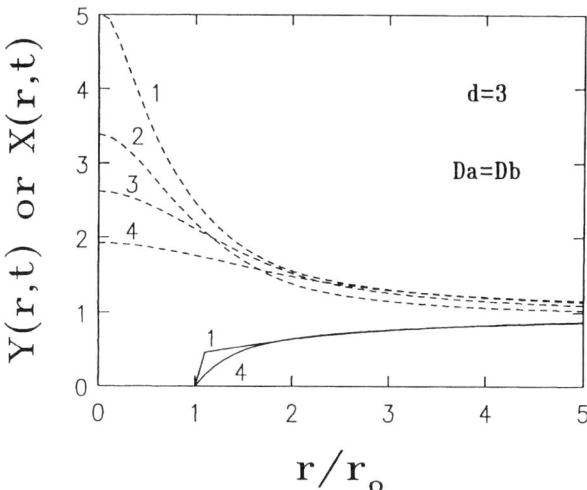

Fig. 7.6. The stationary joint correlation functions as a function of the parameter λ. Curves 1 to 4 correspond to $\lambda = 0; 10^{-1}; 1$ and 10, respectively. $\bar{d} = 3$.

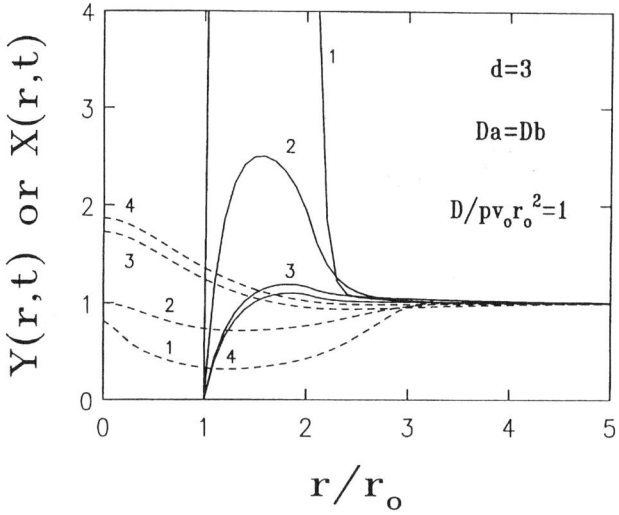

Fig. 7.7. Same as Fig. 7.4 but with $\lambda = 1$.

of the correlation length ξ). In fact, it assumes that parameters δ_A and δ_B are small, but $K(t)$ strives for the steady-state value. At last, expansion in a series in a small parameter λ^{-1} is employed. In reality, in low-dimensional systems defect segregation leads to the decrease of the gradient of the correlation function $Y(r, t)$ at $r < \xi$. The dimensionless reaction rate $K(t)$, being proportional to the gradient of $Y(r, t)$ (see equation (7.1.60)) decreases also monotonously in time which makes the steady-state achievement impossible. This effect is independent from λ value, the latter determines only the distinctive time of segregation.

To illustrate what was said above, in Fig. 7.5 the stationary concentration U_0 for $\bar{d} = 3$ is plotted vs. parameter λ. (There is no macroscopic segregation here.) For the small λ magnitudies (slow diffusion) U_0 strives for its limit known for immobile defects, $U_0 = 1.02$, but as $\lambda \gg 1$, its linear decrease is observed with the slope -0.50. It is in good agreement with equation (7.1.58) of the linear approximation.

The supression of the microscopic defect segregation by strong diffusion is well-observed for $\bar{d} = 3$ – see Fig. 7.6. The main effect is seen for the correlation function of *similar* (rather than dissimilar) defects. At last, a comparison of Figs 7.7 and 7.4 demonstrates an effect of the diffusion upon the correlation functions. It acts to smooth peaks of $X(r,t)$ supressing not only their peculiarity at small r but also peak amplitudes due to the geminate defect correlation.

Note that another consistent approach to the problem of mobile particle accumulation is based on the *field-theoretical formalism* [15, 37, 51]. However by two reasons this approach is not useful for the study of *immobile* particle aggregation: (i) the smallness of the parameter $U(t) = n(t)v_0 \ll 1$ is widely assumed there, (ii) the "free" Hamiltonian turns out to be just zero for immobile particles.

One more formalism used for the solving of this problem is *scaling* [52, 53]. The conclusion was drawn that for $\bar{d} = 1$ and 2 aggregation occurs but it is not true for $\bar{d} = 3$ which contradicts all findings above. The possible explanation is that scaling approach is not sensitive to those non-Poisson fluctuations where a system remains *macroscopically* homogeneous (i.e., the correlation length ξ is finite).

Our findings could be compared with simple estimates made in [54] using the "box" model (to be discussed in detail in Section 7.4) where an annihilation radius is less than lattice spacing and a particle A(B) disappears only if it is created on a site occupied by dissimilar particle B(A); if it landed on some empty site its survival does not depend on types of neighbours surrounding it: reactions occuring on different sites are now *independent* which allows us to get a very simple analytical solution. Since particle creation rate p is the same for A's and B's and these particles appear independently at each site and are uncorrelated in geminate pairs, the mean number of A(B) particles deposited at site is $\langle N_A(t) \rangle = \langle N_B(t) \rangle = N(t) = pt$. The deposition obeys the Poisson statistics; at long times (following the central-limit theorem) the distribution $\eta(N)$ of both N_A and N_B is Gaussian, with the dispersion

$$\sigma^2 = \langle N_A^2 \rangle - \langle N_A \rangle^2 = \langle N_B^2 \rangle - \langle N_B \rangle^2 = N(t). \tag{7.1.64}$$

Since particles A and B do not co-exist simultaneously at one site, the total number N_0 of particles at a site equals $|\delta|$, where $\delta = N_A(t) - N_B(t)$. The total concentration of similar particles, $n(t) = 0.5\langle|\delta|\rangle$. Therefore,

$$n(t) = \left(\frac{\sigma}{\pi}\right)^{1/2} = \left(\frac{pt}{\pi}\right)^{1/2}, \tag{7.1.65}$$

i.e., the predicted *infinite* growth has the algebraic character, with the critical exponent $\alpha = 0.50$. The relevant study of the one-dimensional accumulation of mobile particles yield $\alpha = 0.25$ for non-interacting particles and 0.20 for the screened Coulomb interaction [55, 56]. On the other hand, computer simulations [54] carried out on a one-dimensional discrete lattice have confirmed the algebraic character of the accumulation kinetics with the critical exponent $\alpha \approx 0.30$ obtained for the annihilation radius r_0 ranging between 1 to 20 lattice sites, which is close to the prediction $\alpha = 0.25$ [55] and our finding, $\alpha = 0.20$ (Fig. 7.1) for a continuum model.

To conclude this Section, we would like note that a rigorous theory of many-particle effects in the kinetics of the Frenkel defect accumulation restricted by their annihilation and diffusion has been just presented above. The direct use of Kirkwood's superposition approximation results in the truncation of the infinite sign-alternating series of the correlation functions describing the defect spatial distribution, and thus in considerable underestimate of the defect concentration, as $t \to \infty$. To avoid this shortcoming, a special summation technique has been developed. It is demonstrated that a real saturation of defect concentrations in continuum model takes place after prolonged irradiation only for the spatial dimension $\bar{d} = 3$. It is accompanied by a similar (A–A, B–B) defect clustering in a form of compact aggregates (domains), each containing defects A or B only. Typical size of an aggregate is $10r_0$ only, where r_0 is a recombination radius. Such self-organization allows to accumulate defect concentrations by a factor 3–4 exceeding those for a random AB pair distribution. For low-dimensional systems ($\bar{d} = 1$ and 2) such as surfaces and capillars there is no real saturation which is accompanied by *macroscopic* defect segregation. Diffusion acts to smooth the defect aggregation but does not change qualitatively the pattern (cluster) formation.

The obtained analytical results create a solid basis for the interpretation of experimental data on defect irradiation kinetics in solids of arbitrary nature, as well as in the low-dimensional systems [15, 57].

7.2 DIFFUSION-CONTROLLED PARTICLE AGGREGATION UNDER PERMANENT SOURCE

> There was no "One, two, three, and away," but they began running when they liked and left off when they liked, so that it was not easy to know when the race was over.
>
> Lewis Carroll

The formalism presented in Section 7.1 is generalized here by incorporating the elastic attraction between similar particles (defects) which causes their efficient diffusion-controlled aggregation. It is shown that the aggregation process starts only if both the dose rate p and the elastic attraction energy exceed certain critical values; the aggregation takes place in the limited temperature interval in agreement with experimental data (e.g., known for NaCl crystals). The obtained results are compared with experimental data for this crystal and the mesoscopic theory.

Of special interest in the recent years was the kinetics of defect radiation-induced aggregation in a form of *colloids*; in alkali halides MeX irradiated at high temperatures and high doses bubbles filled with X_2 gas and metal particles with several nanometers in size were observed [58] more than once. Several theoretical formalisms were developed for describing this phenomenon, which could be classified as three general categories: (i) macroscopic theory [59–62], which is based on the rate equations for macroscopic defect concentrations; (ii) mesoscopic theory [63–65] operating with space-dependent local concentrations of point defects, and lastly (iii) discussed in Section 7.1 microscopic theory based on the hierarchy of equations for many-particle densities (in principle, it is infinite and contains complete information about all kinds of spatial correlation within different clusters of defects).

More information about these approaches and their advantages readers could find in [14, 15, 64, 65]; in this Section 7.2 we focus on the further improvement of the *microscopic* approach to the defect aggregation via taking into account *elastic attraction* between point defects.

7.2.1 The basic kinetic equations

We consider in this Section the effects of diffusion-controlled particle aggregation, which occurs when the particles' mobility is high enough which allows us to neglect the contribution of recombination of particles A due to their birth directly within the annihilation sphere around pre-existing dissimilar particles B (and vice versa) compared to the result of diffusive dissimilar-particle drift to each other. Using the dimensionless units $\zeta = \lambda^{-1} = p v_0 r_0^2 / D$, and $r' = r/r_0$, $\mathfrak{D}_\nu = 2D_\nu/D$, $t' = Dt/r_0^2$, and $n'(t) = n(t)v_0$, we arrive at the simple equations for the uncorrelated particle creation when at their birth no geminate (correlated) AB pairs are formed:

$$\left.\frac{dn(t)}{dt}\right|_{cr} = \zeta, \qquad (7.2.1)$$

$$\left.\frac{\partial Y(r,t)}{\partial t}\right|_{cr} = \frac{2\zeta}{n(t)}(1-Y(r,t)), \qquad (7.2.2)$$

$$\left.\frac{\partial X_\nu(r,t)}{\partial t}\right|_{cr} = \frac{2\zeta}{n(t)}(1-X_\nu(r,t)), \quad \nu = A, B. \qquad (7.2.3)$$

The relevant contributions to the kinetic equations (7.2.1) to (7.2.3) from particle diffusion and recombination are (cf. [14] and Chapter 5)

$$\left.\frac{dn(t)}{dt}\right|_{\text{diff+rec}} = -3K(t)n^2(t), \qquad K(t) = \left.\frac{\partial Y(r,t)}{\partial r}\right|_{r=1}, \qquad (7.2.4)$$

$$\left.\frac{\partial Y(r,t)}{\partial t}\right|_{\text{diff+rec}} = \nabla\big(\nabla Y(r,t) + Y(r,t)\nabla\mathfrak{U}(r,t)\big) -$$

$$- 3K(t)n(t)Y(r,t)\sum_{\nu=A,B}J[X_\nu], \qquad (7.2.5)$$

$$\left.\frac{\partial X_\nu(r,t)}{\partial t}\right|_{\text{diff+rec}} = \mathfrak{D}_\nu\nabla\big(\nabla X_\nu(r,t) + X_\nu(r,t)\nabla\mathfrak{U}_\nu(r,t)\big) -$$

$$- 6K(t)n(t)X_\nu(r,t)J[Y]. \qquad (7.2.6)$$

Here, $K(t)$ is a reaction rate and the functional $J[Z]$ for $\bar{d}=3$ is given by equation (5.1.38) whereas the quantities $\mathfrak{U}(r,t)$ and $\mathfrak{U}_\nu(r,t)$ in these equations are the *mean-force potentials* defined by equations (5.1.45) to (5.1.47).

As it was discussed in Chapter 3, neutral point defects in all solids interact with each other by the elastic forces caused by overlap of deformation fields surrounding a pair of defects. These forces are effectively *attractive* for both similar and dissimilar defects (interstitial–interstitial, vacancy–vacancy and interstitial–vacancy, respectively) and decay with the distance between defects as

$$U_{AA}(r) = -\frac{\alpha_A}{r^3}, \qquad U_{AB}(r) = -\frac{\alpha}{r^3}. \qquad (7.2.7)$$

Due to discreteness of a crystalline lattice we have to cut-off elastic interactions at the short relative distances of the order of interatomic distances or r_0

[66]: $U_{AA}(r \leqslant r_0) = U_{AA}(r_0)$, $U_{AB}(r \leqslant r_0) = U_{AB}(r_0)$. Lastly, introducing two distinctive dimensionless parameters

$$r_A = \left(\frac{\alpha_A}{k_B T}\right)^{1/3}, \qquad r_E = \left(\frac{\alpha}{k_B T}\right)^{1/3}, \qquad (7.2.8)$$

the dimensionless interaction potentials, $U(r) \equiv U(r)/(k_B T)$, similar to those used in Section 6.3 finally read as

$$U_{AA}(r) = \begin{cases} -\left(\dfrac{r_A}{r_0}\right)^3, & r \leqslant 1, \\ -\left(\dfrac{r_A}{r_0}\right)^3 \dfrac{1}{r^3}, & r > 1, \end{cases}$$

$$U_{AB}(r) = \begin{cases} -\left(\dfrac{r_E}{r_0}\right)^3, & r \leqslant 1, \\ -\left(\dfrac{r_E}{r_0}\right)^3 \dfrac{1}{r^3}, & r > 1. \end{cases} \qquad (7.2.9)$$

The accumulation kinetics under study is defined by *three* dimensionless parametrs: $\zeta, r_A/r_0$ and r_E/r_0; all three depend on the temperature. The physical meaning of the former parameter has been discussed earlier, in Section 6.3 and Section 7.1. The joint correlation functions X_A, X_B characterize, as before, an aggregation effect of similar particles; their random distribution is taken as the initial condition, $X_A(r,0) = X_B(r,0) \equiv 1$.

In the limiting case of small particle concentrations one can neglect the non-linear terms in equation (7.2.6) which gives us the simplest estimate of the correlation function for similar particles. At relatively long times it reaches the steady-state and gives just the Boltzmann distribution, equation (6.3.4), $X_A^0(r, \infty) \approx \exp(-U_{AA}(r))$. The increasing in time deviation of functions $X_A(r,t)$ above the asymptotic value, observed more than once in previous Chapters means formation of the *non*-Poissonian particle distribution via enriched concentration of pairs of similar particles, AA, or BB, at their short relative distances, which means nothing but their *aggregation*.

7.2.2 Aggregate characterization

To characterize aggregation of similar particles, it is convenient to use not only their correlation functions, but an integral quantity, characterizing the number of particles in aggregates. Indeed, the value of $C_A(r,t) = n_A(t) X_A(r,t)$ has the physical sense of the average concentration of particles A at the distance r from another particle A in the origin of coordinates. Therefore, the quantity $\delta C_A(r,t) = n(t)(X_A(r,t)-1)$ gives us the *surplus* in the concentration with respect to the Poisson (random) particle distribution. This is why the integral quantity

$$\langle N \rangle = \int \delta C_A(r,t)\,\mathrm{d}\vec{r} \qquad (7.2.10)$$

gives us the *average number of particles* A in an aggregate surrounding arbitrary chosen central particle A. Calculations demonstrate a sharp maximum of $X_A(r,t)$ around $r \approx r_0 = 1$ which makes a major contribution to $\langle N \rangle$. Typically, at larger distances X_A drops a little bit *below* its asymptotic value of unity and then slowly approaches it from below; therefore the distinctive distance r^* where $X_A(r^*,t) = 1$ could be defined as the *aggregate boundary*. The quantity of $\langle N \rangle$ is convenient to complement by another distinctive characteristic which is called *the aggregate size*,

$$\langle R \rangle = \frac{1}{\langle N \rangle} \int r \delta C(r,t)\,\mathrm{d}\vec{r}. \qquad (7.2.11)$$

If no aggregation occurs, the joint correlation function coincides with equation (6.3.4) and the value of the integral $\langle R \rangle$, equation (7.2.11), is divergent since as $r \to \infty$, $\delta C(r,t) \propto r^{-3}$ whereas $\mathrm{d}\vec{r} \propto r^2$ only. (Due to a limited integration interval we obtain in actual numerical calculations not infinity but big numbers, of the order of 10^2.) However, as soon as the aggregation process becomes important, the correlation function $X_A(r,t)$ drops much faster than equation (6.3.4), the integration interval shortens down to the values close to the recombination radius r_0 and thus the quantity $\langle R \rangle$ turns out to be a much smaller value, of the order of unity (in units of r_0). Therefore, only the parallel study of the time-development of *both* quantities – $\langle N \rangle(t)$ and $\langle R \rangle(t)$ – gives us a complete picture of the aggregation process.

Analytical calculations of the aggregation process kinetics in terms of this approach are presented in [67, 68]. It is shown there that the aggregation occurs only if the distinctive dimensionless parameter ψ is positive:

$$\psi = \frac{\alpha_A}{k_B T r_0^3} + \frac{1}{4} \ln\left(\frac{p v_0 r_0^2}{D}\right) \geq 0. \tag{7.2.12}$$

It gives the *critical dose rate* (irradiation intensity) p_c as a function of the elastic interaction between similar defects and the temperature:

$$p_c = \frac{3 D_0}{4\pi r_0^5} \exp\left(-\frac{E_I}{k_B T} - \frac{4\alpha_A}{k_B T r_0^3}\right),$$

$$D = D_I + D_V \approx D_I \tag{7.2.13}$$

and $A = v, I$ (vacancy or interstitial).

The critical dose rate p_c necessary for initiating the aggregation process is the smaller, the lower the temperature, the stronger elastic attraction of similar particles and the slower the diffusion (greater the activation energy for hopping). This conclusion is in a complete qualitative agreement with the results obtained recently in terms of a quite different mesoscopic approach [63–65].

To get the reference point for many-particle effects in defect accumulation, the saturation concentration, $n(t \to \infty)$ could be taken from equation (7.1.58) with $U_0 \ll 1$, i.e., in the *linear* approximation neglecting many-particle effects

$$n_0 = n(\infty) = \sqrt{\frac{p}{4\pi D r_0}} \approx \sqrt{\frac{\zeta r_0}{3 r_E}}. \tag{7.2.14}$$

Note that this concentration is governed by a *sum* of diffusion coefficients, $D_I + D_V$, which practically equals D_I since vacancies have much lower mobilities compared to interstitials. In the temperature range of high mobilities the value of r_E usually exceeds r_0 not more than a factor 2–3 and $\xi \ll 1$, that is, $n_0 \approx 1/3\sqrt{\xi} \ll 1$. In the regime of intensive aggregation one can expect n_0 to be considerably lower than this estimate due to disappearance of single defects in their aggregates.

The general theory just discussed will be illustrated below for the particular set of parameters distinctive for NaCl crystals for which a lot of aggregation studies were done (see [60–62] for more detail). However, if one uses, as a test, equation (7.2.12) in order to estimate the just-mentioned reference point – the saturation concentration without the considerable aggregation – he immediately finds for the standard set of NaCl parameters (the activation energy of interstitial hops $E_\mathrm{I} = 0.1$ eV [9, 12], and $D_0 = 10^{-3}$ cm^2 s^{-1}, $r_0 \approx 3$ Å, $p = 7 \times 10^{12}$ cm^{-3} s^{-1}) that at $T = 15°\mathrm{C}$ and $85°\mathrm{C}$ the calculated (theoretical) concentrations are $n_0^\mathrm{th} = 10^{12}$ cm^{-3} and 6.8×10^{11} cm^{-3}, whereas experimental values are much higher [69]: 6×10^{16} cm^{-3} and 10^{16} cm^{-3}, respectively! It is strange that this puzzle has never been discussed in the extensive theoretical literature devoted to the F-centre aggregation [58–62, 69–81]! Recently [65], an explanation of this fact has been suggested assuming that the effective diiffusion coefficient for mobile interstitials is reduced considerably due to their periodical trapping by radiation-induced ($v_\mathrm{a}v_\mathrm{c}$) di-vacancies observed experimentally in [82]. It results in the *effective* activation energy in NaCl of $E_\mathrm{I} = 0.37$ eV and $D_0 = 2.6 \times 10^{-8}$ cm^2 s^{-1}, which approximate very well the experimental data obtained in the non-aggregation regime. (For vacancies $E_\mathrm{V} = 0.9$ eV.) We concentrate now on the F centre aggregation for which much more experimental information is available. As it is known from the calculations for alkali halide crystals parameters [83], typical values of the elastic interaction for two interstitials and interstitial–vacancy pair are: $\alpha_\mathrm{I} = 15$ eV Å^3 and $\alpha = 5$ eV Å^3 respectively. The elastic interaction energy between two vacancies is not exactly known and we accept for our semiqualitative calculation done below the very moderate value of $\alpha_\mathrm{V} \approx 5$ eV Å^3.

7.2.3 Results

Figure 7.8 shows the joint correlation function for vacancies for a fixed dose rate of $p = 10^{17}$ cm^{-3} s^{-1} but varying the temperature from 0°C to 150°C. The two low-temperature curves – at 0°C and 50°C – decay monotonously with r, as it is predicted by equation (6.3.4). However, as the temperature increases further, curves 3 and 4 reveal a new type of behaviour – they fall *below* the asymptotic value of $X = 1$ and then approach it slowly, as $r \to \infty$. Physically it means that the aggregate of similar defects exists which adsorbs intensively similar defects near it due to which their concentration is reduced *below* the statistical level. The distinctive aggregate's size defined as $X(R^*) = 1$, is about $R^* = (1.5\text{–}2.0)r_0$ for the temperature ranging between 100°C and 150°C. Note also that the correlation curve 4 goes *below* curve 3 at short distances which means that at such high temperatures the aggregation of very mobile vacancies is *less* efficient. In other

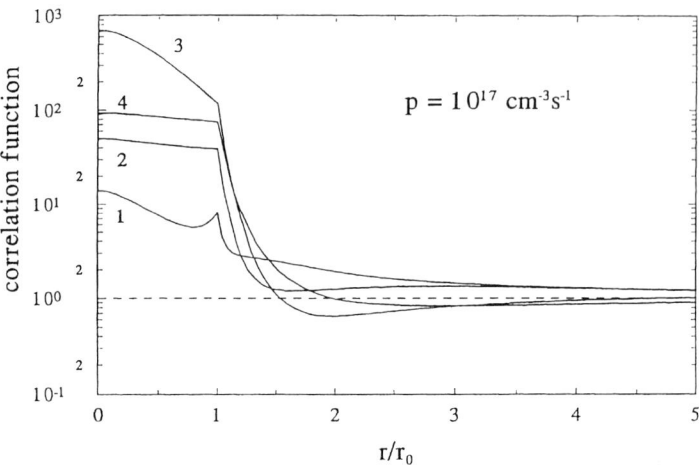

Fig. 7.8. The joint correlation functions $X(r)$ at $t = 10^{10}$ s for similar defects (v–v) in NaCl for the temperatures 0°C (curve 1), 50°C (curve 2), 100°C (curve 3) and 150°C (curve 4). Note that only in curves 3 and 4 the aggregation really takes place: the correlation function crosses the asymptotic value of $X = 1$ (see text). The elastic interaction energy $\alpha_v = 5$ eV Å^3.

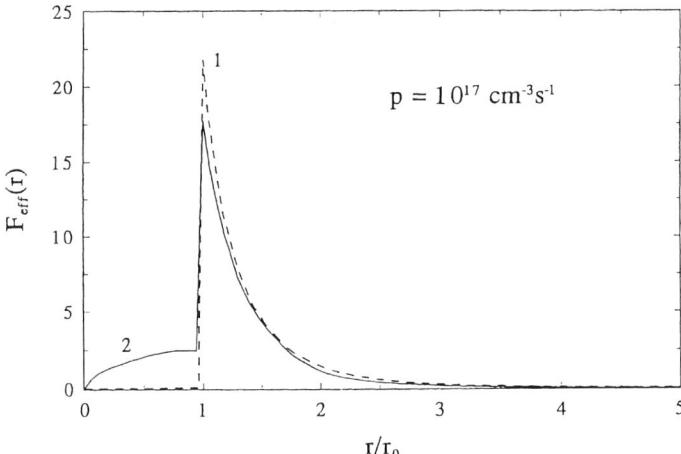

Fig. 7.9. The effective attraction force between vacancies. Curve 1: $T = 0$°C, no aggregation; curve 2: $T = 100$°C, aggregation occurs (note an additional attraction at short distances).

words, there exists the *optimal temperature* for defect aggregation which is discussed in detail below.

Figure 7.9 characterizes the effective attraction between vacancies with and without their aggregation. In the aggregation regime an *additional* short-range force arises (curve 2) which has many-particle nature and is caused by the non-linear term in the mean-force potentials, equations (5.1.45) to (5.1.47), at $r < 1$.

Our microscopic formalism cannot give the distribution function over aggregate *sizes*, but having calculated a mean number of particles within aggregate, equation (7.2.8), we can assume approximate validity of the Poisson distribution (for the moderate aggregation efficiency)

$$\Phi(m) = \exp\left(-\langle N \rangle\right) \frac{\langle N \rangle^m}{m!}, \qquad (7.2.15)$$

and thus calculate concentration of dimers (F_2) defects: $n(F_2) = n(F)\Phi(1)$, trimers; $n(F_3) = n(F)\Phi(2)$ etc. From this point of view, it is reasonable to define as aggregates only those defect clusters where a fraction of dimers is not very small compared to the concentration of single vacancies. Say, if $n(F) = 10^{18}$ cm^{-3}, $n(F_2)$ should be $\geqslant 10^{15}$ cm^{-3}, i.e., $\langle N \rangle \geqslant 10^{-3}$. Such a definition corresponds to a very weak, *marginal* aggregation process which is of our primary interest here.

The F-centre saturation concentrations as a function of the temperature and dose rate are plotted in Fig. 7.10. In a line with general theory, equation (7.2.12), it is higher for large dose rates and low temperatures. What is non-trivial here is the *aggregation region* which lies between the two dashed lines. The left line corresponds to the above-given aggregate definition, $X(R^*) = 1$. The right dashed line corresponds to the second condition discussed above – the concentration of dimer centres, F_2, should not be three and more orders of magnitude less than that for single F centres. This conclusion about the existence of the temperature optimal for the aggregation for a given dose rate is in agreement with our analysis of the joint correlation functions – Fig. 7.8 – and the actual experimental data on NaCl crystals [84].

A large body of experiments (e.g., [70, 73, 74]) deals with the kinetics of *dimer* centre accumulation in alkali halide crystals; it was observed experimentally that $n(F_2)$ increases often as squared single-vacancy concentration, $n^2(F)$. In our formalism, $n(F_2) \approx \langle N \rangle n(F)$. As it is well seen in Fig. 7.11, $\langle N \rangle$ turns out to be a linear function of $n(F)$; in a wide range of defect

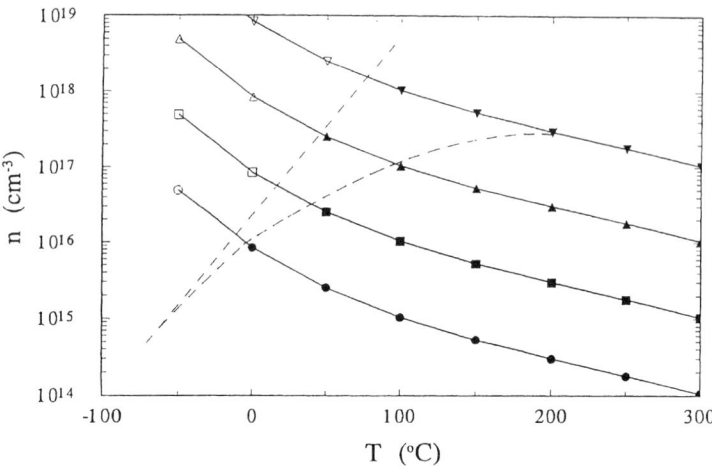

Fig. 7.10. The saturation F centre concentrations as a function of the temperature for four different dose rates: $p = 10^{11}$ cm^{-3} s^{-1} (●), 10^{13} cm^{-3} s^{-1} (■), 10^{15} cm^{-3} s^{-1} (▲) and 10^{17} cm^{-3} s^{-1} (▼). The two dashed lines show the temperature interval in which the aggregation takes place (see text).

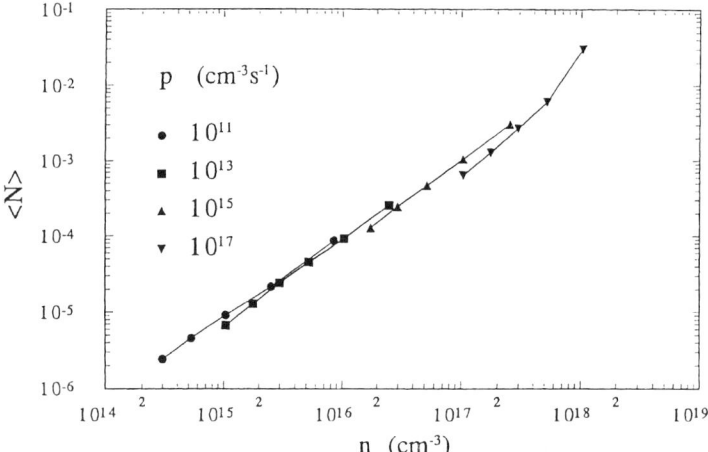

Fig. 7.11. The dependence of the mean number of particles in the aggregate as a function of the single-particle concentration.

Diffusion-controlled particle aggregation under permanent source

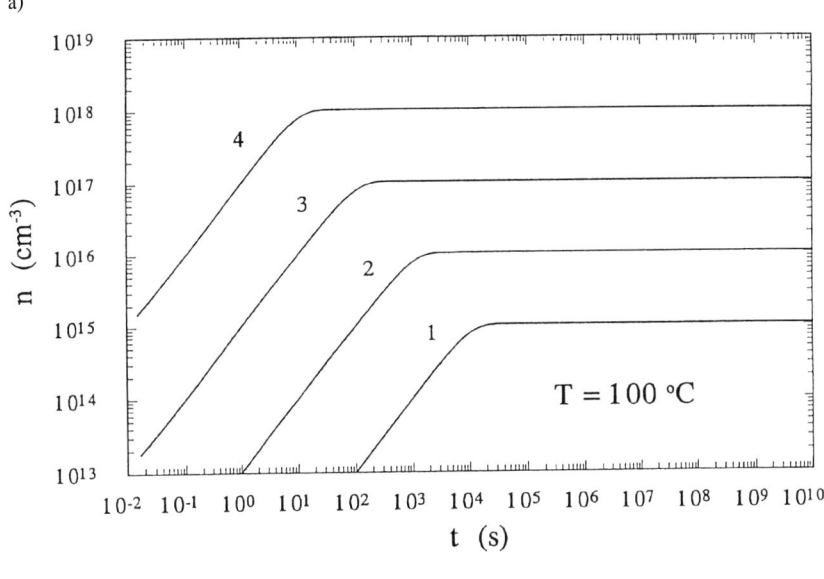

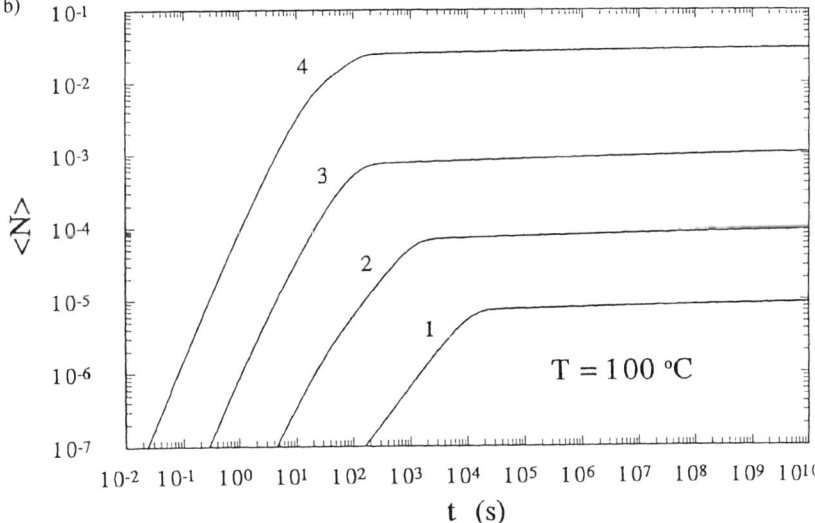

Fig. 7.12. (a) Particle concentration as a function of time for the dose rates $p = 10^{11}$ cm^{-3} s^{-1} (curve 1), 10^{13} cm^{-3} s^{-1} (curve 2), 10^{15} cm^{-3} s^{-1} (curve 3) and 10^{17} cm^{-3} s^{-1} (curve 4). (b) The same for mean numbers of particles in aggregates; note that these curves do not saturate but continue to grow up slowly.

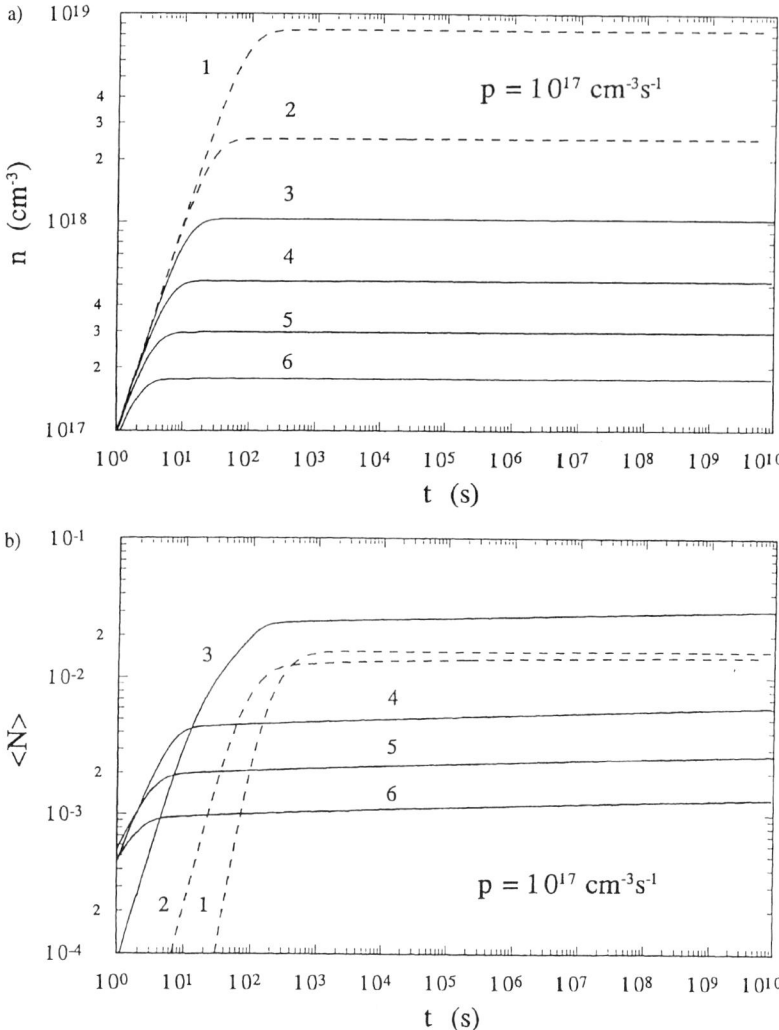

Fig. 7.13. (a) Particle concentration as a function of the temperature: 0°C (curve 1), 50°C (curve 2), 100°C (curve 3), 150°C (curve 4), 200°C (curve 5) and 250°C (curve 6). The dashed line indicates that no aggregation takes place at these temperatures (for a given dose rate of $p = 10^{17}$ cm^{-3} s^{-1}). (b) The same for the mean number of particles in aggregates. Dashed curves are obtained by integrating over sphere with the radius $2r_0$. Note that the aggregation reaches its maximum at 100°C (curve 3). Full curves do not saturate with time-slow aggregation continues.

concentrations and dose rates, $\langle N \rangle \approx (10^{-20} \text{ cm}^{-3}) \times n(F)$. The slope of this curve only slightly depends on the elastic attraction energy α_ν between F centres. Deviation from the quadratic law begins only at quite high defect concentrations, $n \geqslant 10^{18} \text{ cm}^{-3}$. These findings confirm the results of earlier rate-equation model [70–73].

The time development of the F centre concentration, up to saturation, and the mean number of particles in aggregates, are shown in Fig. 7.12 (a) and (b). The n_0 values are well described by equation (7.2.14), valid for not strongly aggregated systems; $n_0 \propto \sqrt{p}$. The concentrations are saturated at time $\tau_0 \propto 1/\sqrt{p}$.

Figure 7.13 shows similar dependence for a fixed dose rate but varying the temperature (F centre mobility). As it is expected from equation (7.2.14), the higher the temperature, the more intensive defect recombination is and thus the lower the saturation concentration. Curve 3 in Fig. 7.13(b) confirms once more that $T \approx 100°C$ is the optimal aggregation temperature for a given dose rate of $10^{17} \text{ cm}^{-3} \text{ s}^{-1}$ whereas at higher temperatures a portion of F_2 centres decreases.

Lastly, a decisive role of the elastic attraction between F centres in their agregation is seen in Fig. 7.14. For small attraction energy α_V (curve 1)

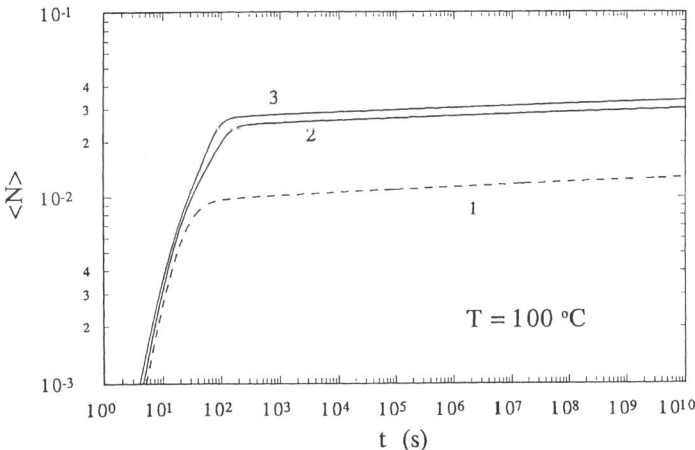

Fig. 7.14. A critical role of the elastic interaction energy between particles: no aggregation occurs for $\alpha = 4 \text{ eV Å}^3$ (dashed curve 1), but it is triggered for larger values of $\alpha = 5 \text{ eV Å}^3$ (curve 2) and 8 eV Å^3 (curve 3).

aggregation does not hold even at such great dose rates as $p = 10^{17}$ cm^{-3} s^{-1}. However, as far as the α_V exceeds some (quite moderate) critical value about 5 eVÅ^3, a number of dimer centres is increased immediately by a factor of 2–3 (curves 2 and 3).

7.2.4 Discussion

As it has been demonstrated in Section 7.2, the microscopic many-point density approach to a system of dynamically interacting particles predicts the existence of the *critical value of elastic attraction* energy between similar particles (which in this particular case are defects – F-centres) – it triggers the aggregation process. The further accumulation kinetics, when the aggregation started to occur, depends only slightly on the interaction energy α_V. Another important conclusion confirming results of other macroscopic studies [60, 62] is that the aggregation occurs only in the *limited temperature range* – at low temperatures it disappears due to very slow mobility of particles whereas at high temperatures relatively weak elastic interaction can no longer keep highly mobile particles together in the aggregate. On the other hand, for a given temperature there exists a *certain critical dose rate* necessary for the aggregation process initialization, and it increases with the temperature increase [73, 81].

All these findings are in agreement with both experimental data for NaCl crystals and the recent *mesoscopic* studies [65, 68] where the critical dose rate was found to be

$$\ln p_c \geqslant \ln D + 2\ln \frac{k_B T}{\Delta_{el}} + \ln\left[\left(1 + \frac{D_V}{D_I}\right)^2 4\pi r_0\right], \qquad (7.2.16)$$

where Δ_{el} is defined by the combination of the elastic interaction energies between similar and dissimilar defects (V–I):

$$\Delta_{el} = \frac{4}{3}\pi\left[\alpha_I + \frac{D_V}{D_I}\alpha_V - \left(1 + \frac{D_V}{D_I}\right)\alpha\right]. \qquad (7.2.17)$$

On the other hand, the present microscopic approach leads to the critical dose rate arising from equation (7.2.13):

$$\ln p_c \geqslant \ln D - \frac{4\alpha_A}{k_B T r_0^3} + \ln \frac{3}{4\pi r_0^5}. \qquad (7.2.18)$$

Both equations (7.2.16) and (7.2.18) have the same dependence on the relative diffusion coefficient, $D = D_\mathrm{I} + D_\mathrm{V}$, but different dependence on the elastic interaction between defects. However, in *both* cases the stronger similar defect attraction, the lower is the critical dose rate. In the mesoscopic approach this effect is less pronounced (logarithmic vs. linear dependence) and here p_c is considerably higher. It seems that this approach is able to detect only those mesoscopic-size aggregates which are already well-developed – unlike the microscopic formalism able to detect even the marginal aggregation effects.

Lastly, the present theory confirms the quadratic dependence between the concentration of dimer F_2 centres (also called M centres [9, 12]) and concentration of single F centres – observed more than once experimentally in alkali halide crystals [70, 73, 75].

It should be noted in conclusion of this Section that preliminary results obtained by means of the discrete lattice formalism are presented in [85]. This study demonstrates clearly the cooperative nature of the aggregation of two kinds of the Frenkel defects, vacancies and interstitials.

7.3 OTHER APPROACHES

> In a theory which has given results like these, there must certainly be a great deal of the truth.
>
> H.A. Lorentz

Alongside the refined many-point density formalism discussed in Sections 7.1 and 7.2, simple combinatoric approaches and computer simulations to be discussed below, in Section 7.4, several other semi-quantitative analytical approaches were presented for the treatment of the $A + B \to 0$ accumulation kinetics with mobile reactants. Using scaling arguments, similar to the pioneering papers by Kang and Redner [86, 87], *macroscopic* segregation was predicted in [52] with the typical aggregate size

$$\xi \propto D^{2-\bar{d}} p^{-2/(2-\bar{d})}, \quad \bar{d} < 2,$$

$$\xi \propto \exp(-\mathrm{const} \cdot D p^{-2}), \quad \bar{d} = 2, \tag{7.3.1}$$

where p is irradiation intensity (dose rate). Note that these aggregates are stable in time. This conclusion has been confirmed in the study [53]. Note that this kind of *scaling* analysis [52, 53] is suitable for observing *macroscopic* aggregation effects only; as noted in [53], "it is not sufficient to determine whether or not observable segregation will occur, but it suggests greater likelihood of segregation for decreasing dimensionality $\bar{d}$". We would like to stress here that our previous analysis (Sections 7.1 and 7.2) also argues for the stronger segregation effects in low dimensions. This is why, the careful analysis of spatial particle correlations, e.g., in terms of the joint correlation functions, is necessary for making a conclusion whether aggregation occurs or not. Otherwise looking at snapshots of particle distribution, readers are invited to decide themselves whether segregation takes place (see, e.g., [88]).

7.3.1 Particle accumulation on fractals

Sometimes, as in the case of *particle segregation on fractals* (e.g., the planar Sierpinski gasket discussed in Section 6.1) this effect indeed is self-evident [88–90]. Its analytical treatment for particle accumulation was presented in [91, 92]; we reproduce here simple mesoscopic estimates following these papers. Particle concentrations obey the kinetic equations

$$\frac{\partial C_A(x,t)}{\partial t} = \mathfrak{D}\Delta C_A(r,t) - R(C_A, C_B) + \varphi_A,$$

$$\frac{\partial C_B(x,t)}{\partial t} = \mathfrak{D}\Delta C_B(r,t) - R(C_A, C_B) + \varphi_B, \qquad (7.3.2)$$

where $C_A(x,t)$, $C_B(x,t)$ are local concentrations of the particles of A and B type, $\mathfrak{D}$ is a *fractal* difusion constant (measured in units of $L^{2+\theta}/T$), Δ the fractal Laplace operator, φ_A, φ_B are flows of produced A and B particles, R is an operator that describes the "particle" mutual annihilation, θ is a anomalous diffusion exponent. The explicit form of the R operator is unknown. Fortunately, this explicit form is not necessary for our estimates.

As it was shown in [16, 93] the concentration of particles is determined mainly by the behaviour of the difference of the species concentrations $Z = C_A - C_B$ which satisfies the linear kinetic equation

$$\frac{\partial Z}{\partial t} = \mathfrak{D}\Delta Z + \varphi, \quad \varphi = \varphi_A - \varphi_B. \qquad (7.3.3)$$

In computer experiments [88, 89] the total numbers of A and B particles created in a system per unit time are equal, so $\langle Z(x,t)\rangle = 0$. Its second moment $\langle Z^2 \rangle$ in the infinite system is given by

$$\langle Z^2 \rangle = \int_0^t d\tau_1 \int_0^t d\tau_2 \int dx\, dx' \times$$
$$\times G(x, t - \tau_1) G(x', t - \tau_2) c(x - x', \tau_1, \tau_2), \tag{7.3.4}$$

where $c(x, t_1, t_2)$ is a two-point correlation function of φ and $G(x, t)$ is the diffusion operator's Green function. In the case of independent production of particles

$$c(x, t_1, t_2) = 2p\delta(x)\delta(t_1 - t_2)\theta(t_1)\theta(t_2),$$
$$p = \langle \varphi_A \rangle = \langle \varphi_B \rangle. \tag{7.3.5}$$

Assuming the full segregation of reactants

$$n(t) = \langle C_A \rangle = \langle C_B \rangle = \frac{1}{2}\langle |Z| \rangle, \quad \langle |Z| \rangle = \sqrt{\frac{2}{\pi}} \langle Z^2 \rangle^{1/2}. \tag{7.3.6}$$

The accuracy of this approximation will be estimated further. Here we shall study the behaviour of the quantity $\langle |Z| \rangle$ only.

The investigation of the behaviour of $Z(r, t)$ in a finite system is a difficult problem due to the fact that the boundary conditions for (7.3.3) are unknown for fractal systems although one can use some scaling arguments based on the knowledge of the properties of the infinite system. In the case of the independent production of different particles we obtain

$$\langle Z^2 \rangle = 2p \int dx \int_0^t G^2(x, t - \tau)\, d\tau. \tag{7.3.7}$$

The growth of $\langle Z^2 \rangle$ (the particle accumulation) is accompanied by the growth of clusters of similar particles (cf. [91]). The characteristic spatial dimension of the cluster grows according to the same law as the mean square displacement of a single particle,

$$l^2 = \delta^{2/(2+\theta)}. \tag{7.3.8}$$

Here δ is a coefficient connected with the diffusion constant $\mathcal{D}$ by the relation,

$$\delta = \lim_{t \to \infty} t^{-2(2+\theta)} \int G(x, t) x^2\, dx. \tag{7.3.9}$$

From the scaling properties of $G(x,t)$ one can derive that $\delta = \text{const}(\bar{d}, \theta)\mathfrak{D}^{\tilde{d}/\bar{d}}$ with $\tilde{d} = 2\bar{d}/(2+\theta)$ the spectral dimension of the fractal. The growth of the cluster's sizes goes on until $l \approx L$ where L is the whole system's size. The further growth of clusters and accumulation of particles stop because the same quantity L is the characteristic scale of a pair of different particles created in the system; according to [91] there is no accumulation effect when particles are created by pairs on fractals of the Sierpinski gasket type.

In a typical case, only two large clusters of the particles of A and B type survive in the system [88, 89]. The steady-state concentration takes a value corresponding to the time needed for the creation of a cluster of such a size. This characteristic time is given by the relation $\delta^{\bar{d}/2} t^{\tilde{d}/2} \approx M/2$ where M is the full volume (number of sites) of the fractal considered. According to equation (7.3.7), at large times one gets

$$\langle C_A \rangle = \langle C_B \rangle \propto p^{1/2} \delta^{-\bar{d}/4} t^{(2-\tilde{d})/4} \tag{7.3.10}$$

for the systems with $\tilde{d} < 2$, and we conclude that the steady-state concentration in a system of volume M (e.g., on the Sierpinski gasket consisting of M sites with lattice spacing $a_0 = 1$) is

$$\langle C_A \rangle = \langle C_B \rangle = \beta p^{1/2} \delta^{-\bar{d}(2/\tilde{d}-1)/4} M^{1/\tilde{d}-1/2}, \tag{7.3.11}$$

where β is a numerical constanst.

The $n_A = n_B = \langle C_A \rangle = \langle C_B \rangle \propto p^{1/2}$ dependence is in good agreement with the results of computer simulations [88, 89]. In [89] it is mentioned that the dependence of n_A, n_B on M is very weak. According to (7.3.11), for the Sierpinski gasket indeed $n_A \propto M^{0.235}$.

Now we can give some numerical estimates. The averaged Green function of the Δ operator is known to be [94]

$$G(x,t) = \frac{1}{2\pi^{\bar{d}/2}} \frac{\Gamma(\bar{d}/2)}{\Gamma(\tilde{d}/2)} \left[(2+\theta)^2 \mathfrak{D}t\right]^{-\tilde{d}/2} \times$$

$$\times \exp\left(\frac{-x^{2+\theta}}{(2+\theta)^2 \mathfrak{D}t}\right). \tag{7.3.12}$$

Using the spherical coordinates

$$dx = \frac{2\pi^{\bar{d}/2}}{\Gamma(\bar{d}/2)} x^{\bar{d}-1} dx \tag{7.3.13}$$

(the coefficient before dx is the surface of a $(\bar{d}-1)$-dimensional sphere) and the relation between $\mathfrak{D}$ and δ derived from (7.3.9), one obtains the following value of the coefficient β:

$$\beta = \frac{\pi^{\bar{d}/4} \bar{d}\Gamma[(\bar{d}+2)/(\theta+2)]^{\bar{d}/4}}{\Gamma(\bar{d}/2)^{1/2} 2^{\bar{d}/4+1/2}}. \tag{7.3.14}$$

Now we can turn to the numerical simulations carried out in [88, 89] for the Sierpinski gasket having $\bar{d} = 1.59$ and $\tilde{d} = 1.36$. The value of δ can be estimated under the assumption that the expression $\langle r^2 \rangle = \delta t^{2/(2+\theta)}$ holds up to the time of a single step. As the particle takes one step of unit length per unit time we can conlude that $\delta = 1$. The total concentration of particles is

$$n_A = n_B = \frac{1}{\sqrt{2\pi}} \beta p^{1/2} \delta^{-\bar{d}(2/\tilde{d}-1)/4} \left(\frac{M}{2}\right)^{1/\tilde{d}-1/2}. \tag{7.3.15}$$

In order to compare this result with the data of computer simulations let us introduce the number RM of particles produced in the whole system per unit time and rewrite (7.3.15) in the form

$$n_A = n_B = \frac{2^{1/(\tilde{d}-1)}}{\sqrt{\pi}} \beta (\mathrm{RM})^{1/2} \delta^{-\bar{d}(2/\tilde{d}-1)/4} M^{1/\tilde{d}-1}. \tag{7.3.16}$$

Using equation (7.3.16) one obtains for $M = 9843$ and RM $= 1, 2$ and 4, $n_A = n_B = 0.026, 0.036$ and 0.052 respectively. These values are only 10% different from those obtained in the actual computer simulations [89]: $n_A = n_B = 0.0246, 0.0340$ and 0.0470 and are even in better agreement with the values given in the earlier work [88] which indicates that the approximations made are not too rough.

In these calculations it was supposed that $n_A = n_B = \frac{1}{2}\langle |Z| \rangle$, i.e., that the cluster with abundance of A (or B) particles consists of A (or B) particles only. Now, in conclusion, we estimate the concentration of guest particles in the cluster. The mean lifetime $\tau(n)$ of a guest particle can be evaluated taking into account the compactness of the volume visited by a particle: it reacts immediately with one of the neighbouring particles of opposite type. The lifetime τ may be determined from the equation

$$\langle l \rangle \approx \gamma \delta^{1/2} \tau(n)^{1/(2+\theta)}, \tag{7.3.17}$$

where $\langle l \rangle = n^{-1/d}$ is the mean interparticle distance, γ is a numerical coefficient of order of unity (taking into account the relation between $\langle l^2 \rangle$ and $\langle l \rangle$ and the mobility of both reacting particles one can take approximately $\gamma = 2,3$ for the Sierpinski gasket). Assuming that half of produced particles hit the cluster of particles of opposite type, we can easily estimate $\delta n \approx p\tau(n)/2$. Under the conditions of the numerical simulation considered δn is less than 10% of A or B so the approximation used is quite valid.

7.3.2 Long-wavelength approximation

One can also carry out systematic treatment of the problem of aggregation of *diffusing* defects (while taking account of their interaction) by the *methods of field theory* with account taken of the fact that the problem contains a small distinctive parameter, $U_0 = nv_0 \ll 1$. The results of this method could be done conveniently presented in the *long-wavelength* approximation, which allows one to obtain graphically practically exact answers for the degree of aggregation of defects (see review paper [15] for more detail). Moreover, it has turned out that the field-theory methods and the long-wavelength approximation at low densities of defects yield results that agree very well. In the stated approximation the density of particles of types A or B can be described by the fluctuating function $C(r,t)$. This quantity is determined exactly if one divides the entire reaction volume into microscopic subvolumes containing a large enough number of particles. At equilibrium in the absence of reactions, the probability density of finding the system in a given $C(r)$ (which differs little from its mean value $\overline{C}$) is given by the Gaussian functional

$$W[C(r)] \propto \exp\left[-\frac{\int (C(r) - \overline{C})^2 d\vec{r}}{2\overline{C}}\right]. \tag{7.3.18}$$

In the presence of diffusion $C(r,t)$ obeys the standard diffusion equation

$$\frac{\partial C(r,t)}{\partial t} = D\Delta C(r,t) + \varphi_0(r,t). \tag{7.3.19}$$

Here $\varphi_0(r,t)$ is the random Gaussian function with the correlator

$$\langle \varphi_0(\vec{r},t)\varphi_0(\vec{r}',t') \rangle = \overline{C}(t)D\delta(t-t')\nabla_r\nabla_{r'}\delta(\vec{r}-\vec{r}'). \tag{7.3.20}$$

As $t \to \infty$, equation (7.3.19) implies equation (7.3.18).

Let us examine now the set of equations controlling the creation and annihilation of neutral A and B particles in the Euclidean space which have equal diffusion coefficients $D_A = D_B = D$ [93]. It has the form of equations (2.2.20) to (2.2.21). Here K is a reaction rate of bimolecular recombination; in particular, it can be equal to $K = 8\pi D r_0$. Also, $\varphi_A(\vec{r}, t)$ and $\varphi_B(\vec{r}, t)$ are the intensities of creation of A and B particles by the external source (irradiation). They have the Poisson statistical properties given by equations (2.2.24) to (2.2.26). Also $\varphi_A(\vec{r}, t)$ and $\varphi_B(\vec{r}, t)$ contain the term with "inner" stochasticity $\varphi_0(\vec{r}, t)$ of (7.3.19) and the correlator of (7.3.20). However, in the long-wavelength approximation and for a small parameter U_0, it always proves to be inessential.

The concentration difference $C_A(\vec{r}, t) - C_B(\vec{r}, t) \equiv Z(\vec{r}, t)$ obeys the equation (2.2.27). The Fourier components $Z(\vec{r}, t)$, $Z(\vec{k}, t)$, are random functions having the correlator $g(\vec{k}, t, t') = \langle Z(\vec{k}, t) Z(-\vec{k}_1, t_1) \rangle$, equation (2.2.28). For small $\vec{k}$, the quantity of $g(\vec{k}, t, t')$ becomes large, which implies separation of the A and B particles over large distances; in the three-dimensional case ($\bar{d} = 3$) this can be illustrated as follows.

For completely mixed particles in large enough volume, the square of the fluctuation of the number of particles $\delta N = N - \overline{N}$ obeys the ordinary statistical relationship

$$\frac{\overline{(\delta N)^2}}{\overline{N}^2} \propto \frac{1}{\overline{N}}. \tag{7.3.21}$$

Yet if the correlator of the number of particles looks like (2.2.28), we have *stronger* level of fluctuations:

$$\frac{\overline{(\delta N)^2}}{\overline{N}^2} \propto \frac{1}{\overline{N}^{1/3}}. \tag{7.3.22}$$

In the two-dimensional case the analogous relationship has the form

$$\frac{\overline{(\delta N)^2}}{\overline{N}^2} \propto \frac{1}{\ln(\overline{N})}. \tag{7.3.23}$$

Lastly, in the one-dimensional case we have

$$\frac{\overline{(\delta N)^2}}{\overline{N}^2} \propto 1. \tag{7.3.24}$$

Equations (7.3.23) and (7.3.24) actually imply that one- and two-dimensional cases actually exhibit already *macroscopic* separation of the system into regions consisting of only A particles and only B particles. This is also confirmed by the fact that the integral over the spectrum of spatial fluctuations diverges in the cases at small k. On the other hand, to find the aggregation of particles in numerical experiments in the *three*-dimensional case we must treat the *deviations* from the Poisson distribution in large volumes. More detailed field-theoretical formalism has confirmed this conclusion [15].

7.3.3 Single-species accumulation kinetics

As earlier in the case of diffusion-controlled concentration decay, *one-species reactions*, $A + A \to A$ and $A + A \to 0$, in one-dimension, could be used as a *proving ground* for the testing theory of particle accumulation, when random particle creation is added. For the former reaction the master equation could be easily derived in the form [95]

$$\frac{\partial w(x,t)}{\partial t} = 2D \frac{\partial^2 w(x,t)}{\partial x^2} - pxw(x,t), \tag{7.3.25}$$

with the boundary conditions

$$w(0,t) = 1, \qquad w(\infty, t) = 0. \tag{7.3.26}$$

$w(x,t)$ is probability that an interval of length x is empty (i.e., the nearest particle A in the randomly chosen point r is at a distance greater than x). This function is simply related to the time-dependent concentrations,

$$n_A(t) = -\frac{\partial w(x,t)}{\partial x}\bigg|_{x=0} \tag{7.3.27}$$

and the interparticle distribution function is

$$f(x,t) = \frac{1}{n_A} \frac{\partial^2 w(x,t)}{\partial x^2}. \tag{7.3.28}$$

The general solution of equation (7.3.25) may be expressed as a sum of the terms $e^{-\lambda t} w_\lambda(x)$ with the eigenfunctions $w_\lambda(x)$ satisfying

$$-\lambda w_\lambda(x) = 2D \frac{\partial^2 w_\lambda(x)}{\partial x^2} - pxw_\lambda(x). \tag{7.3.29}$$

This eigenvalue problem is solved by inspection – this is just Airy's equation. The properly normalized stationary ($\lambda = 0$) solution is

$$w_0(x) = \frac{\text{Ai}\left[\left(\frac{p}{2D}\right)^{1/3} x\right]}{\text{Ai}[0]}, \qquad (7.3.30)$$

where $\text{Ai}[z]$ is Airy's function, satisfying $\text{Ai}''[z] = z\text{Ai}[z]$, and, in particular, $\text{Ai}[0] = 0.35502\ldots$. The rest of the spectrum is

$$\lambda_n = |a_n|(2Dp^2)^{1/3}, \quad n = 1, 2, \ldots, \qquad (7.3.31)$$

with a_n the nth zero of Airy's function. For example, $-a_1 = 2.3381\ldots$, $-a_2 = 4.0879\ldots$, etc. The higher eigenfunctions, now labeled by n, are

$$w_n(x) = \text{Ai}\left[\left(\frac{p}{2D}\right)^{1/3} x + a_n\right]. \qquad (7.3.32)$$

The stationary concentration $n_s(p, D)$ of the single species coagulation process with random particle production is, from equation (7.3.27)

$$n_s(p, D) = \left(\frac{\text{Ai}'[0]}{\text{Ai}[0]}\right)\left(\frac{p}{2D}\right)^{1/3} = (0.72901\ldots)\left(\frac{p}{2D}\right)^{1/3}, \qquad (7.3.33)$$

with $\text{Ai}'[0] = -0.25881\ldots$. The stationary distribution function calculated from equations (7.3.27) and (7.2.30) is

$$f_s(x) = \left(\frac{p}{2D}\right)^{1/3} \frac{\text{Ai}''\left[\left(\frac{p}{2D}\right)^{1/3} x\right]}{|\text{Ai}'[0]|}. \qquad (7.3.34)$$

These results agree well with what was said above about the A + B → 0 reaction (see Fig. 7.5) – the larger reactant diffusities and/or smaller irradiation intensity, the smaller saturation concentrations n_s. The Monte Carlo simulations [95] very well confirm these results. These simulations were performed on a lattice of 10^5 sites, by the "direct" simulation method. The interparticle probability density was also measured in the simulations, and the results are compared with theory: the agreement is excellent.

The same approach could be applied to the similar *one*-particle reaction A + A → 0 [96]. In this case one expects smaller saturation concentration since two particles dissapear in their accounters. Indeed, as it was shown [97] in this reaction

$$n'_s = \frac{n_s}{2^{2/3}}, \qquad (7.3.35)$$

where n_s is equation (7.3.33). The steady-state distribution function differs from that for the $A + A \to A$ reaction; these results are confirmed by the Monte Carlo simulations [97].

The single-species $A + A \to A$ reaction allows us also to test the applicability of the shortened and the complete Kirkwood superposition approximations, (2.3.62) and (2.3.64) [98]. The calculated quantities of the saturation concentrations

$$n_s^{sh} = \frac{1}{2^{1/3}} \left(\frac{p}{2D}\right)^{1/3} = (0.7937\ldots)\left(\frac{p}{2D}\right)^{1/3} \qquad (7.3.36)$$

and

$$n_s^{com} = \left(\frac{3}{8}\right)^{1/3} \left(\frac{p}{2D}\right)^{1/3} = (0.7211\ldots)\left(\frac{p}{2D}\right)^{1/3} \qquad (7.3.37)$$

differ from the exact result, equation (7.3.28), by 9% and 1% respectively.

7.4 PROBABILISTIC MODELS AND COMPUTER SIMULATIONS

> There are nine and sixty ways of constructing tribal lays, and-every-single-one-of-them-is-right.
>
> R. Kipling

7.4.1 Simple models

As it has been said above, accumulation of radiation-induced (Frenkel) defects takes place in all kinds of solids irrespective which of the two basic recombination mechanisms – annihilation or tunnelling recombination – occurs [9, 11, 13, 17–20, 39, 42, 99–107]. In a good approximation this process could be considered as the $A + B \to 0$ reaction with the particle input. In many cases strong arguments exist for the clustering of radiation-induced similar defects. We discussed earlier that for evaluating the results of analytical calculations which use highly different (often implicit) assumptions and for making a comparison between them and the results of computer simulations, the dimensionless quantity $U_0 = n_0 v_0$ ($n_0 = n(\infty)$) is the steady-state concentration of accumulated defects of the similar type at saturation could be used. The probability of recombination of an interstitial atom i in the sphere of spontaneous recombination around a vacancy will be henceforth considered to be uniform within the limits of the sphere and zero outside it. For the continuum model it is clear from dimensional considerations that its magnitude U_0, if it exists, is a *universal* characteristic of the problem that does not depend on the magnitude of U_0. Table 7.1 collects various estimates of this quantity, which will be discussed in greater detail below.

The problem of the kinetics of particle accumulation is very complicated owing to its multiparticle character; clusters of similar defects that arise initially owing to statistical fluctuations behave differently with regard to recombination from the way individual defects do. Thus, for example, while the recombination of an isolated vacancy with an individual interstitial diminishes the effective recombination volume by the amount of the volume of spontaneous recombination of the vacancy, the recombination of an interstitial with *a cluster* consisting of vacancies with overlapping recombination volumes (below, in discussing accumulation, an important role is played precisely by these clusters, rather than simply by the accumulation of vacancies with nonoverlapping recombination spheres), can lead to a quite insignificant change in the effective recombination volume.

In studying processes of accumulation of the Frenkel defects, one uses three different types of simple models: the box, continuum, and discrete (lattice) models. In the simplest, box model, which was proposed first in [22], one studies the accumulation of complementary particles in boxes having a certain capacity, with walls impenetrable for diffusion of particles among the boxes. The continuum model treats respectively a continuous medium; the intrinsic volume of similar defects at any point of the space is not bounded. In the model of a discrete medium a single cell (e.g., crystalline lattice site) cannot contain more than one defect (v or i).

The box *model* amounts to a system of N boxes, into which one randomly and successively throws white (A) and black (B) balls, with the total numbers of thrown balls n_A and n_B equal to one another at any instant (apart from a single ball). In the absence of annihilation the colour of a ball has no

TABLE 7.1
Various estimates of the dimensionless concentration of defects $U_0 = n_0 v_0$ at saturation (without taking account of correlation in genetic pairs) in the case of immobile defects in the one-, two- and three-dimensional cases [31]

U_0	Notes	References
0.5	Kirkwood's superposition approximation	[34, 41, 122]
$\ln 2 \approx 0.69$	Neglect of correlation of similar particles	[34]
$\geqslant 1$	Simulation taking account of tunnelling recombination	[35, 105, 106]
0.46–3.6	$1\bar{d}$ simulation with number of lattice sites varied from 2 to 700	[107, 109, 112]
1.36 and 1.08	Analytical theory using the Kirkwood approximation in the linear approximation for the $1\bar{d}$ and $3\bar{d}$ cases	[31, 111]
1.02, $\bar{d} = 3$	Analytical theory using the improved Kirkwood approximation, no saturation in $1\bar{d}$ case occurs	[68]
0.33–2.77	Probability estimates for 2–100 sites in the recombination sphere	[21, 22, 23]
3.4; 3.2	Analogously, for the $1\bar{d}$ and $3\bar{d}$ cases	[25–28]
4.2; 2.07; 1.04	Simulation in the continuum approximation for $1\bar{d}$, $2\bar{d}$ and $3\bar{d}$ respectively	[118]
1.36	$2\bar{d}$ simulation for 400 lattice sites in the recombination sphere	[113]
∞	Model allowing an infinite local density of defects	[24]

meaning and the distribution of balls between the boxes is described by the binomial law usual for random events:

$$W_k(\gamma) = C_\gamma^k P^k (1-P)^{\gamma-k}. \tag{7.4.1}$$

Here $W_k(\gamma)$ is a probability of filling of an arbitrarily chosen box with k spheres after γ throws, $\gamma = n_A + n_B$ denotes also the total number of thrown balls, and the C_γ^k are binomical coefficients, $P = 1/N$ is a probability that a ball is thrown into an arbitrary box. As $P \to 0$ and for finite $s \equiv \gamma P$, the distribution of (7.4.1) goes over into the Poisson equation

$$W_k(\gamma) = \frac{s^k \exp(-s)}{k!}, \qquad (7.4.2)$$

while when k is sufficiently close to s, into the Gaussian distribution:

$$W_k(\gamma) = (2\pi s)^{-1/2} \exp\left(-\frac{(k-s)^2}{2s}\right). \qquad (7.4.3)$$

Now let us introduce the process of recombination: the entry of a white and a black ball into a single box leads to their annihilation. When $N = 1$, the number of balls in the box is determinate and is equal for alternate throwing to 0 or 1. When $N > 1$, the situation is altered in principle owing to the random character of entry of A and B into the boxes. When recombination is included the distributions (7.4.1) to (7.4.3) no longer hold; the probability of finding m balls of a chosen color in an arbitrary box after 2γ throws (γ A balls and γ B balls) is

$$W_m(2\gamma) = \sum_{k=0}^{\gamma-m} C_\gamma^k P^k (1-P)^{\gamma-k} C_\gamma^{k+m} P^{k+m} (1-P)^{\gamma-k-m}. \qquad (7.4.4)$$

We can also study the case in which the capacity of each box is restricted to a certain number M. Here the pairs of throws in which even one of two balls enters a box filled to the limit with balls of the same colour are not realizable and are not taken into account in calculating the number γ. For small values of P, large M, and $\gamma \to \infty$, equation (7.4.4) is approximately reduced to

$$W_m(2\gamma) = \exp(-2s) \lim_{\gamma \to \infty} \sum_{k=0}^{\gamma-m} \frac{s^{2k+m}}{k!(k+m)!} = \exp(-2s) I_m(2s).$$

Here I_m is the Bessel function of imaginary argument. The asymptotic representation of I_m leads, when $\gamma \gg N, M$, to a probability $W_m(2\gamma)$ that does not depend on m: $W_m(2\gamma) \propto 1/2\sqrt{\pi s}$. Here the means are $\overline{m} \approx (M+1)/2$, $\overline{m^2} \approx M(M+1)/3$, while we have

$$\delta_m = \frac{\sqrt{\overline{m^2}}}{\overline{m}} \approx 1.16. \tag{7.4.5}$$

Thus the random distribution of the particles among the boxes in the generation-recombination process leads to an *equiprobable* distribution, cardinally different from the Poisson distribution, of "boxons" (sets of a given number of particles of the same colour) over the boxes and to the macroscopic magnitude of the fluctuations of the number of particles in the boxes of (7.4.5).

These characteristic features are manifested also in the generation-recombination processes of the Frenkel defects in real crystals. However, in contrast to the box model, in a crystal statistical screening of the recombing particles occurs in coordinate space that leads to a complex spatial distribution of vacancies v and interstitial atoms i. This distribution depends on the law whereby the probability of recombination varies with the distance r between complementary particles. Usually one approximates this law by the step-distribution $\sigma(r) = 1$ $(r \leqslant r_0)$, $\sigma(r) = 0$ $(r > r_0)$, r_0 is a recombination radius.

The first attempts to seek a closed expression for $n(t)$ did not include an analysis of pair correlation of defects, even at the level of pair densities, and seem at least to be ambiguous. Such an approach based on simple probabilistic considerations was first used in [24] (see also [108]). Since here we are not treating explicitly the relationship between two-particle and higher-particle densities, it is difficult to make a correct estimate of the effect of overlap of the forbidden volumes of several closely-lying defects. This leads to the need to introduce some *a priori* assumptions. A characteristic example is [24], where the implicit assumption of a chaotic distribution of defects through the reaction volume (along with partially taking it into account) led to a physically incorrect result – the *absence* of an effect of saturation of concentration doses (see Table 7.6 below and comments to it).

In the above-mentioned paper [25, 26, 28], which assumes the existence of steady-state concentration of accumulated defects in a one-dimensional model, and which takes the distribution of A and B defects separately in the form of periodically distributed clusters (groups) of identical dimension, it was found from simple probabilistic considerations that the mean number of defects per cluster is

$$\overline{N} = 4\exp\left(\frac{r_0}{\overline{r}}\right) - 1. \tag{7.4.6}$$

Here $\bar{r}$ is a mean distance between particles in the cluster (an AB pair instaneously recombines if the distance r between A and B is $r \leqslant r_0$). The values were obtained in the model being discussed of $r_0/\bar{r} = 3.43$, $\overline{N} \approx 143$. We can record the results of this study by using the parameter a_0. Thus, we easily note that

$$U_0 = \frac{r_0 L/\bar{r}}{L + (a+b)}. \tag{7.4.7}$$

Here we have $L = \overline{N}\bar{r}$, $a = [2\bar{r}(\bar{r} + r_0)]^{1/2}$, $b = r_0 - (\bar{r}/2)$. From (7.4.7) we have (as $\overline{N} \gg 1$)

$$U_0 = \frac{r_0}{\bar{r}} = 3.43. \tag{7.4.8}$$

A discrete case has been studied in [27]. Then, in contrast to equation (7.4.6), one has

$$\overline{N}_L = 4\left(1 - \frac{n_0}{\gamma}\right)^{-\gamma r_0} - 1. \tag{7.4.9}$$

Here γ is concentration of traps (localization sites). The calculation of the dependence of n_0 on $\log(2r_0\gamma)$ that was performed agrees with the data of computer simulation, with the deviation appreciable only in the region of small values of $\log(2r_0\gamma)$, where only two sites exist in the recombination sphere. In the discrete model with the recombination radius r_0 any site of the crystal is surrounded by a sphere of radius $r_0/2$ within which only defects of the same type exist or there are none at all. One can call such a region of the crystal *a homogeneity interval* and characterize each site of the crystal with the number k of similar defects contained within its homogeinity interval; m_k is a number of such sites, M is a number of sites in the homogeinity interval. In the steady-state the average over the volume of the crystal of the concentration of defects is

$$\frac{km_k}{M} = \left(\frac{\partial P_k}{\partial t}\right)_g \tau_k. \tag{7.4.10}$$

Here $(\partial P_k/\partial t)_g$ is a rate of generation of defects, and τ_k is lifetime of a defect with respect to recombination. It was assumed in [109] as an approximation that τ_k is determined by the concentration of defects solely within the homogeneity interval (the recombination region). Here one can derive recursion relationships between the m_k, which lead to

$$m_k = m_0 \prod_{j=1}^{k}\left(1 + \frac{1-j}{M}\right), \qquad (7.4.11)$$

and, together with the condition of constancy of the number of sites

$$N = m_0 + 2\sum_{k=1}^{M} m_k$$

also to the steady-state concentration values and

$$n_0 = N_v = N_i = \frac{N}{(1 + 2q_M)},$$

$$q_M = \sum_{k=1}^{M}\sum_{j=1}^{k}\left(1 + \frac{1-j}{M}\right). \qquad (7.4.12)$$

At low enough values of M the calculated values of the parameter U_0 agree well with the results of computer experiments. For large M the relationships of (7.4.11) lead to a rapid growth in $U_0(M)$, which qualitatively agrees with experiment. However, the n_0 calculated by (7.4.12) increases with increasing M *more rapidly* than in the computer calculations, as explained by the mentioned approximation for determining τ_k.

We stress an important feature of the generation-recombination process being discussed. The lifetime of a vacancy v, owing to the competition for capture of an i that has fallen at some site α with other vacancies that also lie within the sphere of recombination of i, depends on the concentration of vacancies $N_\gamma(\alpha)$ in the sphere with its centre at the site α. Yet the quantity $N_\gamma(\alpha)$ depends analogously on $N_\gamma(\alpha')$ for the sites α' of the recombination sphere with its centre at α', etc. Thus an effective interaction arises between immobile point defects caused by the generation–recombination process, which leads to "cleaning out" of the volume of the crystal of a single defect and small clusters and their aggregation into large clusters.

If both components of the Frenkel pair (v, i) are mobile and no other reaction but annihilation occur, the tendency to generation–recombination clustering is preserved, but the effect of clustering is less sharply marked. However in real systems usually only capture reactions of mobile v and i by various sinks and by one another are effective; while the complexes that are formed here (e.g., divacancies, tetravacancies, vacancy + impurity atom) lose mobility. The possibility of effective accumulation of clusters of similar particles in such systems is determined by the relationships among the capture cross Sections of particles of opposite type. There are a number of causes that lead to a systematic difference between the capture cross Sections of v and i by various defects (*preference*) that gives rise to effective separation of them with respect to weak recombination. A consequence of this is a low radiation stability of the materials being irradiated. The effect of statistical screening is manifested also in this case. However, even in the absence of preference clustering of similar defects occurs, completely due to the statistical screening. The latter statement is well confirmed by the results of computer simulations discussed below.

It is important to note that it was K. Dettmann [34] who first created the solid basis of a rigourous theory of the kinetics of accumulation of the Frenkel defects in crystals. He showed that in the absence of diffusion the accumulation curve, $n = n(t)$, is determined by the *infinite* sum of the correlation functions $\rho_{m,0}$, $m = 2, \ldots \infty$ that describe the spatial correlations of all orders of *similar* defects *only*.

The accumulation equation can be written in the form similar to equations (7.1.16):

$$\frac{dn(t)}{dt} = p(1 - 2\delta) \qquad (7.4.13)$$

with

$$\delta = \sum_{m=1}^{\infty} \frac{(-1)^{m-1}}{m!} \times$$

$$\times \int \ldots \int \theta(\vec{r}_1) \ldots \theta(\vec{r}_m) \rho_{m,0}(\vec{r}_1, \ldots, \vec{r}_m; t) d\vec{r}_1 \ldots d\vec{r}_m. \qquad (7.4.14)$$

Here p is a creation rate (dose rate) of stable pairs of defects, while the quantity δ is a fraction of the reaction volume overlapped by recombination spheres, i.e., the ratio of the effective recombination volume to the entire volume of the crystal. The problem of constructing the kinetic equation of accumulation for $n(t)$ could be solved if one could obtain δ in explicit form, including actually all the information on aggregation (see Section 7.1). This would mean the solution of the multiparticle problem in the accumulation kinetics. But in practice one must restrict the treatment to certain handable approximations. Thus, in the simplest case, upon neglecting the correlation between similar defects (i.e., assuming that $\rho_{m,0} \approx \rho_{1,0}^m$), the following result was obtained [34]:

$$\delta = 1 - \exp(-U_0). \tag{7.4.15}$$

The corresponding saturation concentration of defects is $U_0 = \ln 2 \approx 0.69$.

7.4.2 Computer simulations

In this Section we discuss results of extensive computer simulations performed mainly in the former Soviet Union in the 70s and 80s and thus almost unknown in the West (see the review article [15] for more detail).

7.4.2.1 Quasi-continuum model in particle accumulation

To establish the fundamental laws of the process of the kinetics of particle accumulation in solids, initially the *quasicontinuum* model of a crystal was studied [107]. A one-dimensional "crystal" was represented in the form of a segment containing L cells with periodic boundary conditions (the ends of the segment are closed). The simulation was conducted for different dimensions L of the crystals and magnitudes l of the recombination region. The results of the simulation are given in Table 7.2.

The simulation shows that: 1) the curve of the dependence of the number of accumulated defects on the total number of defects created reveals a tendency to saturation; 2) when $L = 100$–400 and $l = 5$–20, *complete separation* was attained of the "crystal" into a region of interstitials and a region of accumulation of vacancies; the location of the regions changes upon further generation of defects; 3) the sum of the lengths of the wells filled with vacancies for $L/l > 2$ considerably exceeds the length L (when $L = 2000$, $l = 5$, $U_0 = n_0 l \approx 5$, where n_0 is the concentration of *accumulated* defects, in total about 10^6 pairs of defects was generated). It should be stressed once more that for small values of L/l the magnitude of U_0 depends on the ratio L/l.

Inspection of a series of successive patterns for $L = 200$ and $l = 20$ [15, 107] demonstrates that *complete separation* was established of the "crystal"

TABLE 7.2
Results of calculations for a quasi-continuum one-dimensional model [107, 15]

L	l	U_0
40	20	1.16
200	100	1.16
100	20	1.32
200	20	2.5
400	20	3.5
2000	5	5.0
2000	10	5.0

into regions containing only vacancies or interstitials (cf. [52, 53]). With time such regions of similar defects move through the "crystal" and sometimes break up into smaller ones. As a result, for small values of L/l, the regions of similar defects "crawl" through the "crystal" owing to the fitting at the ends of the regions of similar defects. For larger values of L/l regions of similar defects also exist in a form of *clusters*. However, in this case the separation into two regions was not observed. Evidently, as long as L/l is small, there is not enough room for arrangement of the accumulated defects in the "crystal", and hence they are collected in such a way that the greatest number of them is accommodated. This will happen precisely upon division of the entire "crystal" into two regions occupied by similar defects. When $L/l \geqslant 200$ one no longer finds a dependence of $n_0 l$ on L/l.

We recall here that estimates on the basis of a simple model of the mean number of defects in a cluster [25, 26, 28] gave $U_0 = r_0/\bar{r} = 3.43$, where $\bar{r}$ is a mean distance between defects. The mean number of particles in a cluster is $\overline{N} = 120$. These values correlate with the values of U_0 from the computer experiment, which obtained $U_0 \simeq 5$ and a mean number of defects in a cluster, respectively, of about 100. (As follows from the pattern of accumulation for $L = 2000$ and $l = 5$ with a total number of creation events of 5×10^5 [107].)

7.4.2.2 The discrete model

To simulate better a real experiment, a computer simulation was performed also for a *discrete* one-dimensional model described in detail in [107, 110].

TABLE 7.3
Dependence of the steady state value of the concentration of defects U_0 on the number v_p of sites in the recombination sphere (one-dimensional case)

v_p	[109, 110]	[21–23]
2	0.464	0.333
4	0.636	0.500
10	0.922	0.830
24	1.248	1.323
50	1.625	1.937
100	2.25	2.77
150	2.3±0.2	
200	2.5	
300	3.0	
400	3.2	
500	3.5	
700	3.6	
∞	4.12	

In contrast to the quasi-continuum model, the "crystal" is divided into $2N$ cells of two types: at the initial instant of time the cells with odd numbers are occupied by atoms, and the rest are empty. A vacancy can appear only in a cell with an odd number, and an interstitial atom can lie in a cell with an even number. Each cell contains not more than one defect.

It has been established in [107, 110] the existence of saturation of the concentration n_0 and the dependence of the number of lattice sites v_p within the recombination sphere v_0 (Table 7.3); $V = 2M$ in the theory [22, 23].

Paper [109] determined the value of U_0 upon approach to the steady state from above. One-dimensional crystals were simulated of length from 8×10^3 to $2 \times 10^4 a_0$ (a_0 is a lattice constant; the spatial correlation in genetic pairs is neglected). The limiting values $U_0 = 3.5$–3.6 for 500 and 700 sites in the recombination sphere (Table 7.3, third column) are close to the value 3.43 obtained in the continuum approximation by an approximate method [22] and considerably exceed the estimate 1.36 implied by the approach based on many-point densities in the *linear* approximation [31, 111]; remember that this formalism based on improved Kirkwood's approximation [33] predicts no saturation of defect concentrations at all for $\bar{d} = 1$ (Section 7.1).

To determine the existence of stable steady state, a model [109] was studied of the destruction of clusters in the case $v_p = 700$. At the initial instant of time 10 uniformly distributed clusters of 300 vacancies each, were put into the "crystal", and interstitial atoms in the intervals between them ($U_0 \approx 10$). Then pairs of randomly distributed defects of different types were created in the "crystal". The newly generated defects break up the orginally existing clusters and the concentration of defects *declines* to a steady-state value. The values of U_0 were obtained by averaging a region of the curve of length 2.5×10^4 events of defect creation. The result unambiguously implies the existence of *stable steady state* in the problem of accumulation of point defects and in the problem of breakup of clusters.

A connection was established by simulation of accumulation [110] between n_0 and the ratio of the number of active interstitials to the total number of them in the crystal

$$U_0 = 1 - 2\beta. \tag{7.4.16}$$

Here β is a function of v_p, the number of sites active toward recombination in the recombination sphere. (In [110] the concentration and recombination volumes were expressed in units of the volume v_0 of the unit cell, and n_0 coincides with the fraction of sites or interstiatial sites occupied respectively by vacancies or interstitial atoms.) We note that, in the model being discussed, the cell itself in which a vacancy occurs is considered inactive with respect to recombination of an interstitial on it.

For a qualitative study of the aggregation effect in the defect accumulation, also *pair correlation functions* have been defined for similar and dissimilar defects. The distribution function of similar defects at small values of r shows a region of enrichment of their concentration ($X(r) > 1$) at $r \leqslant 2r_0$, which indicates the clustering of defects. The distribution function Y of dissimilar particles in the same region has a value smaller than unity. (These results agree well with the analytical theory in the continuum approximation – Section 7.1.) It was established by simulation that clustering increases with increasing number of lattice nodes v_p in the recombination sphere. Thus, when $v_p = 4$, the clustering effect is practically absent: $X \approx 1$ for all relative distances. The absence of the clustering effect at low v_p indicates the possibility of applying the ordinary superposition approximation in such cases.

In the paper [112] the following accumulation equation was proposed on the basis of results of simulating the accumulation of point defects in a one-dimensional crystal in the saturation stage:

$$\frac{dn(t)}{dt} = p\left(1 - \frac{2\alpha(t)n(t)v_0}{1 - n(t)v^*}\right), \qquad (7.4.17)$$

(in the notations [110] $v_0 = v_p v^*$, where v_p is the number of sites in the recombination sphere), $\alpha(t)$ is a coefficient of recombination efficiency; $\alpha(0) = 1$ characterizes the efficiency of recombination of a single defect but with increasing the degree of overlap of the recombination spheres, $\alpha(t)$ declines. It was assumed for $\alpha(t)$ that [112]

$$\alpha(t) = 1 - [1 - \alpha(\infty)]\frac{n(t)}{n_0}. \qquad (7.4.18)$$

Here $\alpha(\infty) = \alpha\,(t \to \infty)$ is the value in the saturation stage.

To test the hypotheses (7.4.17) and (7.4.18), the kinetics of accumulation was simulated on a computer by the method described in [110]. For each of the values $v_p = 10, 16, 24$, and 50, the process of accumulation was performed independently 200 times until the stage of steady-state values of n_0 was reached. The relationships $n(N)$, $N = pt$, and $\alpha(n)$ were constructed from the mean values obtained in this series. It was shown that within the limits of error of computer experiment ($\approx 5\%$), the slowly varying function $\alpha(n)$ can be well approximated by the linear dependence of (7.4.18), which confirms the suitability of this approach for describing the accumulation of point defects in the discrete model. Analogous results are obtained for $v_p = 16$ and 50 for which the values were found respectively, of 1.092 and 1.625 for n_0 and 0.463 and 0.478 for $\beta(\infty) = \alpha(\infty)v_0 n_0$.

To estimate the role of spatial correlations in *genetic pairs* of created defects, the case was simulated in which a created interstitial lies at various distances l from the edge of the recombination sphere of its vacancy [109]. Table 7.4 shows the values of U_0 for different l and $v_p = 10$. As it is well seen from these results, only when $l/a < 5$ (strong initial correlation), do these correlations substantially alter U_0; for small l the magnitude of U_0 declines owing to the suppression of the clustering effect by correlations within the pairs being created. This is in agreement with analytical theory presented in Section 7.1.

TABLE 7.4
Dependence of the concentration of defects at saturation on the degree of correlation within geminate pairs [107]

l/a	U_0
1	0.75
2	0.81
3	0.85
5	0.89
30	0.92

The distribution functions of similar and dissimilar defects also have a distinctive form. Thus, in the steady-state distribution function of dissimilar defects one observes a *maximum* at small distances (in the region of relative distances corresponding to the interpair correlation). On the other hand, the distribution function of similar defects in the region of small r values takes respectively on *smaller* values than in the case of absence of interpair correlations. This also agrees well with the analytical calculations for the continuum model [30, 31, 34] discussed in Section 7.1.

The accumulation was also simulated in [113] for a *two-dimensional* "crystal" represented by a square lattice with the lattice constant a_0 and dimensions $L \times L$ (in units of a_0). In the initial state $L \times L$ atoms are placed at lattice sites, while $L \times L$ interstitial states are free. The recombination region amounts to a square containing $l \times l$ sites. Otherwise the simulation is analogous to the one-dimensional case. (Periodic conditions are imposed on the boundaries of the crystal.) The number N of accumulated defects of a single type was determined, together with the dimensionless parameter U_0 (the multiplicity of the coverage of the crystal by the sum of the areas of the instability zones of the accumulated defects). A steady state is established in the cases studied, as a rule, after 4×10^4 events of pair creation. To obtain the steady-state values of U_0 and N, averaging was performed over the last 5×10^4 creation events with a total number of them of 10^5. The results are given in Table 7.5, which shows the results of simulation for a set of parameters L, l and L/l. In all cases we observe a strongly marked effect of aggregation of similar defects. As $L/l \geqslant 10$, the steady-state value of U_0 no longer depends on L.

TABLE 7.5
Results of computer simulations of defect accumulation in a two-dimensional crystal of length L. The area of the recombination zone is $l \times l$, the number of accumulated defects is N [113]

$L \times L$	L/l	$l \times l$	U_0	N
50×50	5	10×10	0.89 ± 0.02	22.34
100×100	10	10×10	1.2	120.7
150×150	15	10×10	1.2	270.0
250×250	25	10×10	1.2	752.4
100×100	10	20×20	1.0	25.3
500×500	25	20×20	1.36	322.3

In the two-dimensional case the value of U_0 is *smaller* than for the one-dimensional case at the same magnitudes of the recombination region (thus, when $v_p = 400$, $U_0 = 3.2$ in the one-dimensional case). However, the aggregation effect is expressed rather strongly.

The quantity U_0 characterizes the degree of aggregation only on the average. Therefore it is important in each particular case to analyze also the spatial distribution of the defects. Thus, the low-temperature accumulation of the Frenkel defects in the two- and three-dimensional cases was simulated in [36, 114] and the obtained values of U_0 for $\bar{d} = 2$ considerably exceed the same in [113]. In contrast to the latter, the authors of [36, 114] used a *circle* as the recombination region. In its turn, the values of u_0 obtained in [115] are considerably larger than those found in [114]. We note that it was assumed in [115] that, when an interstitial atom occurs at a site where the recombination spheres of several vacancies overlap, it recombines with the *closest* vacancy. This demonstrates very well how any details, insignificant at first glance, can affect considerably the accumulation kinetics.

In the three-dimensional case [114] the saturation concentration is *smaller* than for $\bar{d} = 1$ and 2; the maximum value is $U_0 = 1.02$ (for $v_p = 266$ sites in the recombination sphere), and its dependence on v_p is also weaker. Both the magnitude of U_0 and this trend are in complete agreement with analytical theory presented in Section 7.1.

A two-dimensional simulation has also been performed of the accumulation of defects with an *asymmetric* recombination region [113] chosen in the form of a rectangle with $v_p = a \times b$. The anisotropic case in which the larger sides of all the rectangles are oriented in parallel was studied. For $v_p = 6 \times 16$ and $L \times L = 150 \times 150$, the steady-state value was obtained of $U_0 = 1.35$. The greater lengths of the sides of the rectangle enhance the aggregation effect in this direction. Thus, in a simulation with $v_p = 6 \times 16$, *extended* clusters were observed, which divided the crystals into strips [15].

A simulation of accumulation with account taken of *diffusion* was first performed for $\bar{d} = 2$ in [116, 117]. Evidently the inclusion of diffusion smears out the clusters – the more efficiently the greater is the mobility of the defects.

Lastly, in [118] the accumulation kinetics was simulated for immobile particles for $\bar{d} = 1, 2$ and 3 in a continuum model. The saturation concentrations obtained, $U_0 = 4.2, 2.07$ and 1.04 respectively, agree with results presented above.

7.4.2.3 Taking account of tunnelling recombination of defects

A simulation has been carried out [105, 106, 119] of the process of accumulation of the immobile Frenkel defects restricted by *tunnelling recombination* of dissimilar defects, as it is observed in many solid insulators. As follows from Chapter 3, in contrast to the ionic process of instant annihilation of close pairs of the vacancy–atom type, it is characterized by a broad spectrum of recombination times. Thus, the probability for a pair of chosen defects that lie at the relative distance r to survive for τ seconds is

$$W(\tau) = \exp\left(-\sigma_0 \tau \exp\left(-\frac{r}{r_0}\right)\right). \qquad (7.4.19)$$

For small r the lifetime of the pair is minimal: $\tau \geqslant \sigma_0^{-1}$ (usually $\tau \geqslant 10^{-5}$–10^{-8} s), and increases *exponentially* with the distance r between the defects.

The simulation was performed for the three-dimensional case with imposition of the periodic boundary conditions on the cube in which the defects are being created. The initial distribution function of genetic defects was chosen in the form

$$f(r) = \frac{\exp(-r/r^*)}{4\pi r^2 r^*}, \qquad \int f(r)\mathrm{d}\vec{r} = 1. \qquad (7.4.20)$$

The results of the simulation confirmed the hypothesis expressed in [105, 120, 121] that tunnelling recombination is a secondary reaction that leads

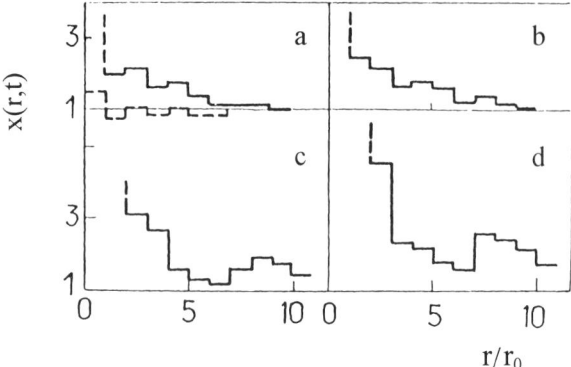

Fig. 7.15. The joint correlation function of similar particles under their accumulation restricted by the tunnelling recombination [106]. The dose rate p: 5×10^{20} cm^{-3} s^{-1} (curve a), 6.7×10^{19} (curve b), 1.2×10^{18} (curve c), 2.2×10^{16} (curve d). The dotted line around the abscissa axis in (a) shows the case when there is no recombination.

to *saturation* of the concentration of defects upon prolonged irradiation of many insulators.

Figure 7.15 shows the joint correlation function of similar defects in the region of concentration saturation as calculated from the results of the simulation. We see that the fraction of the close Frenkel defects (of the type of dimer F_2-centres) exceeds by approximately *threefold* the value expected in the Poisson distribution, which agrees well with the analytical theory presented in Section 7.1 for the annihilation mechanism (see also [31, 111]) and with actual experiments for alkali halide crystals [13].

In contrast to the case of annihilation, for the tunnelling recombination not only the dose ($= pt$, i.e., number of all created defects) but also the dose rate p plays a substantial role since, in view of the relatively large lifetimes of close-lying defects, the appearance of a third defect influences their recombination. We see from Fig. 7.15 that, the smaller is the dose rate p the closer the similar defects lie, and the better the clustering is marked. We can interpret the curves of Fig. 7.15 (c) and (d) as the creation of periodically arranged similar defects.

Figure 7.16 demonstrates the obtained probability density of finding *nearest* neighbours of the same (a) or opposite (b) types as a function of the relative distance. We see that smaller irradiation intensity leads to increase

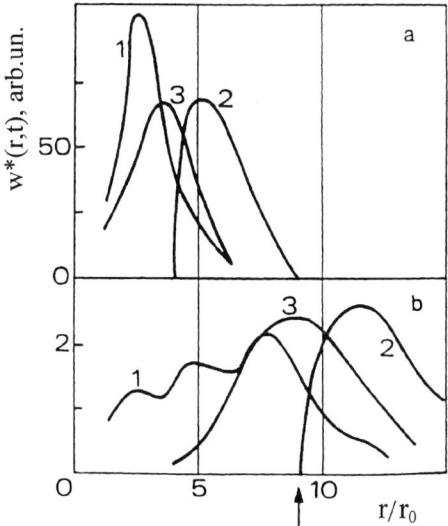

Fig. 7.16. Probability density of finding a closest particle of the same kind (1) and of the opposite kind (2) at the distance r from a given particle at the coordinate origin [106]. Curve 3 shows absence of the recombination. The dose rate p: 5×10^{20} cm^{-3} s^{-1} (curves a), 2.2×10^{16} (curves b).

of the minimal relative distance between dissimilar defects; at smaller distances the created AB pairs efficiently recombine. Owing to the tunnelling recombination, similar defects prove to lie closer, whereas dissimilar defects farther than in the case of their equal-probability (Poisson) distribution throughout the volume (curves 3 in Fig. 7.16).

Since greater irradiation intensities lead to creation of closer dissimilar defects (see Fig. 7.16), we can expect that this should faciliate attainment also of larger defect concentration at saturation. The curves and the inset in Fig. 7.17 illustrate well what we have said.

We can also easily conclude from Fig. 7.17 that, the stronger is the degree of correlation in genetic pairs ($r^* \to 0$), the smaller is the concentration n_0 at saturation. This agrees well also with the analytical theories (Section 7.1 and [41, 122]). Figure 7.17 (curves 1 and 3) implies that, in the saturation region, the mean distance between similar defects $\bar{r} = 0.554/n_0^{1/3}$ amounts to be about 50 Å, whereas the distance, even in weakly correlated pairs ($r^*/r_0 = 10$) [see equation (7.4.20)] is of the order of 25 Å. That is, genetic

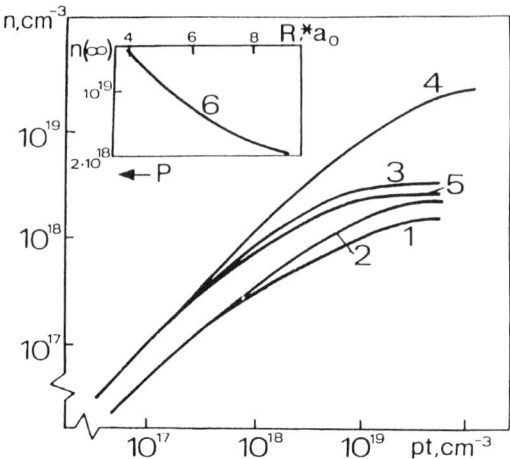

Fig. 7.17. Accumulation kinetics restricted by the tunnelling recombination. Curve 1 shows correlated pairs ($r^* = 10a$), $p = 7 \times 10^{16}$ cm^{-3} s^{-1}, curve 2 – same with $p_2 = 10p_1$, curve 3 – uncorrelated particles with $p_3 = p_1$, curve 4 – uncorrelated particles with $p_4 = 5 \times 10^{20}$ cm^{-3} s^{-1}, curve 5 – as curve 4 with $p_5 = 2.2 \times 10^{16}$, curve 6 – dependence of the saturation concentration on the dose rate p.

pairs do not mix, even in the saturation region, which leads curves 1 and 3 to reach saturation at the *different* concentrations.

The dependence of the low-temperature accumulation curve on the intensity p of irradiation is a characteristic feature of the tunnelling recombination; It has been observed in the most varied solid matrices: alkali-halide crystals [17], glasses [121, 123] etc.

7.4.3 Analysis of the accumulation equations

Some analytic expressions are collected in Table 7.6 that have been used in literature to describe the experimental accumulation kinetics, $n(t)$, or dn/dt, the rate of concentration accumulation. Experimentally such kinetics have been studied, both in the alkali-halide crystals [13, 17] and in many metals [43–45] in a wide temperature interval, starting with low (liquid-helium, 4 K) temperatures. Since often a succesful approximation of the accumulation curve is associated with better understanding of a micromechanism of defect formation (see, e.g., [40]) and with other important physical conclu-

TABLE 7.6
Different forms of the approximations of the accumulation curve $n(t)$ and the rate of accumulation, $dn(t)/dt$ [35]

Kinetic equation	Notes	References
IMMOBILE PARTICLES		
1. $\frac{dn}{dt} = p(1-\gamma)(2e^{-nv_0} - 1) - \frac{p\gamma}{2nv_0}(2nv_0 - 1 + 4e^{-nv_0}(1-nv_0) - 3e^{-2nv_0})$	Neglecting correlation of similar defects, $\gamma = \int_0^{r^*} f(r)\,d\vec{r}$	[41]
2. $\frac{dn}{dt} = p(2e^{-nv_0} - 1)$	The same, neglecting correlation of genetic defects under creation ($\gamma = 0$), $n_0 = \ln 2/v_0 \approx 0.69/v_0$	[41, 122]
3. $\frac{dn}{dt} = p(1 - 2nv_0) = pe^{-2pv_0 t}$, $n = n_0[1 - \exp(-2pv_0 t)]$	Two terms of the expansion (2) in $nv_0 \ll 1$, $n_0 = 1/(2v_0)$ which coincides with the result of the superposition approximation	[20, 34, 45, 122]
4. $\frac{dn}{dt} = p[1 - 2v_0 n + (nv_0)^2] = p(1-nv_0)^2$	Three terms of the expansion (2) in $nv_0 \ll 1$	[44–46, 167]
5. $\frac{dn}{dt} = p[1 - 2nv_0 + (nv_0)^2 - \frac{2}{3}(nv_0)^3]$	Four terms of the expansion (2) in $nv_0 \ll 1$. Note that in [41] the coefficient for the cubic term is pointed out erroneously as 0.01 instead of 2/3	[41, 43, 44]
6. $n = \frac{1}{4v_0}[\ln(4pv_0 t - 1) + 1]$	For temperatures at which defects are immobile, and for moderate radiation doses. Owing to incompletely taking account of the correlation of similar defects, $n_0 \to \infty$ as $t \to \infty$	[41]

TABLE 7.6
(Continued)

Kinetic equation	Notes	References
	MOBILE PARTICLES	
7. $n = \frac{1}{v_0} \ln(pv_0 t + 1)$	At moderate temperatures at which interstitial atoms are mobile	[24, 84]
8. $\frac{dn}{dt} = p(4pv_0 t + 1)^{-1}$	At high temperature the coefficient $1/4v_0$ is replaced by $1/v_0$	[41]
9. $n = \sum_k A_k(1 - e^{-\beta_k t}) + Dt$	Has been applied preferentially in the high-temperature region	[125]
10. $n = n_0[1 - \exp(2Kt)]^{1/2}$	Empirical equation in which K is not necessarily equal to pv_0	[126]
11. $n = n_0\left(1 - \frac{1}{(bt+1)^{1/2}}\right)$	Empirical relationship with the two fitting parameters: b and n_0	[127]
12. $\frac{dn}{dt} = p\left(1 - \frac{2\alpha n v_p}{1 - n v_0}\right)$	It is assumed that the efficiency of recombination $\alpha = \text{const} = 0.85$. v_p is number of sites in recombination sphere	[108]
13. $\frac{dn}{dt} = p\left\{1 - 2\left[1 - \left(1 - \frac{1 - n_0 v^*}{2n_0 v_0}\right)\frac{n(t)}{n_0}\right]\frac{n(t)v_0}{1 - n(t)v^*}\right\}$	Obtained in an approximation of the results of computer simulation, $\alpha(0) = 1$; $\alpha(t) < 1$; $v^* = v_0/v_p$	[107, 112]

sions, it is reasonable to discuss briefly the region of applicability and the substantiation of the expressions that have been used. Hereafter numbers of equations mentioned refer to those in Table 7.6.

Equation (1) is the most general and rigorous analytical result. However, it does not take account of the aggregation of similar defects and hence it is applicable only at not very large irradiation doses (up to a concentration of defects $\leqslant (1/2)n_0$, where n_0 is concentration at saturation). In fact, here the existence of clusters of similar defects is allowed, but it is actually assumed that these clusters are statistical fluctuations of the Poisson distribution of similar defects, which does not reflect a real pattern of cluster formation with a substantially *non*-Poisson spectrum of fluctuations. It is assumed implicitly in equation (1) that, after each event of creating a new pair of defects, the entire system of defects is stirred to attain the Poisson distribution. In the case of the absence of the defect correlation in genetic pairs we arrive at equation (2).

Equation (3) is the most widely used in analyzing experimental curves, since its form is intuitively clear: the rate of the defect accumulation is determined by the fraction of free volume of the crystal not occupied by previously created defects, without taking account of the overlap of the annihilation volumes of similar defects. Evidently it is applicable only in the initial stage of accumulation kinetics at relatively low concentrations of defects, $nv_0 \ll 1$. As it was stated in Section 7.1, the superposition approximation corresponds to the first two terms of expansion (2) in powers of nv_0.

In their turn, the equations (4) and (5) correspond to taking account of two more terms of this expansion respectively. (It was erroneously stated in [41] that the coefficient of the cubic term of equation (5) equals -0.01, which was then copied in all the experimental studies that used this approximation.) However, it is not always acknowledged that these two equations also assume the absence of aggregation of defects; the authors [41] propose using of equation (5) up to the saturation concentration. Therefore it is not justified to extrapolate the rate of accumulation of defects to a zero value for obtaining n_0 from equations (3)–(5) (cf. [43, 44, 46]). Actually the saturation concentration can exceed several times the value predicted by these equations.

Equation (6) predicts a logarithmic growth of the concentration of defects without any saturation. To obtain the saturation effect, which has been experimentally observed by the authors themselves [17], a hypothetical secondary reaction was introduced. The defects of this model were discussed earlier, while we note only that in the initial stage the relationship (6) qualitatively resemble the previous ones. This also has been used in interpreting the experimental data obtained for alkali-halide crystals [40], and it yielded a very big number of sites within the recombination sphere ($v_p = 3000$ anion sites around a vacancy), which explicitly indicates a tunnelling mechanism of defect recombination in alkali halides.

The kinetic equations for *diffusion-controlled* defect accumulation are presented in the rest of the Table 7.6. Equations (7)–(11) are approximations of the kinetics of accumulation that are not substantiated theoretically in any way but give curves with saturation, which qualitatively resemble the form of equations (1) to (5).

Equation (12) proposed in [108] yields $U_0 = 0.59$. This evidently too low value is obtained owing to the unsubstantiated assumption of the independence of the efficiency of recombination α of v_p and of the irradiation dose, as well as the ill-substantial choice of the value $\alpha = 0.85$.

Finally, the simple empirical equation (13) was presented in [107, 112], which describes well the computer simulation results. It was shown that the efficiency of recombination depends on the accumulation time and on the number of lattice sites within the recombination sphere, v_p. Therefore, this equation could serve as a solid basis for analyzing the actual experimental data. However, in this case, first of all, we must convince ourselves by comparison of the experimental and theoretical accumulation curves of their qualitative resemblance. For example, we easily note that convex relationships of the type of Fig. 1 in [43, 44] require the use of more complex models than those discussed above, where the rate of accumulation declines monotonically.

An approximate analysis using equation (13) for description of the accumulation curve for defects in Cu (Fig. 7 in [45]) yields the value $U_0 = n_0 v_0 \approx 1.2$, which clearly indicates an aggregation effect [35]. This result was obtained on the basis of estimates of values of p and v_p obtained in the same paper with the assumption of the absence of the correlation in genetic pairs. (Evidently taking account of correlation would have led to an even larger value of the U_0 parameter.)

The experimental kinetics of accumulation of the Frenkel defects – F centres in alkali-halide crystals at liquid-helium temperatures – was studied in [17] and [40] within the framework of a model that yields a logarithmic dependence of the concentration of defects on the irradiation dose – equation (6) of Table 7.6). Although we criticized this relationship above, at low radiation doses it can be represented as a polynomial in powers of $n_0 v_0$ resembling equations (3) to (5). At the same time cogent arguments exist favoring the idea that in alkali halide crystals the accumulation of F-centres is restricted by their tunnelling recombination with the complementary H-centres (see, e.g., Fig. 7.17). This recombination gives rise to its products-pairs of charged F^+- and I-centres, which influence the kinetics of accumulation and complicate the very simple model of accumulation being discussed.

We note in conclusion that taking account of correlation of defects in genetic pairs, formation of pairs of new defects (e.g., owing to the tunnelling mechanism of recombination), and of radiation-induced disclocation loops, etc., substantially complicate the development of rigorous and universal theory of the kinetics of defect accumulation. The *temperature dependence* of the efficiency of defect accumulation contains substantial information on the correlation within genetic pairs and on the nature of their interaction [119, 124] and is also of great theoretical importance.

The analysis conducted in this Chapter dealing with different theoretical approaches to the kinetics of accumulation of the Frenkel defects in irradiated solids (the bimolecular $A + B \to 0$ reaction with a permanent particle source) with account taken of many-particle effects has shown that all the theories confirm the *effect of low-temperature radiation-stimulated aggregation* of similar neutral defects and its substantial influence on the spatial distribution of defects and their concentration at saturation in the region of large radiation doses. The aggregation effect must be taken into account in a quantitative analysis of the experimental curves of the low-temperature kinetics of accumulation of the Frenkel defects in crystals of the most varied nature – from metals to wide-gap insulators; it is universal, and does not depend on the micro-mechanism of recombination of dissimilar defects – whether by annihilation of atom–vacancy pairs (in metals) or tunnelling recombination (charge transfer) in insulators.

The kinetics of defect accumulation requires further theoretical studies at temperatures at which they become mobile. The creation under the action of radiation of an ordered structure from the accumulation of radiation defects is an effect akin to those discussed in the theory of self-organization of structures – *synergetics* [14].

References

[1] E. Sonder and W.A. Sibley, in: Point Defects in Solids, Vol. 1, eds J.H. Crawford and L.M. Slifkin (Plenum, New York, 1972) p. 201.

[2] W. Schilling and K. Sonnenberg, J. Phys. F: 3 (1973) 322.
[3] K. Schröder, Diffusion-Controlled Reactions of Point Defects, Report 1083-FF, Jülich, 1974.
[4] K. Schröder, Point Defects in Metals (2), Springer Tracts Mod. Phys. 87 (1980) 71.
[5] N. Itoh, Adv. Phys. 31 (1982) 49.
[6] N. Itoh, Semicond. Insul. 5 (1983) 165.
[7] M. Ueta, H. Kanzaki, K. Kobayashi, Y. Toyozawa and E. Manamura, Excitonic Processes in Solids (Springer, Berlin, 1986).
[8] V.V. Emtsev, T.V. Machovets, V.V. Mikhnovich and N.A. Vitkovskii, Radiat. Eff. Defects Solids 111/112 (1989) 99.
[9] Ch. Lushchik and A. Lushchik, Decay of Electronic Excitations into Defects in Solids (Nauka, Moscow, 1989).
[10] N. Itoh and K. Tanimura, J. Phys. Chem. Solids 51 (1990) 717.
[11] V. Kirsanov, A. Suvorov and Yu. Trushin, Processes of Radiation-Induced Defect Producting in Metals (Energoatomisdat, Moscow, 1985).
[12] M.N. Kabler, in: Point Defects in Solids (1), ed. J. Crawford (Plenum Press, New York, 1972) Ch.4.
[13] B.F. Faraday and W.D. Compton, Phys. Rev. Sect. A: 138 (1965) 893.
[14] V.N. Kuzovkov and E.A. Kotomin, Rep. Prog. Phys. 51 (1988) 1479.
[15] V.L. Vinetsky, Yu.H. Kalnin, E.A. Kotomin and A.A. Ovchinnikov, Sov. Phys. Usp. 33 (1990) 793.
[16] Ya.B. Zeldovich and A.S. Mikhailov, Sov. Phys. Usp. 30 (1987) 977.
[17] L.W. Hobbs, A.E. Hughes and D. Pooley, Proc. R. Soc. London Ser. A: 332 (1973) 167.
[18] G. Lück and R. Sizmann, Phys. Status Solidi 5 (1964) 683.
[19] G. Lück and R. Sizmann, Phys. Status Solidi 17 (1966) K61.
[20] G. Lück and R. Sizmann, Nucleonik 8 (1966) 256.
[21] V.L. Vinetsky, Sov. Phys. Solid State 25 (1983) 1159.
[22] V.L. Vinetsky and A.B. Kondrachuk, Ukr. Fiz. Zh. 27 (1982) 383 (Ukrainian J. Phys., in Russian).
[23] V.L. Vinetsky and O.M. Kirshon, Latv. PSR Zinat. Akad. Vestis Fiz. Teh. Zinat Ser. 1 (1988) 36 (Proc. Latv. Acad. Sci. Phys. Technol. Ser., in Russian).
[24] A.E. Hughes and D. Pooley, J. Phys. C: 4 (1971) 1963.
[25] V.V. Antonov-Romanovskii, Sov. Phys. Solid State 23 (1981) 1393.
[26] V.V. Antonov-Romanovskii, Sov. Phys. Solid State 25 (1983) 341; 25 (1983) 1896.
[27] V.V. Antonov-Romanovskii, Phys. Status Solidi B: 121 (1983) 133.
[28] V.V. Antonov-Romanovskii, Sov. Phys. Solid State 27 (1985) 674.
[29] Yu.H. Kalnin, J. Lumin. 16 (1976) 311.
[30] V.N. Kuzovkov and E.A. Kotomin, Phys. Status Solidi B: 105 (1981) 789.
[31] V.N. Kuzovkov and E.A. Kotomin, J. Phys. C: 17 (1984) 2283.
[32] H. Schnörer, V. Kuzovkov and A. Blumen, J. Chem. Phys. 93 (1990) 7148.
[33] V.N. Kuzovkov and E.A. Kotomin, Phys. Scr. 47 (1993) 585.
[34] K. Dettmann, Phys. Status Solidi 10 (1965) 269.

[35] Yu.H. Kalnin and E.A. Kotomin, Probl. At. Nauki Tekh. Kharkov Fiz. Tekh. Inst. 29 (1984) 18 (Probl. Atom. Sci. Technol., Kharkov Phys. Tech. Inst., in Russian).
[36] F.V. Pirogov and E.I. Palagashvili, Latv. PSR Zinat. Akad. Vestis Fiz. Teh. Zinat Ser. 4 (1984) 46 (Proc. Latv. Acad. Sci. Phys. Technol. Ser., in Russian).
[37] S.F. Burlatsky and A.A. Ovchinnikov, Sov. Phys. JETP 65 (1987) 908.
[38] V.V. Gromov, Zh. Fiz. Khim. 55 (1981) 1377 (Sov. J. Phys. Chem., in Russian).
[39] M.A. Elango, Elementary Inelastic Radiation Processes (Nauka, Moscow, 1988).
[40] E. Mercier, G. Guillot and A. Nouilhat, Phys. Rev. B: 20 (1979) 1578.
[41] K. Dettmann, G. Leibfried and K. Shröder, Phys. Status Solidi 22 (1967) 423.
[42] Ch.B. Lushchik, I.K. Vitol and M.A. Elango, Sov. Phys. Usp. 122 (1977) 233.
[43] M. Nakagawa, Phys. Rev. 16 (1977) 5285.
[44] M. Nakagawa, J. Nucl. Mater. 108/109 (1982) 194.
[45] M. Nakagawa, W. Mansel and K. Boning, Phys. Rev. B: 19 (1979) 742.
[46] G. Duesing, W. Sassin et al., Cryst. Lattice Defects 1 (1969) 55.
[47] A.B. Doktorov and E.A. Kotomin, Phys. Status Solidi B: 114 (1982) 9.
[48] Yu.H. Kalnin, Latv. PSR Zinat. Akad. Vestis Fiz. Teh. Zinat Ser. 3 (1977) 70 (Proc. Latv. Acad. Sci. Phys. Technol. Ser., in Russian).
[49] V.L. Vinetsky, I.I. Yaskovets and I.V. Kelman, in: Physical Processes in Crystals with Defects (Naukova Dumka, Kiev, 1972) p. 111.
[50] F.V. Pirogov, PhD Thesis (Salaspils, 1981).
[51] A.M. Gutin, A.S. Mikhailov and V.V. Yashin, Sov. Phys. JETP 65 (1987) 533.
[52] Zhang Yi-Cheng, Phys. Rev. Lett. 59 (1987) 1726.
[53] K. Lindenberg, B.J. West and R. Kopelman, Phys. Rev. Lett. 60 (1988) 1777; Phys. Rev. A: 42 (1990) 2279.
[54] A. Blumen, S. Luding and I.M. Sokolov, J. Stat. Phys. 65 (1991) 849.
[55] I.M. Sokolov and A. Blumen, Europhys. Lett. 21 (1993) 885.
[56] I.M. Sokolov, P. Argyrakis and A. Blumen, J. Phys. Chem. 98 (1994) 7256.
[57] E.A. Kotomin and V.N. Kuzovkov, Rep. Prog. Phys. 55 (1992) 2079.
[58] K.K. Schwartz and Yu.H. Ekmanis, Dielectric Materials: Radiation-Induced Processes and Stability (Zinatne, Riga, 1989).
[59] A.B. Lidiard, Philos. Mag. 39 (1979) 647.
[60] U. Jain and A.B. Lidiard, Philos. Mag. 35 (1977) 245.
[61] J. Seinen, J.C. Groote, J.R.W. Weerkamp and H.W. Hartog, Radiat. Eff. 124 (1992) 325.
[62] W. Soppe, J. Phys. C: 5 (1993) 3519.
[63] G. Martin, in: Diffusion in Materials, eds A.L. Lasker, J.L. Bocquet, G. Brebec and C. Monty (Kluwer, Dordrecht, 1990) p. 129, and references therein.
[64] E.A. Kotomin and M. Zaiser, Appl. Phys. 57 (1993) 117.
[65] E.A. Kotomin, M. Zaiser and W. Soppe, Philos. Mag. 70 (1994) 313.
[66] V.N. Kuzovkov and E.A. Kotomin, J. Chem. Phys. 98 (1993) 9107.
[67] E.A. Kotomin, V.N. Kuzovkov, W. Frank and A. Seeger, J. Phys. A: 27 (1994) 1453.
[68] V.N. Kuzovkov and E.A. Kotomin, Phys. Scr. 50 (1994) 720; E.A. Kotomin, V.N. Kuzovkov, M. Zaiser and W. Soppe, Radiat. Eff. Def. Solids, 136 (1995) 209.

[69] E.R. Hodgson, A. Delgado and J.L. Alvarez Rivas, Solid State Commun. 16 (1975) 785.
[70] E.R. Hodgson, A. Delgado and J.L. Alvarez Rivas, J. Phys. C: 12 (1979) 1239; 12 (1979) 4393.
[71] E.R. Hodgson, A. Delgado and J.L. Alvarez Rivas, Phys. Rev. B: 18 (1978) 2911.
[72] E.R. Hodgson, A. Delgado and J.L. Alvarez Rivas, J. Phys. C: 14 (1981) 337.
[73] E.R. Hodgson, A. Delgado and J.L. Alvarez Rivas, Radiat. Eff. 74 (1983) 193.
[74] J.D. Comins and B.O. Carragher, Phys. Rev. B: 24 (1981) 283.
[75] F. Agullo-Lopez, F.J. Lopez and F. Jaque, Cryst. Lattice Defects Amorph. Mater. 9 (1982) 227.
[76] F. Agullo-Lopez and F. Jaque, J. Phys. Chem. Solids 34 (1973) 1949.
[77] M. Aguilar, F. Jaque and F. Agullo-Lopez, Radiat. Eff. 61 (1982) 215.
[78] E. Sonder, G. Bassagnani and P. Camagni, Phys. Rev. 180 (1969) 882.
[79] J. Rubio, J. Hermandes, H. Murrieta and S. Ramos, Phys. Rev. B: 34 (1986) 5820.
[80] E. Sonder, Phys. Rev. B: 5 (1972) 3259.
[81] P.W. Levy, J. Phys. Chem. Solids 52 (1991) 319.
[82] A.V. Gektin and N.V. Shiran, Phys. Solid State 30 (1988) 870; A.V. Gektin, DSc thesis (Riga, 1990).
[83] K. Bachmann and H. Peisl, J. Phys. Chem. Solids 31 (1970) 1525.
[84] A.E. Hughes, Radiat. Eff. 74 (1983) 57; 97 (1986) 161.
[85] V.N. Kuzovkov and E.A. Kotomin, J. Phys. Cond. Matter 7 (1995) L481.
[86] K. Kang and S. Redner, Phys. Rev. Lett. 52 (1984) 955.
[87] K. Kang and S. Redner, Phys. Rev. A: 32 (1985) 435.
[88] L.W. Anacker and R. Kopelman, Phys. Rev. Lett. 58 (1987) 289.
[89] L.W. Anacker and R. Kopelman, J. Phys. Chem. 91 (1987) 5335.
[90] E. Clement, R. Kopelman and L.M. Sander, Chem. Phys. 146 (1990) 343.
[91] I.M. Sokolov, Sov. Phys. JETP 67 (1988) 1846.
[92] I.M. Sokolov, Phys. Lett. A 139 (1989) 403.
[93] A.A. Ovchinnikov and S.F. Burlatsky, Sov. Phys. JETP Lett. 43 (1986) 638.
[94] B. O'Shaughnessy and I. Procaccia, Phys. Rev. A: 54 (1985) 455.
[95] C.R. Doering and D. Ben-Avraham, Phys. Rev. Lett. 62 (1989) 2563.
[96] Z. Ra'cz, Phys. Rev. Lett. 55 (1985) 1707.
[97] D. Ben-Avraham and C.R. Doering, Phys. Rev. A: 37 (1988) 5007.
[98] J.-C. Lin, C.R. Doering, and D. Ben-Avraham, Chem. Phys. 146 (1990) 355.
[99] N. Itoh, in: Defects in Insulating Crystals, eds V. Tuchkevich and K. Shvarts (Springer, Berlin, 1981) p. 341.
[100] M.I. Klinger, Ch.B. Lushchik, T.V. Mashovets et al., Sov. Phys. Usp. 28 (1985) 994.
[101] V.M. Koshkin and Yu.R. Zabrodskii, Sov. Phys. Dokl. 21 (1976) 203.
[102] Yu.R. Zabrodskii and V.M. Koshkin, Sov. Phys. Solid State 18 (1976) 1669.
[103] E.A. Kotomin and A.B. Doktorov, Phys. Status Solidi B: 114 (1982) 287.
[104] P. Dean, Prog. Solid State Chem. 8 (1973) 1.
[105] I.A. Tale, D.K. Millers and E.A. Kotomin, J. Phys. C: 8 (1975) 2366.
[106] E.A. Kotomin, PhD thesis (Inst. of Physics, Latv. Acad. Sci., Riga, 1975).

[107] Yu.H. Kalnin and Yu.Yu. Krikis, Latv. PSR Zinat. Akad. Vestis Fiz. Teh. Zinat Ser. 1 (1983) 104 (Proc. Latv. Acad. Sci. Phys. Technol. Ser., in Russian).
[108] A. Van den Bosch, Radiat. Eff. 8 (1971) 63.
[109] Yu.H. Kalnin, V.L. Vinetskii, O.M. Kirshon and Yu.Yu. Krikis, Abstracts of the All-Union Seminar on Simulation of Radiation Defects (Ioffe Institute of the Acad. Sci. USSR, Leningrad, 1986) p. 46.
[110] Yu.H. Kalnin and Yu.Yu. Krikis, Probl. Radiat. Sci. Technol. (3) (1981) 21.
[111] E.A. Kotomin, V.N. Kuzovkov and I.A. Tale, Latv. PSR Zinat. Akad. Vestis Fiz. Teh. Zinat Ser. 4 (1984) 114 (Proc. Latv. Acad. Sci. Phys. Technol. Ser., in Russian).
[112] Yu.H. Kalnin, Latv. PSR Zinat. Akad. Vestis Fiz. Teh. Zinat Ser. 5 (1982) 3 (in Russian).
[113] Yu.H. Kalnin, O. Perevalova and M. Ogrinsh, in: Computer Simulation of Structural Defects in Crystals (Ioffe Institute of the Acad. Sci. USSR, Leningrad, 1988) p. 145.
[114] F.V. Pirogov and E.I. Palagashvili, Latv. PSR Zinat. Akad. Vestis Fiz. Teh. Zinat Ser. 2 (1985) 27 (Proc. Latv. Acad. Sci. Phys. Technol. Ser., in Russian).
[115] A.B. Kondrachuk and S.S. Yatsyuk, Ukr. Fiz. Zh. 32 (1987) 305 (Ukranian J. Phys., in Russian).
[116] Yu.H. Kalnin and F.V. Pirogov, Latv. PSR Zinat. Akad. Vestis Fiz. Teh. Zinat Ser. 2 (1980) 29 (Proc. Latv. Acad. Sci. Phys. Technol. Ser., in Russian).
[117] F.V. Pirogov and Yu.A. Ekmanis, Latv. PSR Zinat. Akad. Vestis Fiz. Teh. Zinat Ser. Phys., 2 (1981) 7 (Proc. Latv. Acad. Sci. Phys. Technol. Ser., in Russian).
[118] S.D. Baranovskii, E.L. Ivachenko and B.I. Shklovskii, Sov. Phys. JETP 65 (1987) 1260.
[119] E.A. Kotomin, Dr. Habil. Thesis (Inst. of Physics, Latv. Acad. Sci., Salaspils, 1988).
[120] Ya.R. Bogans, Ya.A. Valbis et al., Bull. Acad. Sci. USSR Phys. Ser. 37 (1973) 49.
[121] A.I. Mikhailov, Dokl. Akad. Nauk SSSR 197 (1971) 136; 197 (1971) 223.
[122] E.A. Kotomin and V.N. Kuzovkov, Solid State Commun. 39 (1981) 351.
[123] V.V. Voevodskii, Physics and Chemistry of Elementary Chemical Processes (in Russian) (Khimiya, Moscow, 1970).
[124] E.A. Kotomin and I.I. Fabrikant, Radiat. Eff. 46 (1980) 85; 46 (1980) 91.
[125] M.A. Elango, Proc. Eesti NSV Tead. Akad. Toim. 43 (1974) 175.
[126] A.B. Lidiard, Solid State Commun. 8 (1978) 70.
[127] F. Hermann and P. Pinard, J. Phys. C: 3 (1970) 1037.

Chapter 8

Systems Under Birth and Death Conditions: Lotka and Lotka–Volterra Models

8.1 A NOVEL CRITERION OF THE MARGINAL COMPLEXITY INSURING SELF-ORGANIZATION IN THE COURSE OF CHEMICAL REACTIONS

> Essences should not be multiplied without necessity.
> (Entia non sunt multiplicanda sine necesitate)
>
> Ockam

8.1.1 Autowaves in non-equilibrium extended systems

As it is well known [1–4] the extended systems of quite different nature consisting of a large number of interacting elements and being far from the thermodynamic equilibrium, demonstrate existence of the *auto-wave processes* [5]. Such self-supported processes are observed in active non-linear systems, their properties remain constant due to the energy source distributed in the medium. At long enough time their characteristics (oscillation period, wavelength, its velocity, amplitude etc.) become independent on the initial conditions and are defined entirely by *local properties* of the active medium. Autowave propagation in space has a diffusion character.

According to the classification [5], the spatial structures existing in homogeneous active (in particular, chemical) media are as follows:

1. Switch-like wave motion (fire spreading);
2. Propagation of a single wave (impulse);
3. Autonomous localized wave sources [4];
4. Standing waves;
5. Reverberator (a spiral wave [4]);
6. Synchronous spatial auto-oscillations;
7. Quasi-stochastic waves;
8. Spatially inhomogeneous distributions stationary in time.

Of special interest is the so-called *Belousov–Zhabotinsky* class of similar reactions [4, 6–12]. This system can serve as an extremely successful example of *self-organisation*: proper mixing of several liquids in a given proportion and at certain temperature demonstrates practically *all* kinds of the autowave processes just mentioned.

Numerous versions of the Belousov–Zhabotinsky system differ by chemical compounds used. The typical reaction involves oxidation of some organic compound by bromate ion (BrO_3^-) occurring in acid medium with metal catalyst (Ce^{3+}, Mn^{2+}, as well as complexes of Fe^{2+}, Ru^{2+}). As an example, a particular reaction [4] could be mentioned, where an organic reductor is malonic acid $CH_2(COOH)_2$ and Ce^{3+} ions serve as a catalyst. In this reaction a solution changes periodically its colour due to oscillations in Ce^{3+} concentration. Generally speaking, the reaction consists of two stages. At the first one metal is oxidized

$$Ce^{3+} \xrightarrow[H^+]{BrO_3^-} Ce^{4+}, \qquad (8.1.1)$$

but at the second stage it is reduced by the malonic acid

$$Ce^{4+} \xrightarrow{MA} Ce^{3+}. \qquad (8.1.2)$$

Despite the fact that from a *principal* point of view a problem of concentration oscillations could be considered as solved [4], satisfactory theoretical descriptions of experimentally well-studied particular reactions are practically absent. Due to very complicated reaction mechanism (in order to describe the Belousov–Zhabotinsky reaction even in terms of standard chemical kinetics *several tens* of concentration equations for intermediate products should be written down and solved numerically [4, 9, 10]) these equations contain large number of ill-defined parameters – reaction rates [10].

8.1.2 Basic models

Following an idea of Zhabotinsky [4] that chemical oscillating systems could expediently be treated as a "black box" and that the relevant mathematical semi-phenomenological model has to focus on the basic reactions only neglecting those less important, Vasiliev, Romanovsky and Yakhno [5] suggested a concept of the *basic model*. These simplified models of an extended active medium could be obtained either by a reduction of pre-existing (but too complicated for the detailed analysis) sets of higher-order equations or using some *a priori* information about physico-chemical mechanisms of the processes under study. A hybrid of these two approaches is used for chemical oscillatory reactions – starting from a simplified reaction mechanism a set of kinetic equations is written down, which then is shortened using certain criteria (thus passing from the elementary reaction stages to "macrokinetic" stages [7]). The readers can find numerous examples in [4, 6, 9, 10, 13]. Different versions of the so-called *Oregonator* model [10, 14, 15] fall into this category of basic models.

Finally, the basic model could be also constructed *ad hoc* just to reproduce the kinetic phenomena observed experimentally in time and in space; the well-known examples are the Brusselator or *Prigogine–Lefever* model (see [2]) and the model by Smoes [7]). Practically *any* basic model is oriented for a simplest and transparent description of a particular kind of the autowave processes.

Therefore, often main attention in studying chemical oscillations is paid to their formal description on the macroscopic level rather than to an attempt to understand in detail the micromechanism of oscillations. It often results in necessity to make a choice between several *alternative* models suggested for a particular chemical system. It is difficult to restrict ourselves in theory to a definite *universal* basic model since it can turn out to be either too complicated for studying a particular kind of the autowave processes or, on the other hand, of a limited use due to its inability to reproduce all types of auto-wave processes.

8.1.3 Systems with a complete stirring

Field, Kórös and Noyes [13] suggested to use as the basic model for the Belousov–Zhabotinsky system a rather complicated set of chemical reactions with seven intermediate products. Its more global analysis based on macrokinetic stages and retaining still the principal features of this reaction [14] has led to the simplified scheme with *three* intermediate products only. This model called *Oregonator* [9, 15] is described by the following equations:

$$\begin{aligned} &A + B \to P, \\ &E + A \to 2A + C, \\ &F + B \to A, \\ &A + A \to Q, \\ &C \to F + B. \end{aligned} \quad (8.1.3)$$

Here E and F are initial reactants whereas P and Q are final products, A, B and C are intermediate compounds: $HBrO_2$, Br^- and Ce^{4+}. Concentrations n_E and n_F of the initial reactants are assumed to be constant in an *open system* under study due to stationary matter source. Under well-stirring condition, the kinetic law of mass action leads to a set of the ordinary differential equations

$$\frac{dn_A(t)}{dt} = k_1 n_F n_B(t) - k_2 n_A(t) n_B(t) + k_3 n_E n_A(t) - k_4 n_A(t)^2, \quad (8.1.4)$$

$$\frac{dn_B(t)}{dt} = -k_1 n_F n_B(t) - k_2 n_A(t) n_B(t) + f k_5 n_C(t), \quad (8.1.5)$$

$$\frac{dn_C(t)}{dt} = k_3 n_E n_A(t) - k_5 n_C(t). \quad (8.1.6)$$

Along with this model [13], other basic models were presented which involve three intermediate products [10, 16, 17], as well as four products [10, 18].

A numerical solution of the basic equations demonstrated their ability to reproduce concentration oscillations. At the same time, for the systems possessing three and more intermediate products the standard method to prove existence of periodical solutions, using a phase portrait of a system (Section 2.1.1) *fails*. An additional reduction in a number of differential equations, e.g., using an idea that one of concentrations, say, $[BrO_3^-]$, serves as a rapid variable and thus the relevant kinetic equation (8.1.5) could be solved as the stationary [10], cannot be always justified due to uncertainty in the kinetic coefficients k_i.

As it was mentioned in Section 2.1.1, the concentration oscillations could be simulated quite well by a set of even *two* ordinary differential equations of the first order but paying the price of giving up the rigid condition imposed on interpretation of mechanisms of chemical reactions; namely that they are based on mono- and bimolecular stages only (remember the Hanusse theorem [19])! An example of what Smoes [7] called the *heuristic–topological* model is the well-known *Brusselator* [2]. Its scheme was discussed in Section 2.1.1; see equations (2.1.33) to (2.1.35).

8.1.4 Role of diffusion

Kinetic coefficients k_i entering equations for the well-stirred systems like (8.1.4) to (8.1.6) are defined, in principle, by mutual diffusive approach of reactants but in the framework of such a formally-kinetic statement of a problem diffusion begins to play a role only when we turn to the inhomogeneous spatially extended systems. Such diffusive systems are described by equations (2.1.40) containing functions non-linear in concentrations. An idea of these equations came from the theory of well-stirred systems (Section 2.1.1). Of common knowledge is the basic model of the *extended Brusselator* [2]. Its basic equations are nothing but the generalized equations (2.1.34) and (2.1.35):

$$\frac{\partial C_A(\vec{r},t)}{\partial t} = D_A \nabla^2 C_A(\vec{r},t) + k_1 n_E -$$
$$- (k_2 n_C + k_4) C_A(\vec{r},t) + k_3 C_A(\vec{r},t)^2 C_B(\vec{r},t), \qquad (8.1.7)$$

$$\frac{\partial C_B(\vec{r},t)}{\partial t} = D_B \nabla^2 C_B(\vec{r},t) + k_2 n_C C_A(\vec{r},t) -$$
$$- k_3 C_A(\vec{r},t)^2 C_B(\vec{r},t). \qquad (8.1.8)$$

When deriving (8.1.7) and (8.1.8), it was assumed that the concentrations n_E and n_C of initial reactants remain to be homogeneously distributed in space.

The extended *Brusselator* [2, 5], *Oregonator* [5, 10] and other similar systems [4, 7] demonstrate other autowave processes whose distinctive spatial and temporal properties are independent on initial concentrations, boundary conditions and often even on geometrical size of a system. As it was noted by Zhabotinsky [4], Vasiliev, Romanovsky and Yakhno [5], a number of well-documented results obtained in the theory of autowave processes is much less than a number of problems to be solved. In fact, mathematical methods for *analytical solution* of the autowave equations and for analysis of their stability are practically absent so far.

The principal role of diffusion in these processes could be established considering rather simple examples [2]. If the kinetic equations for a well-stirred system are able to reproduce self-oscillations (the limit cycle), the *extended system* could be presented as a set of non-linear oscillators continuously distributed in space. Diffusion acts to conjunct these local oscillations and under certain conditions it can result in the *synchronisation* of oscillations. Thus, autowave solutions could be interpreted as a result of a *weak* coupling (conjunction) of local oscillators when they are not synchronised completely. The stationary spatial distributions in an initially homogeneous systems can also arise due to diffusion, which makes homogeneous solutions unstable.

8.1.5 The criterion of the marginal complexity of self-organized systems

As it follows from the above-said, nowadays any study of the autowave processes in chemical systems could be done on the level of the basic models only. As a rule, they do not reproduce real systems, like the Belousov–Zhabotinsky reaction in an implicit way but their solutions allow to study experimentally observed *general* kinetic phenomena. A choice of models is defined practically uniquely by the mathematical formalism of standard chemical kinetics (Section 2.1), generally accepted and based on the law of mass action, i.e., reaction rates are proportional just to products of reactant concentrations.

As it was mentioned in Section 2.1.1, according to the Hanusse, Tyson and Light theorem [19, 20], the limit cycle (auto-oscillating regime) can arise in a system only with *three and more* intermediate products (degrees of freedom) emerging during the mono- and bimolecular reaction stages. The well-known *Oregonator*, equations (8.1.3) to (8.1.6), could be treated as a *pragmatic illustration* of this theorem, which puts a rigid limit on the critical (marginal) complexity of a system to be self-organised.

The limitations of this theorem are reduced also if *trimolecular* reaction stages are involved [2]. A distinctive illustration is the above-discussed *Brusselator*, equations (2.1.33) to (2.1.35). Note that models with only two intermediate products cannot describe chaotic oscillations.

However, if we give up the macroscopic approach, Section 2.1, and pass to the *microscopic language* in treating the stochastic processes, Section 2.3.2, we might expect the marginal criterion of complexity to be considerably changed. Indeed, as it was shown earlier, in Chapter 2, application of the stochastic microscopic formalism to homogeneous system has led us formally to standard chemical kinetics equations but with the reaction rates which are no longer constants but functions of time and concentrations. Therefore, the very proof of the Hanusse, Tyson and Light theorem [19, 20] assuming essentially that all non-linearity in the relevant kinetic equations arises due to concentrations but all coefficients (reaction rates) are constants, *becomes now quite questionable*. Moreover, reaction rates in such microscopic approach are coupled to equations for the *correlation dynamics* describing spatial distribution of particles. As it was demonstrated in Chapters 2, 6, 7, spatially homogeneous systems nevertheless can reveal the effect of *microscopic self-organisation* [21] which was described in terms of the joint correlation functions. These new variables (the correlation functions) introduce into the problem nothing but a *continuum* of degrees of freedom; respectively, qualitatively *new types of motion* can also emerge here.

Staying within a class of mono- and bimolecular reactions, we thus can apply to them safely the technique of many-point densities developed in Chapter 5. To establish a new criterion insuring the self-organisation, we consider below the autowave processes (if any) occurring in the simplest systems – the Lotka and Lotka–Volterra models [22–24] (Section 2.1.1). It should be reminded only that standard chemical kinetics *denies* their ability to self-organisation either due to the absence of undamped oscillations (the Lotka model) or since these oscillations are unstable (the Lotka–Volterra model).

Since the many-point density formalism in its practical applications assumes *macroscopically homogeneous* system, we will restrict ourselves to a particular class of *microscopically* self-organized autowave processes. Without investigating in Chapter 8 all possible kinds of autowave processes, we are aimed to answer a *principal* question – whether these two models under question could be attributed to the basic models useful for the study of autowave processes.

To treat the stochastic Lotka and Lotka–Volterra models, we have now to extend the formalism presented in Section 2.2.2, where "collective" variables-numbers of particles N_A and N_B were used to describe reactions. The point is that this approach neglects *local density fluctuations* in small element volumes. To incorporate both these fluctuations and their correlations due to diffusive conjunction, we are in position now to reformulate these models in terms of the *diffusion-controlled processes* – in contrast to the rather primitive birth–death formalism used in Section 2.2.2. It permits also to demonstrate in the non-trivial way a role of diffusion in the autowave processes. The main results of this Chapter are published in [21, 25].

8.2 THE LOTKA–VOLTERRA MODEL

Do not claim too much without necessity.
(Pluralitas non est ponenda sine necessitate)

Ockam

8.2.1 The stochastic Lotka–Volterra model

Let us reformulate the standard Lotka–Volterra model [23, 24] described by the set (2.1.27) in terms of the *diffusion-controlled process* as it was suggested for the first time by Kuzovkov [21, 25–27]. Its basic elements are as follows.

(i) We consider a set of point particles A and B undergoing hopping migration with the diffusion coefficients D_A and D_B.

(ii) In the first autocatalytic stage $E + A \to 2A$ concentration of initial matter is assumed to be constant and spatially homogeneous, therefore creation of new A particles could be described as a spontaneous process induced by another (catalytic) particle A nearby. This creation is characterized by the production rate α and by the probability density $\mu_A(r)$ that a new A particle emerges at the distance r from the existing catalyst particle A.

(iii) The second autocatalytic stage $A + B \to 2B$ consists of two elements: trapping of particle A by some B with its further transformation into a new particle B. This newly-created particle B could emerge at some distance from the catalyst B which is characterized by the probability density $\mu_B(r)$.

A *biological interpretation* of this problem in terms of population dynamics could be as follows: both prey animals A and predators B living on them are reproduced by division in a medium with a spontaneous production of food E for them. With these studies in mind the model has to be slightly modified: any new predator B is born after killing the prey animal A with the probability $\zeta < 1$, which needs in our statement of the problem just a trivial replacement of $\mu_B(r)$ for $\zeta \mu_B(r)$.

(iv) Death rate of particles (animals) B is characterized by the constant rate β.

Given all these quantities, the master equation could be written for a set of reactions (2.1.27) as the Markov process. (Actual choice of the functions $\sigma(r)$, $\mu_A(r)$, $\mu_B(r)$ is discussed below.)

8.2.2 Kinetic equations

To formulate this stochastic model in terms of concentrations and joint correlation functions only, i.e., in a manner we used earlier in Chapters 2, 4 and 5, it is convenient to write down a master equation of the Markov process under study in a form of the infinite set of coupled equations for many-point densities $\rho_{m,m'}$. Let us write down the first equations for indices $(m + m') = 1$:

$$\frac{\partial \rho_{1,0}}{\partial t} = D_A \nabla_1^2 \rho_{1,0} + \alpha \int \mu_A(|\vec{r}_1 - \vec{r}|) \rho_{1,0}(\vec{r};t) \, d\vec{r} - \int \sigma(|\vec{r}_1 - \vec{r}_1'|) \rho_{1,1} \, d\vec{r}_1', \tag{8.2.1}$$

$$\frac{\partial \rho_{0,1}}{\partial t} = D_B \nabla_1'^2 \rho_{0,1} - \beta \rho_{0,1} + \int \mu_B \left(|\vec{r}_1' - \vec{r}_2'| \right) d\vec{r}_2' \times$$
$$\times \int \sigma \left(|\vec{r}_1 - \vec{r}_2'| \right) \rho_{1,1}(\vec{r}_1; \vec{r}_2'; t) \, d\vec{r}. \tag{8.2.2}$$

Hereafter on r.h.s. of the equations "standard" arguments of the many-point densities are omitted: $\rho_{m,m'} \equiv \rho_{m,m'}(\vec{r}_1, \ldots, \vec{r}_m; \vec{r}_1', \ldots, \vec{r}_{m'}'; t)$ whereas only those of principal importance are given. Diffusion and recombination terms in (8.2.1) are the same as we had in Chapter 5. Now we have an additional integral term proportional to α and describing particle A creation at $\vec{r}_1$, which is induced by the catalyst A at the point $\vec{r}$. Similarly, in (8.2.2) we have integral term for creation of a new has trapped B at $\vec{r}_1'$, provided the catalyst B at $\vec{r}_2'$ the particle A at $\vec{r}_1$. The integration in (8.2.1), (8.2.2) runs over all relative distributions of both particles B and A.

Similar detailed interpretation could be given also for further equations of a set with $(m + m') = 2$ which read as

$$\frac{\partial \rho_{2,0}(\vec{r}_1; \vec{r}_2; t)}{\partial t}$$
$$= D_A \nabla_1^2 \rho_{2,0} + D_A \nabla_2^2 \rho_{2,0} +$$
$$+ \alpha \mu_A \left(|\vec{r}_1 - \vec{r}_2| \right) \left(\rho_{1,0}(\vec{r}_1; t) + \rho_{0,1}(\vec{r}_2; t) \right) +$$
$$+ \alpha \int \mu_A \left(|\vec{r}_1 - \vec{r}| \right) \rho_{2,0}(\vec{r}, \vec{r}_2; t) \, d\vec{r} +$$
$$+ \alpha \int \mu_A \left(|\vec{r}_2 - \vec{r}'| \right) \rho_{2,0}(\vec{r}', \vec{r}_1; t) \, d\vec{r}' -$$
$$- \int \sigma \left(|\vec{r}_1 - \vec{r}_1'| \right) \rho_{2,1} \, d\vec{r}_1' - \int \sigma \left(|\vec{r}_2 - \vec{r}_1'| \right) \rho_{2,1} \, d\vec{r}_1', \tag{8.2.3}$$

$$\frac{\partial \rho_{2,0}(\vec{r}_1'; \vec{r}_2'; t)}{\partial t}$$
$$= D_B \nabla_1'^2 \rho_{2,0} + D_B \nabla_2'^2 \rho_{0,2} - 2\beta \rho_{0,2} +$$
$$+ \mu_B \left(|\vec{r}_1' - \vec{r}_2'| \right) \int \sigma \left(|\vec{r}_1 - \vec{r}_1'| \right) \rho_{1,1}(\vec{r}_1; \vec{r}_1'; t) \, d\vec{r}_1 +$$
$$+ \mu_B \left(|\vec{r}_1' - \vec{r}_2'| \right) \int \sigma \left(|\vec{r}_1 - \vec{r}_2'| \right) \rho_{1,1}(\vec{r}_1; \vec{r}_2'; t) \, d\vec{r}_1 +$$

$$+ \int \mu_B(|\vec{r}_1' - \vec{r}''|)\,d\vec{r}'' \int \sigma(|\vec{r}_1 - \vec{r}''|) \rho_{1,2}(\vec{r}_1;\vec{r}'',\vec{r}_2';t)\,d\vec{r}_1 +$$

$$+ \int \mu_B(|\vec{r}_2' - \vec{r}''|)\,d\vec{r}'' \int \sigma(|\vec{r}_1 - \vec{r}''|) \rho_{1,2}(\vec{r}_1;\vec{r}'',\vec{r}_1';t)\,d\vec{r}_1, \qquad (8.2.4)$$

$$\frac{\partial \rho_{1,1}(\vec{r}_1;\vec{r}_1';t)}{\partial t}$$

$$= D_A \nabla_1^2 \rho_{1,1} + D_B \nabla_1'^{2} \rho_{1,1} - \int \sigma(|\vec{r}_1 - \vec{r}_2'|)\rho_{1,2}\,d\vec{r}_2' -$$

$$- \beta \rho_{1,1} - \sigma(|\vec{r}_1 - \vec{r}_1'|)\rho_{1,1} + \alpha \int \mu_A(|\vec{r}_1 - \vec{r}|)\rho_{1,1}(\vec{r};\vec{r}_1';t)\,d\vec{r} +$$

$$+ \int \mu_B(|\vec{r}_1' - \vec{r}''|)\,d\vec{r}'' \int \sigma(|\vec{r}_2 - \vec{r}''|)\rho_{2,1}(\vec{r}_1,\vec{r}_2;\vec{r}'';t)\,d\vec{r}_2. \qquad (8.2.5)$$

Unlike the elementary A + B → 0 reaction considered in Chapter 5, now a set of equations contains more integrals including multiple integrals.

Since our principal aim in studying the Lotka–Volterra model is to clarify whether the limit cycle or chaotic regime could arise for this model, let us specify now the functions $\mu_A(r)$, $\mu_B(r)$ and $\sigma(r)$ in a way simplifying the integral terms in (8.2.1) to (8.2.5).

Keeping in mind the biological interpretation of the autocatalytic stages as the reproduction of animals by division, let us choose

$$\mu_\nu(r) = \frac{1}{\gamma_{\bar{d}}} r^{1-\bar{d}} \delta(r - r^*), \quad \nu = A, B \qquad (8.2.6)$$

with the further limiting transition $r^* \to 0$. Here δ is the Dirac function, $\bar{d}$ is space dimension, the factor $\gamma_{\bar{d}}$ was introduced in (4.1.50).

Trapping of particles A by B is described as earlier in terms of the black sphere model (3.2.16). A model of particle reproduction by division (8.2.6) along with a simplification of integral terms has also the following advantage. Creation of particles, as it was shown in Chapter 7 leads usually to the problem of the proper account of free volume available for particles A; the superposition approximation is valid here only for small dimensionless particle concentrations. In our treatment of the reproduction this problem does not arise since prey animals A appear near other A's which are outside the

The Lotka–Volterra model

recombination spheres around the predators B and vice versa. In other words, the model (8.2.6) assumes that predators trap their prey animals only due to diffusion of A's into the recombination spheres around B's. This permits also to treat carefully, using the Lotka–Volterra model as a basic one, the principal *role of diffusion* in the autowave processes.

It is convenient to perform the limiting transition $r^* \to 0$ at two stages. At the first stage let us do it in all terms of (8.2.1) to (8.2.5), where functions $\mu_A(r)$ and $\mu_B(t)$ are under integrals. Thus the integration accompanied with the limiting transition allows us to get rid of multiple integrals. Using equations (8.2.1) and (8.2.5) we get

$$\frac{\partial \rho_{1,0}(\vec{r}_1;t)}{\partial t} = D_A \nabla_1^2 \rho_{1,0} + \alpha \rho_{1,0} - \int \sigma(|\vec{r}_1 - \vec{r}_1'|) \rho_{1,1} \, d\vec{r}_1', \tag{8.2.7}$$

$$\frac{\partial \rho_{0,1}(\vec{r}_1';t)}{\partial t} = D_B \nabla_1'^2 \rho_{0,1} - \beta \rho_{0,1} + \int \sigma(|\vec{r}_1 - \vec{r}_2'|) \rho_{1,1} \, d\vec{r}_1. \tag{8.2.8}$$

Now the terms on r.h.s. do not differ principally from those considered in Chapters 5, 6, 7 for the $A + B \to 0$ reaction.

Separating now the singular terms of equations $(\mu_\nu(r) = \mu(r))$ containing δ-functions, we arrive at the set of equations for the joint densities:

$$\frac{\partial \rho_{2,0}(\vec{r}_1;\vec{r}_2;t)}{\partial t}$$
$$= D_A \nabla_1^2 \rho_{2,0} + D_A \nabla_2^2 \rho_{2,0} + 2\alpha \rho_{2,0} +$$
$$+ \alpha \mu(|\vec{r}_1 - \vec{r}_2|)(\rho_{1,0}(\vec{r}_1;t) + \rho_{0,1}(\vec{r}_2;t)) +$$
$$+ \int \sigma(|\vec{r}_1 - \vec{r}_1'|) \rho_{2,1} \, d\vec{r}_1' - \int \sigma(|\vec{r}_2 - \vec{r}_1'|) \rho_{2,1} \, d\vec{r}_1', \tag{8.2.9}$$

$$\frac{\partial \rho_{0,2}(\vec{r}_1';\vec{r}_2';t)}{\partial t}$$
$$= D_B \nabla_1'^2 \rho_{0,2} + D_B \nabla_2'^2 \rho_{0,2} - 2\beta \rho_{0,2} +$$
$$+ \mu(|\vec{r}_1' - \vec{r}_2'|) \int \sigma(|\vec{r}_1 - \vec{r}_1'|) \rho_{1,1}(\vec{r}_1;\vec{r}_1';t) \, d\vec{r}_1 +$$
$$+ \mu(|\vec{r}_1' - \vec{r}_2'|) \int \sigma(|\vec{r}_1 - \vec{r}_2'|) \rho_{1,1}(\vec{r}_1;\vec{r}_2';t) \, d\vec{r}_1 +$$
$$+ \int \sigma(|\vec{r}_1 - \vec{r}_1'|) \rho_{1,2} \, d\vec{r}_1 + \int \sigma(|\vec{r}_1 - \vec{r}_2'|) \rho_{1,2} \, d\vec{r}_1, \tag{8.2.10}$$

$$\frac{\partial \rho_{1,1}(\vec{r}_1;\vec{r}_1';t)}{\partial t}$$

$$= D_A \nabla_1^2 \rho_{1,1} + D_B {\nabla'}_1^2 \rho_{1,1} - \int \sigma(|\vec{r}_1 - \vec{r}_2'|)\rho_{1,2}\,\mathrm{d}\vec{r}_2' -$$

$$- \beta\rho_{1,1} - \sigma(|\vec{r}_1 - \vec{r}_1'|)\rho_{1,1} + \alpha\rho_{1,1} +$$

$$+ \int \sigma(|\vec{r}_2 - \vec{r}_1'|)\rho_{2,1}\,\mathrm{d}\vec{r}_2. \tag{8.2.11}$$

Therefore the singular terms with $\mu(r)$ enter equations for the joint densities of *similar* particles only. We show below how these terms modify the relevant boundary conditions at the coordinate origin.

8.2.3 Superposition approximation

The infinite set of kinetic equations could be decoupled by using the superposition approximation (2.3.62), similarly as it was done in Chapter 5. Let us introduce now the dimensionless variables (primes are omitted below):

$$r' = \frac{r}{r_0}, \qquad t' = \frac{Dt}{r_0^2}, \qquad D'_\nu = \frac{2D_\nu}{D},$$

$$a' = \frac{\alpha r_0^2}{D}, \qquad \beta' = \frac{\beta r_0^2}{D},$$

$$N_a = n_A \gamma_{\bar{d}} r_0^{\bar{d}}, \qquad N_b = n_B \gamma_{\bar{d}} r_0^{\bar{d}},$$

where $D = D_A + D_B$.

First equations (8.2.7) and (8.2.8) of the set are not affected by the superposition approximation and thus yield the *exact* equations for the time development of the dimensionless macroscopic densities (concentrations):

$$\frac{\mathrm{d}N_a(t)}{\mathrm{d}t} = \alpha N_a(t) - K(t)N_a(t)N_b(t), \tag{8.2.12}$$

$$\frac{\mathrm{d}N_b(t)}{\mathrm{d}t} = K(t)N_a(t)N_b(t) - \beta N_b(t). \tag{8.2.13}$$

A set of equations (8.2.12) and (8.2.13) for the concentration dynamics is formally similar to the standard statement of the Lotka–Volterra model given

The Lotka–Volterra model

by equations (2.1.28) and (2.1.29). However, a new and principal feature of these equations is that (2.1.28), (2.1.29) the parameter K is just a constant value, whereas in newly derived equations (8.2.12), (8.2.13) the "reaction rate" $K(t)$ depends on the time. It is defined in the same way as for the elementary $A + B \to 0$ reaction combined with the black sphere model

$$K(t) = \left. \frac{\partial Y(r,t)}{\partial r} \right|_{r=1}. \qquad (8.2.14)$$

In other words, $K(t)$ is a *functional* of the joint correlation function of similar particles. In this respect, a set of equations (8.2.12) and (8.2.13) is similar to the stochastic treatment of the Lotka–Volterra model (equations (2.2.68) and (2.2.69)) considered in Section 2.3.1 using the similar time-dependent reaction rate (2.2.67).

The quantity $K(t)$ is defined by equations for the *correlation dynamics* complementing equations (8.2.12) and (8.2.13)

$$\frac{\partial X_A(r,t)}{\partial t} = D_A \Delta X_A(r,t) - 2K(t) N_b(t) X_A(r,t) J[Y] +$$

$$+ \frac{2\alpha}{N_a(t)} r^{1-\bar{d}} \delta(r - r^*), \qquad (8.2.15)$$

$$\frac{\partial X_B(r,t)}{\partial t} = D_B \Delta X_B(r,t) + 2K(t) N_a(t) X_B(r,t) J[Y] +$$

$$+ \frac{2K(t) N_a(t)}{N_b(t)} r^{1-\bar{d}} \delta(r - r^*), \qquad (8.2.16)$$

$$\frac{\partial Y(r,t)}{\partial t} = \Delta Y(r,t) +$$

$$+ K(t) Y(r,t) \{ N_a(t) J[X_A] - N_b(t) J[X_B] \}. \qquad (8.2.17)$$

These two kinds of dynamics – for particle correlations and concentrations – become coupled through the reaction rate. The functionals $J[Z]$ in (8.2.15) to (8.2.17) were defined in Chapter 5 (5.1.36) to (5.1.38) for different space dimensions $\bar{d} = 1, 2, 3$. They emerge in those terms of (8.2.9) to (8.2.11) which are affected by the superposition approximation. It should be stressed that in the case of the Lotka–Volterra model it is the only approximation used for deriving the equations of the basic model.

A comparison with the correlation dynamics of the $A + B \to 0$ reaction, equations (5.1.33) to (5.1.35), shows their similarity, except that now several terms containing functionals $J[Z]$ have changed their signs and several singular correlation sources emerged. The accuracy of the superposition approximation in the diffusion-controlled and static reactions was recently confirmed by means of large-scale computer simulations [28]. It was shown to be quite correct up to large reaction depths $\Gamma = 3$ studied.

8.2.4 Boundary conditions

The uncorrelated particle distribution (4.1.12) is used, as standard initial conditions for the correlation dynamics. After the transient period the solution (for the stable regime) becomes independent on the initial conditions. For both the joint correlation functions boundary conditions at large distances $X_\nu(\infty, t) = Y(\infty, t) = 1$ has to be fulfilled due to the correlation weakening. The black sphere model imposes the additional boundary condition (5.1.39) for the correlation function $Y(r, t)$.

Strictly speaking, equations for the joint densities of similar particles have to be solved with the boundary condition (5.1.40) imposed at the coordinate origin. However, the singular terms $\delta(r - r^*)$ with $r^* \to 0$ have to modify it. To illustrate this point, let us consider the generalized diffusion equation with the singular term

$$\frac{\partial g(r, t)}{\partial t} = D\nabla^2 g(r, t) - q(r, t, \{g\})g(r, t) + \\ + \lambda r^{1-\bar{d}}\delta(r - r^*), \qquad (8.2.18)$$

corresponding to equations (8.2.15) and (8.2.16). The question is, in what respect the singular term (8.2.18) affects the use of standard numerical methods developed for such equations. The so-called standard conservative difference scheme [29] is derived by integrating (8.2.18) over thin spherical layers. Introducing a discrete mesh through $t_i = i\Delta t$, $r_m = m\Delta r$, $g(r_m, t_i) = g_m^i$ with intermediate coordinate points $r_{m+1/2} = (m+1/2)\Delta r$, one can take the integration running over a thin spherical layer in the limits $r_{m-1/2}$ to $r_{m+1/2}$. The boundary condition at the coordinate origin is used here for integrating over sphere with the radius $r_{1/2} = \Delta r/2$. In doing so, let us consider the integral emerging from (8.2.18),

$$\int_0^{\Delta r/2} r^{\bar{d}-1}\, dr \left(D\nabla^2 g + \lambda r^{1-\bar{d}}\delta(r-r^*) - qg - \frac{\partial g}{\partial t} \right) = 0. \quad (8.2.19)$$

Since the integration of two last terms in brackets is the standard procedure [29], let us concentrate on the first two terms:

$$\int_0^{\Delta r/2} r^{\bar{d}-1}\, dr \left(D\nabla^2 g + \lambda r^{1-\bar{d}}\delta(r-r^*) \right)$$

$$= D \left(r^{\bar{d}-1} \frac{\partial g(r,t)}{\partial r} \right) \Big|_{r=0}^{r=r_{1/2}} + \lambda. \quad (8.2.20)$$

When deriving (8.2.20) we used the fact that in the limit $r^* \to 0$ an argument r^* falls into the integration region and thus the basic property of the Dirac δ-function could be used. The term in (8.2.20) proportional to D is estimated by means of the boundary condition (5.1.40) (zero flux through the coordinate origin which permits to put the term to zero at the lower integration limit). Finally, instead of (8.2.20) we get

$$D \left(\frac{\Delta r}{2} \right)^{\bar{d}-1} \frac{\partial g(r,t_i)}{\partial r} \bigg|_{r=r_{1/2}} + \lambda = D \frac{(\Delta r)^{\bar{d}-2}}{2^{\bar{d}-1}} (g_1^i - g_0^i) + \lambda. \quad (8.2.21)$$

(The derivative is replaced here by the difference.) Therefore the final kinetic equation with the boundary condition modified due to the singular term is quite similar to (5.1.40) discussed above. Therefore the singular terms in equations for the joint densities (8.2.15) and (8.2.16) for similar particles could be also omitted:

$$\frac{\partial X_A(r,t)}{\partial t} = D_A \Delta X_A(r,t) - 2K(t)N_b(t)X_A(r,t)J[Y], \quad (8.2.22)$$

$$\frac{\partial X_B(r,t)}{\partial t} = D_B \Delta X_B(r,t) + 2K(t)N_a(t)X_B(r,t)J[Y]. \quad (8.2.23)$$

The numerical methods for solving equations like (8.2.17), (8.2.22) and (8.2.23) are discussed in Section 5.1. In practice the conservative difference schemes are widely used for solving differential equations with the accuracy of the order $O(\Delta t + \Delta r^2)$ [21, 26, 27] used as well $O(\Delta t^2 + \Delta r^2)$ [25]. Unlike mathematically similar equations for the $A + B \to 0$ reaction (Section 5.1), where the correlation functions vary monotonously in time, the

Lotka–Volterra model reveals different kind of autowave processes with the non-monotonous behaviour of the correlation functions accompanied by their great spatial gradients and rapid change in time. Due to this fact the space increment $\Delta r \ll 1$ was used whereas the time increment Δt was variable to ensure that the relative change of any variable in the kinetic equations does not exceed a given small value. The difference schemes described above were absolutely stable and a choice of coordinate and time mesh was controlled by additional calculations with reduced mesh.

8.2.5 Standing waves

The Lotka–Volterra equations written in the dimensionless parameters contain only several control parameters: birth and death rates α, β and the ratio of diffusion coefficients $\kappa = D_A/(D_A + D_B)$, $0 \leqslant \kappa \leqslant 1$, i.e., $D_A = 2\kappa$, $D_B = 2(1-\kappa)$ whereas their sum is constant, $D_A + D_B = 2$. Lastly, it is also the space dimension $\bar{d}$ determining the functionals $J[Z]$, equations (5.1.36) to (5.1.38), the Laplace operator (3.2.8) as well as the boundary condition (8.2.21) for the correlation functions of similar particles. Before discussing the results of the joint solution of the complete set of the kinetic equations, let us consider first the following statements.

Statement 1. Provided $K(t) = K = \text{const}$, i.e., neglecting change in time of the correlation functions, equations (8.2.12) and (8.2.13) of the concentration dynamics describe undamped concentration oscillations with the frequencies $\omega < \omega_0 = \sqrt{\alpha\beta}$, dependent on the initial conditions. The dependence $\omega = \omega(K)$ is weak. This statement is based on the analysis of the Lotka–Volterra model by both topological and analytical methods (see Section 2.1.1).

Statement 2. Substitution into the concentration dynamics (equations (8.2.12) and (8.2.13)) of the reaction rate $K = K(N_a, N_b)$, dependent on the *current* concentrations, changes the nature of the singular point. In particular, a centre (neutral stability) could be replaced by stable or unstable focus. This conclusion comes easily from the topological analysis; its illustrations are well-developed in biophysics (see, e.g., a book by Bazikin [30]).

Statement 3. If the concentrations are fixed, $N_a = \text{const}$, $N_b = \text{const}$, the set of kinetic equations (8.2.17), (8.2.22) and (8.2.23) as functions of the control parameter demonstrates two kinds of motions: for $\kappa \geqslant \widehat{\kappa}$ the stationary (quasi-steady-state) solution holds, whereas for $\kappa < \widehat{\kappa}$ a regular (quasi-regular) oscillations in the correlation functions like *standing waves* are observed. The critical value $\widehat{\kappa} = \widehat{\kappa}(\alpha, \beta, N_a, N_b)$ increases when any of two concentrations decreases.

This statement is not self-evident and needs some comments. A role of concentration degrees of freedom in terms of the formally-kinetic description was discussed in Section 2.1.1. Stochastic approach adds here a set of equations for the correlation dynamics where the correlation functions are field-type values. Due to very complicated form of the complete set of these equations, the analytical analysis of the stationary point stability is hardly possible. In its turn, a numerical study of stability was carried out independently for the correlation dynamics with the fixed particle concentrations.

First of all, the stationary solution was sought for. It is shown that it exists for $\bar{d} = 3$ only; two types of *qualitatively* different behaviours take place depending on the value of parameter κ. If $\kappa \geqslant \widehat{\kappa}$, the correlation functions change in space monotonously; the correlation function for dissimilar reactants Y increases from zero to its asymptotic value of the unity, whereas that for similar particles X approaches to unity from above. However, for $\kappa < \widehat{\kappa}$ such monotonous behaviour disappears and these correlation functions begin to oscillate in space.

The solution of the first kind is stable and arises as the limit, $t \to \infty$, of the non-stationary kinetic equations. Contrary, the solution of the second kind is unstable, i.e., the solution of non-stationary kinetic equations oscillates periodically in time. The joint density of similar particles remains monotonously increasing with coordinate r, unlike that for dissimilar particles. The autowave motion observed could be classified as the *non-linear standing waves*. Note however, that by nature these waves are not standing waves of concentrations in a real $3\bar{d}$ space, but these are more the waves of the joint correlation functions, whose oscillation period does not coincide with that for concentrations. Speaking of the auto-oscillatory regime, we mean first of all the asymptotic solution, as $t \to \infty$. For small t the transient regime holds depending on the initial conditions.

As it was said above, there is *no* stationary solution of the Lotka–Volterra model for $\bar{d} = 1$ (i.e., the parameter $\widehat{\kappa}$ does not exist), whereas for $\bar{d} = 2$ we can speak of the quasi-steady state. If the calculation time $t_{\max}$ is not too long, the marginal value of $\widehat{\kappa} = \widehat{\kappa}(\alpha, \beta, N_a, N_b, t_{\max})$ could be also defined. Depending on κ, at $t < t_{\max}$ both oscillatory and monotonous solutions of the correlation dynamics are observed. At long t the solutions of non-steady-state equations for correlation dynamics for $\bar{d} = 1$ and $\bar{d} = 2$ are qualitatively similar: the correlation functions reveal oscillations in time, with the oscillation amplitudes slowly increasing in time.

Statement 4. Results similar to the Statement 3 hold, if in a set of the kinetic equations (8.2.17), (8.2.22) and (8.2.23) concentrations are no longer fixed but the values of $N_a = \beta/K$, $N_b = \alpha/K$ corresponding to the solution of the concentration dynamics are used. The relevant marginal parameter is $\kappa_0 = \kappa_0(\alpha, \beta)$. This statement could be proved in the same way as above. Substitution of $N_a = \beta/K$, $N_b = \alpha/K$ changes slightly a form of the non-linear kinetic equations (taking into account the K definition (8.2.14). All peculiarities of the $\bar{d} = 1, 2$ class remain valid.

8.2.6 Concentration dynamics

The performed calculations demonstrate that a *type* of the asymptotic solution of a complete set of the kinetic equations is independent of the initial particle concentrations, $N_a(0)$ and $N_b(0)$. Variation of parameters α and β does not also result in new asymptotic regimes but just modifies there boundaries (in t and κ). In the calculations presented below the parameters $N_a(0) = N_b(0) = 0.1$ and $\alpha = \beta = 0.1$ were chosen. The basic parameters of the diffusion-controlled Lotka–Volterra model are space dimension $\bar{d}$ and the ratio of diffusion coefficients κ. The basic results of the developed stochastic model were presented in [21, 25–27].

Let us consider a projection of the complex many-dimensional motion (which variables are both concentrations and the correlation functions) onto the "phase plane" (N_a, N_b). It should be reminded that in its classical formulation the trajectory of the Lotka–Volterra model is a closed curve – Fig. 2.3. In Fig. 8.1 a change of the phase trajectories is presented for $\bar{d} = 3$ when varying the diffusion parameter κ. (For better understanding logarithms of concentrations are plotted there.)

(a) The situation shown in Fig. 8.1 corresponds to the "fast prey animals" ($\kappa = 0.9$). The phase portrait reveals the distinctive features of the unstable focus. It emerges under the following conditions: for large κ values the inequality $\kappa > \kappa_0$ holds, simultaneously $\kappa > \widehat{\kappa}$ takes place for chosen initial conditions within a wide time interval. In a line with Statements 3 and 4 it means that assuming absence of the concentration motion, the correlation functions strive for the certain stationary states dependent effectively on concentrations. In the steady state the reaction rate could be calculated by means of (8.2.14) as a function of concentrations, $K = K(N_a, N_b)$. Despite the fact that concentrations in fact change, for chosen parameters α and β their oscillation period $T > 2\pi/\omega_0 > 10$ exceeds greatly the distinctive time

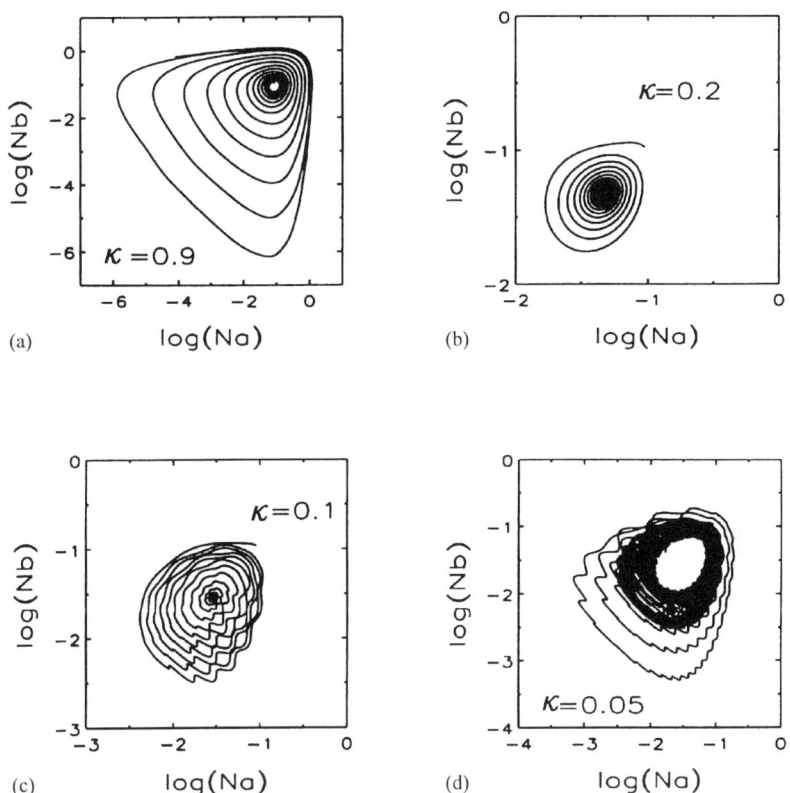

Fig. 8.1. Phase portraits of the Lotka–Volterra model for $\bar{d}=3$: (a) Unstable focus ($\kappa=0.9$); (b) Stable focus ($\kappa=0.2$); (c) Concentration oscillations during the steady-state formation ($\kappa=0.1$); (d) Chaotic regime ($\kappa=0.05$). The values of the distinctive parameter are shown.

for the stationary state formation (when the correlation functions approach very close the stationary distributions). If the concentration motion is slow, the correlation motion has to follow it. The reaction rate could be presented as $K(t) = K(N_a(t), N_b(t))$, i.e., as function of the current concentrations and indirectly of time. According to the Statement 2, the dependence of the reaction rate on concentrations change the *character* of the singular point; namely, unstable focus arises here instead of a centre. Therefore, despite the fact that the Lotka–Volterra equations have the stationary solution, it is in fact unstable due to the time dependence of concentration variables.

As it follows from the phase trajectory, at long times the concentration amplitudes are essentially limited unlike the *minimal* concentrations which decay systematically. In the biophysical interpretation of the Lotka–Volterra model such trajectories are known as *catastrophes* [30], since reduction of the biological population below certain limit in a finite-size system means nothing but their disappearance. However, the phase trajectory at long times does not reflect novel effect-emergence of the important particular dependence of the reaction rate on time. In a line with the Statement 3, the marginal value of $\widehat{\kappa}$ depends on the concentrations, increasing with their decay. In the unstable focus regime the concentrations fall into the region where $\kappa < \widehat{\kappa}$ holds. The correlation dynamics is changed here for the auto-oscillations (standing waves) and the reaction rate $K(t)$ oscillations have their *own* period independent on the period of the concentration motion. Since the concentration oscillations weakly depend on the K values (Statement 1), the oscillations $K(t)$ are not practically seen on the phase diagram. For the greater times these oscillations in $K(t)$ might change the catastrophe's type which could be pronounced well for the lower-dimensional systems discussed below.

(b) The case of $\kappa = 0.2$ shown in Fig. 8.1 demonstrates the case of "slow prey animals". Unlike the case above, this dependence of the reaction rate $K(t) = K(N_a(t), N_b(t), t)$ leads to the *stable focus regime*. A complete set of the kinetic equations has a single stable stationary solution achieved as $t \to \infty$.

(c) Further reduction of control parameter down to $\kappa = 0.1$ demonstrates quite well that the correlation motion *no longer* follows the concentration motion. As earlier, a complete system of the kinetic equations reveals a stable steady-state achieved as $t \to \infty$. However, κ value is so small that for concentrations close to the stationary value the condition $\kappa > \widehat{\kappa}$ is violated (but $\kappa > \kappa_0$ still holds). It results in the appearance of the standing wave regime for the correlation dynamics with so large amplitudes of the $K(t)$ oscillations that they affect (with a certain damping) the concentration oscillations.

(d) Lastly, for very small parameter $\kappa = 0.05$ both inequalities, $\kappa < \kappa_0$, and $\kappa < \widehat{\kappa}$, simultaneously take place. In a line with the Statements 3 and 4, the stationary stable solution is absent here due to the correlation motion. It results in the distinctive chaotic motion with self-intersections of the phase trajectory, so that at long times phase trajectories fill in densely a limited phase region. Such a complicated motion arises by two reasons. For fixed initial concentrations the standing wave regime for the correlation functions results in a perfect periodic motion characterised by an intrinsic oscillation period (Statements 3, 4). On the contrary, according to the Statement 1, the concentration motion has no definite oscillation period; for any fixed K the phase trajectory and the relevant oscillation period are defined entirely by the *current* particle concentrations, which play a role of initial conditions for the further motion.

Therefore, oscillations of $K(t)$ result in the transition of the concentration motion from one stable trajectory into another, having also another oscillation period. That is, the concentration dynamics in the Lotka–Volterra model acts as a *noise*. Since along with the particular time dependence $K = K(t)$ related to the standing wave regime, it depends also effectively on the current concentrations (which introduces the damping into the concentration motion), the concentration passages from one trajectory onto another have the *deterministic* character. It results in the limited amplitudes of concentration oscillations. The phase portrait demonstrates existence of the distinctive range of the allowed periods of the concentration oscillations.

For a given set of parameters the period of concentration oscillations (or its average for a periodic motion) exceeds greatly the period of the correlation motion. For the slow concentration motion not only the period of the standing wave oscillations but also their amplitudes and, consequently, the amplitude in the $K(t)$ oscillations depend on the current concentrations $N_a(t)$ and $N_b(t)$. In other words, the oscillations of the reaction rate *are modulated by the concentration motion*. Respectively, the influence of the time dependence $K = K(t)$ upon the concentration dynamics has irregular, aperiodic character. A noise component modulates the autowave component (the standing waves) but the latter, in its turn, due to back-coupling causes transition to new noise trajectories. What we get as a result is *aperiodic motion (chaos)*. The mutual influence of the concentration and correlation motions and vice versa is illustrated in Fig. 8.2, where time developments of both the concentrations and reaction rates are plotted.

The conclusion could be drawn from Fig. 8.2 that the peaks in $K(t)$ produce a fine structure in the concentration curves. Despite the fact that these oscillations in $K(t)$ have two orders of magnitude, the fine structure is not of a primary importance. In its turn, the concentration oscillations modulate oscillations of the reaction rate $K(t)$.

A slightly different motion is observed in the low-dimensional systems – $\bar{d} = 2$ (Fig. 8.3(a)) and $\bar{d} = 1$ (Fig. 8.3(b)). Irrespective of the parameter κ the concentration oscillations result in the fatal decay of the concentration

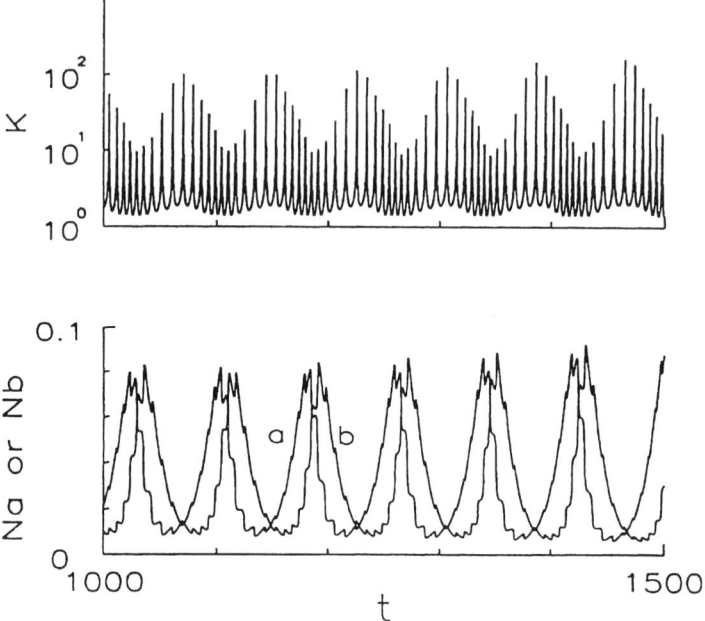

Fig. 8.2. Chaotic oscillations in the Lotka–Volterra model. Parameter $\kappa = 0.05$, $\bar{d} = 3$.

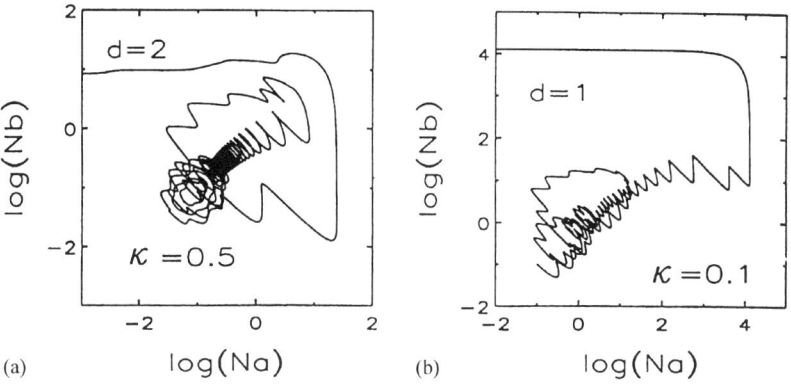

Fig. 8.3. Catastrophes in the Lotka–Volterra model: (a) Parameters $\kappa = 0.5$, $\bar{d} = 2$; (b) $\kappa = 0.1$, $\bar{d} = 1$.

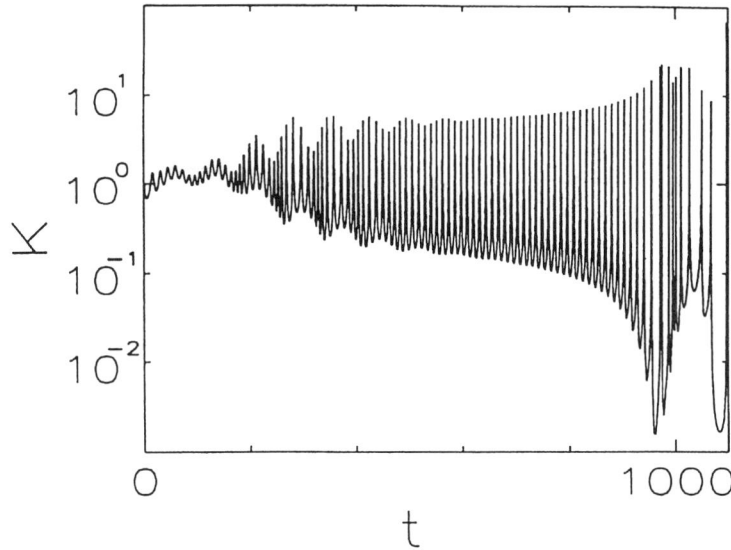

Fig. 8.4. The reaction rate in the Lotka–Volterra model. Parameters $\kappa = 0.5$, $\bar{d} = 2$.

of A's ("prey animals"). The actual value of κ affects the catastrophe's time only.

Such catastrophic motions arise first of all due to the peculiarities of the correlation dynamics in the low-dimensional systems (Statement 3). The equations of the correlation dynamics with fixed particle concentrations demonstrate here neither the stable stationary state nor the regular auto-oscillating regime. For any κ values the standing waves arise with time with increasing amplitudes. Respectively the amplitude of the oscillations in $K(t)$ is also changed. As follows from Fig. 8.4, it can be changed by several orders of magnitude. During one of such well-pronounced maximums in $K(t)$, a whole system's volume turns out to be completely overlapped by the "recombination" regions around predators B and thus the population of the "prey animals" A disappears.

8.2.7 Correlation dynamics

Study of the concentration motion could be supplemented with that for the *correlation functions*. The particle reproduction by division leads to the distinctive peculiarities of the correlation functions for similar particles at short distances: they have a maximum at the coordinate origin. Note that earlier we observed qualitatively similar behaviour of the correlation functions for the simplest bimolecular $A + B \to 0$ reaction discussed in Chapters 5, 6, 7. Such an effect could be reasonably interpreted as the aggregation of similar particles; i.e., in terms of the population dynamics *both predators and prey animals are grouped in "packs"*. Their sizes are defined by the scales (correlation lengths) ξ_A and ξ_B at which the correlation functions greatly exceed their asymptotic values; $X_\nu(\xi_\nu, t) \gg 1$. The mean particle (animal) density in a "pack" changes also with time.

Trapping of "preys" A occurs mainly on the boundary of its pack encircled by predators. Death of the particle A results in the emergence of a new particle B, which is not necessarily accompanied by an increase of the whole volume covered by the reaction trapping volumes around B's; the situation depends essentially on the diffusion parameter κ. Thus, for large $\kappa \sim 1$ particles B are practically immobile and their aggregates exist mainly due to a continuous flux of particles A. On the contrary, as κ becomes small, the aggregates of A's are only slowly expanding due to the particle diffusion. Their reproduction (characterized – according to the Maltus law – by the birth rate α) leads to the formation of dense aggregates of A's. Mobile particles B approaching these aggregates, transform them autocatalyticaly and rapidly into similar aggregates but now consisting of particles B. A formation of this dense B aggregates due to the absence of new preys A nearby excludes many particles B from the reaction (the so-called *effect of statistical screening*); only particles B on the aggregate's periphery can compete for new preys A. These structure rearrangements result in the oscillations of the reaction rate $K(t)$.

The behaviour of the correlation functions shown in Fig. 8.5 corresponds to the regime of unstable focus whose phase portrait was earlier plotted in Fig. 8.1. For a given choice of the parameter $\kappa = 0.9$ the correlation dynamics has a stationary solution. Since a complete set of equations for this model has no stationary solution, the concentration oscillations with increasing amplitude arise; in its turn, they create the "passive" standing waves in the correlation dynamics. These latter are characterized by the monotonous behaviour of the correlations functions of similar and dissimilar particles. Since both the amplitude and oscillation period of concentrations increase in time, the standing waves do not reveal a periodical motion. There are two kinds of particle distributions distinctive for these standing waves. Figure 8.5 at $t = 295$ demonstrates the structure at the maximal concentration

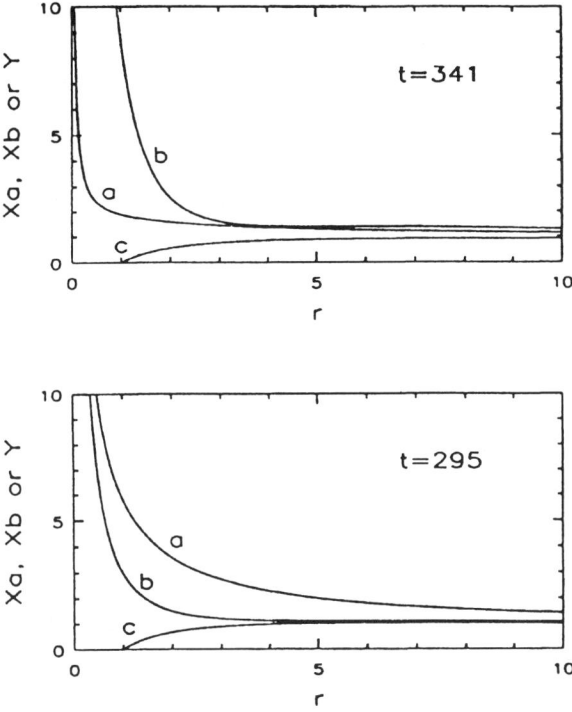

Fig. 8.5. The joint correlation functions in the Lotka–Volterra model in the unstable focus regime. (a) – $X_A(r,t)$, (b) – $X_B(r,t)$, (c) – $Y(r,t)$. $\kappa = 0.9$, $\bar{d} = 3$.

$N_a(t)$ whereas at $t = 341$ the structure with the minimal $N_a(t)$ is plotted. In the former-case at $r \sim 1$, $X_A(r,t) \gg X_B(r,t)$ holds (small aggregates of B's restrict large aggregates of A's) whereas in the latter case $X_A(r,t) \ll X_B(r,t)$ (the opposite situation) takes place.

More complicated case of standing waves emerges in the regime of chaotic oscillations. Here the equations for the correlation dynamics are able to describe auto-oscillations (for $\bar{d} = 3$). However, a noise in concentrations changes stochastically the amplitude and period of the standing waves. It results finally in the correlation functions with *non-monotonous* behaviour. Despite the fact that the motion of both concentrations and of the correlation functions is aperiodic, the time evolution of the correlation functions reveals several distinctive distributions shown in Fig. 8.6.

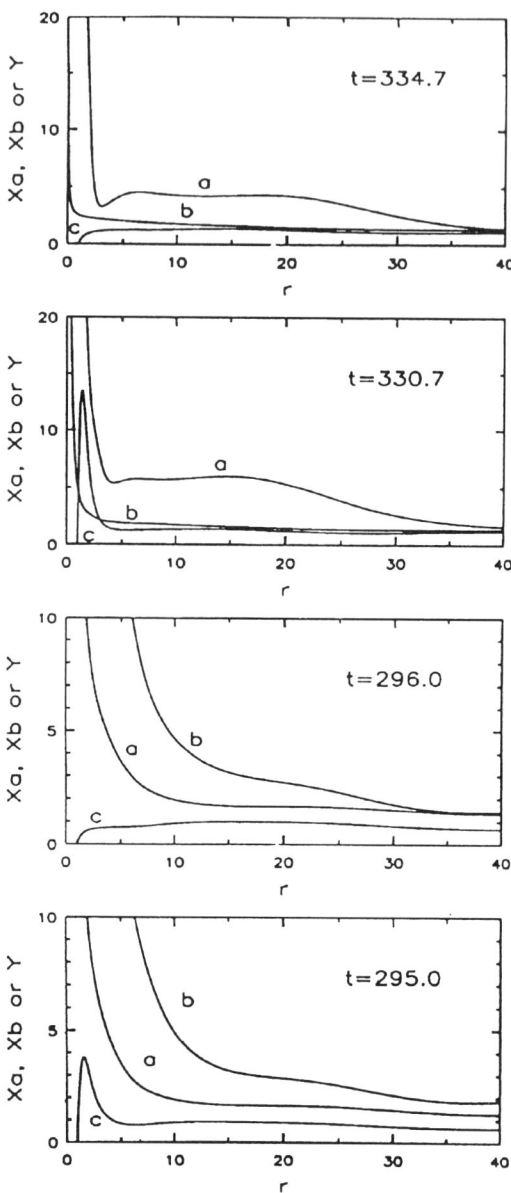

Fig. 8.6. The joint correlation functions in the Lotka–Volterra model: (a) – $X_A(r,t)$, (b) – $X_B(r,t)$, (c) – $Y(r,t)$. Parameters $\kappa = 0.05$, $\bar{d} = 3$.

These curves show four kinds of structures which are dependent on the current particle concentrations and the oscillation phases of the reaction rate $K(t)$. The moment of time $t = 295.0$ corresponds to the $K(t)$ maximum whose concentration $N_a(t)$ is close to its minimum value. The behaviour of the correlation functions reminds that shown in Fig. 8.5 but the function for the dissimilar particles has now maximum. After a short time interval, at $t = 296.0$, despite very small change of concentrations and the correlation functions for similar particles, the maximum in the correlation functions for *dissimilar* particles completely disappeared ($K(t)$ has a minimum).

The greater changes of the correlation functions are observed near $N_a(t)$ maximum shown in Fig. 8.6. The pattern for $t = 330.7$ corresponds to the $K(t)$ maximum; the correlation functions of dissimilar particles $Y(r,t)$ has a considerable peak. In its turn, at $t = 334.7$ we observe $K(t)$ *minimum* and no peak of Y. When particle concentrations change, so does the correlation function of similar particles. These changes as it is seen in Fig. 8.5 demonstrate the correspondence of the behaviuor of the correlation functions to the current particle concentrations.

A role of other parameters of the model is investigated by Kuzovkov [26]. It is demonstrated that an increase of the ratio α/β for a fixed $\omega_0 = (\alpha\beta)^{1/2}$ and the control parameter κ acts to accelerate a change of the focal regime for chaotic. Simultaneously, the amplitudes of oscillations in concentration for particles of different kinds are no longer close. A study of the stochastic Lotka–Volterra model performed here shows that irregular concentration motion observed experimentally in the Belousov–Zhabotinsky systems [8] indeed could take place in a system with mono- and bimolecular stages and two intermediate products only.

8.3 THE LOTKA MODEL

> What can be explained by means of simple arguments should not be expressed using stronger efforts.
> (Frustra fit per plura quod potest fieri per pauciora)
>
> Ockam

8.3.1 The stochastic Lotka model

The Lotka–Volterra model [23, 24] considered in the preceding Section 8.2 involves two autocatalytic reaction stages. Their importance in the self-organized chemical systems was demonstrated more than once [2]. In this

Section we show that presence of two such intermediate stages is more than enough for the self-organization manifestation. Lotka [22] was the first to demonstrate theoretically that the concentration oscillations could be in principle described in terms of a simplest kinetic scheme based on the law of mass action [4]. Its scheme given by (2.1.21) is similar to that of the Lotka–Volterra model, equation (2.1.27). The only difference is the mechanism of creation of particles A: unlike the reproduction by division, $E + A \to 2A$, due to the autocatalysis, a simpler reproduction law $E \to A$ with a constant birth rate of A's holds here. Note that analogous mechanism was studied by us above for the $A + B \to B$ and $A + B \to 0$ reactions (Chapter 7).

Following [21, 25], let us visualize the scheme (2.1.21) of the Lotka model in terms of a simple stochastic model as follows.

(i) Particle motion is characterized by the diffusion coefficients D_A and D_B;

(ii) Particles A are created with the constant rate p per unit volume and time;

(iii) The autocatalytic stage $A + B \to 2B$ consists of two elements: trapping of particles A near some particle B characterized by the rate $\sigma(r)$ and further creation of a new particle B at arbitrary distance r from an initial catalyst particle B (with the probability density $\mu_B(r)$);

(iv) Decay of particles B ($B \to P$) is characterized by a constant rate β.

8.3.2 Kinetic equations

Analogously the Lotka–Volterra model, let us write down the fundamental equation of the Markov process in a form of the infinite hierarchy of equations for the many-point densities. Thus equations for the single densities $(m + m') = 1$ read:

$$\frac{\partial \rho_{1,0}}{\partial t} = D_A \nabla_1^2 \rho_{1,0} + p - \int \sigma(|\vec{r}_1 - \vec{r}_1'|) \rho_{1,1} \, d\vec{r}_1', \qquad (8.3.1)$$

$$\frac{\partial \rho_{0,1}}{\partial t} = D_B \nabla_1'^2 \rho_{0,1} - \beta \rho_{0,1} +$$

$$+ \int \mu_B(|\vec{r}_1' - \vec{r}_2'|) \, d\vec{r}_2' \int \sigma(|\vec{r}_1 - \vec{r}_2'|) \rho_{1,1}(\vec{r}_1; \vec{r}_2'; t) \, d\vec{r}_1. \quad (8.3.2)$$

Equations for the joint densities $(m + m') = 2$ are

$$\frac{\partial \rho_{2,0}(\vec{r}_1, \vec{r}_2; t)}{\partial t}$$
$$= D_A \nabla_1^2 \rho_{2,0} + D_A \nabla_2^2 \rho_{2,0} + p(\rho_{1,0}(\vec{r}_1; t) + \rho_{0,1}(\vec{r}_2; t)) -$$
$$- \int \sigma(|\vec{r}_1 - \vec{r}_1'|) \rho_{2,1} \, d\vec{r}_1' - \int \sigma(|\vec{r}_2 - \vec{r}_1'|) \rho_{2,1} \, d\vec{r}_1', \qquad (8.3.3)$$

$$\frac{\partial \rho_{0,2}(\vec{r}_1', \vec{r}_2'; t)}{\partial t}$$
$$= D_B \nabla_1'^2 \rho_{0,2} + D_B \nabla_2'^2 \rho_{0,2} - 2\beta \rho_{0,2} +$$
$$+ \mu_B(|\vec{r}_1' - \vec{r}_2'|) \int \sigma(|\vec{r}_1 - \vec{r}_1'|) \rho_{1,1}(\vec{r}_1; \vec{r}_1'; t) \, d\vec{r}_1 +$$
$$+ \mu_B(|\vec{r}_1' - \vec{r}_2'|) \int \sigma(|\vec{r}_1 - \vec{r}_2'|) \rho_{1,1}(\vec{r}_1; \vec{r}_2'; t) \, d\vec{r}_1 +$$
$$+ \int \mu_B(|\vec{r}_1' - \vec{r}''|) \, d\vec{r}'' \int \sigma(|\vec{r}_1 - \vec{r}''|) \rho_{1,2}(\vec{r}_1; \vec{r}'', \vec{r}_2'; t) \, d\vec{r}_1 +$$
$$+ \int \mu_B(|\vec{r}_2' - \vec{r}''|) \, d\vec{r}'' \int \sigma(|\vec{r}_1 - \vec{r}''|) \rho_{1,2}(\vec{r}_1; \vec{r}'', \vec{r}_1'; t) \, d\vec{r}_1, \qquad (8.3.4)$$

$$\frac{\partial \rho_{1,1}(\vec{r}_1; \vec{r}_1'; t)}{\partial t}$$
$$= D_A \nabla_1^2 \rho_{1,1} + D_B \nabla_1'^2 \rho_{1,1} - \int \sigma(|\vec{r}_1 - \vec{r}_2'|) \rho_{1,2} \, d\vec{r}_2' -$$
$$- \beta \rho_{1,1} - \sigma(|\vec{r}_1 - \vec{r}_1'|) \rho_{1,1} + p\rho_{0,1} +$$
$$+ \int \mu_B(|\vec{r}_1' - \vec{r}''|) \, d\vec{r}'' \int \sigma(|\vec{r}_2 - \vec{r}''|) \rho_{2,1}(\vec{r}_1, \vec{r}_2; \vec{r}''; t) \, d\vec{r}_2. \qquad (8.3.5)$$

Due to a similarity of reaction stages in the Lotka and Lotka–Volterra models the equations for the $\rho_{0,1}$ and $\rho_{0,2}$ remain the same as in Section 8.2. Other kinetic equations are slightly simplified, a number and multiplicity of integrals are reduced.

As it was done in Section 8.2, to simplify further the derived kinetic equations let us consider the particular choice of the function $\mu_B(r) = \mu(r)$ in the form of equation (8.2.6), whereas particle trapping is considered in terms of the black sphere model (3.2.16). Despite the fact that model (8.2.6) permits to simplify the integral terms, unlike the Lotka–Volterra model, it cannot in principle be reduced to the diffusion-control problem. Creation of particles A with the constant rate means that along with the diffusive approach of A–B particles, some fraction of particles A is produced directly inside trapping (reaction) spheres around particles B. It results in an old problem of the calculation of a fraction of the system's volume available for a new particle creation discussed in Chapter 7. The use of the superposition approximation is restricted by some conditions which could be reformulated as certain limitations imposed on the model parameters. The limiting transition $r^* \to 0$ could be done in the way discussed in Section 8.2. Use of the model (8.2.6) affects equations (8.3.2), (8.3.4) and (8.3.5) only:

$$\frac{\partial \rho_{0,1}(\vec{r}_1';t)}{\partial t} = D_B \nabla_1'^2 \rho_{0,1} - \beta \rho_{0,1} + \int \sigma(|\vec{r}_1 - \vec{r}_2'|) \rho_{1,1} \, d\vec{r}_1, \tag{8.3.6}$$

$$\frac{\partial \rho_{0,2}(\vec{r}_1', \vec{r}_2';t)}{\partial t}$$
$$= D_B \nabla_1'^2 \rho_{0,2} + D_B \nabla_2'^2 \rho_{0,2} - 2\beta \rho_{0,2} +$$
$$+ \mu(|\vec{r}_1' - \vec{r}_2'|) \int \sigma(|\vec{r}_1 - \vec{r}_1'|) \rho_{1,1}(\vec{r}_1; \vec{r}_1'; t) \, d\vec{r}_1 +$$
$$+ \mu(|\vec{r}_1' - \vec{r}_2'|) \int \sigma(|\vec{r}_1 - \vec{r}_2'|) \rho_{1,1}(\vec{r}_1; \vec{r}_2'; t) \, d\vec{r}_1 +$$
$$+ \int \sigma(|\vec{r}_1 - \vec{r}_1'|) \rho_{1,2} \, d\vec{r}_1 + \int \sigma(|\vec{r}_1 - \vec{r}_2'|) \rho_{1,2} \, d\vec{r}_1, \tag{8.3.7}$$

$$\frac{\partial \rho_{1,1}(\vec{r}_1; \vec{r}_1';t)}{\partial t}$$
$$= D_A \nabla_1^2 \rho_{1,1} + D_B \nabla_1'^2 \rho_{1,1} - \int \sigma(|\vec{r}_1 - \vec{r}_2'|) \rho_{1,2} \, d\vec{r}_2' -$$
$$- \beta \rho_{1,1} - \sigma(|\vec{r}_1 - \vec{r}_1'|) \rho_{1,1} + p\rho_{0,1} + \int \sigma(|\vec{r}_2 - \vec{r}_1'|) \rho_{2,1} \, d\vec{r}_2. \tag{8.3.8}$$

That is, the singular terms containing $\mu(r)$ enter equation (8.3.7) for $\rho_{0,2}$, (the joint density of similar particles) only. As earlier, these terms are used to modify the boundary conditions at the coordinate origin.

8.3.3 Superposition approximation

A set of kinetic equations is cut off by means of the Kirkwood approximation (2.3.62). Let us introduce now dimensionless variables (primes are omitted below):

$$r' = \frac{r}{r_0}, \quad t' = \frac{Dt}{r_0^2}, \quad D'_\nu = \frac{2D_\nu}{D}, \quad p' = p\gamma_{\bar{d}} \frac{r_0^{2+\bar{d}_0}}{D},$$

$$\beta' = \beta \frac{r_0^2}{D}, \quad N_a = n_A \gamma_{\bar{d}} r_0^{\bar{d}}, \quad N_b = n_B \gamma_{\bar{d}} r_0^{\bar{d}},$$

where $D = D_A + D_B$.

The first equations (8.3.1) and (8.3.6) of a set define the dimensionless macroscopic densities of particles

$$\frac{dN_a(t)}{dt} = p - \widetilde{K}(t) N_a(t) N_b(t), \tag{8.3.9}$$

$$\frac{dN_b(t)}{dt} = \widetilde{K}(t) N_a(t) N_b(t) - \beta N_b(t). \tag{8.3.10}$$

They are accompanied by the equation for the reaction rate

$$\widetilde{K}(t) = K(t) + \frac{p}{\bar{d} N_a(t)}. \tag{8.3.11}$$

Here the first term arises from the *diffusive* approach of reactants A into trapping spheres around B's; it is nothing but the standard expression (8.2.14). The second term arises due to the direct production of particles A inside the reaction spheres (the forbidden for A's fraction of the system's volume). Unlike the Lotka–Volterra model, the reaction rate is defined by an approximate expression (due to use of the Kirkwood superposition approximation), therefore first equations (8.3.9) and (8.3.10) of a set are also approximate.

In deriving (8.3.11) we took into account the fact that despite the reaction rate is still defined through the integral (4.1.19), its magnitude depends greatly on the equation for the correlation function for dissimilar particles. Making use of the definition of the functional (5.1.5), let us write down equation (8.3.8) (in "usual" dimensional units) modified by the Kirkwood superposition approximation (2.3.62)

$$\frac{\partial Y(r,t)}{\partial t} = D\nabla^2 Y(r,t) - \sigma(r)Y(r,t) + \frac{p}{n_A(t)}\left(1 - Y(r,t)\right) +$$
$$+ Y(r,t)\{n_A(t)I[X_A] - n_B(t)I[X_B]\}. \tag{8.3.12}$$

This equation differs from (5.1.4) which served us as an example for calculating the reaction rate in the black sphere model. Introducing the function $h(r,t) = \sigma(r)Y(r,t)$, and taking into account (3.2.16), we arrive at

$$\sigma_0^{-1}\frac{\partial h(r,t)}{\partial t} = \theta(r_0 - r)\left(D\nabla^2 Y(r,t) + \frac{p}{n_A(t)}\right) - h(r,t) +$$
$$+ \sigma_0^{-1} h(r,t)\{n_A(t)I[X_A] - n_B(t)I[X_B]\} -$$
$$- \sigma_0^{-1}\frac{p}{n_A(t)} h(r,t). \tag{8.3.13}$$

In the limit of an instant recombination, $\sigma_0 \to \infty$, the degenerate equation defining $h(r,t)$

$$h(r,t) = \theta(r_0 - r)\left(D\nabla^2 Y(r,t) + \frac{p}{n_A(t)}\right) \tag{8.3.14}$$

differs from (5.1.24) by an additional term having non-diffusion nature. The use of (8.3.14) in calculating the reaction rate in a way demonstrated above, equation (5.1.26), leads directly to (8.3.11).

Equation (8.3.14) is not an asymptotically exact result for the black sphere model due to the superposition approximation used. When deriving (8.3.14), we neglected in (8.3.11) small terms containing functionals $I[Z]$, i.e., those terms which came due to Kirkwood's approximation. However, the study of the immobile particle accumulation under permanent source (Chapter 7) has demonstrated that direct use of the superposition approximation does not reproduce the exact expression for the volume fraction covered by the reaction spheres around B's. The error arises due to the incorrect estimate of the order of three-point density $\rho_{2,1}$ for a large parameter σ_0: at some relative distances ($|\vec{r}_1 - \vec{r}_1'| < r_0$, $|\vec{r}_2 - \vec{r}_1'| > r_0$) the superposition approximation is correct, $\rho_{2,1} \propto \sigma_0^{-1}$, however, it gives a wrong order of magnitude $\rho_{2,1} \propto \sigma_0^{-2}$ instead of the exact $\rho_{2,1} \propto \sigma_0^{-1}$ (if $|\vec{r}_1 - \vec{r}_1'| < r_0$, $|\vec{r}_2 - \vec{r}_1'| < r_0$). It was noted in Chapter 7 that the superposition approximation gives only first terms of expansion in powers of the dimensionless concentration, i.e., its range of validity is limited by small concentrations. Its generalization in the way demonstrated in Chapter 7 for mobile particles leads to very complicated equations which hardly could be used for real calculations.

The reaction rate $\widetilde{K}(t)$ is defined by the following equations for the correlation dynamics complementary to (8.3.9) and (8.3.10):

$$\frac{\partial X_A(r,t)}{\partial t} = D_A \Delta X_A(r,t) - 2K(t) N_b(t) X_A(r,t) \widetilde{J}[Y], \quad (8.3.15)$$

$$\frac{\partial X_B(r,t)}{\partial t} = D_B \Delta X_B(r,t) + 2K(t) N_a(t) X_B(r,t) \widetilde{J}[Y] +$$
$$+ \frac{2K(t) N_a(t)}{N_b(t)} r^{1-\bar{d}} \delta(r - r^*), \quad (8.3.16)$$

$$\frac{\partial Y(r,t)}{\partial t} = \Delta Y(r,t) + \frac{p}{N_a(t)} (1 - Y(r,t)) +$$
$$+ K(t) Y(r,t) \{ N_a(t) \widetilde{J}[X_A] - N_b(t) \widetilde{J}[X_B] \}. \quad (8.3.17)$$

A general form of the functionals $\widetilde{J}[Z]$ entering (8.3.15) to (8.3.17) is

$$\widetilde{J}[Z] = J[Z] + \frac{p}{K(t) \bar{d} N_a(t)} \hat{J}[Z]. \quad (8.3.18)$$

The functionals $J[Z]$ arise due to the diffusion approach of reactants and are defined by (5.1.36) to (5.1.38). Second term in (8.3.18) appears due to particle creation, the functional $\hat{J}[Z]$ reads there

$$\hat{J}[Z] = \bar{d} \int_0^1 \left(\widetilde{Z}(r^*,t) - 1 \right) r'^{\bar{d}-1} \, dr', \quad (8.3.19)$$

where $\widetilde{Z}(r^*,t)$ is defined in (5.1.8). The two functionals, $J[Z]$ and $\hat{J}[Z]$, numerically are close for smooth functions $Z(r,t)$ and could be approximated as $(Z(r,t) - 1)$.

As it follows from the definition (8.3.18) the relative contributions of particle generation and diffusion into the functional $\widetilde{J}[Z]$ depend on the parameter $\delta(t) = p/(K(t) \bar{d} N_a(t))$. In an analogous way the reaction rate $\widetilde{K}(t)$

in (8.3.11) could be presented in a form $\widetilde{K}(t) = K(t)(1+\delta(t))$. Taking into account the limited range of the applicability of the superposition approximation and complexity of functionals (8.3.19) involved (numerical calculation of two-fold integrals is rather time-consuming), we restrict ourselves hereafter by the regime of diffusion control where co-factors of small parameter $\delta(t)$ are omitted. In its turn, smallness of $\delta(t)$ could be guaranteed by a proper choice of the control parameters and of the initial conditions.

8.3.4 Basic equations

Let us consider as the basic equations those of the diffusion-controlled stochastic Lotka model which are derived in the superposition approximation, thus neglecting terms having a small parameter $\delta(t)$.

The relevant equations of the concentration dynamics read

$$\frac{dN_a(t)}{dt} = p - K(t)N_a(t)N_b(t), \qquad (8.3.20)$$

$$\frac{dN_b(t)}{dt} = K(t)N_a(t)N_b(t) - \beta N_b(t). \qquad (8.3.21)$$

Equations of the correlation dynamics are

$$\frac{\partial X_A(r,t)}{\partial t} = D_A \Delta X_A(r,t) - 2K(t)N_b(t)X_A(r,t)J[Y], \qquad (8.3.22)$$

$$\frac{\partial X_B(r,t)}{\partial t} = D_B \Delta X_B(r,t) + 2K(t)N_a(t)X_B(r,t)J[Y] +$$

$$+ \frac{2K(t)N_a(t)}{N_b(t)} r^{1-\bar{d}} \delta(r-r^*), \qquad (8.3.23)$$

$$\frac{\partial Y(r,t)}{\partial t} = \Delta Y(r,t) + \frac{p}{N_a(t)}(1 - Y(r,t)) +$$

$$+ K(t)Y(r,t)\{N_a(t)J[X_A] - N_b(t)J[X_B]\}. \qquad (8.3.24)$$

As it was shown in Section 8.2, the singular term in (8.3.23) could be omitted due to the transformation of the boundary condition at coordinate origin for the correlation function $X_B(r,t)$,

$$\frac{\partial X_B(r,t)}{\partial t} = D_B \Delta X_B(r,t) + 2K(t) N_a(t) X_B(r,t) J[Y]. \tag{8.3.25}$$

Numerical methods of solving these equations do not differ from those discussed in Section 8.2.

8.3.5 Standing waves

Independent control parameters of the Lotka model are p and β, describing reproduction of particles A and decay of B's, as well as the relative diffusion parameter $\kappa = D_A/(D_A+D_B)$, and the space dimension $\bar{d}$. Before discussing the solution of a complete set of the kinetic equations (8.3.20) to (8.3.24), let us formulate several statements.

Statement 1. The solution of the concentration dynamics (8.3.20) and (8.3.21) provided $K(t) = K = $ const, i.e., neglecting the correlation function motion, strives for the steady-state $N_a(\infty) = \beta/K$ and $N_b(\infty) = p/K$ irrespective of the magnitudes of parameters p, β and K. As $pK < 4\beta^2$, these equations describe damped concentration oscillations (a stable focus). As $pK > 4\beta^2$, the relaxation motion (a stable node) occurs.

This statement comes from analytical and topological studies [4]. Unlike the Lotka–Volterra model where due to the dependence of the reaction rate $K(t)$ on concentrations N_A and N_B, the nature of the critical point varied, in the Lotka model the concentration motion is always decaying. Autowave regimes in the Lotka model can arise under quite rigid conditions. It is easy to show that not any time dependence of $K(t)$ emerging due to the correlation motion is able to lead to the principally new results. For example, the reaction rate of the $A + B \to 0$ reaction considered in Chapter 6 was also time dependent, $K(t) \propto t^{1-\bar{d}/4}$ but its monotonous change accompanied by a strong decay in the concentration motion has resulted only in a monotonous variation of the quasi-steady solutions of (8.3.20) and (8.3.21): $N_a(t) \sim \beta/K(t)$ and $N_b(t) \sim p/\beta = $ const.

Deviation from standard chemical kinetics described in (Section 2.1.1) can happen only if the reaction rate $K(t)$ reveals its own *non-monotonous* time dependence. Since $K(t)$ is a functional of the correlation functions, it means that these functions have to possess their own motion, practically independent on the time development of concentrations. The correlation functions characterize the intermediate order in the particle distribution in a spatially-homogeneous system. Change of such an intermediate order could be interpreted as a series of structural transitions.

Statement 2. A set (8.3.22) to (8.3.24) with fixed concentrations $N_a = \beta/K$ and $N_b = p/\beta$ has two kinds of motions dependent on the value of parameter κ. As $\kappa \geqslant \kappa_0$, the stationary (quasi-steady-state) solution occurs, whereas for $\kappa < \kappa_0$ the correlation functions demonstrate the regular (quasi-regular) oscillations of the *standing wave type*. The marginal magnitude is $\kappa_0 = \kappa_0(p, \beta)$.

This statement could be proved in the manner similar to that used in Section 8.2. It is important to note that the correlation dynamics of the Lotka and Lotka–Volterra model do not differ qualitatively. A stationary solution exists for $\bar{d} = 3$ only. Depending on the parameter κ, different regimes are observed. For $\kappa \geqslant \kappa_0$ the correlation functions are changing monotonously (a stable solution) but as $\kappa < \kappa_0$, the spatial oscillations of the correlation functions (unstable solution) are observed. In the latter case a solution of non-steady-state equations of the correlation dynamics has a form of the non-linear standing waves. In one- and two-dimensional cases there are no stationary solutions of the Lotka model.

In a system with strong damping of the concentration motion the concentration oscillations are constrained; they follow oscillations in the correlation motion. As compared to the Lotka–Volterra model, where the concentration motion defines essentially the autowave phenomena, in the Lotka model it is less important being the result of the correlation motion. This is why when plotting the results obtained, we focus our main attention on the correlation motion; in particular, we discuss in detail oscillations in the reaction rate $K(t)$.

8.3.6 Concentration dynamics and oscillations of $K(t)$

A type of the asymptotic solution of a complete set of the kinetic equations does not also depend on the initial concentrations $N_a(0)$ and $N_b(0)$. Therefore, in calculations [21, 25] these values were fixed: $N_a(0) = N_b(0) = 0.1$, $p = 0.01$ and $\beta = 0.1$. Note once more that of primary importance of the diffusion-controlled Lotka model are the space dimension $\bar{d}$ and the relative diffusion parameter κ.

Figure 8.7 shows the dependence of the reaction rate $K(t)$ for different κ values. For the space dimension $\bar{d} = 3$ the obtained results could be easily interpreted as follows: there exists the marginal value of κ_0 (Statement 2). For $\kappa = 0.05$ the inequality $\kappa > \kappa_0$ holds. The stable stationary solution exists for the correlation dynamics and due to *decay* of the concentration motion

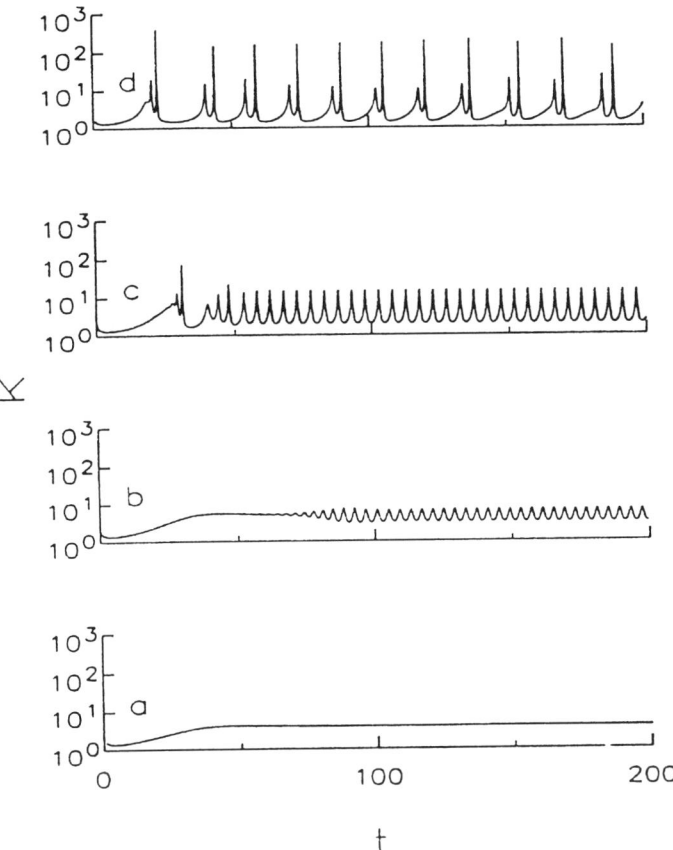

Fig. 8.7. Time dependence of the reaction rate $K(t)$ for $\bar{d} = 3$. κ: (a) – 0.05, (b) – 0.04, (c) – 0.02, (d) – 0.005.

(Statement 1) a stable stationary solution of a complete set of equations of the Lotka model holds. At long t the reaction rate $K(t)$ strives for the stationary value. Time development of concentrations obeys standard chemical kinetics, Section 2.1.1.

(b) The magnitude $\kappa = \kappa_0 \approx 0.04$ is marginal, the first bifurcation in the solution of a set of the Lotka equations arises. The standing waves emerge in the correlation motion, the oscillations $K(t)$ are developed also. For the long times these oscillations are perfectly periodical which means that *a limit cycle* takes place in the system. Note that oscillations of $K(t)$ are defined entirely by the correlation dynamics. In the region $\kappa < \kappa_0$ equations of the correlation dynamics give oscillatory solutions for fixed concentrations also. A change of the parameter κ leads to a series of bifurcation solutions of a set of kinetic equations, i.e., to more complicated motion which is demonstrated in Fig. 8.7 (c) and (d).

(c) Solution with $\kappa = 0.02$ for time interval $t < 200$ is quasi-periodic (neglecting a short transient period when a new structure of spatial particle distribution is formed). "Quasi" means that a real periodic solution arises after some transient period τ only. This τ is defined by the diffusion mechanism of the correlation function changes. Note specially a formation of the steady state for the correlation function $X_A(r,t)$, equation (8.3.22). It takes the time of the order of $\tau \sim L^2/D_A$ necessary to reach the optimal function at the distinctive distance L whose time development is described by the non-linear diffusion equation with the diffusion coefficient D_A. Since $D_A = 2\kappa$, we have $\tau \propto \kappa^{-1}$. Therefore, one more singular point arises naturally at $\kappa = 0$ at which the diffusion mechanism of the structure stabilisation disappears. When parameter κ decreases the formation time of the limit cycle increases.

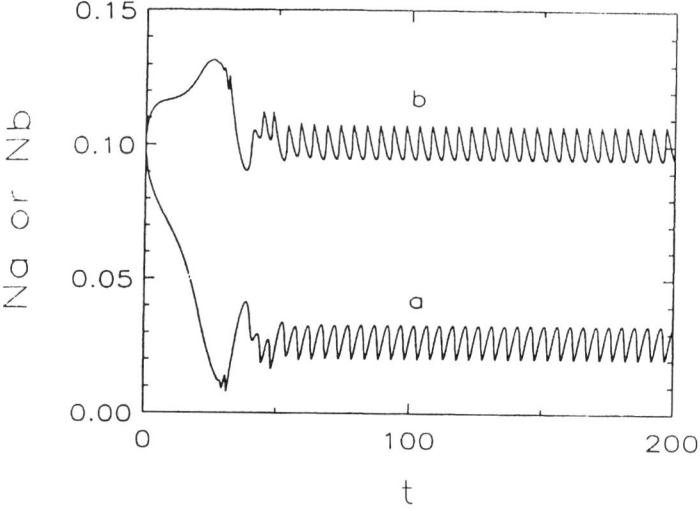

Fig. 8.8. Constrained concentration oscillations in the Lotka model: (a) – $N_a(t)$, (b) – $N_b(t)$, Parameters $\kappa = 0.02$, $\bar{d} = 3$.

It results in a number of transient regimes to be discussed below. At $\kappa = 0$ another (quasi-chaotic) kind of solution arises, which has another asymptotic behaviour.

Solution for a very small parameter $\kappa = 0.005$ demonstrates a distinctive transient regime with oscillations considerably different at the very beginning from what has been discussed in this Chapter.

Figure 8.8 shows constrained concentration oscillations for the parameter $\kappa = 0.02$, which corresponds to the oscillations of $K(t)$ in Fig. 8.7(c).

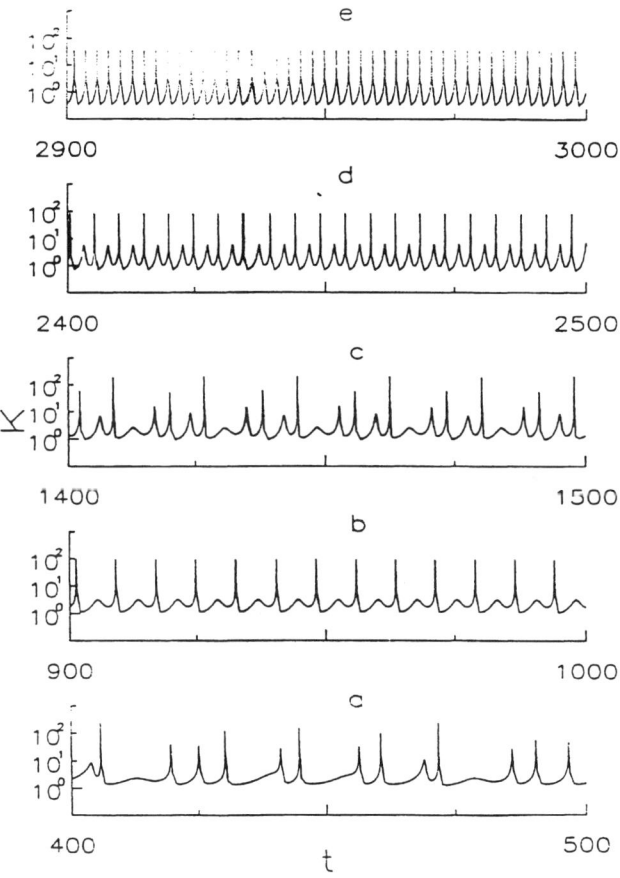

Fig. 8.9. Transient regimes in the Lotka model. Parameters $\kappa = 0.005$, $\bar{d} = 3$.

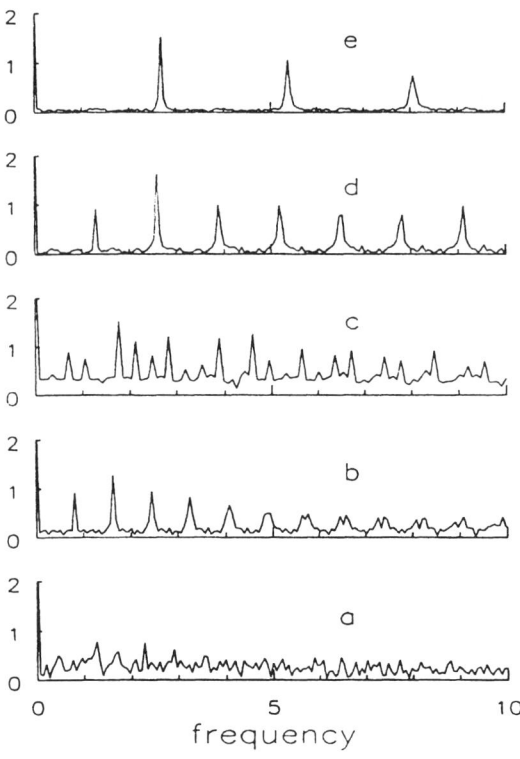

Fig. 8.10. The Fourier spectrum of the reaction rate $K(t)$. Parameters $\kappa = 0.005$, $\bar{d} = 3$. Enumeration of curves (a) to (e) corresponds to the time intervals presented in Fig. 8.9.

The amplitude of the oscillations $K(t)$ is relatively small, the same is true for the amplitudes of the concentration oscillations. The concentration $N_a(t)$ oscillates around the value of $b/\overline{K}$ where $\overline{K}$ is a mean value of $K(t)$ whereas $N_b(t)$ oscillates around p/β respectively.

Figure 8.9 serves as a good illustration of different possible *transient* regimes arising as κ is reduced. As stabilization time increases, $\tau \propto \kappa^{-1}$, the Lotka model reveals a series of quasi-periodic motions, separated by chaotic transient phases. The main trend seen from the analysis of results, is emergence of the periodic motion with a minimal period. To get some important properties of the transient irregular regimes, such as the presence of main frequencies or a white noise, it is useful to analyze the Fourier spec-

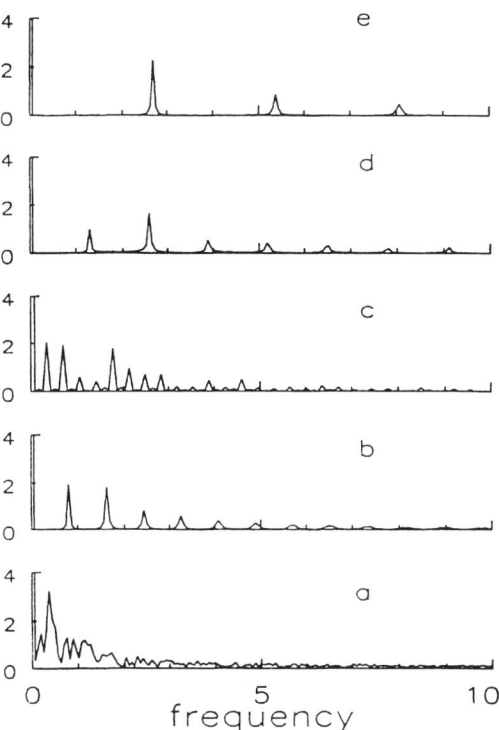

Fig. 8.11. The Fourier spectrum of concentration $N_a(t)$ (in relative units). Parameters $\kappa = 0.005$, $\bar{d} = 3$. Enumeration of curves (a) to (e) corresponds to the time intervals presented in Fig. 8.9.

trum [8]. The Fourier analysis of solutions (Figs. 8.10 and 8.11) shows that if we neglect a smooth change of frequencies due to the diffusion formation of the correlations in the system, the main reason of the quasi-periodic motion disappearance is a doubling of the main frequency. It is seen very well while comparing curves (d) and (e). This mechanism is pronounced better at long t when the distinctive profile of the correlation functions is formed already. The noise component (being well pronounced at short times) disappears gradually from the Fourier spectrum – curves (a). However, there exist another effect observed if we compare curves (b) to (d) – it is caused by the main frequency reduction in time. It is accompanied by a chaotisation of a motion and by the noise component increase.

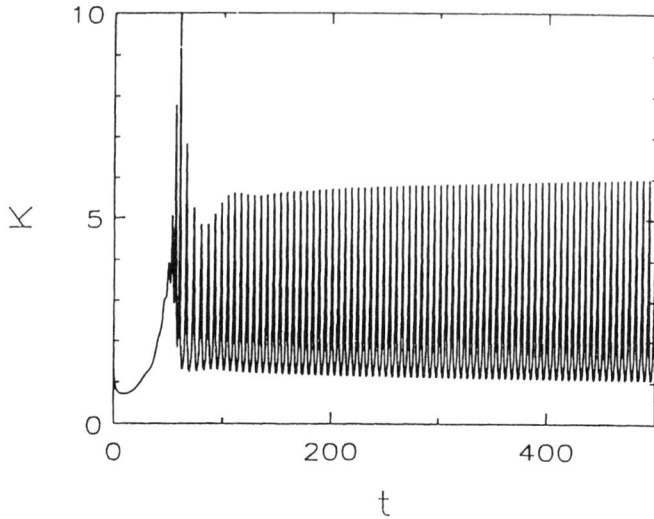

Fig. 8.12. Time dependence of the reaction rate for $\bar{d} = 2$. $\kappa = 0.1$.

The Fourier spectra of concentrations and of the reaction rate are quite similar. The difference is that the strong concentration decay suppresses the Fourier amplitudes at high frequencies. For the concentration motion the change in time of the $K(t)$ acts as an external noise from which the concentration motion "selects" the main frequencies.

The main peculiarity of low dimensions, $\bar{d} = 1, 2$ is that here a set of kinetic equations do not reveal neither a stationary solution, nor a limit cycle-type solution. However, for $\bar{d} = 2$ a solution has, however, some features similar to the three-dimensional problem. When restricting the calculation time by some finite value $t_{\max}$, two-dimensional case allows to introduce the marginal magnitude of $\kappa_0 = \kappa_0(p, \beta, t_{\max})$. As $\kappa > \kappa_0$, the quasi-steady solution is settled with the reaction time slowly varying in time, and the concentration $N_a(t) \approx \beta/K(t)$ follows the magnitude of $K(t)$. If $\kappa < \kappa_0$, a set of the correlation functions describes the standing waves. This motion, however, is not asymptotically periodic: both its amplitudes and periods change in time – see Fig. 8.12. The change of transient regimes as the diffusion parameter κ decreases is also observed for $\bar{d} = 2$.

The largest deviation is observed for $\bar{d} = 1$ – see Fig. 8.13. Here irrespectively of the value of parameter κ the oscillatory regime arises rather rapidly

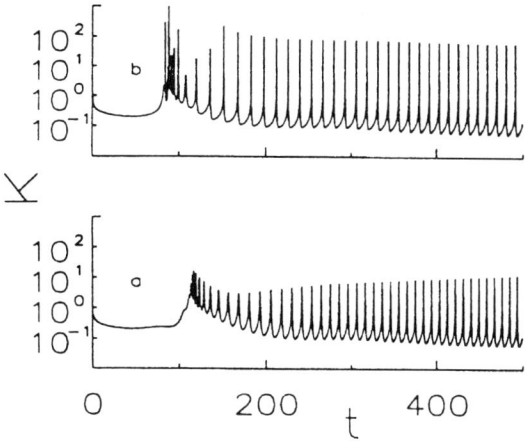

Fig. 8.13. Same for $\bar{d} = 1$. κ: (a) – 0.9, (b) – 0.1.

with the slowly increasing amplitudes of the concentration oscillations and $K(t)$. Change of κ does not result in qualitatively new results. Standard chemical kinetics (Section 2.1.1) loses here completely its applicability.

8.3.7 Correlation dynamics

The correlation functions for the Lotka model in the auto-oscillating regime are presented in Figs 8.14 and 8.15. The value of the parameter $\kappa = 0.02$ corresponds to the curves plotted in Figs 8.7(c) and 8.8. The correlation functions' motion is completely periodic, the results shown here correspond to the the minimum and maximum of $K(t)$.

The behaviour of the correlation function $X_B(r, t)$ is defined by the autocatalytic reaction stage: the probability to find some particle B near another B is rather high if they are reproduced by a division. For the short relative distances r the function $X_B(r, t)$ has a singularity which is, however, weakly pronounced, i.e., particles B are quasi-randomly distributed in space. For a chosen parameter $\kappa = 0.02$ the relative diffusion coefficients is large; $D_B = 2(1 - \kappa)$, $D_B \gg D_A$. The aggregates emerging under reproduction of B's are spread out rapidly due to the diffusion.

Of the greater interest is to understand the reason of the singularity of the correlation function for similar particles $X_A(r, t)$ at small r values. In

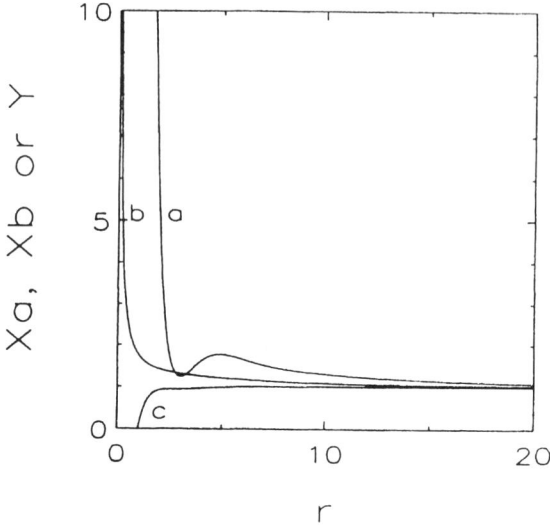

Fig. 8.14. The joint correlation functions for the Lotka model. (a) – $X_A(r,t)$, (b) – $X_B(r,t)$, (c) – $Y(r,t)$. Parameters $\kappa = 0.02$, $\bar{d} = 3$. These curves correspond to the $K(t)$ minimum.

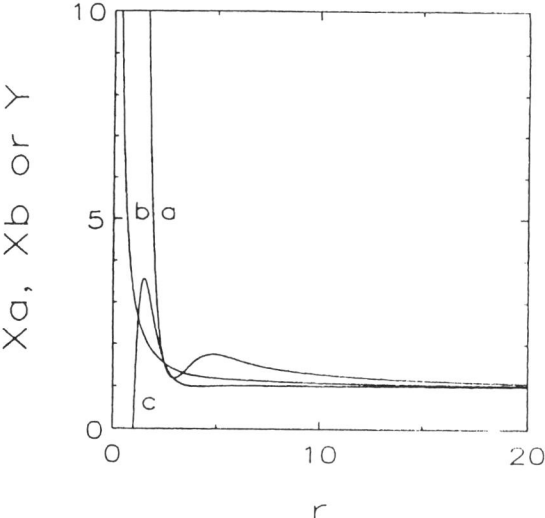

Fig. 8.15. Same as Fig. 8.14 for the $K(t)$ maximum.

the Lotka–Volterra model (Section 8.2) it was also defined by the auto-catalytic stage. In the Lotka model this stage is replaced by a more simple production stage for particles A; E → A. At this stage there is formally no clear correlation in the distribution of particles A in space. In reality, however, particles A are accumulated only in domains free of particles B, which makes A's to aggregate. For a given density $N_b(t)$ its value defines practically the mean sizes L of domains occupied by particles A. Consequently, the correlation function $X_A(r,t)$ differs from its unity asymptotics only at $r < L$. The effective aggregation of non-interacting particles A observed in the Lotka model reminds strongly the statistical effect of the Frenkel defect aggregation under permanent source (Chapter 7). However, the mentioned particle aggregation in domains is important only when mobility of particles is slow. This gives an explanation to the smallness of the marginal value of κ_0. As $\kappa > \kappa_0$, the diffusive smoothing of particles A turns out to be sufficient for destroying the auto-oscillating processes in the system.

In Fig. 8.14 particles B are distributed quasi-randomly, $X_B(r,t) \gg 1$ only at $r \leqslant 1$. The distinctive sizes of the aggregates of A's are of the order of several trapping radii, $L \sim 1$. Aggregates of A's have small amount of B's since on their boundaries the correlation function of dissimilar reactants $Y(r,t) \leqslant 1$.

The structural changes in the system due to reaction of particles B with aggregates A are shown in Fig. 8.15. The auto-catalytic stage $A + B \to 2B$ transforms particles A into aggregates of B's which are rapidly distributed in volume due to their diffusion. It results in an increase of $X_B(r,t)$ and decrease of $X_A(r,t)$. For the dense aggregates of A's such an auto-catalytic transformation of the boundary particles A into B's enriches greatly a number of AB pairs at the relative distances of the order of the aggregate's size. Respectively, at $r \approx 1$ the correlation function of dissimilar particles is large, $Y(r,t) \gg 1$. Therefore, emergence of the reaction rate maximum reflects the distinctive structural change – an increase of a number of AB pairs capable to participate in the reaction, i.e., separated by the distance of the order of the trapping radius.

After particles B destroyed several aggregates of A's, they leave in the same region the aggregates of B's smoothing out rapidly due to the diffusion. After some distinctive time particles B dying out under a deficiency of A's turn out to be randomly distributed in the system's volume. Simultaneously, particle A are accumulated in domains. Then the distinctive cycle of the structural change is repeated thus leading to the emergence of the self-oscillations. Note that the structural changes in the passage between Fig. 8.14 to Fig. 8.15 are not considerable. As a result, the amplitudes of the concentration oscillations, shown in Fig. 8.8, are also not large.

8.3.8 Conclusions

Irregular behaviour of concentrations and the correlation functions observed in the chaotic regime differ greatly from those predicted by law of mass action (Section 2.1.1). Following Nicolis and Prigogine [2], the stochastic Lotka–Volterra model discussed in this Section, could be considered as an example of *generalized turbulence*.

Therefore, the study of the stochastic Lotka and Lotka–Volterra models carried out in Chapter 8, has demonstrated that the traditional estimates of the complexity of the system necessary for its self-organisation *are not correct*. Incorporation of the fluctuation effects and thus introduction of a continuous number of degrees of freedom prove their ability for self-organisation and thus put them into a class of the *basic models* for the study of the autowave processes.

References

[1] W. Ebeling, Structurbildung bei Irreversiblen Prozessen (Teubner, Leipzig, 1976).
[2] G. Nicolis and I. Prigogine, Self-Organization in Non-Equilibrium Systems (Willey, New York, 1977).
[3] H. Haken, Synergetics (Springer, Berlin, 1978).
[4] A.M. Zhabotinsky, Concentration Self-Oscillations (Nauka, Moscow, 1974) (in Russian).
[5] V.A. Vasilev, Yu.M. Romanovsky and V.G. Yakhno, Autowave Processes (Nauka, Moscow, 1987) (in Russian).
[6] D. Edelson, R.M. Noyes and R.J. Field, Int. J. Chem. Kinet. 225 (1970) 535.
[7] M.L. Smoes, in: Dynamics of Synergetic Systems, ed. H. Haken (Springer, Berlin, 1980) p. 80.
[8] C. Vidal, in: Chaos and Order in Nature, ed. H. Haken (Springer, Berlin, 1981) p. 68.
[9] R.J. Field, in: Oscillations and Traveling Waves in Chemical Systems, eds R.J. Field and M. Burger (Willey, New York, 1985).
[10] J.J. Tyson, in: Oscillations and Traveling Waves in Chemical Systems, eds R.J. Field and M. Burger (Willey, New York, 1985).
[11] R. Feistel and W. Ebeling, Evolution of Complex Systems. Self-Organization, Entropy and Development (Kluwer, Boston, 1988).
[12] C. Vidal and H. Lemarchard, La Réaction Créatrice. Dynamique des Systémes Chimiques (Harmann, Paris, 1988).

[13] R.J. Field, E. Kórös and R.M. Noyes, J. Am. Chem. Soc. 94 (1972) 8649.
[14] R.J. Field and R.M. Noyes, J. Chem. Phys. 60 (1974) 1877.
[15] W.C. Troy, in: Oscillations and Traveling Waves in Chemical Systems, eds R.J. Field and M. Burger (Willey, New York, 1985).
[16] R.J. Field, J. Chem. Phys. 63 (1975) 2289.
[17] K. Tomita, A. Ito and T. Ohta, J. Theor. Biol. 68 (1977) 459.
[18] S. Schmidt and P. Ortoleva, J. Chem. Phys. 74 (1981) 4488.
[19] P. Hanusse, C. R. Acad. Sci. Ser. C: 274 (1972) 1245.
[20] J.J. Tyson and J.C. Light, J. Chem. Phys. 59 (1973) 4164.
[21] V.N. Kuzovkov and E.A. Kotomin, Rep. Prog. Phys. 51 (1988) 1579.
[22] A.J. Lotka, J. Phys. Chem. 14 (1910) 271.
[23] A.J. Lotka, J. Am. Chem. Soc. 27 (1920) 1595.
[24] V. Volterra, Leçons sur la Theorie Mathematique de la Lutte Pour la Vie (Paris, 1931).
[25] V.N. Kuzovkov, Latvian J. Phys. Techn. Sci. 2 (1991) 3.
[26] V.N. Kuzovkov, Teor. Eksp. Khim. 24 (1988) 1 (Sov. Theor. Exp. Chem., in Russian).
[27] V.N. Kuzovkov, Latv. PSR Zinat. Akad. Vestis Fiz. Teh. Zinat. Ser. 1 (1988) 17 (Proc. Latv. Acad. Sci. Phys. Technol. Ser., in Russian).
[28] E.A. Kotomin and V.N. Kuzovkov, Rep. Prog. Phys. 55 (1992) 2079.
[29] A.A. Samarsky, Theory of Difference Equations (Nauka, Moskow, 1983) (in Russian).
[30] A.D. Bazikin, Mathematical Biophysics of Interacting Populations (Nauka, Moskow, 1985) (in Russian).

Chapter 9

Catalytic Reactions on Solid Surfaces

9.1 A STOCHASTIC MODEL FOR SURFACE REACTIONS WITHOUT ENERGETIC INTERACTIONS

> Big battalions are always right.
> (Les gros bataillons ont toujours raíson)
> Napoleon

9.1.1 A general stochastic model of surface reactions

9.1.1.1 Introduction

In this Chapter we introduce a stochastic *ansatz* which can be used to model systems with surface reactions. These systems may include mono- and bimolecular steps, like particle adsorption, desorption, reaction and diffusion. We take advantage of the Markovian behaviour of these systems using *master equations* for their description. The resulting infinite set of equations is truncated at a certain level: in a small lattice region we solve the exact lattice equations and connect their solution to continuous functions which represent the behaviour of the system for large distances from a reference point. The stochastic *ansatz* is used to model different surface reaction systems, such as the oxidation of CO molecules on a metal (Pt) surface, or the formation of NH_3.

Surface reactions are very important in both theoretical and applied research. Experimental information on the individual reaction steps is difficult to obtain and the interpretation of the data is not easy. However, investigation of individual steps of a surface reaction can be obtained by using theoretical models, where discrete lattices are used to represent the surface. Depending on the number of different particles and on the adsorption and reaction steps, the models are classified as $A + A \to 0$, $A + B \to 0$, $A + \frac{1}{2}B_2 \to 0$, etc., reactions. Such lattice systems can be handled by computer simulations. A large variety of the Monte Carlo and cellular automata simulations has been done in the recent years. Some of them are in reasonable agreement with experimental observations [1]. The most prominent systems which have been studied are: the CO oxidation on a Pt catalyst [2, 3] and the catalytic formation of NH_3 [4]. The computer simulations have the disadvantage that the reaction steps should not be too complicated because of the tremendous amount of computer time which is needed. Therefore, theoretical (analytical) approaches are very important as an alternative for the description of these lattice models. Many different attempts have been done to handle these systems. Most of them focused on the very simple $A + A \to 0$ reaction because in this model many aspects can be solved analytically, at least in one dimension [5–9].

More complex systems which model real systems cannot be solved using purely analytical methods. For this reason we want to introduce in this Chapter a novel formalism which is able to handle complex systems using analytical and numerical techniques and which takes explicitly structural aspects into account. The *ansatz* can be formulated following the theory described below. In the present stochastic *ansatz* we make use of the assumption that the systems we will handle are of the Markovian type. Therefore these systems are well suited for the description in terms of master equations.

In such a representation of an infinite set of master equations for the distribution functions of the state of the surface and of pairs of surface sites (and so on) will arise. This set of equations cannot be solved analytically. To handle this problem practically, this hierarchy must be truncated at a certain level. In such an approach the numerical part needs only a small amount of computer time compared to direct computer simulations. In spite of very simple theoretical descriptions (for example, mean-field approach for certain aspects) structural aspects of the systems are explicitly taken here into account. This leads to results which are in good agreement with computer simulations. But the stochastic model successfully avoids the main difficulty of computer simulations: the tremendous amount of computer time which is needed to obtain good statistics for the results. Therefore more complex systems can be studied in detail which may eventually lead to a better understanding of such systems.

Definitions

The simulation takes place on a discrete lattice with coordination number z. Each lattice site is given a lattice vector l. The state of the site l is represented by the lattice variable σ_l, which may depend on the state of the catalyst site (e.g., promoted or not) and on the coverage with a particle. Let us assume we deal with the simple case in which all catalyst sites are equal and therefore σ_l depends only on its coverage. Another case is explained in the paper [10]. Therefore $\sigma_l = \{0, A, B, \ldots\}$ where 0 represents a vacant catalyst site, A is a site which is occupied by an A particle and so on. We also want to define a variable $\alpha_{l,n}$ which is unity for the case where l and n are the nearest neighbour sites on the lattice and zero otherwise.

Monomolecular steps

Monomolecular steps are processes which depend only on a single lattice site. Examples for such steps are the creation of a particle ($0 \to A$), the annihilation ($A \to 0$) or the particle transformation ($A \to B$). The prominent feature of these steps is the fact that the neighbourhood of the actual site plays no role here. Therefore these steps can be described by the following equation:

$$\sigma_l \overset{p}{\Rightarrow} \sigma_l', \qquad p \equiv P(\sigma_l \to \sigma_l'), \tag{9.1.1}$$

where p is the transition probability from state σ_l to σ_l'. All these transition probabilities are listed in the matrix P.

The transition probabilities are of course independent of l, only the initial and the final state are important. Normally many matrix elements of P are zero because not every transition is allowed.

Bimolecular steps

A more complicated situation of bimolecular steps arises if the step depends on two lattice sites l and n. Examples are reaction process ($A + B \to 0 + 0$), diffusion ($A + 0 \to 0 + A$) or a pair creation ($0 + 0 \to B + B$) which is useful for the description of dissociative adsorption events. All these processes can be formulated by

$$\sigma_l \sigma_n \overset{k}{\Rightarrow} \sigma_l' \sigma_n', \qquad k \equiv K(\sigma_l \sigma_n \to \sigma_l' \sigma_n')/z. \tag{9.1.2}$$

Here we have introduced a factor $1/z$ in order to simplify the equations which will arise.

i-Point probabilities

We introduce in this Section the probabilities $\rho^{(i)}$ which depend on i lattice sites. For $i = 1$ we obtain simple density of particles on the lattice. For these probabilities the condition

$$\sum_{\sigma_l} \rho^{(1)}(\sigma_l) = 1 \tag{9.1.3}$$

holds. We use the abbreviation

$$C_\lambda = \rho^{(1)}(\sigma_l), \quad \lambda = \sigma_l \tag{9.1.4}$$

yielding $\sum_\lambda C_\lambda = 1$.

The two-point probabilities $\rho^{(2)}(\sigma_l \sigma_m)$ depend on two states (σ_l and σ_m) respectively and on the coordinate difference $r = l - m$ of the lattice sites. For these probabilities the following sum rule holds:

$$\sum_{\sigma_m} \rho^{(2)}(\sigma_l \sigma_m) = \rho^{(1)}(\sigma_l). \tag{9.1.5}$$

Later we prefer to use the *correlation functions* $F_{\lambda\mu}(l - m)$, where $\lambda = \sigma_l$ and $\mu = \sigma_m$. With the help of $F_{\lambda\mu}$, $\rho^{(2)}$ can be expressed by

$$\rho^{(2)}(\sigma_l \sigma_m) = C_\lambda C_\mu F_{\lambda\mu}(l - m). \tag{9.1.6}$$

In the models discussed here $F_{\lambda\mu}$ depends only on the difference $r = l - m$. As $|r| \to \infty$, the correlations between the particles vanish:

$$\lim_{|r| \to \infty} F_{\lambda\mu}(r) = 1. \tag{9.1.7}$$

From the definitions made above we see that a product $c_\mu = C_\mu F_{\mu\lambda}(r)$ represents the mean density of μ-particles, if the central site is in the state λ. We will use this product later to characterize the different particle phases which appear in this model.

9.1.1.2 Master equations

Now we write down the master equations in the form of an infinite set of equations for the many-point probabilities.

A) Equation of motion for the one-point probabilities.

For the one-point probabilities we obtain

$$\frac{d\rho^{(1)}(\sigma_l)}{dt} \equiv \frac{dC_\lambda}{dt} = \left.\frac{dC_\lambda}{dt}\right|_{in} - \left.\frac{dC_\lambda}{dt}\right|_{out}, \tag{9.1.8}$$

where the term with the index 'in' describes the creation of a state σ_l and the term with the index 'out' the annihilation of a state σ_l:

$$\left.\frac{dC_\lambda}{dt}\right|_{in} = \left.\frac{d\rho^{(1)}(\sigma_l)}{dt}\right|_{in} = \sum_{\sigma'_l} P(\sigma'_l \to \sigma_l)\rho^{(1)}(\sigma'_l) +$$

$$+ \sum_n \frac{\alpha_{l,n}}{z} \sum_{\sigma'_l \sigma'_n \sigma_n} K(\sigma'_l \sigma'_n \to \sigma_l \sigma_n)\rho^{(2)}(\sigma'_l \sigma'_n), \quad (9.1.9)$$

and

$$\left.\frac{dC_\lambda}{dt}\right|_{out} = \left.\frac{d\rho^{(1)}(\sigma_l)}{dt}\right|_{out}$$

$$= \sum_{\sigma'_l} P(\sigma_l \to \sigma'_l)\rho^{(1)}(\sigma_l) +$$

$$+ \sum_n \frac{\alpha_{l,n}}{z} \sum_{\sigma'_l \sigma'_n \sigma_n} K(\sigma_l \sigma_n \to \sigma'_l \sigma'_n)\rho^{(2)}(\sigma_l \sigma_n). \quad (9.1.10)$$

In these equations monomolecular steps need one-point probabilities and bimolecular steps need two-point probabilities for their description. In order to simplify the equation we set $\sigma_l = \lambda$, $\sigma'_l = \lambda'$, $\sigma_n = \nu$, $\sigma'_n = \nu'$. The summation over n leads to z equal terms because all the correlation functions are taken for $|r| = 1$ (nearest neighbours). Therefore we can rewrite the last two equations in the form:

$$\left.\frac{dC_\lambda}{dt}\right|_{in} = \sum_{\lambda'} P(\lambda' \to \lambda)C_{\lambda'} +$$

$$+ \sum_{\lambda' \nu' \nu} K(\lambda' \nu' \to \lambda \nu)C_{\lambda'} C_{\nu'} F_{\lambda' \nu'}(1) \quad (9.1.11)$$

and

$$\left.\frac{dC_\lambda}{dt}\right|_{out} = \sum_{\lambda'} P(\lambda \to \lambda')C_\lambda +$$

$$+ \sum_{\lambda' \nu' \nu} K(\lambda \nu \to \lambda' \nu')C_\lambda C_\nu F_{\lambda \nu}(1). \quad (9.1.12)$$

In the last equation all terms have a cofactor of C_λ. Therefore one can introduce a very simple representation of the density equations:

$$\frac{dC_\lambda}{dt} = A_\lambda[C, F] - B_\lambda[C, F]C_\lambda, \qquad (9.1.13)$$

where $A_\lambda[C, F]$ and $B_\lambda[C, F]$ are simple positive functions (polynomials) of the densities C and of the correlation functions F.

B) Equation of motion for the two-point probabilities.

Among the processes which take place on two different lattice sites, l and m, the following ones are possible:

(1) l and m are not the nearest neighbours on the lattice.

(a) The state σ_l can be created or annihilated independently of σ_m. Only the neighbourhood of l is important.

(b) The state σ_m can be created or annihilated independently of l. Only the neighbourhood of m is important.

(2) If l and m are the nearest neighbours on the lattice additional terms must be taken into account which represent the bimolecular steps.

A *diagrammatic* description for these processes could be introduced:

$$\frac{d\rho^{(2)}(\sigma_l \sigma_m)}{dt} = \blacksquare\!\!-\!\!\underset{l}{\square} + \underset{m}{\square}\!\!-\!\!\blacksquare + \alpha_{l,m}\{\blacksquare\blacksquare\}. \qquad (9.1.14)$$

The symbol $\square$ denotes the site which plays no role for the determination of the state of the other site. On the site $\blacksquare$ a process takes place. Therefore the first two terms correspond to the cases (1a) and (1b) whereas the third term represents the bimolecular step. For the latter term we obtain

$$\{\blacksquare\blacksquare\} = \frac{1}{z} \sum \left(\{\blacksquare\blacksquare\}_{\text{in}} - \{\blacksquare\blacksquare\}_{\text{out}}\right)$$

$$= \frac{1}{z} \sum_{\lambda'\mu'} \left(K(\lambda'\mu' \to \lambda\mu)\rho^{(2)}(\sigma'_l\sigma'_m) - \right.$$

$$\left. - K(\lambda\mu \to \lambda'\mu')\rho^{(2)}(\sigma_l\sigma_m)\right) \qquad (9.1.15)$$

or

$$\alpha_{l,m}\{\blacksquare\blacksquare\} = \alpha_{l,m}\frac{1}{z}\sum_{\lambda'\mu}\left(K(\lambda'\mu' \to \lambda\mu)C_{\lambda'}C_{\mu'}F_{\lambda'\mu'}(1) - \right.$$

$$\left. - K(\lambda\mu \to \lambda'\mu')C_\lambda C_\mu F_{\lambda\mu}(1)\right). \qquad (9.1.16)$$

The diagrams of type (1a) and (1b) are of the form

$$\blacksquare_l - \square_m = \left(\blacksquare_{in\ l} - \square_m - \blacksquare_{out\ l} - \square \right). \tag{9.1.17}$$

The first term on the right hand side represents all processes which create the state σ_l. It differs from equation (9.1.9) in having the additional condition that the site m is in the state σ_m. Therefore we can write:

$$\left. \frac{d\rho^{(2)}(\sigma_l \sigma_m)}{dt} \right|_l^{in}$$

$$= \blacksquare_{in\ l} - \square_m$$

$$= \sum_{\sigma'_l} P(\sigma'_1 \to \sigma_l) \rho^{(2)}(\sigma'_l \sigma_m) +$$

$$+ \sum_n \frac{\alpha_{l,n}}{z} \sum_{\sigma'_l \sigma'_n \sigma_n} K(\sigma'_1 \sigma'_n \to \sigma_l \sigma_n) \rho^{(3)}(\sigma'_l \sigma'_n \sigma_m). \tag{9.1.18}$$

The corresponding 'out'-term is analogous to this expression. In the last equation the three-point probability arises for which the following sum rule holds:

$$\sum_{\sigma_m} \rho^{(3)}(\sigma_l \sigma_n \sigma_m) = \rho^{(2)}(\sigma_l \sigma_n). \tag{9.1.19}$$

9.1.1.3 The superposition approximation

We must truncate the infinite set of master equations in order to obtain a finite system of non-linear equations. To this end we use the lattice form of the superposition approximation introduced by Kirkwood [11] (Section 2.3).

In this approximation the three-point probability is expressed via two-point probabilities:

$$\rho^{(3)}(\sigma_l \sigma_n \sigma_m) \Rightarrow C_\lambda C_\nu C_\mu F_{\lambda\nu}(l - n) \times$$
$$\times F_{\nu\mu}(n - m) F_{\mu\lambda}(m - l). \tag{9.1.20}$$

This superposition approximation fulfills all necessary conditions:
(1) It is a function of C and F.

(2) If the lattice sites l, n and m are far away from each other, their state should be independent which means

$$\rho^{(3)}(\sigma_l \sigma_n \sigma_m) \approx \rho^{(1)}(\sigma_l)\rho^{(1)}(\sigma_n)\rho^{(1)}(\sigma_m) = C_\lambda C_\nu C_\mu.$$

This holds because of equation (9.1.20).

(3) If two sites are near each other (e.g., l and n) and m is far away, the state σ_m should be independent of the others. This means that
$\rho^{(3)}(\sigma_l \sigma_n \sigma_m) \approx \rho^{(2)}(\sigma_l \sigma_n)\rho^{(1)}(\sigma_m) = C_\lambda C_\nu C_\mu F_{\lambda\nu}(l-n).$

This approximation fails if all three points are situated near each other. In this case the equation (9.1.20) is only an approximation, but an approximation which has proved to be very successful [12].

Improvement of the superposition approximation

There is a problem in the use of the superposition approximation which arises in the case of lattice systems (the case which we are interested in): for $r = $ const we obtain a matrix for the correlation function $F_{\lambda\mu}(r)$ with $\lambda, \mu = 0, 1, \ldots, s-1$ suggesting that there are s different lattice states. Because of the symmetry $(F_{\lambda\mu} = F_{\mu\lambda})$ we obtain a number of functions which is smaller than s^2; the sum rule (9.1.5) leads to

$$\sum_\mu c_\mu = C_\mu F_{\mu\lambda}(r) = 1. \tag{9.1.21}$$

This means that we have s conditions ($\lambda = 0, \ldots, s-1$) and $s-1$ dependent correlation functions of the form

$$F_{\lambda\lambda}(r) = \frac{1}{C_\lambda}\left(1 - \sum_{\mu \neq \lambda} C_\mu F_{\lambda\mu}(r)\right). \tag{9.1.22}$$

This may lead to numerical difficulties because solutions can arise which do not fulfill the condition

$$F_{\lambda\mu} > 0. \tag{9.1.23}$$

This problem arises from the following fact: from equation (9.1.14) the correct equation for the one-point probabilities follows. It is easy to show that

A stochastic model for surface reactions without energetic interactions

$$\frac{d\rho^{(1)}(\sigma_l)}{dt} = \sum_{\sigma_m} \frac{d\rho^{(2)}(\sigma_l \sigma_m)}{dt}. \tag{9.1.24}$$

This equation is based on equation (9.1.19). But if we shall use the *ansatz* (9.1.20), we do not obtain equation (9.1.24). The reason lies in the bimolecular steps.

To avoid the problem with the dependent correlation functions we should seek another *ansatz* in which all correlation functions can be handled as independent ones.

In order to save equation (9.1.24) we break the symmetry of the equation (9.1.20) with respect to l, n and m. To this end we may introduce the strong condition – equation (9.1.19) (discussed later) – or we can use another, weak condition. To find this weak condition, let us look at the diagram (9.1.14). The three-point probabilities can be determined from the first diagram. To obtain equation (9.1.24) we have to compute two sums:

$$\sum_{\sigma_m} \left(\blacksquare\!\!-\!\!\square \atop l \quad m\right) + \sum_{\sigma_m} \left(\square\!\!-\!\!\blacksquare \atop l \quad m\right). \tag{9.1.25}$$

The second sum is given via equation (9.1.17) by

$$\sum_{\sigma_1} \left(\blacksquare\!\!-\!\!\square \atop l \quad m\right) = \sum_{\sigma_1} \left(\blacksquare \text{ in} \!\!-\!\!\square \atop l \quad m} - \blacksquare \text{ out} \!\!-\!\!\square \atop l \quad m}\right). \tag{9.1.26}$$

The sum takes the value of zero independent of the *ansatz* for the three-point probabilities because the in- and out-terms lead to the same value. Therefore we want to compute the first sum in equation (9.1.25) using an *ansatz* which is independent of the choice of the matrix $K(\sigma_l \sigma_n \to \sigma'_l \sigma'_n)$. To this end we look at equation (9.1.18) where we find a factor which includes a three-point probability and does not depend on $K(\sigma_l \sigma_n \to \sigma'_l \sigma'_n)$:

$$\sum_n \frac{\alpha_{l,n}}{z} \rho^{(3)}(\sigma_l \sigma_n \sigma_m). \tag{9.1.27}$$

(Here we have neglected the prime sign.) In this sum $n \neq m$ because l, m and n are different lattice sites. In addition we set

$$\rho^{(3)}(\sigma_l \sigma_n \sigma_m) \equiv 0 \tag{9.1.28}$$

for $l = n$, $n = m$ and $m = l$. A similar definition can be made for the two-point probabilities and the correlation functions:

$$\rho^{(2)}(\sigma_l \sigma_m) = 0 \qquad (9.1.29)$$

for $l = m$, or $F_{\lambda\mu}(r) = 0$ for $r = 0$. With these definitions we can compute the sum exactly:

$$\sum_{\sigma_m} \left(\sum_n \frac{\alpha_{l,n}}{z} \rho^{(3)}(\sigma_l \sigma_n \sigma_m) \right) = \left(1 - \frac{\alpha_{l,m}}{z} \right) C_\lambda C_\nu F_{\lambda\nu}(1). \qquad (9.1.30)$$

Instead of using condition (9.1.19) we assume that only the weak condition (9.1.30) holds. In this way we can derive equation (9.1.24) and we will not encounter the numerical difficulties mentioned above. For example, we can derive equation (9.1.21) automatically and we can handle all correlation functions as independent ones.

Using now this weak condition, we have to calculate the new resulting *ansatz*. To this end we have to make closer look at the Kirkwood approximation. The latter has a high symmetry with respect to the lattice vectors, but in the kinetic equations the three-point probabilities appear only in the form of equation (9.1.27) where the lattice sites play different roles: l and n are then nearest neighbours and in nearly all terms $|l - n| = 1$, $|l - m|$ and $|n - m| > 1$ holds. Therefore we can suggest that the sites l and n are in all cases more important than the site m. (The case $|l - m| = 1$ is an exception which we do not study in detail.) Therefore we can introduce another *ansatz* instead of (9.1.20) which is non-symmetric:

$$\rho^{(3)}(\sigma_l \sigma_n \sigma_m) \Rightarrow G_{\lambda\nu}(l - m) C_\lambda C_\nu C_\mu F_{\lambda\nu}(l - n) \times$$
$$\times F_{\nu\mu}(n - m) F_{\mu\lambda}(m - l). \qquad (9.1.31)$$

In this equation a correction factor $G_{\lambda\nu}(l - m)$ appears. We can order the sites appearing in this equation with respect to their importance: the site l is the main site (first order), n is the second site and m is a site which is far away. The matrix $G_{\lambda\mu}$ now becomes non-symmetric. The equation must fulfill all the conditions which are fulfilled by the *ansatz* (9.1.20). To this end $G_{\lambda\nu}(l - m)$ must be unity for large distances $|l - m|$. We use equation (9.1.30) and define a lattice average:

$$\widehat{F}_{\nu\mu}(l-m) = \sum_n \frac{\alpha_{l,n}}{z} F_{\nu\mu}(n-m). \qquad (9.1.32)$$

This leads to the condition

$$G_{\lambda\nu}(l-m)\left(\sum_\mu C_\mu \widehat{F}_{\nu\mu}(l-m) F_{\mu\lambda}(l-m)\right) = \left(1 - \frac{\alpha_{l,m}}{z}\right). \qquad (9.1.33)$$

With this condition the coefficients $G_{\lambda\mu}$ are easy to determine.

We use *ansatz* (9.1.31) with equation (9.1.18) (and corresponding terms) and define the temporal evolution as

$$\frac{d\rho^{(2)}(\sigma_l \sigma_m)}{dt} = \frac{d}{dt}\left(C_\lambda C_\mu F_{\lambda\mu}(l-m)\right). \qquad (9.1.34)$$

This leads to the system of non-linear equations of the general form

$$\frac{d}{dt} F_{\lambda\mu}(\boldsymbol{r}) = A_{\lambda\mu}[C, F] - B_{\lambda\mu}[C, F] F_{\lambda\mu}(\boldsymbol{r}). \qquad (9.1.35)$$

We are able to solve the set of equation (9.1.13) and equation (9.1.35) together. This *ansatz* avoids the main numerical difficulties described above.

Another *ansatz* different from (9.1.31) is also possible. In the direct *ansatz* using the strong condition (9.1.19) one can describe the sites l and n as being of the same order (they are the nearest neighbours) and the site m as a site which is far away. The resulting *ansatz* reads:

$$\rho^{(3)}(\sigma_l \sigma_n \sigma_m) \Rightarrow H_{\lambda\nu}(l-m, l-n) C_\lambda C_\nu C_\mu F_{\lambda\nu}(l-n) \times$$
$$\times F_{\nu\mu}(n-m) F_{\mu\lambda}(m-l). \qquad (9.1.36)$$

The correction factor $H_{\lambda\nu}(l-m, l-n)$ is a symmetric function which depends on the distance $(l-m)$ (main axis) and on the orientation to a neighbour (z possibilities). This factor can be defined by using equation (9.1.19):

$$H_{\lambda\nu}(l-m, l-n)\left(\sum_\mu C_\mu F_{\nu\mu}(n-m) F_{\mu\lambda}(m-l)\right) = 1. \qquad (9.1.37)$$

This *ansatz* is feasible but will lead to a lot of work because for every vector $(n-m)$ z functions must be calculated instead of the easy calculation of the matrix $G_{\lambda\mu}(l-m)$.

We have calculated several models and have compared the results with computer simulations to check the validity of the *ansatz* (9.1.31). In all models the results are in very good agreement with the corresponding computer simulations. These models will be discussed below.

9.1.1.4 Method of solution

The main problem in solving the obtained set of equations is connected with the solution of an infinite system of non-linear differential equations for a given type of lattice. To solve this problem in practice the following approximation is used [10]. A threshold value m_0 is introduced. For $|r| < m_0$ the lattice equations are solved for all non-equivalent points of the lattice ($m_0 = 5$ was used). This first area determines several coordination spheres in which the lattice aspect of the problem is important. In the second area all properties change quasi-continuously when the distance $|r|$ increases. Therefore we can use a continuum approximation introducing the coordinates $r = |r|$ and substituting instead of the correlation function $F(r)$ the radial one $F(r)$. Because of the weakness of the correlation we can use $F_{\lambda\mu}(\infty) = 1$ as the correct boundary condition (for $r \to \infty$). As the left (or inner) boundary condition (circumference of the circle with radius m_0) the solution within the first area at $|r| \leqslant m_0$ is used. When evaluating equation (9.1.32) some space point l is chosen. We assign to this point the average value of the function in the first coordination sphere (the nearest neighbours of l). The continuum analogue of this operation is

$$\widehat{F}_{\lambda\mu}(r) = \frac{1}{\pi}\int_{-1}^{+1} F_{\lambda\mu}(r')\frac{ds}{\sqrt{1-s^2}},$$

$$r' = \sqrt{r^2 + 1 - 2rs}. \qquad (9.1.38)$$

After these transformations the model can be solved effectively by numerical methods. As the initial condition, we have to specify the concentration of adsorbed particles and the pair correlation function. For example, for non-correlated distributed pairs we set $F_{\lambda\mu}(r) = 1$.

9.1.2 The $A + \frac{1}{2}B_2 \to 0$ surface reaction: island formation and complete particle segregation

9.1.2.1 Particular bimolecular reaction

In this Section we introduce a stochastic model for the $A + \frac{1}{2}B_2 \to 0$ reaction on a square lattice. Reaction between an A and a B particles occurs if they are the nearest neighbours on the lattice. To this system which includes adsorption and reaction steps we add the effect of A-diffusion and A-desorption. We describe the model in terms of master equations exploiting the Markovian behaviour of the system. The equations are truncated at a certain level via the modified Kirkwood approximation. In this Section the reaction between particles occur when they are the nearest neighbours on the lattice. This approach which is different from the standard one [13] requires a special treatment of the stochastic equations and the relevant correlation functions. In particular the Kirkwood superposition approximation has to be modified. The resulting system of lattice equations is solved in a small region around a reference point. The solution is connected to continuous functions which describe the system behaviour for larger distances. This system shows kinetic phase transitions which separate the reactive regime from two non-reactive states where the lattice is completely covered (*poisoned*) by A or B. The location and the character of the phase transitions was studied in detail in [2]. With the help of correlation functions we identify the different phases of particles on the lattice. Island formation and segregation of the particles on the lattice are found to be dominant processes. It is established that finite lattices which have to be used in simulations can be seriously inadequate and miss physical processes. This problem does not appear in the *ansatz* presented here.

We want to study a very important and interesting reaction model which takes place on a lattice. The particular reaction scheme reads as follows:

$$A(\text{gas}) + 0 \to A(\text{ad}) \quad \text{or} \quad 0 \xrightarrow{p} A, \quad p = p_A, \tag{9.1.39}$$

$$B_2(\text{gas}) + 20 \to 2B(\text{ad}) \quad \text{or} \quad 00 \xrightarrow{k} BB, \quad k = p_B/z, \tag{9.1.40}$$

$$A(\text{ad}) + B(\text{ad}) \to 20 \quad \text{or} \quad AB \xrightarrow{k} 00, \quad k = R/z, \tag{9.1.41}$$

$$A(\text{ad}) \to A(\text{gas}) + 0 \quad \text{or} \quad A \xrightarrow{p} 0, \quad p = k_A, \tag{9.1.42}$$

$$A0 \xrightarrow{k} 0A, \quad k = D/z, \tag{9.1.43}$$

where 0 denotes a free lattice site and A and B are reactant particles of our interest. This model (the first 3 equations) historically has been introduced by *Ziff, Gulari* and *Barshad* and is therefore called *the ZGB-model* [2]. Many investigations of this system have been performed because of its similarity to surface reaction systems. It has been shown [1] that some aspects of the behaviour of this system are in agreement with the heterogeneously catalyzed oxidation of CO. In this case A stands for a CO molecule, B_2 for an O_2 molecule and O for a free metal site which represents the catalyst. Equation (9.1.39) then shows the event of a CO-adsorption on a free surface site, equation (9.1.40) the dissociative adsorption of O_2 which requires two adjacent vacant sites and equation (9.1.41) the reaction of adsorbed CO and O to form CO_2 which desorbs immediately from the surface. Equation (9.1.42) is an extension of the original ZGB-model and describes the desorption of a CO molecule.

The most prominent feature of the original ZGB-model ($k_A \equiv 0$, $D = 0$) is the existence of *kinetic phase transitions*. Denoting the mole fraction of CO in the gas phase by Y_{CO} (and therefore $Y_{O_2} = 1 - Y_{CO}$), one finds a reactive interval $0.395 \pm 0.005 = y_1 < Y_{CO} < y_2 = 0.525 \pm 0.005$ [2] in which *both* particle types are coexisting on the surface. For $Y_{CO} < y_1$ and for $Y_{CO} > y_2$ the surface is completely covered by O_2 or CO, respectively. The phase transitions are found to be of the second order at y_1 and of the first order at y_2.

If one allows CO to diffuse (by the nearest neighbour hopping – equation (9.1.43)), which is a very common process at the normal temperatures of the CO oxidation, the value of y_2 is increased and reaches the stoichiometric value of 2/3 in the limit of an infinite diffusion rate [3]. Because of the irreversible character of the model (9.1.39) to (9.1.41) a completely covered surface (by one kind of species) means that the system can not escape from this absorbing state and therefore no further reactions can take place. In terms of the catalyzed CO oxidation, this is a state which describes a poisoned catalyst. The value of $y_2 = 2/3$ is in agreement with experimental observations [14].

The additional aspect of CO desorption, equation (9.1.42), leads to the disappearance of the CO-poisoned state [15] because at every value of Y_{CO} adsorbed CO molecules are able to leave the surface.

Many other investigations of reactions under various conditions have been performed, like energetic particle interactions [16], physisorption and reaction via the Eley–Rideal mechanism [17]. Most of the investigations of this model have been performed by computer simulation methods (i.e., the Monte Carlo and cellular automata). These methods are well suited for the study of this kind of system but they have two serious disadvantages: (1) They need a large amount of computing time in order to obtain good statistics. The necessary computing time can be reduced by using cellular automata instead of the Monte Carlo simulations [18] but the general problem remains. (2) All simulations can deal only with finite lattices. Especially at the phase transition of the second order at y_1 very large correlations appear which can only be described up to a certain degree correctly by using finite lattices.

Purely analytic models, on the other hand, are ruled out because these models are too complex for such an approach. Very simple theoretical approaches like mean-field *ansatz* which neglect correlations, fail in the prediction of the system behaviour at the phase transition point: the second-order phase transition is completely missing [3]. In addition, island formation and reactant segregation at the first-order phase transition are incorrectly described. The latter effect is namely a topic of this Section. In order to obtain a good description of the ZGB-model which avoids the difficulties described above and which can be practically handled, we have introduced a stochastic *ansatz* which describes the system by using the master equations (under the assumption that the system is Markovian) [10, 13, 19]. This *ansatz* takes explicitly into account correlations in the particle distribution on the lattice, which is necessary for a correct reaction description. Within this *ansatz* an infinite hierarchy of master equations appears which must be truncated at a certain level, say using SA. However, this approximation has a serious limitation for the original ZGB-model in which reaction occurs between the nearest neighbours with an infinite reaction rate R (see below). In [13] a variation of the ZGB-model was studied in which reaction occurs when a CO molecule jumps onto a lattice site which is already occupied by an O atom. We called this a *local model* in which the reaction rate R equals the (limited) diffusion rate D – in contrast to the original ZGB-model (*distant model*) where $R = \infty$. The results of the former model turned out to be in overall good agreement with the ZGB-model but some differences are existing due to the different reaction mechanism.

In this Section we want to introduce a stochastic model for the ZGB-model with diffusion in which reaction occurs between the nearest neighbours with an infinite reaction rate. This model shows many similarities to our previous model discussed above. Here we want to study only the new and different aspects. For more detail of the previous model see therefore Section 9.1.1. In addition we study the effect introduced by desorption and the segregation.

9.1.2.2 Basic equations

We introduce the stochastic equivalent to the ZGB-model with an infinite reaction rate R between an A and B particle which are the nearest neighbours on the lattice. First, we study the master equations for our system without diffusion and desorption in order to demonstrate how difficulties in using the Kirkwood approximation arise and why this approximation should be modified for the ZGB-model. We denote hereafter by $p_A = Y_{CO}$ and $p_B = 2(1-Y_{CO})$ the rate of creation of A and B, respectively. We focus our interest on the reaction system on a square lattice with $z = 4$ nearest neighbours (an extension to other discrete lattices is simple). We normalize the rates to a single lattice site in order to obtain simple single-point expressions. In configurations with two sites a normalizing factor of $1/z$ appears (for the case of the square lattice this factor is 1/4). The state of a lattice site is denoted by 0, A or B, which means an empty site, and a site occupied by A or B, respectively.

Let us now write down the temporal evolution of the A- and B-densities (equation (9.1.8)):

$$\frac{d}{dt}C_A = p_A C_0 - R C_A C_B F_{AB}(1) \tag{9.1.44}$$

and

$$\frac{d}{dt}C_B = p_A C_0^2 F_{00}(1) - R C_A C_B F_{AB}(1), \tag{9.1.45}$$

where $F_{\lambda\mu}(1)$ is a correlation function between the nearest neighbours. The first terms in these two equations represent the creation of an A particle or a BB pair, the second term shows the reaction between the nearest neighbours. Due to this term in which reaction rate tends to infinity ($R \to \infty$), serious problems arise. In order to show this we study the temporal evolution of AB pairs (AB = $\rho^{(2)}$(AB) = $C_A C_B F_{AB}(1)$):

$$\frac{d}{dt}\mathbf{AB} = p_A \mathbf{OB} + \frac{1}{4}p_B \left(\mathbf{AOO} + 2\mathbf{A}\overset{O}{\mathbf{O}}\right) - \frac{R}{4}\mathbf{AB} -$$
$$- \frac{R}{4}\left(\mathbf{BAB} + 2\mathbf{A}\overset{B}{\mathbf{B}}\right) - \frac{R}{4}\left(\mathbf{ABA} + 2\mathbf{A}\overset{A}{\mathbf{B}}\right). \tag{9.1.46}$$

We have used bold face letters for the sites of interest, in the equation above these are the ones on which an AB pair is formed or removed from the lattice. To denote the lattice we use a section of a grid accommodating larger configurations. This diagrammatic description is very easy to understand: the first term shows the creation of an AB pair by adsorption of an A in a 0B configuration. The second term shows the creation of AB by adsorption of B_2. The kinetic coefficients follow from the number of all possible configurations. The third term represents direct annihilation of an AB pair and the fourth and fifth terms describe an indirect annihilation processes, respectively.

In the distant model R goes to infinity and therefore $\mathbf{AB} \equiv 0$, but a product $R\,\mathbf{AB} = \mathbf{ab} \neq 0$ remains finite and plays a very important role. For example, the temporal evolution of the A-density reads:

$$\frac{d}{dt}\mathbf{A} = p_a \mathbf{O} - R\,\mathbf{AB} = p_a \mathbf{O} - \mathbf{ab} \quad \text{for} \quad R \to \infty. \tag{9.1.47}$$

If we would use the Kirkwood approximation, equation (9.1.20), directly in equation (9.1.46), we would neglect all three-particle terms. In particular, we would obtain

$$R(\mathbf{ABA}) = R\rho^{(3)}(\mathbf{ABA})$$
$$\Rightarrow R\frac{\rho^{(2)}(\mathbf{AB})\rho^{(2)}(\mathbf{BA})\rho^{(2)}(\mathbf{AA})}{\rho^{(1)}(\mathbf{A})\rho^{(1)}(\mathbf{A})\rho^{(1)}(\mathbf{B})} = \frac{1}{R}\frac{\mathbf{ab}\,\mathbf{ab}\,\rho(2)(\mathbf{AA})}{C_A C_A C_B} \to 0$$
$$\text{for} \quad R \to \infty. \tag{9.1.48}$$

Let us re-write equation (9.1.46) in the form

$$\frac{d}{dt}\mathbf{AB} = \frac{1}{R}\frac{d}{dt}\mathbf{ab} = 0 \quad \text{as} \quad R \to \infty \tag{9.1.49}$$

which is an exact result and holds for all other distribution functions. Therefore in the Kirkwood approximation we obtain

$$\mathbf{ab} = 4p_A \mathbf{OB} + p_B \left(\mathbf{AOO} + 2\mathbf{A}\overset{\mathbf{O}}{\underset{\mathbf{O}}{}} \right), \tag{9.1.50}$$

because all distribution functions $\rho^{(i)}$ which have more A than B particles (or more B than A's) in the neighbourhood play no role – in the Kirkwood approximation the product $R\rho^{(i)} = 0$ as $R \to \infty$ (see equation (9.1.48)). Therefore the direct use of the Kirkwood approximation is not suitable for our model. To solve this problem we study what will be called *virtual* distributions employed earlier in Chapter 7. Virtual distributions are those which arise from a normal particle distribution ($\rho^{(i)} \neq 0$) by creating an A particle or a BB pair in the empty sites of the configuration so that this results in the creation of one or more AB pairs. It has been shown that for such distribution functions the product $R\rho^{(i)} = \text{const}(t)$ holds [20]. Therefore we must study all these virtual distributions in detail.

First we write equation (9.1.46) in the form

$$\frac{1}{R}\frac{d}{dt}\mathbf{ab} = p_A \mathbf{OB} + \frac{1}{4}p_B \left(\mathbf{A}\overset{O}{\mathbf{O}} + 2\mathbf{A}\overset{O}{\mathbf{O}}\right) -$$
$$-\frac{1}{4}\mathbf{ab} - \frac{1}{4}\left(\mathbf{bab} + 2\overset{b}{\mathbf{ab}}\right) - \frac{1}{4}\left(\mathbf{aba} + \overset{a}{\mathbf{ab}}\right) = 0, \quad (9.1.51)$$

where $\mathbf{aba} = R(\mathbf{ABA})$. In this equation the virtual distributions play an important role. We must search for all these distributions which affect equation (9.1.51). They are characterized by the fact that their time derivatives go to zero. For example such a virtual distribution is given by equation

$$\frac{1}{R}\frac{d}{dt}\left(\overset{b}{\mathbf{ba}}\right) \Rightarrow 0 = p_A\overset{B}{\mathbf{BO}} - \frac{2}{4}\overset{b}{\mathbf{ba}} - \frac{1}{4}\left(\overset{b}{\mathbf{bab}} + \overset{b}{\underset{b}{\mathbf{ba}}}\right). \quad (9.1.52)$$

To obtain the temporal evolution of this virtual distribution (defined by the left hand side of this equation) we must analyse in which way it can be created and annihilated. The first term on the right hand site describes the creation due to an A-adsorption event. It can be annihilated by a direct (second term) and by indirect reaction events (third and fourth terms). The factor of 2/4 in the second term on the right hand side of the equation written above comes from the fact that here there are two possibilities to annihilate the A particle. The events written on the right hand side are all possibilities to create or annihilate this virtual distribution. Now we list all other virtual distributions which affect the temporal evolution of the AB pairs (equation (9.1.51)). With the help of all the virtual distributions we are able to express all virtual distributions through normal ones in equation (9.1.51). To this end we list all virtual distributions which affect the evolution of $\mathbf{ab}$ and solve it as a set of linear equations for the virtual distributions. The solution will be inserted in equation (9.1.51) in order to obtain an exact and handable equation. First, we study other virtual distributions with an A particle in the center and B particles in the neighbourhood. They are formed by A-adsorption in an appropriate configuration of B particles. In the last equation the A particle has two B particles as its neighbours. Now we write down the other possible virtual distributions with two and more B particles surrounding an A particle:

$$\frac{1}{R}\frac{d}{dt}(\mathbf{bab}) \Rightarrow 0 = p_A \mathbf{BOB} - \frac{2}{4}\mathbf{bab} - \frac{2}{4}\left(\mathbf{b}\atop\mathbf{bab}\right), \qquad (9.1.53)$$

$$\frac{1}{R}\frac{d}{dt}\left(\mathbf{b}\atop\mathbf{bab}\right) \Rightarrow 0 = p_A \mathbf{BOB}\atop - \frac{3}{4}\mathbf{b}\atop\mathbf{bab} - \frac{1}{4}\left(\mathbf{b}\atop\mathbf{bab}\atop\mathbf{b}\right), \qquad (9.1.54)$$

$$\frac{1}{R}\frac{d}{dt}\left(\mathbf{b}\atop\mathbf{bab}\atop\mathbf{b}\right) \Rightarrow 0 = p_A \mathbf{BOB}\atop\mathbf{B} - \frac{4}{4}\left(\mathbf{b}\atop\mathbf{bab}\atop\mathbf{b}\right). \qquad (9.1.55)$$

The equations listed above represent all possible virtual distributions with an A particle in the center surrounded by B particles.

Let us now study another set of virtual distributions with a B particle in the center and A particles in the neighbourhood. They are given by

$$\frac{1}{R}\frac{d}{dt}\left(\mathbf{a}\atop\mathbf{ab}\right) \Rightarrow 0 = p_B \frac{2}{4}\mathbf{AOO}\atop - \frac{2}{4}\mathbf{a}\atop\mathbf{ab} - \frac{2}{4}\left(\mathbf{a}\atop\mathbf{aba}\right), \qquad (9.1.56)$$

$$\frac{1}{R}\frac{d}{dt}(\mathbf{aba}) \Rightarrow 0 = p_B \frac{2}{4}\mathbf{AOA}\atop\mathbf{O} - \frac{2}{4}\mathbf{aba} - \frac{2}{4}\left(\mathbf{a}\atop\mathbf{aba}\right), \qquad (9.1.57)$$

$$\frac{1}{R}\frac{d}{dt}\left(\mathbf{a}\atop\mathbf{aba}\right) \Rightarrow 0 = p_B \frac{1}{4}\mathbf{AOA}\atop\mathbf{O} - \frac{3}{4}\mathbf{a}\atop\mathbf{b} - \frac{1}{4}\left(\mathbf{a}\atop\mathbf{aba}\atop\mathbf{a}\right). \qquad (9.1.58)$$

The last configuration where a B particle is surrounded by four A particles is not possible to obtain by creations of BB pairs and therefore it is zero.

With these exact equations we obtain an exact equation for **ab**:

$$\mathbf{ab} = p_A\left(4\,\mathbf{OB} - 2\,\mathbf{BOB} - 4\,\mathbf{BO}\atop\mathbf{B} + 4\mathbf{BOB}\atop\mathbf{B} - \mathbf{BOB}\atop\mathbf{B}\right) +$$

$$+ p_B\left(\mathbf{AOO} + 2\,\mathbf{AO}\atop\mathbf{O} - 2\mathbf{AOA} - \mathbf{AOO}\atop\mathbf{A} + \mathbf{AOA}\atop\mathbf{A}\right). \qquad (9.1.59)$$

Comparing now this exact equation with equation (9.1.50) we see that the Kirkwood approximation takes only a few terms into account. Next we must express equation (9.1.59) in terms of the single particle densities C_λ ($\lambda = 0, A, B$) and pair correlation functions $F_{\lambda\mu}(r)$. We see that if a rate $R_{\lambda\mu}$ is infinite, the corresponding distribution function $\rho^{(2)}$ is zero and

$F_{\lambda\mu}(1) = 0$. Therefore we should not write down the temporal evolution of $F_{\lambda\mu}(1)$ and should not solve it with the Kirkwood approximation. Instead we must calculate the product $R_{\lambda\mu}\rho^{(2)} = \Psi_{\lambda\mu} \neq 0$. To this end we use an extension of the idea of Kirkwood which takes into account all the probabilities up to $\rho^{(1+m)}$ (a central site plus $m = 1, \ldots, z$ neighbour sites). Here we use the so-called *Mamada–Takano approximation* [21]:

$$\rho^{(1+m)}(\sigma_l, \sigma_1, \ldots, \sigma_m) \Rightarrow \rho^{(1)}(\sigma_l) \prod_{i=1}^{m} \frac{\rho^{(2)}(\sigma_l, \sigma_i)}{\rho^{(1)}(\sigma_l)}. \tag{9.1.60}$$

Using this approximation in equation (9.1.59) we obtain

$$\mathbf{ab} = p_A C_0 g(\beta) + p_B C_0^2 F_{00}(1) h(\alpha) \tag{9.1.61}$$

with $\beta = C_B F_{0B}(1)$ and $\alpha = C_A F_{0A}(1)$ and

$$g(\beta) = 4\beta - 6\beta^2 + 4\beta^3 - \beta^4 \equiv 1 - (1-\beta)^z, \tag{9.1.62}$$

$$h(\alpha) = 3\alpha - 3\alpha^2 + \alpha^3 \equiv 1 - (1-\alpha)^{z-1}. \tag{9.1.63}$$

Because of the sum rule for the correlation functions, equation (9.1.21), $\alpha, \beta \leqslant 1$ and $h(\alpha), g(\alpha)$ are monotonic functions with $g(0) = h(0) = 0$ and $g(1) = h(1) = 1$.

To get better understanding of these results, let us study the temporal evolution of the A- and B-density:

$$\frac{d}{dt} C_A = p_A C_0 \big(1 - g(\beta)\big) - p_B C_0^2 F_{00}(1) h(\alpha), \tag{9.1.64}$$

$$\frac{d}{dt} C_B = p_B C_0^2 F_{00}(1) \big(1 - h(\alpha)\big) - p_A C_0 g(\beta). \tag{9.1.65}$$

For $R = \infty$ this means that number of A particles can increase only due to creation of an A particle if no B particles exist in their nearest neighbourhood. Otherwise reaction occurs and the number of A particles remains unchanged. The term $1 - g(\beta)$ represents the probability that in the nearest neighbourhood of an empty site there is no B particles and $g(\beta)$ stands for the probability that in the nearest neighbourhood of an empty site there is at least one B particle. The quantity $h(\alpha)$ has the corresponding meaning for a pair of empty sites and A particles in the neighbourhood. From the physical meaning it is clear that all these four probabilities should be positive. But in the Kirkwood approximation we obtain for the case $C_B \approx 1$ and $C_A \ll 1$ (which means $\alpha \ll 1$ and $\beta \approx C_B \approx 1$) that $g(\beta) = 4\beta$. That is for $\beta > 0.25$ a crude approximation because $1 - g(\beta) < 0$. But our new *ansatz* leads for this case to $g(\beta) \approx 1$ and $1 - g(\beta) \approx 0$ which is physically a correct result.

9.1.2.3 The diffusion and desorption processes

Now we have to introduce the process of diffusion and desorption of the A particles. It is a straightforward procedure; **ab** is changed by the diffusion to $\mathbf{ab} = \mathbf{ab}|_\text{creation} + \mathbf{ab}|_\text{diffusion}$ with

$$\mathbf{ab}\big|_\text{diffusion} = D\left(\overset{A}{AOB} + 2\overset{B}{OB} - 3\overset{B}{AOB} + \underset{A}{BOB} \right). \tag{9.1.66}$$

These terms represent jumps of an A particle into the central empty site forming one or more AB pairs. Using equation (9.1.60) we obtain

$$\mathbf{ab}\big|_\text{diffusion} = DC_A C_0 F_{0A}(1) h(\beta). \tag{9.1.67}$$

The A-desorption affects the A-density. We must only add simple terms like $-k_A C_A$ in the equation for the temporal evolution of the A-density. Here k_A is a desorption constant, equation (9.1.42).

Let us write down the temporal evolution of the A- and B-density:

$$\frac{d}{dt} C_A = p_A C_0 (1 - g(\beta)) - p_B C_0^2 F_{00}(1) h(\alpha) - \\ - DC_A C_0 F_{0A}(1) h(\beta) - k_A C_A, \tag{9.1.68}$$

$$\frac{d}{dt} C_B = p_B C_0^2 F_{00}(1)(1 - h(\alpha)) - p_A C_0 g(\beta) - \\ - DC_A C_0 F_{0A}(1) h(\beta). \tag{9.1.69}$$

The temporal evolution equations for $F_{\lambda\mu}(r)$, $\lambda\mu \neq AB$, which are not shown in this Section are the same as in Section 9.1.1.

9.1.2.4 Results

We discuss the model described above firstly neglecting both diffusion and desorption [22], which makes it analogous to the original ZGB-model. We obtain for the critical points $y_1 = 0.395 \pm 0.005$ and $y_2 = 0.565 \pm 0.005$,

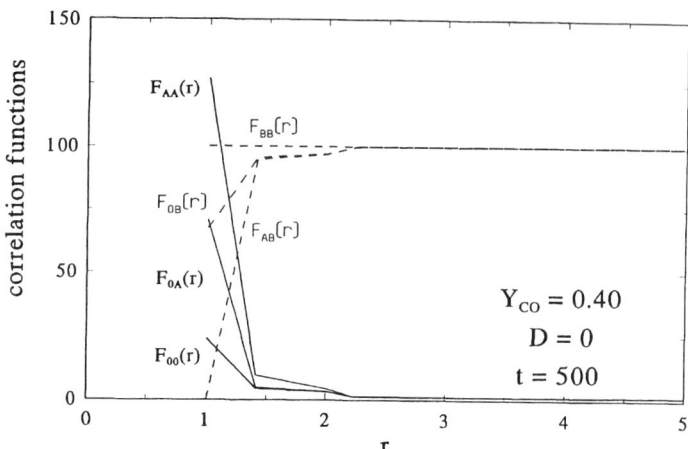

Fig. 9.1. The correlation functions for $Y_{CO} = 0.40$ slightly above the critical point y_1. Diffusion and desorption are neglected. The correlation functions with the non-bold letter F are multiplied by a factor of 100.

which is in good agreement with the results obtained by the Monte Carlo simulation of Ziff et al. [2] for y_1 and only fair for y_2 ($y_1 = 0.395 \pm 0.005$ and $y_2 = 0.525 \pm 0.005$). It is especially remarkable that the phase transition point y_1 is in such good agreement because at this point a phase transition of the second order is located. Models neglecting correlations in particle distributions (like mean-field models) lead to $y_1 = 0$ showing the importance of long-range correlations. The disagreement in y_2 will be explained below.

In Fig. 9.1 the correlations are shown for $Y_{CO} = 0.40$ which is slightly above the critical point y_1. (The non-bold characters indicate that the corresponding correlation functions are scaled by a factor of 100 to obtain a better representation.) Here $C_B = 0.986$ and $C_0 = 0.014$ are stationary coverages. $F_{0A}(1)$ is large because only empty sites are present in the neighbourhood of an A particle. $F_{AA}(1)$ is also quite large thus showing that the survival probability of several A particles clustered together is larger than for a single A particle. $F_{00}(1)$ is large but finite.

From equation (9.1.44) and (9.1.45) it follows for the stationary state, $p_A = Y_{CO}$, $p_B = 2(1 - Y_{CO})$,

$$Y_{CO} = 2(1 - Y_{CO})C_0 F_{00}(1). \tag{9.1.70}$$

A stochastic model for surface reactions without energetic interactions

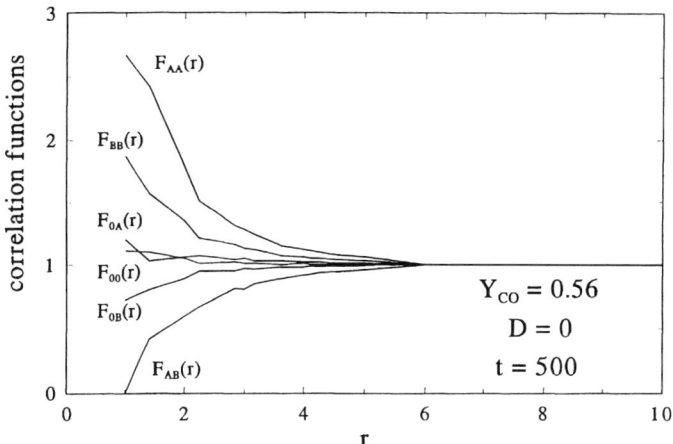

Fig. 9.2. The correlation functions for the reactive state at $Y_{CO} = 0.56$. Diffusion and desorption are neglected.

Rewriting the last equation in the form

$$C_0 = \frac{Y_{CO}}{2(1 - Y_{CO})F_{00}(1)}, \qquad (9.1.71)$$

we see that a phase transition (at which $C_0 \to 0$) with $y_1 \neq 0$ is only possible as $F_{00}(1) \to \infty$ for $t \to \infty$. From the results for $Y_{CO} = 0.4$ presented above it follows that we find ourselves $Y_{CO} = 0.4$ above the critical value of y_1. F_{BB} is nearly unity which means that the surface is nearly completely filled by B particles. F_{AB} is smaller than unity for very short distances ($r < \xi \approx 2$) showing that the A-clusters are very small.

Figure 9.2 shows the correlation functions for the reactive state just below the phase transition point ($Y_{CO} = 0.56 < y_2$). The steady state is reached in a very short time. The particle composition on the surface is a mixture of A and B particles with many empty sites in between.

In the case $Y_{CO} = 0.57 > y_2$ the poisoning with A occurs (Fig. 9.3), the stationary point is reached very slowly. The correlation functions show the extreme tendency of the particles towards clustering. The correlation length ξ of the clusters increases with increasing time. For $r < \xi$ the A particles form a matrix having almost no empty sites ($F_{AA} \approx 1$ and $F_{0A} \approx 0$).

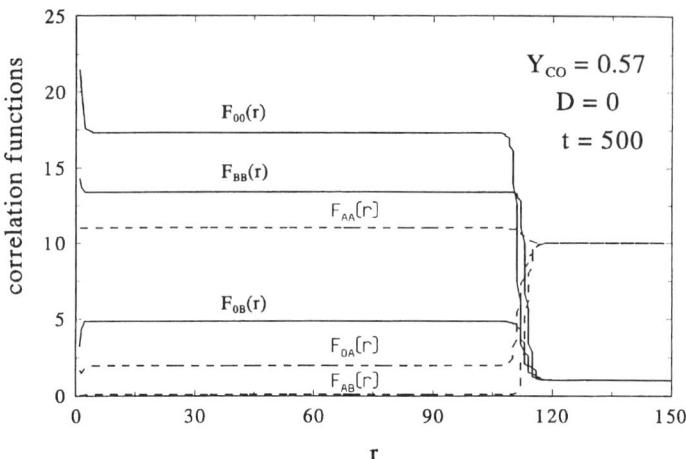

Fig. 9.3. The correlation functions for $Y_{CO} = 0.57$ where poisoning with A occurs. The correlation functions with a non-bold letter F are multiplied by a factor of 10.

The B-clusters are not so compact which means that there are many holes in these clusters ($F_{0B} \gg 1$). The B particles and the empty sites together form a cluster which comes from the fact that $F_{00} \gg 1$. This cluster does not contain A particles ($F_{AB} \approx 0$). The correlation length ξ increases with time but this does not mean that the B-clusters are growing (C_B decreases). It means only that the smaller clusters are removed first from the surface due to their higher reactive surface (ξ represents the minimal size of the remaining clusters).

It should be noticed that the above (correct) results are very difficult or impossible to obtain by using direct simulation methods for the case of a large correlation length ($\xi > 100$). Normally simulations are restricted to lattices which are smaller than 256×256 sites. This is in contrast to the method discussed here where we have used on (virtual) infinite lattice. Therefore a difference in the value of the critical points obtained by these two different procedures is understandable.

In Fig. 9.4 the temporal evolution of the concentrations C_A and C_B is shown for $Y_{CO} = 0.56$ (curve 1) and $Y_{CO} = 0.57$ (curve 2). For $Y_{CO} < y_2$ (curve 1) the relaxation time is very short. For $Y_{CO} > y_2$ (curve 2) the system needs much more time to reach the stationary point.

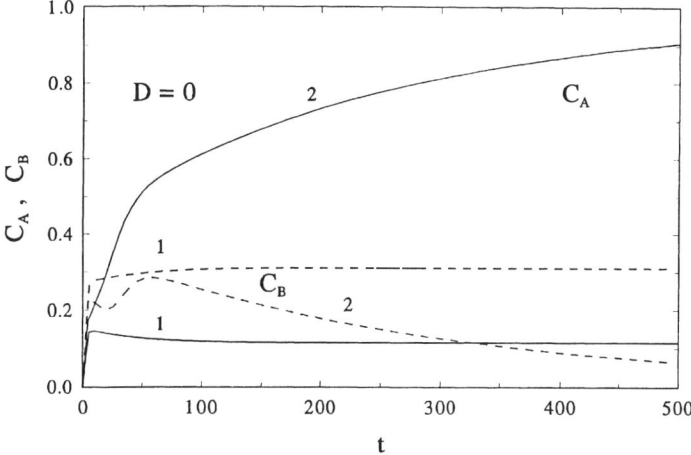

Fig. 9.4. The temporal evolution of C_A (solid line) and C_B (dashed line) for $Y_{CO} = 0.56$ (curve 1) and for $Y_{CO} = 0.57$ (curve 2).

9.1.2.5 The extended ZGB-model incorporating diffusion and desorption processes

At the next to our previous steps we want to study the additional aspect caused by A-diffusion [22]. In Fig. 9.5 the coverages of A and B and the reaction rate R_{CO_2} are shown as a function of the mole fraction of A (or CO) in the gas phase for the different diffusion rates $D = 0, 1, 10$, and 100. One can see that the phase transition at y_1 is not influenced by the A-diffusion because at this value of Y_{CO} there are only few A particles on the lattice. The value of y_2 increases with increasing D. The character of the phase transitions is not changed by the influence of the diffusion. This is also in agreement with the Monte Carlo simulations where y_2 approaches in the case of very fast diffusion the value of 2/3 [3].

We turn to the additional effect of the desorption of A particles characterized by the rate k_A. In Fig. 9.6 the particle concentrations and the reaction rate R_{CO_2} are shown as a function of Y_{CO} for $D = 0$ (no diffusion) and for three different desorption rates ($k_A = 0, 0.05$, and 0.20). For $k_A > 0$ the phase transition point y_2 vanishes because the complete coverage of the lattice by A particles cannot occur. For $Y_{CO} > y_2$ ($D = 0$) we obtain two different reactive regimes: (1) A stationary state for which the correlation length ξ for the A- and B-clusters remains finite for $t \to \infty$. In Fig. 9.6 we

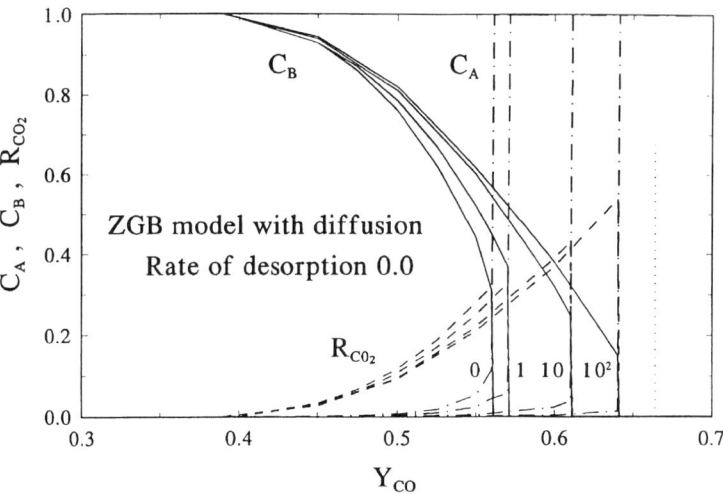

Fig. 9.5. The coverages and the reaction rate as a function of Y_{CO} for different diffusion rates ($D = 0, 1, 10$ and 100).

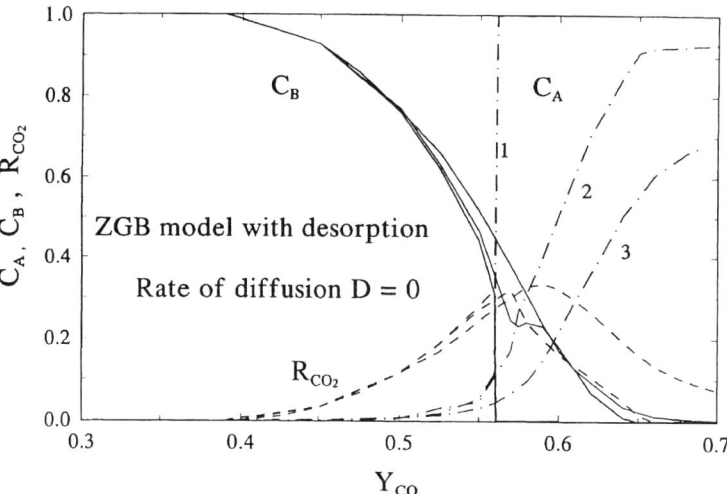

Fig. 9.6. The coverages and the reaction rate as a function of Y_{CO} for $D = 0$ (no diffusion) and for different desorption rates $k_A = 0(1), 0.05(2)$ and 0.20 (3).

A stochastic model for surface reactions without energetic interactions

obtain such states for $k_A = 0.05$ for $Y_{CO} < 0.57$ as well as for $Y_{CO} > 0.64$ and also for $k_A = 0.20$ for $Y_{CO} < 0.62$ as well as $Y_{CO} > 0.63$. The larger k_A the larger is the possible parameter regime for reaching such a state. (2) A very interesting non-stationary state in which ξ increases up to infinity as $t \to \infty$ occurs for $k_A = 0.05$ in the interval $0.57 < Y_{CO} < 0.64$ and for $k_A = 0.20$ when $0.62 < Y_{CO} < 0.63$. The larger k_A the shorter is the interval of Y_{CO} to reach this state. For diffusion rates $D > 10$ and for larger k_A the system cannot reach such a non-stationary state at all. The maximum for C_B in curve 2 should be noted. Here the system has reached a non-stationary state in which the correlation length increases with increasing time (see also Fig. 9.7). We obtain complete segregation in which particles form large islands. This increases the adsorption probability of the B particles because vacant lattice pairs can be found easier compared to the case of a random particle distribution. This effect leads to a larger coverage of B particles than for the stationary states. R_{CO} decreases also sharply if the system approaches the non-stationary state showing the segregation of the particles which leads to decreased reactivity because reaction can only occur at the borders of the formed clusters. For $k_A = 0.20$ we do not observe a maximum in C_B

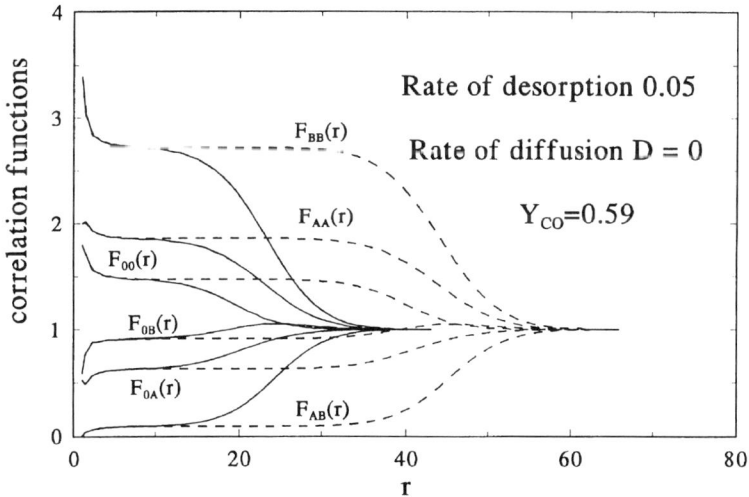

Fig. 9.7. The correlation functions for $t = 100$ (solid curve) and for $t = 200$ (dashed curve). $Y_{CO} = 0.59$, $D = 0$ and $k_A = 0.05$.

because the non-stationary state occurs at large values of Y_{CO} where C_B is very small.

In Fig. 9.7 the correlation functions for a non-stationary state are shown for $t = 100$ (solid curve) and for $t = 200$ (dashed curve). The most prominent feature of this state is that the correlation length increases with increasing time but the concentrations are constant over the whole time ($C_A = 0.404$ and $C_B = 0.232$). We use the quantity c_λ ($\lambda = 0, A,$ and B) to characterize the appearing phases. For example, $c_\lambda = C_\lambda F_{\lambda\mu}(r)$ gives the mean density of λ particles at a distance r from a central site occupied by μ particle. We observe three different phases for $r < \xi$: (1) a phase of A particles in which single B particles occur ($\mu = A$, $c_A = 0.75$ and $c_B = 0.02$), (2) a phase which consists completely of B particles ($\mu = B$, $c_A = 0.04$ and $c_B = 0.63$) and lastly, (3) a mixture of particles A, B and empty lattice sites ($\mu = 0$, $c_A = 0.54$ and $c_B = 0.25$). With increasing time the phases (1) and (2) evolve and phases (3) vanishes as $t \to \infty$. The final state of a system consists of phases (1) and (2) and a transition region between them. A complete covering does not occur because of the empty sites and the particles of the other type inside the phases (1) and (2). We studied the system under different initial conditions by occupying the lattice with particles which represents phase (1) or (2), respectively. After a short time the system evolves to the same state described above (Fig. 9.7). This means that phases (1) and (2) are existing only together on the lattice, i.e., we have a two phase region in an irreversible reaction. This corresponds to the complete particle segregation.

9.1.2.6 Conclusion

We have introduced in this Section a stochastic model for the $A + \frac{1}{2}B_2 \to 0$ reaction which is equivalent to the ZGB-model [2] and thus remedies a deficiency of a previously presented model [13]. In this model we obtain for the case of no diffusion for the phase transition points $y_1 = 0.395$ and $y_2 = 0.565$, which are in good or fair agreement with the results of the ZGB-model ($y_1 = 0.395$ and $y_2 = 0.525$). In the model [13] where the reaction occurs only if A particle jumps to active site occupied by a B particle, we obtain $y_1 = 0.27$ and $y_2 = 0.65$ (for $D = 10$). Because the reaction occurs only due to diffusion, we cannot directly compare this model with the ZGB-model in which no diffusion exists. But the value of y_2 is in agreement with computer simulations of the extended ZGB-model including diffusion ($y_2 = 0.65$ for a high diffusion rate) [3]. The value of y_1 should not be influenced by the additional aspect of A-diffusion because too few A particles are on the surface at this value of Y_{CO}. Therefore the result of our previous model for the phase transition point at y_1 is not in good agreement with the ZGB-model. The model presented in this Section is similar to the ZGB-model (reaction between the nearest neighbours with probability equal to unity), it leads to results which are in very good agreement with the pioneering computer simulation of Ziff, Gulari and Barshad [2]. A small difference remains for the value of y_2. This discrepancy arises due to the finite lattice size used in the simulations. These finite lattices are not able to describe correctly long-range correlations which may appear at phase transition points.

We have seen that the width of the reaction interval $y_1 < Y_{CO} < y_2$ is given by the size of the B-clusters (see Fig. 9.3). If they have reached a size $L > L_0 = L_0(y_{CO}, C_A(0))$ the state remains stable. Here $C_A(0) = C_A(t = 0)$ is the starting A particle concentration. If $C_A(0) > 0$ it is more difficult for the B particles to form large clusters and therefore the critical point y_2 is lowered [2, 3]. During the poisoning process by A the B-clusters are removed from the surface starting with the smallest clusters which have the highest reactivity (largest reactive surface). In a finite system it is impossible to obtain a complete distribution over all cluster sizes. That means that the value of y_2 obtained in a finite system is smaller than the value obtained by this stochastic model which gives the limit value for an infinite system.

The additional effect of A-diffusion shifts the value of y_2 to larger values of Y_{CO} which is in agreement with simulation methods [3]. The value of y_1 is not affected by A-diffusion and desorption because at this point of Y_{CO} too few A particles are on the surface.

Because of A-desorption the complete poisoning of the surface by A cannot occur. Instead we observe a non-stationary phase above y_2 (see Fig. 9.6 and 9.7). In a mixture of A and B particles two phases arise. One phase consists of A particles with empty sites in between whereas another consists of B particles. These two phases can only co-exist together. Starting with the first or second phase we obtain in every case the same final result which consists of mixture of both phases. The density of particles is constant but the correlation length ξ increases with time and thus we obtain as a result a completely segregated system. This state is limited to a small parameter interval of Y_{CO} which is controlled by the desorption and the diffusion rate. An increase of the desorption and diffusion rate narrows the interval of Y_{CO}. For $D > 10$ and for $k_A > 0.2$ the system cannot reach such an unusual, non-stationary state. The long-range correlations cannot be correctly described by simple theories like a mean-field *ansatz* which neglects the spatial particle correlations. Simulations are also limited due to the use of finite lattices. The stochastic model presented here avoids these problems and leads therefore to a realistic formalism reproducing results of the ZGB-model.

9.1.3 The influence of surface disorder on the $A + \frac{1}{2}B_2 \to 0$ reaction

9.1.3.1 Problem statement

We study here the $A + \frac{1}{2}B_2 \to 0$ reaction upon a *disordered* square lattice on which only a certain fraction S of lattice sites can be accessed by the particles (the so-called *active sites*). We study the system behaviour as a function of the mole fractions of A and B in the gas phase and as a function of a new parameter S. We obtain reactive states for $S > S_0$ where S_0 is the kinetically defined percolation threshold which means existence of an infinite cluster of active sites. For $S < S_0$ we obtain only finite clusters of active sites exist. On such a lattice all active sites are covered by A and B and no reaction takes place as $t \to \infty$.

The behaviour of surface reaction is strongly influenced by structural variations of the surface on which the reaction takes place [23]. Normally theoretical models and computer simulations for the study of surface reaction systems deal with perfect lattices such as the square or the triangular lattice. However, it has been shown that fractal-like structures give much better description of a real surface [24]. In this Section we want to study the system (9.1.39) to (9.1.42).

It should be reminded here that the most interesting feature of the original ZGB-model is the existence of kinetic phase transitions. Denoting the mole fraction of CO in the gas phase by Y_{CO} (and therefore $Y_{CO_2} = 1 - Y_{CO}$), one finds a reactive interval $0.395 = y_1 < Y_{CO} < y_2 = 0.525$ [2] in which both particle types are coexisting on the surface. For $Y_{CO} < y_1$ and for $Y_{CO} > y_2$ the surface is completely covered by O_2 or CO, respectively. The phase transitions are found to be of the second order at y_1 and of the first order at y_2. Because of the irreversible character the model describes a poisoned state from which the system cannot escape and reaction comes to a stop.

We have studied the system (9.1.39) to (9.1.41) by means of the Monte Carlo method on a disordered surface where the active sites form a percolation cluster built at the percolation threshold and also above this threshold [25]. Finite clusters of active sites were removed from the surface to study only the effect of the ramification of the infinite cluster. The phase transition points show strong dependence on the fraction of active sites and on the lattice size used in the simulations. If S approaches the percolation threshold from above, the phase transition points approach each other decreasing a reactive interval. The main problem of the computer simulations is the use of finite lattice: due to the finite-size effect and to the relatively short simulation time (which is strongly limited by the performance of the available computers) we found at $S = S_c$ a short reactive interval, $[0.305, 0.424]$. These restrictions of the computer simulations will be removed in theory presented in this Section, using the *ansatz* described in [19]. In the ZGB-model a difficulty arises due to the fact that the reaction rate R goes to infinity which means that the reaction probability of an AB pair on the lattice equals unity. For this case we need an improvement of the superposition approximation which is discussed in [26]. The formal description for the case when not all lattice sites are accessible to the particles is described in detail in [10]. In this Section we do not repeat the theoretical description but concentrate on the results of this model.

9.1.3.2 Effects of disorder

We study the system for different densities S of lattice sites which are accessible (active) to the reacting particles. For the case of $0 < S < 1$ we obtain a lattice in which the active sites form smaller or larger islands. These islands are separated on the lattice in such a way that the particles occupying one island cannot interact with particles sitting on another island. In contrast to the computer simulations performed for such a system [25], we do not remove here finite clusters from the surface.

If S is smaller than the site percolation threshold for the square lattice $S_c = 0.59275$ we obtain a system which consists only of finite clusters. In principle these clusters can be completely occupied by one-kind species. For the case that no desorption is allowed this represents a poisoned state for which the production rate R_{CO_2} goes to zero as $t \to \infty$. Then the whole system consists of finite clusters poisoned by particles A or B. For this state the condition $C_A + C_B = S$ holds where C_λ is the density of particles of type λ ($C_A + C_B + C_0 \equiv S$).

For $S > S_c$ we obtain an infinite cluster for which in principle a reactive state exists. We use this fact to define the percolation threshold in a kinetic way for the particular reaction at hand as the transition point from the reactive ($R_{CO_2} > 0$) to the non-reactive ($R_{CO_2} = 0$) state. As we have shown above, this transition happens in such a way that the kinetic phase transition points of y_1 and y_2 are approaching each other if $S \to S_c$ [25]. At $S = S_c$ the condition $y_1 = y_2$ holds. This is a very stringent test for our system. We have obtained as the kinetic percolation threshold $S = S_0 = 0.63$ which is larger than S_c. The discrepancy arises partly due to our approximations used in this model, but the major part of the discrepancy is expected to arise from the requirements of the B_2 adsorption process (two-site requirement). If a site does not have four nearest neighbours which is a typical case on a percolation cluster, the B_2 adsorption is strongly hindered compared to the case of the square lattice. This leads to the fact that S_0 exceeds S_c. A more compact percolation cluster is required to give a reactive state. Less compact clusters, i.e., the percolation clusters at S_c always lead to the poisoning via A particles which have only a one-site requirement for adsorption. The present value of S_0 is thus the kinetic percolation threshold for the $A + \frac{1}{2}B_2 \to 0$ reaction.

In Fig. 9.8 the ratio of the number of particles to the number of active sites and the production rate R_{CO_2} (multiplied by a factor of 10) are shown as a

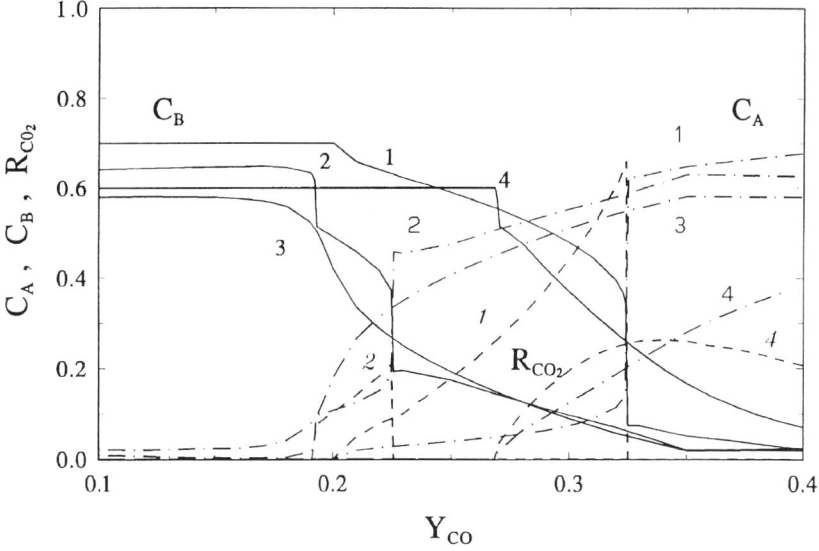

Fig. 9.8. The density of particles (C_B is given by a full line, C_A – by a dash-dot line) and the production rate R_{CO_2} multiplied by a factor of 10 (dashed line) are shown as a function of Y_{CO} for $S = 0.70$ (curves 1), $S = 0.65$ (curves 2) and $S = 0.60$ (curves 3). Curves 4 show the same quantities as curves 3 but adding A-desorption with the rate $k_A = 0.1$.

function of Y_{CO} for $S = 0.70$ (curves 1), $S = 0.65$ (curves 2) and $S = 0.60$ (curves 3). In contrast to the curves 1–3, curves 4 show the system behaviour with the additional A-desorption with the corresponding rate $k_A = 0.1$ and $S = 0.60$.

For $S = 0.70$ which is well above the percolation threshold we observe the second-order phase transition at $y_1 = 0.200$ and the first-order phase transition at $y_2 = 0.324$. For $Y_{CO} < y_1$ all active sites are covered by B particles ($C_B = S$). A complete occupation with A does not exist ($C_A < S$) thus showing that clusters which are poisoned by B are still remaining on the lattice (otherwise these clusters would be occupied by A). We obtain a reactive interval between y_1 and y_2 with $R_{CO_2} > 0$.

For the case $S = 0.65$ (curves 2) which is very close S_0 we obtain the first-order phase transitions at $y_1 = 0.193$ and at $y_2 = 0.225$; the reaction interval is very narrow. As for the case discussed above we do not have a state where the lattice is completely covered by A.

For $S_0 > S = 0.60$ the state of the system is characterized by the condition $C_A + C_B = S$ and a reactive state does not exist. Therefore the production rate R_{CO_2} which is equal to zero is not shown in this figure either.

In curves 4 the system behaviour is shown under the conditions of curves 3 but now including the effect of A-desorption, equation (9.1.42). In this case we obtain a reactive state. We observe a phase transition of the first order at $y_1 = 0.268$. For $Y_{CO} < y_1$ the lattice is completely occupied by particles B. A phase transition point y_2 does not exist but we obtain a smooth transition over a wide range of Y_{CO}. Due to the desorption the state poisoned by A does not exist.

Next we study the correlation functions at $S = S_0$. In Fig. 9.9 they are shown for $Y_{CO} = 0.19$ which is in the region in which surface poisoning with B particles occurs. The corresponding particle densities are $C_A = 0.045$ and $C_B = 0.585$, i.e., nearly all lattice sites are occupied by B. The A particles form small clusters which in order to survive are separated by empty sites. This explains the large values of $F_{0A}(1)$ (which is the nearest neighbour correlation between an empty site and a site occupied by A) and the large value of $F_{AA}(1)$. Very interesting is also the behaviour of $F_{00}(r)$: for $r = 1$ the correlation function has an infinite value. For $r = \sqrt{2}$ and for $r = 2$ it takes the value of zero. This means that only pairs of empty sites are existing; around these pairs no other free sites are existing. This is a necessary condition for the system to achieve a state which is completely covered by B particles [26]. In these pairs of free sites, B_2 particles can adsorb. If more

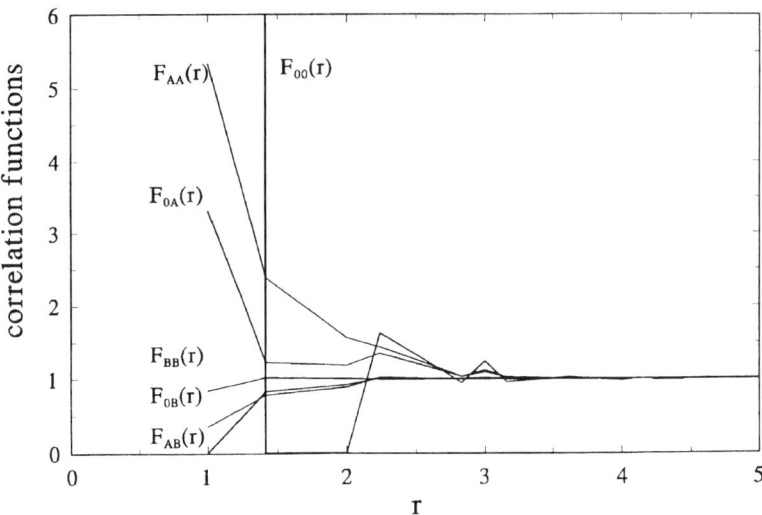

Fig. 9.9. The correlation functions for $S = 0.63$ and $Y_{CO} = 0.19$.

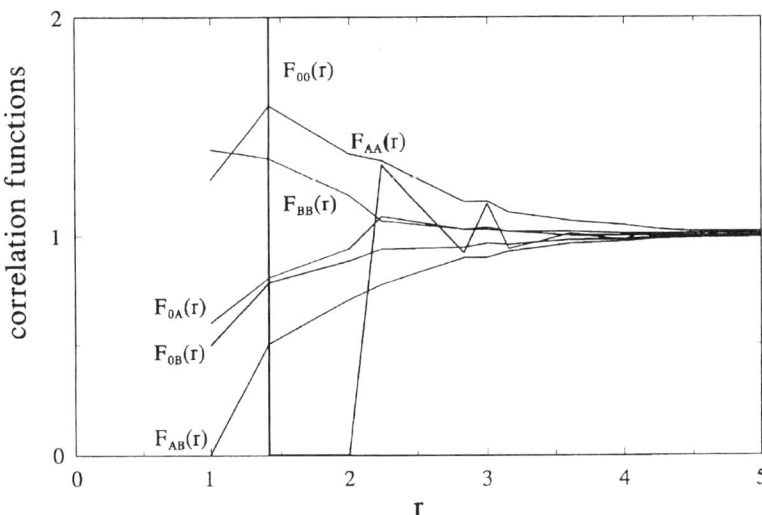

Fig. 9.10. The correlation functions for $S = 0.63$ and $Y_{CO} = 0.20$.

A stochastic model for surface reactions without energetic interactions   549

empty sites existed around such a pair, A particles could easily adsorb and we would not obtain a state which is completely occupied by B. For larger values of r smaller oscillations can be observed in the $F_{00}(r)$ behaviour. This shows that for $r > 2$ empty nearest-neighbour pairs must exist in order to reach a state poisoned by B. Around these pairs other vacant sites are not allowed. Of course the amplitude of the oscillations decreases with increasing r showing that the correlations decrease from a reference site.

In Fig. 9.10 the behaviour is shown at $Y_{CO} = 0.20$ which is slightly above the point $y_1 = y_2$. The particle densities are $C_A = 0.330$ and $C_B = 0.300$. For $r < 5$ the correlation functions $F_{AA}(r)$ and $F_{BB}(r)$ are larger than unity which shows a segregated state in which A and B form islands; surface poisoning cannot occur. $F_{00}(r)$ shows behaviour similar to the one discussed above.

9.1.3.3 Summary

We have studied above a model for the surface reaction $A + \frac{1}{2}B_2 \to 0$ on a disordered surface. For the case when the density of active sites S is smaller than the kinetically defined percolation threshold S_0, a system has no reactive state, the production rate is zero and all sites are covered by A or B particles. This is quite understandable because the active sites form *finite* clusters which can be completely covered by one-kind species. Due to the natural boundaries of the clusters of active sites and the irreversible character of the studied system (no desorption) the system cannot escape from this case. If one allows desorption of the A particles a reactive state arises, it exists also for the case $S > S_0$. Here an infinite cluster of active sites exists from which a reactive state of the system can be obtained. If S approaches S_0 from above we observe a smooth change of the values of the phase-transition points which approach each other. At $S = S_0$ the phase transition points coincide ($y_1 = y_2$) and no reactive state occurs. This condition defines kinetically the percolation threshold for the present reaction (which is found to be 0.63). The difference with the percolation threshold of $S_c = 0.59275$ is attributed to the reduced adsorption probability of the B_2 particles on percolation clusters compared to the square lattice arising from the two site requirement for adsorption, to balance this effect more compact clusters are needed which means S_0 exceeds S_c. The correlation functions reveal the strong correlations in the reactive state as well as segregation effects.

9.1.4 A stochastic model for complex surface reaction systems: Application to the NH_3 catalytic formation

9.1.4.1 Introduction

In this Section we focus our attention on the development of the formalism for complex reactions with application to the formation of NH_3. The results obtained (phase transition points and densities of particles on the surface) are in good agreement with the Monte Carlo and cellular automata simulations. The stochastic model can be easily extended to other reaction systems and is therefore an elegant alternative to the above-mentioned methods.

Real catalytic reactions upon solid surfaces are of great complexity and this is why they are inherently very difficult to deal with. The detailed understanding of such reactions is very important in applied research, but rarely has such a detailed understanding been achieved neither from experiment nor from theory. Theoretically there are *three basic approaches*: kinetic equations of the mean-field type, computer simulations (Monte Carlo, MC) and cellular automata CA, or stochastic models (master equations).

The kinetic equations are useful as a fitting procedure although their basis – the homogeneous system – in general does not exist. Thus they cannot deal with segregation and island formation which is frequently observed [27]. Computer simulations incorporate fluctuation and correlation effects and thus are able to deal with segregation effects but so far the reaction systems under study are oversimplified and contain only few aspects of a real system. The use of computer simulations for the study of surface reactions is also limited because of the large amount of computer time which is needed. Especially MC simulations need so much computer time that complicated aspects (e.g., the dependence of the results on the distribution of surface defects) in practice cannot be studied. For this reason CA models have been developed which run very fast on parallel computers and enable to study more complex aspects of real reaction systems. Some examples of CA models which were studied in the past years are: the NH_3 formation [4] and the problem of the universality class [18]. However, CA models are limited to systems which are suited for the description by a purely parallel *ansatz*.

Master equations present a very powerful and general approach because they also maintain correlations, fluctuations and structural informations. Their use for the complex reactions which take place on a lattice is only limited by the fact that analytically they can be solved up to now only for very simple systems in one dimension [5]. But it is possible to introduce sensible

A stochastic model for surface reactions without energetic interactions   551

approximations and/or solve the equations numerically, e.g., certain aspects of the problem can be solved analytically (instead of a purely numerical approach) and one does not attempt to include structural effects and fluctuations at large distances where they are negligible. Many different approaches have been developed to handle such systems; some important references are cited in [6–9].

In this Section we introduce a stochastic alternative model for surface reactions. As an application we will focus on the formation of NH_3 which is described below, equations (9.1.72) to (9.1.76). It is expected that these stochastic systems are well-suited for the description via master equations using the Markovian behaviour of the systems under study. In such a representation an infinite set of master equations for the distribution functions describing the state of the surface and of pairs of surface sites (and so on) arises. As it was told earlier, this set cannot be solved analytically and must be truncated at a certain level. The resulting equations can be solved exactly in a small region and can be connected to a mean-field solution for large distances from a reference point. This procedure is well-suited for the description of surface reaction systems which includes such elementary steps as adsorption, diffusion, reaction and desorption. The numerical part needs only a very small amount of computer time compared to MC or CA simulations.

In spite of simple theoretical formalism (for example, mean-field descriptions of certain aspects) structural aspects of the systems are still explicitly taken into account. This leads to the results which are in a good agreement with computer simulations. But the stochastic model avoids the main difficulty of computer simulations: the tremendous amount of computer time which is needed for obtaining good statistics for the reliable results. Therefore more complex systems can be studied in detail which may eventually lead to better understanding of real systems. In the theory discussed below we deal with a disordered surface. This additional complication will be handled in terms of the stochastic approach. This is also a very important case in catalytic reactions.

The equations are written for the specific reaction of NH_3 formation as a fully general approach would be unwieldy. The modification of the approach to other reactions is not trivial but could be done following the outline below. Another application of the formalism to a very complex reaction system ($CO + O_2$ on a Pt/Sn disordered catalyst) is demonstrated, as well as the generality of the stochastic *ansatz* [28].

The knowledge of the details of catalytic reactions is in general of great industrial importance; in particular much effort has been undertaken to understand the reaction mechanism of the NH_3 synthesis. But even today this system is not well understood because it turned out to be very complex. With the help of modern surface spectroscopy some aspects of the reaction mechanism [29] and some details of the role of structural and energetic promoters on the surface have been clarified but no consensus has been reached on several important points. The NH_3 synthesis is industrially performed with a *promoted iron catalyst*, where the promoters are structural ones such as Al_2O_3 and electronic ones such as K_2O. In the laboratory single crystals of iron, tungsten or rhenium are used. H_2 adsorbs dissociatively with a large sticking coefficient and N_2 adsorbs molecularly with an activation barrier towards dissociation into atoms which together lead to a low sticking coefficient of about 10^{-7}. The dissociative adsorption is the rate limiting step. The sticking coefficient is defined as the probability that a molecule impinging on the surface is adsorbed (dissociatively).

The reaction occurs between adsorbed atoms via the steps $N + H \to NH$, $NH + H \to NH_2$ and finally $NH_2 + H \to NH_3$, which desorbs then after formation. Computer simulations are another tool which may be helpful for understanding certain aspects of the behaviour of this system. Complex kinetic and thermodynamic calculations have been introduced by Stoltze and Norskov [30, 31] who took many aspects of this system into account. Their results are in very good agreement with experimental observations of the reaction rate. Due to the use of many experimental data their model becomes involved and thus one cannot understand the evolution of the reaction system in detail.

In [4] we have introduced a CA model for the NH_3 formation which accounts only for a few aspects of the reaction system. In our simulations the surface was represented as a two-dimensional square lattice with periodic boundary conditions. A gas phase containing N_2 and H_2 with the mole fraction of y_N and $y_H = 1 - y_N$, respectively, is above this surface. Because the adsorption of H_2 is dissociative an H_2 molecule requires two adjacent vacant sites. The adsorption rule for the N_2 molecule is more difficult to be described because experiments show that the sticking coefficient of N_2 is unusually small (10^{-7}). The adsorption probability can be increased by high energy impact of N_2 on the surface. This process is interpreted as tunnelling through the barrier to dissociation [32].

Another possibility to increase the adsorption probability is via electronic promoters (K_2O). It is believed that this promoter, which is enriched on the surface, leads to a larger binding energy of molecular N_2 via an increased metal π-electron backbonding. Connected with this is a lowering of the activation energy for dissociation [33, 34]. Because of the presence of promotors we introduce two different adsorption sites: activated ones, S_1, and normal ones, S_2. We suppose that the dissociative adsorption of N_2 occurs on a pair of neighbouring vacant sites from which at least one must be activated ($S_1 - S_{1,2}$) where $S_{1,2}$ means a surface site of type one (activated) or two (non activated).

It is unnecessary to require two neighbouring activated sites for the dissociative adsorption of N_2. The effect of the promotor is not a strictly localized one but influences also the neighbourhood. At the typical concentrations of K^+ it is rather unlikely that two neighbour sites are both activated. Thus our rules take into account the fact that an activated site also influences the energetic behaviour of the neighbouring sites. The dissociative adsorption of H_2 can take place on every pair of free sites S_1–S_1, S_1–S_2 or S_2–S_2. The concentration of S_1 is a measure for the concentration of K^+ on the surface. If N atom is the nearest neighbour of a H atom reaction occurs to $HN-S_{1,2}$. Via further reaction steps the product molecule NH_3 is formed which desorbs immediately after formation. We neglect recombination reactions. Therefore the basic reaction steps are

$$N_2(gas) + S_1 - S_{1,2} \to N-S_1 + N-S_{1,2}, \tag{9.1.72}$$

$$H_2(gas) + S_{1,2} - S_{1,2} \to 2N-S_{1,2}, \tag{9.1.73}$$

$$N-S_{1,2} + H-S_{1,2} \to HN-S_{1,2} + S_{1,2}, \tag{9.1.74}$$

$$HN-S_{1,2} + H-S_{1,2} \to H_2N-S_{1,2} + S_{1,2}, \tag{9.1.75}$$

$$H_2N-S_{1,2} + H-S_{1,2} \to HN(gas) + 2S_{1,2}. \tag{9.1.76}$$

We neglected in our model the *back* reactions, in order to simplify the already quite complicated process. In some of the steps this does not correspond to the reality with the exception of the last step because NH_3 is removed from the surface and the reactor. We do not believe that the inclusion of the back reactions will alter significantly the conclusions of the present very simplified reaction model except via a reduced reaction rate. The surface coverages should remain essentially unchanged and they are the prominent information available from our model. Further on the removal of NH_3 introduces a "drag" on the reaction process in the direction of smaller importance of the back reactions. As a result of our model we found in the case when all sites are activated the first-order kinetic phase transition for the coverages as a function of y_N. For $y_N < y_1 = 0.262$ the surface is nearly completely covered by H and nearly no reaction takes place. With decreasing y_N, the coverage of H(θ_H) increases up to unity. For the value $y_N = 0$ itself only H_2 molecules adsorb and no reaction event occurs. Therefore θ_H is equal to 0.88 which is the maximum coverage for the adsorption of dimers. For the case that not all sites are activated, a change in the character of the phase transition from the first to the second-order is observed.

9.1.4.2 The stochastic model

The state of the lattice site l (activated or unactivated) is represented by the lattice variable γ_l with

$$\gamma_l = \begin{cases} 0 & \text{if the site is unactivated (u)}, \\ 1 & \text{if the site is activated (a)}. \end{cases} \tag{9.1.77}$$

We define $S = \langle \gamma_l \rangle$ as the mean value of the activity of the catalyst and this is independent of l.

We introduce a variable for the particles $\zeta_l \in \{0, H, N, A, B\}$ where 0 represents a vacant site, A represents a NH particle and B a NH_2 particle. The state of a lattice point σ_l consists of the state of the catalyst (activated or unactivated) and its coverage with a particle. This leads to the following possible states:

$$\sigma_l = \{\zeta_l \gamma_l\} \equiv 0u, 0a, Hu, Ha, \ldots, Bu, Ba. \tag{9.1.78}$$

The H and N particles are created on the surface by adsorption from of the gas phase with the rates

$$p_{H_2} = y_H \quad \text{and} \quad p_H = 2p_{H_2} = 2y_H \tag{9.1.79}$$

for the H particles and

$$p_{N_2} = 1 - y_H \quad \text{and} \quad p_N = 2(1 - y_H) \tag{9.1.80}$$

for the N particles. y_H and y_N are the mole fraction of H_2 and N_2 in the gas phase, with $y_H + y_N = 1$. By these adsorption steps pairs of HH and NN particles are created because of the dissociative character of the adsorption.

For the adsorption of N_2 at least one of the lattice sites must be activated: $\gamma_l + \gamma_m > 0$.

At normal temperatures H atoms are very mobile on metal surfaces. We take this into account by the possibility of diffusion steps for the H atoms. A H atom jumps with rate D/z onto the nearest neighbour sites on the lattice. If this site is occupied by N, reaction occurs and an A particle (NH particle) is formed. The same holds if the site is occupied by A or B (NH_2) where the products B or $NH_3 = 0$ are formed, respectively. NH_3 desorbs immediately from the surface and an empty site is formed. That is we deal with the *diffusion-limited* reaction system. It is important to note that all the reaction steps discussed above (with the exception of the N_2-adsorption) are independent of γ_l and γ_m.

Next we define the distribution function $\rho^{(i)}$ of order i characterizing the state of the surface. For $i = 1$ one gets

$$\rho^{(1)}(\sigma_l) \equiv \rho^{(1)}(\zeta_l \gamma_l) = C_\sigma \equiv C_\zeta^\gamma(t), \qquad (9.1.81)$$

where $C_0^u, C_0^a, \ldots, C_B^a$ are the lattice densities which are independent of l because of the translation invariance of the lattice. We define

$$\theta_\zeta = \sum_\gamma C_\zeta^\gamma(t) = C_\zeta^u(t) + C_\zeta^a(t). \qquad (9.1.82)$$

With the help of the definitions of Section 9.1 made above we are able to write down the equations for the temporal evolution of the distribution functions.

The temporal evolution of the B density is given by

$$\frac{\partial}{\partial t} C_B^\gamma = D \left(C_A^\gamma \sum_{\gamma'} C_H^{\gamma'} F_{A\gamma H \gamma'}(1) - C_B^\gamma \sum_{\gamma'} C_H^{\gamma'} F_{B\gamma H \gamma'}(1) \right). \qquad (9.1.83)$$

$F_{\sigma\sigma'}(1)$ is a correlation function for the nearest neighbours. The other densities are given by the following equations of motion.

A-density:

$$\frac{\partial}{\partial t} C_A^\gamma = D \left(C_N^\gamma \sum_{\gamma'} C_H^{\gamma'} F_{N\gamma H\gamma'}(1) - \right.$$

$$\left. - C_A^\gamma \sum_{\gamma'} C_H^{\gamma'} F_{A\gamma H\gamma'}(1) \right). \tag{9.1.84}$$

N-density:

$$\frac{\partial}{\partial t} C_N^\gamma = p_N C_0^\gamma \sum_{\substack{\sigma' \\ (\gamma+\gamma'>0)}} C_0^{\gamma'} F_{0\gamma 0\gamma'}(1) -$$

$$- D C_N^\gamma \sum_{\gamma'} C_H^{\gamma'} F_{N\gamma H\gamma'}(1). \tag{9.1.85}$$

H-density:

$$\frac{\partial}{\partial t} C_H^\gamma = p_H C_0^\gamma \sum_{\gamma'} C_0^{\gamma'} F_{0\gamma 0\gamma'}(1) + D C_0^\gamma \sum_{\gamma'} C_H^{\gamma'} F_{0\gamma H\gamma'}(1) -$$

$$- D C_H^\gamma \left(\sum_{\gamma'} C_0^{\gamma'} F_{H\gamma 0\gamma'}(1) + \sum_{\gamma'} C_N^{\gamma'} F_{H\gamma N\gamma'}(1) + \right.$$

$$\left. + \sum_{\gamma'} C_A^{\gamma'} F_{H\gamma A\gamma'}(1) + \sum_{\gamma'} C_B^{\gamma'} F_{H\gamma B\gamma'}(1) \right). \tag{9.1.86}$$

0-density:

$$\frac{\partial}{\partial t} C_0^\gamma = -p_H C_0^\gamma \sum_{\gamma'} C_0^{\gamma'} F_{0\gamma 0\gamma'}(1) - p_N C_0^\gamma \sum_{\substack{\gamma' \\ (\gamma+\gamma'>0)}} C_H^{\gamma'} F_{0\gamma H\gamma'}(1) -$$

$$- D C_0^\gamma \left(\sum_{\gamma'} C_0^{\gamma'} F_{H\gamma 0\gamma}(1) + \sum_{\gamma'} C_N^{\gamma'} F_{H\gamma N\gamma'}(1) + \right.$$

A stochastic model for surface reactions without energetic interactions

$$+ \sum_{\gamma'} C_A^{\gamma'} F_{H\gamma A\gamma'}(1) + \sum_{\gamma'} C_B^{\gamma'} F_{H\gamma B\gamma'}(1) \Big) +$$

$$+ DC_B^{\gamma} \sum_{\gamma'} C_H^{\gamma'} F_{B\gamma H\gamma'}(1) - DC_0^{\gamma} \sum_{\gamma'} C_H^{\gamma'} F_{0\gamma H\gamma'}(1). \quad (9.1.87)$$

In our calculation as the initial condition an empty surface is chosen, $C_\zeta^\gamma(t=0) = 0$ for $\gamma \in \{H, N, A, B\}$. The activated sites of the catalyst are distributed with $C_0^a(0) = S$ and $C_0^u(0) = 1 - S$ where S is the mean activity of the surface.

9.1.4.3 The system behaviour for $S = 1$

First of all, we want to study the case in which all surface sites are activated ($S = 1$). This means that a N_2 molecule can adsorb at every pair of free surface sites. Figure 9.11 represents the behaviour of the surface coverages θ_i of the various chemical species i as a function of the mole fraction of N_2 in the gas phase, y_N, for the case $S = 1$ and $D = 1$. The most prominent feature in this figure is the kinetic phase transition of the second order at $y_N = y_1 \approx 0.21$. For $y_N < y_1$ the surface is nearly completely covered by

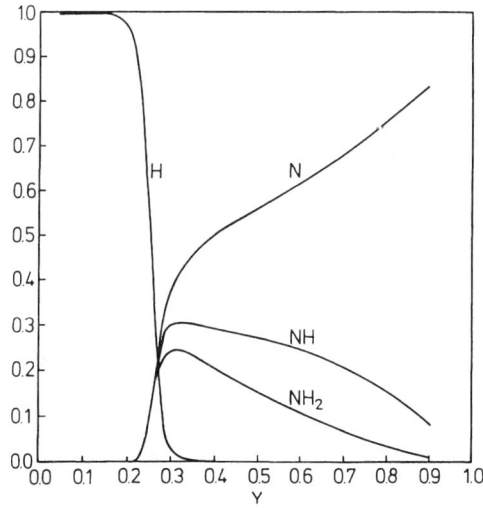

Fig. 9.11. Phase diagram for the case $S = 1$ and $D = 1$.

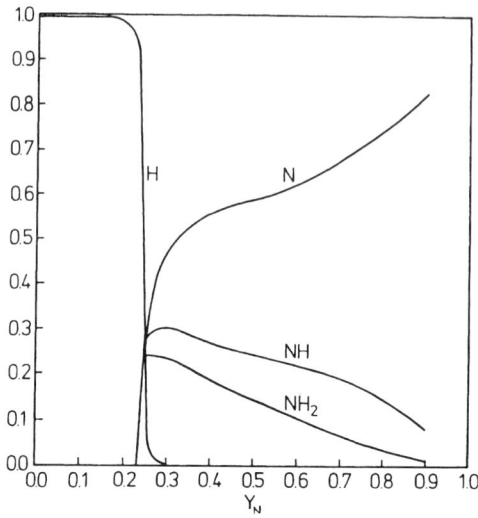

Fig. 9.12. Phase diagram for the case $S = 1$ and $D = 10$.

H and nearly no reaction takes place. For $y_N \geqslant y_1$, θ_H drops to zero and the coverages of N, NH and NH$_2$ increase. For $y_N \to 1$, θ_N increases up to 0.88 which is the maximum coverage for the adsorption of dimers. The coverages of NH and NH$_2$ increase rapidly for $y_N > y_1$ because the composition is nearly stoichiometric on the surface. For larger y_N these coverages decrease because there are no enough adsorbed H atoms on the surface for their formation (the coverage of N atoms increases). It must be noted that the value of y_1 is very close to the stoichiometric ratio of 0.25 but not identical with it. This difference arises from the small diffusion rate of the H atoms. Therefore the system is not well-stirred and adsorbate clusters are formed. This invalidates a description by a mean-field model. The reaction rate obtained for the stochastic and the CA models are in all cases close. It is a nonlinear function of y_N rising steeply at the phase transition point and leveling off for larger values of y_N. We do not represent it in figures.

Figure 9.12 shows similar data of the surface coverages as in Fig. 9.11 but now we use a larger diffusion rate of $D = 10$. It can be seen that the value of y_1 is shifted to a larger value of y_N ($y_1 \approx 0.23$) which is closer to the stoichiometric ratio of 0.25. The kinetic phase transition sharpens also and it is now nearly of the first order. This can easily be understood from the

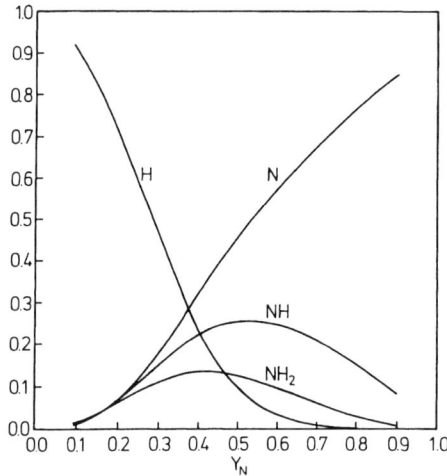

Fig. 9.13. Phase diagram for the case $S = 1$ and $D = 0.1$.

fact that the fast moving H atoms react with N, NH, and NH$_2$ molecules at the border of the adsorbate clusters. Therefore this system is more of mean-field type. The coverages of NH and NH$_2$ are a little bit smaller compared to the case of $D = 1$ which shows the more reactive character of the fast moving H atoms. Just above y_1 the coverage of N slightly exceeds that for the case $D = 1$. This arises from the fact that at this point more vacant pairs of surface sites are present which enlarge the probability of the dissociative adsorption of N$_2$.

An example of a very low diffusion rate of $D = 0.1$ is shown in Fig. 9.13. For this case the phase transition at y_1 vanishes. Because of the very small reactivity of the H atoms large cluster structures of particles are formed. This allows the simultaneous appearance of H and N atoms on the surface.

9.1.4.4 The system behaviour for $S = 1/8$

Let us discuss now a more realistic case in which not all surface sites are activated. This means a reduction of the adsorption probability of the N$_2$ molecules. In the following we assume that $S = 1/8$.

In Fig. 9.14 the coverages for $D = 1$ are shown. The kinetic phase transition is of the second order and the value of y_1 is shifted to larger values of y_N compared to the analogous case above. Over a whole interval of y_N, θ_H is

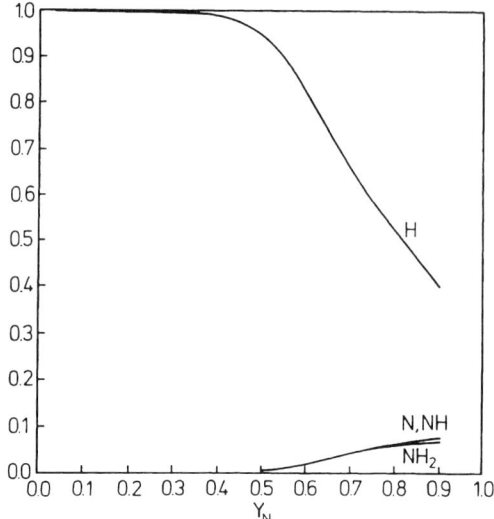

Fig. 9.14. Phase diagram for the case $S = 1/8$ and $D = 1$.

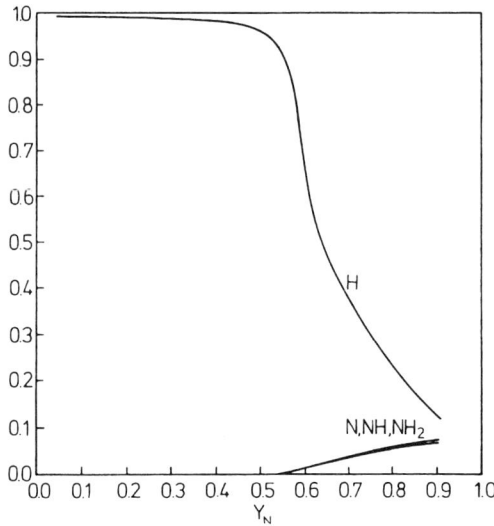

Fig. 9.15. Phase diagram for the case $S = 1/8$ and $D = 10$.

significantly larger than 0.2. The concentrations of N, NH and NH$_2$ increase with increasing y_N but they are smaller than 0.1. This behaviour can easily be understood from the fact that N$_2$ molecules can only adsorb on the vacant nearest neighbour pairs, and at least one of these sites must be activated. During the reaction many activated sites will be blocked via occupation with NH and NH$_2$ because reaction occurs by the jump of H atoms to sites which are covered by N, NH and NH$_2$. After the reaction in many cases the product covers an activated site. Another process for the blocking the activated sites is the adsorption of H$_2$ which is very important for small values of y_N.

If we enlarge the diffusion rate of the H atoms up to $D = 10$ as shown in Fig. 9.15, the value of the phase transition point y_1 and the coverages of N, NH and NH$_2$ are almost unaffected by this change, only the concentration of H atoms drops more rapidly for $y_N > y_1$. This behaviour shows increased reactivity of H atoms which are now more mobile. The coverages of the other particles depend only on the concentration of the activated sites of the surface but not on the reactivity of the H atoms.

9.1.4.5 The correlation functions

We have also studied the joint correlation functions of the adsorbed particles which contain a variety of informations on the adsorbate spatial structure. In contrast to the CO oxidation model (Section 9.2), a large number (55) of different pair correlations appear in the present reaction model. We observe that all of these are very small with the exception for H–H and N–N correlations between the nearest neighbour sites. This correlation results from the dissociative adsorption of the molecules, but within a very short distance (3 or 4 sites) the correlation vanishes. This is clear because of the occurrence of many different pairs which results from the large number of different particles and the state of the surface. In the present model large N-clusters cannot appear because of the formation of NH and NH$_2$ particles during the reaction which immediately breaks off N-clusters. The H atoms are distributed randomly over the surface because of their mobility and large reactivity. For $S < 1$ the additional effect of randomly distributed active surface sites leads to a random distribution of N$_2$-adsorption events. Therefore it is clear that in this model large correlations cannot appear. This gives a justification for the statement made following equation (9.1.20).

9.1.4.6 Discussion

In this Section a stochastic model for heterogeneous catalytic reactions have been introduced and applied to the formation of NH$_3$. We took into account the occurrence of activated sites for the adsorption of N_2. Earlier approach was based on the CA formalism [4]. The reaction step in the CA model is defined in other way than that in the model described here; namely, reaction takes place if two reactive particles are the nearest neighbours on the surface lattice. The effect of the diffusion of the H atoms was not included which is in some respect unrealistic. Differences which occur between these two different models arise due to the reaction and diffusion events. In the stochastic model reaction occurs by a hopping process of H atoms to sites which are occupied by dissimilar particles reacting with H atoms. The product is formed on the site to which the H atom hops. This means that in this case the probability to block an activated site is *larger* in the stochastical model than in the CA model in which the reaction product occupies one of the two sites, which is randomly chosen. Therefore the probability of the blocking of activated sites is reduced. This results in a larger coverage of N atoms for the case $S < 1$.

The reaction models are different in the CA and the stochastic approaches because of the so far insurmountable difficulties with the nearest-neighbour reaction in the stochastic equations. However, a detailed investigation of a simple model of the oxidation of CO [13], where diffusion was included in both approaches, leads to the conclusion that the results of the two models are quite close: the CO poisoning of the lattice is of the same order of magnitude and occurs at the same point. Also the oxygen poisoning and the order of the phase transition agrees in both the approaches, but the transition point is shifted to a somewhat lower value. But on the whole the agreement is very close. We thus can expect that also in the present model there will be only small changes due to the introduction of the different reaction rule.

For the case $S = 1$ and $D = 1$ the results of the stochastic model are in good agreement with the CA model ($y_1 = 0.262$). This is understandable because the different definition of the reaction which leads to a difference in the blocking of activated sites cannot play significant role because all sites are activated. The diffusion rate of $D = 10$ leads nearly to the same reactivity as if we define the reaction between the nearest-neighbour particles. If the diffusion rate is considerably lowered ($D = 0.1$), the behaviour of the system changes completely because of the decrease of the reaction probability. This leads to the disappearance of the kinetic phase transition at y_1 because different types of particles may reside on the surface as the nearest neighbours without reaction, a case which does not occur at all in the CA approach.

The results obtained from the CA model for $S = 1/8$ are in quite good agreement with the results obtained from the stochastic *ansatz* for $D = 1$ and $D = 10$; the value of the phase transition point y_1 is found to be $y_1 \approx 0.4$ in both models. In the stochastic model the density of adsorbed N atoms is smaller compared to the CA model because of the blocking effect of activated sites which arises from the different reaction mechanism.

We have also performed calculations for higher diffusion rates ($D = 100$) and for the triangular lattice (coordination number $z = 6$). The qualitative behaviour is in complete agreement with the calculation presented here. For the case $S = 1$ the increase of the diffusion rate or change of the lattice structure leads to a very small shift of the phase transition point y_1 to higher values of y_N. This trend is clear because the reactivity of the H atoms is increased by the larger mobility. For $S < 1$ nearly no effect is observed which means that the system's behaviour is mainly dominated by the number of activated sites. The correlation of the adsorbed particles are rather small as expected for $S < 1$.

9.1.4.7 Conclusion

The results obtained for the stochastic model show that surface reactions are well-suited for a description in terms of the master equations. Since this infinite set of equations cannot be solved analytically, numerical methods must be used for solving it. In previous Sections we have studied the catalytic oxidation of CO over a metal surface with the help of a similar stochastic model. The results are in good agreement with MC and CA simulations. In this Section we have introduced a much more complex system which takes into account the state of catalyst sites and the diffusion of H atoms. Due to this complicated model, MC and in some respect CA simulations cannot be used to study this system in detail because of the tremendous amount of required computer time. However, the stochastic *ansatz* permits to study very complex systems including the distribution of special surface sites and correlated initial conditions for the surface and the coverages of particles. This model can be easily extended to more realistic models by introducing more aspects of the reaction mechanism. Moreover, other systems can be represented by this *ansatz*. Therefore, this stochastic model represents an *elegant alternative* to the simulation of surface reaction systems via MC or CA simulations.

9.2 A STOCHASTIC MODEL FOR SURFACE REACTIONS INCLUDING ENERGETIC PARTICLE INTERACTIONS

> The system existed under them not because they were able to establish order but because they were unable to destroy it.
>
> V. Klyuchevsky, old Russian historian

9.2.1 A general stochastic model

9.2.1.1 Introduction

In Section 9.1.1 we have introduced a stochastic model for the description of surface reaction systems which takes correlations explicitly into account but neglects the energetic interactions between the adsorbed particles as well as between a particle and a metal surface. We have formulated this by master equations upon the assumption that the systems are of the Markovian type. In the model an infinite set of master equations for the distribution functions of the state of the surface and of pairs of surface sites (and so on) arise. This chain of equations cannot be solved analytically. To handle this problem practically this hierarchy was truncated at a certain level. The resulting equations can be solved numerically exactly in a small region and can be connected to a mean-field solution for large distances from a reference point.

Therefore the model avoids two main difficulties: the large amount of computer time which is normally needed for simulations and the loss of structural information which occurs in simple theoretical models (mean-field models) which do not take into account the structural aspects of the adsorbate layer. Mean-field-kind models fail in the prediction of phase transitions of the second order because at these points the long-range correlations appear. They also fail in describing the system's behaviour in the neighbourhood of the point of first-order kinetic phase transition.

We have applied our model to the $A + \frac{1}{2} B_2 \rightarrow 0$ reaction and compared it in Section 9.1.2 with results of computer simulations. We found that the results are in very good agreement with each other. Disordered surfaces were treated within the stochastic approach in Sections 9.1.3 and 9.1.4. Lastly, in this Section we introduce energetic interactions into the model defined earlier. We define a standard model in order to compare different surface reactions which are modeled using this theoretical *ansatz*. We show that in the case when energetic interactions are neglected, the model reduces to that presented earlier in Section 9.1.1.

9.2.1.2 The model

Definitions. We consider a lattice with coordination number z. To each lattice site is given a lattice vector l. The state of the site l is represented by the lattice variable σ_l which may depend on the state of the catalyst site (e.g., promoted or not) and on its coverage with a particle. Here we deal only with the simple case in which all sites are identical and therefore σ_l depends only on its coverage (the other case is explained in Section 9.1.4. Therefore $\sigma_l \in \{0, A, B, \ldots\}$ where 0 represents a vacant site, A is a site which is occupied by an A particle and so on. Next we want to define a variable $\alpha_{l,n}$ which is unity for the case that l and n are the nearest neighbour sites on the lattice and zero otherwise. Sometimes we will use the abbreviation $\sigma_l = \lambda$, $\sigma'_l = \lambda'$, $\sigma_n = \nu$ and $\sigma'_n = \nu'$. The states of the neighbourhood (z sites) of site l are denoted by $\{\sigma\}_l^z$.

Monomolecular steps. We study the processes which depends only on one lattice site. Examples for such steps are the creation of a particle ($0 \to A$), the annihilation ($A \to 0$) or the transformation ($A \to B$). These steps can be described by equation

$$\sigma_l \stackrel{p}{\Rightarrow} \sigma'_l, \qquad p \equiv \mathfrak{P}(\sigma_l \to \sigma'_l \mid \{\sigma\}_l^z), \tag{9.2.1}$$

where p is a transition rate from the state σ_l to σ'_l. For the square lattice $z = 4$. All these transition rates are listed in the matrix $\mathfrak{P}$. The transition probabilities do not depend on a chosen site l (due to the spatial homogeneity) but they depend on the neighbourhood of the site l. This is different from the model without energetic interactions.

Bimolecular steps. A more complicated situation arises if the step depends on *two* lattice sites, l and n. Examples are reaction processes ($AB \to 00$), diffusion processes ($A0 \to 0A$) or a pair creation ($00 \to BB$); the latter being useful for the description of dissociative adsorption events. All these processes can be formulated by equation

$$\sigma_l \sigma_n \stackrel{k}{\Rightarrow} \sigma'_l \sigma'_n, \qquad k \equiv \mathfrak{K}(\sigma_l \sigma_n \to \sigma'_l \sigma'_n \mid \{\sigma\}_l^{z-1}; \{\sigma\}_n^{z-1})/z. \tag{9.2.2}$$

For these processes the neighbourhood is given by the $(z-1)$ nearest neighbour sites of site l (with the exception of site n) and by the $(z-1)$ neighbours of a site n (with the exception of site l). We study only such lattices where $\{\sigma\}_l^{z-1}$ is not equal to $\{\sigma\}_n^{z-1}$. This holds for the square but not for the triangular lattice. We have introduced a factor $1/z$ in order to simplify the arising equations.

9.2.1.3 Master equations

Next we introduce the probabilities $\rho^{(i)}$ which depend on i lattice sites (Section 9.1.1). We write down the master equations in the form of an infinite set of equations for the many-point probabilities.

For the one-point probabilities we obtain equation (9.1.8), where the term with the index 'in' describes the creation of a state σ_l and the term with the index 'out' the annihilation of a state σ_l:

$$\left.\frac{dC_\lambda}{dt}\right|_{in} = \left.\frac{d\rho^{(1)}(\sigma_l)}{dt}\right|_{in} = \sum_{\{\sigma\}_l^z}\sum_{\sigma_l'}\mathfrak{P}(\sigma_l' \to \sigma_l \mid \{\sigma\}_l^z)\rho^{(1+z)}(\sigma_l'; \{\sigma\}_l^z) +$$

$$+ \sum_n \frac{\alpha_{l,n}}{z} \sum_{\{\sigma\}_l^{z-1}}\sum_{\{\sigma\}_n^{z-1}}\sum_{\sigma_l'\sigma_n'\sigma_n}\mathfrak{K}(\sigma_l'\sigma_n' \to \sigma_l\sigma_n \mid \{\sigma\}_l^{z-1}; \{\sigma\}_n^{z-1}) \times$$

$$\times \rho^{(2z)}(\sigma_l'\sigma_n'; \{\sigma\}_l^{z-1}; \{\sigma\}_n^{z-1}) \tag{9.2.3}$$

and

$$\left.\frac{dC_\lambda}{dt}\right|_{out} = \left.\frac{d\rho^{(1)}(\sigma_l)}{dt}\right|_{out} = \sum_{\{\sigma\}_l^z}\sum_{\sigma_l'}\mathfrak{P}(\sigma_l \to \sigma_l' \mid \{\sigma\}_l^z)\rho^{(1+z)}(\sigma_l; \{\sigma\}_l^z) +$$

$$+ \sum_n \frac{\alpha_{l,n}}{z} \sum_{\{\sigma\}_l^{z-1}}\sum_{\{\sigma\}_n^{z-1}}\sum_{\sigma_l'\sigma_n'\sigma_n}\mathfrak{K}(\sigma_l\sigma_n \to \sigma_l'\sigma_n' \mid \{\sigma\}_l^{z-1}; \{\sigma\}_n^{z-1}) \times$$

$$\times \rho^{(2z)}(\sigma_l\sigma_n; \{\sigma\}_l^{z-1}; \{\sigma\}_n^{z-1}). \tag{9.2.4}$$

In these equations monomolecular steps need $(1+z)$-point probabilities and bimolecular steps need $2 + 2(z-1) = 2z$-point probabilities for their description. We can divide the terms of equation (9.2.4) e.g., into mono- and bimolecular steps:

$$\left.\frac{dC_\lambda}{dt}\right|_{out} = \left.\frac{dC_\lambda}{dt}\right|_{out}^{mono} + \left.\frac{dC_\lambda}{dt}\right|_{out}^{bi} \tag{9.2.5}$$

with a similar splitting for the 'in' term. We study the first term in the equation written above:

$$\left.\frac{dC_\lambda}{dt}\right|_{\text{out}}^{\text{mono}} = \sum_{\{\sigma\}_l^z} \sum_{\sigma_l'} \mathfrak{P}(\sigma_l \to \sigma_l' \mid \{\sigma\}_l^z) \rho^{(1+z)}(\sigma_l; \{\sigma\}_l^z). \tag{9.2.6}$$

In principle many different approaches can be used to handle this equation; we use a *correspondence principle*. In the case of no energetic interactions this model should be equivalent to the model discussed in Section 9.1.1. This results in the condition

$$\sum_{\{\sigma\}_l^z} \rho^{(1+z)}(\sigma_l; \{\sigma\}_l^z) = \rho^{(1)}(\sigma_l), \tag{9.2.7}$$

where we have used equation (9.1.10). In order to avoid the explicit appearance of high-order correlation functions we make use of the Mamada–Takano approximation [21], equation (9.1.60) for $m = z$, which automatically gives the condition (9.2.7). For the square lattice with $z = 4$ we can rewrite this equation in the form

$$\rho^{(1+z)}(\sigma_l; \{\sigma\}_l^z) \Rightarrow C_\lambda \prod_{i=1}^{4} \left(C_{\nu_i} F_{\lambda \nu_i}(1)\right). \tag{9.2.8}$$

From this the monomolecular steps follow like

$$\left.\frac{dC_\lambda}{dt}\right|_{\text{out}}^{\text{mono}} = \sum_{\sigma_l'} P(\sigma_l \to \sigma_l') \rho^{(1)}(\sigma_l), \tag{9.2.9}$$

which is formally identical to the model without energetic interaction. Here P (which replaces $\mathfrak{P}$) is defined as a mean value over the nearest neighbourhood

$$P(\sigma_l \to \sigma_l') = \sum_{\{\sigma\}_l^z} \mathfrak{P}(\sigma_l \to \sigma_l' \{\sigma\}_l^z) \prod_{i=1}^{z} \frac{\rho^{(2)}(\sigma_l \sigma_i)}{\rho^{(1)}(\sigma_l)}. \tag{9.2.10}$$

Thus we do not take into account the explicit configuration of the neighbours, which is certainly a sensible approximation. By doing so, we obtain for the first term in equation (9.2.5)

$$\left.\frac{dC_\lambda}{dt}\right|_{\text{out}}^{\text{mono}} = \sum_{\lambda'} P(\lambda \to \lambda') C_\lambda, \tag{9.2.11}$$

where the mean value $P(\lambda \to \lambda')$ is a homogeneous polynomial function of order z in C_μ and in $F_{\lambda\mu}(1)$. The monomolecular 'in'-term has a very similar form

$$\left.\frac{dC_\lambda}{dt}\right|_{in}^{mono} = \sum_{\lambda'} P(\lambda' \to \lambda) C_{\lambda'}. \qquad (9.2.12)$$

For the bimolecular steps we proceed in a similar way as follows.

$$\left.\frac{dC_\lambda}{dt}\right|_{out}^{bi} = \sum_n \frac{\alpha_{l,n}}{z} \sum_{\{\sigma\}_l^{z-1}} \sum_{\{\sigma\}_n^{z-1}} \sum_{\sigma'_l \sigma'_n \sigma_n} \mathfrak{K}\left(\sigma_l \sigma_n \to \sigma'_l \sigma'_n \mid \{\sigma\}_l^{z-1}; \{\sigma\}_n^{z-1}\right) \times$$
$$\times \rho^{(2z)}\left(\sigma_l \sigma_n; \{\sigma\}_l^{z-1}; \{\sigma\}_n^{z-1}\right). \qquad (9.2.13)$$

Here we use the sum rule

$$\sum_{\{\sigma\}_l^{z-1}} \sum_{\{\sigma\}_n^{z-1}} \rho^{(2z)}\left(\sigma_l \sigma_n; \{\sigma\}_l^{z-1}; \{\sigma\}_n^{z-1}\right) = \rho^{(2)}(\sigma_l \sigma_n). \qquad (9.2.14)$$

Using the Mamada–Takano [21] approximation we get

$$\rho^{(2z)}\left(\sigma_l \sigma_n; \{\sigma\}_l^{z-1}; \{\sigma\}_n^{z-1}\right)$$
$$\Rightarrow \rho^{(2)}(\sigma_l \sigma_n) \prod_{i=1}^{z-1} \frac{\rho^{(2)}(\sigma_l \sigma_i)}{\rho^{(1)}(\sigma_l)} \prod_{j=1}^{z-1} \frac{\rho^{(2)}(\sigma_n \sigma_j)}{\rho^{(1)}(\sigma_n)}. \qquad (9.2.15)$$

Using the last expression we obtain

$$\left.\frac{dC_\lambda}{dt}\right|_{out}^{bi} = \sum_n \frac{\alpha_{l,n}}{z} \sum_{\sigma'_l \sigma'_n \sigma_n} K\left(\sigma_l \sigma_n \to \sigma'_l \sigma'_n\right) \rho^{(2)}(\sigma_l \sigma_n), \qquad (9.2.16)$$

which is identical to the model without energetic interactions. We have used as a mean value over the neighbourhood

$$K(\sigma_l \sigma_n \to \sigma'_l \sigma'_n)$$
$$= \sum_{\{\sigma\}_l^{z-1}} \sum_{\{\sigma\}_n^{z-1}} \mathfrak{K}\left(\sigma_l \sigma_n \to \sigma'_l \sigma'_n \mid \{\sigma\}_l^{z-1}; \{\sigma\}_n^{z-1}\right) \times$$

$$\times \prod_{i=1}^{z-1} \frac{\rho^{(2)}(\sigma_l \sigma_i)}{\rho^{(1)}(\sigma_l)} \prod_{j=1}^{z-1} \frac{\rho^{(2)}(\sigma_n \sigma_j)}{\rho^{(1)}(\sigma_n)}. \tag{9.2.17}$$

We can rewrite this result in the form

$$\left. \frac{dC_\lambda}{dt} \right|_{\text{out}}^{\text{bi}} = \sum_{\lambda' \nu' \nu} K(\lambda \nu \to \lambda' \nu') C_\lambda C_\nu F_{\lambda \nu}(1). \tag{9.2.18}$$

Here the effective transition rate $K(\lambda \nu \to \lambda' \nu')$ is a function of the density C_μ and the joint correlation functions $F_{\lambda \mu}(1)$ and $F_{\nu \mu}(1)$. For the 'in'-term of the bimolecular step we obtain

$$\left. \frac{dC_\lambda}{dt} \right|_{\text{in}}^{\text{bi}} = \sum_{\lambda' \nu' \nu} K(\lambda' \nu' \to \lambda \nu) C_{\lambda'} C_{\nu'} F_{\lambda' \nu'}(1). \tag{9.2.19}$$

The temporal evolution of the densities can be written in the form of equation (9.1.13), where $A_\lambda[C, F]$ and $B_\lambda[C, F]$ are simple positive functions (polynomials) of the densities C and the correlation functions F.

9.2.1.4 Equation of motion for the two-point probabilities

Let us study the next processes which take place on two different lattice sites l and m. The following processes are possible (Section 9.1.1):

(1) l and m are not the nearest neighbours on the lattice.

   (a) The state σ_l can be created or annihilated independently of m. Only the neighbourhood of l is important.
   (b) The state σ_m can be created or annihilated independently of l. Only the neighbourhood of m is important.

(2) If l and m are the nearest neighbours on the lattice additional terms must be taken into account which represent the bimolecular steps.

We want to introduce a diagrammatic description of these processes, equation (9.1.14). A process takes place on the site ■, whereas the symbol □ means that this site plays no role for the determination of the state of another site. Therefore the first two terms correspond to the cases (1a) and (1b) and the third term represents the bimolecular step. The diagrams of type (1a) and (1b) are of the form (9.1.17), the second term on the right hand side represents all processes which annihilate the state σ_l. The difference to equation (9.2.4) lies in the additional condition imposed that the site m is in the state σ_m. Therefore we can write down (cf. equation (9.2.4)):

$$\left.\frac{d\rho^{(2)}(\sigma_l\sigma_m)}{dt}\right|_l^{\text{out}}$$

$$= \blacksquare \left.\frac{\text{out}}{l}\right.\square_m = \sum_{\{\sigma\}_l^z}\sum_{\sigma_l'}\mathfrak{P}(\sigma_l \to \sigma_l' \mid \{\sigma\}_l^z)\rho^{(2+z)}(\sigma_l\sigma_m;\{\sigma\}_l^z) +$$

$$+ \sum_n \frac{\alpha_{l,n}}{z} \sum_{\{\sigma\}_l^{z-1}}\sum_{\{\sigma\}_n^{z-1}}\sum_{\sigma_l'\sigma_n'\sigma_n} \mathfrak{K}(\sigma_l\sigma_n \to \sigma_l'\sigma_n' \mid \{\sigma\}_l^{z-1};\{\sigma\}_n^{z-1}) \times$$

$$\times \rho^{(2z+1)}(\sigma_l\sigma_n\sigma_m;\{\sigma\}_l^{z-1};\{\sigma\}_n^{z-1}). \tag{9.2.20}$$

This equation is valid only for $|l - m| > 1$. The corresponding case when l and m are the nearest neighbours will be treated below. Let us first discuss the monomolecular step for $|l - m| > 1$. Using equation (9.1.60), we obtain

$$\rho^{(2+z)}(\sigma_l\sigma_m;\{\sigma\}_l^z) \Rightarrow \rho^{(2)}(\sigma_l\sigma_m) \prod_{i=1}^{z} \frac{\rho^{(2)}(\sigma_l\sigma_i)}{\rho^{(1)}(\sigma_l)}. \tag{9.2.21}$$

From this equation it follows that

$$\left.\frac{d[C_\lambda C_\mu F_{\lambda\mu}(l - m)]}{dt}\right|_l^{\substack{\text{mono}\\\text{out}}}$$

$$= \left.\frac{d\rho^{(2)}(\sigma_l\sigma_m)}{dt}\right|_l^{\substack{\text{mono}\\\text{out}}} = \sum_{\lambda'} P(\lambda \to \lambda')C_\lambda C_\mu F_{\lambda\mu}(l - m). \tag{9.2.22}$$

For the 'in'-term a similar expression can be derived in the same way.

Let us now treat the terms where a process on site l occurs without m being affected, but l and m are the nearest neighbours. For $|l - m| = 1$ (the nearest neighbours) we change the expression for $\rho^{(2+z)}$ for

$$\rho^{(1+z)}(\sigma_l\sigma_m;\{\sigma\}_l^{z-1})$$

and $\rho^{(2z+1)}$ for

$$\rho^{(2z)}(\sigma_l\sigma_n\sigma_m;\{\sigma\}_l^{z-2};\{\sigma\}_n^{z-1}),$$

respectively.

Again we use equation (9.2.15). But now we get a non-complete mean-value because one site is in the state σ_m. Therefore we introduce the definition

$$P(\sigma_l \to \sigma'_l \mid \sigma_m) = \sum_{\{\sigma\}_l^{z-1}} \mathfrak{P}(\sigma_l \to \sigma'_l \mid \{\sigma\}_l^z) \prod_{i=1}^{z-1} \frac{\rho^{(2)}(\sigma_l\sigma_i)}{\rho^{(1)}(\sigma_l)},$$

$$\{\sigma\}_l^z = \{\sigma\}_l^{z-1}, \sigma_m. \tag{9.2.23}$$

For the nearest neighbours follows

$$\left.\frac{d[C_\lambda C_\mu F_{\lambda\mu}(1)]}{dt}\right|_l^{\text{mono}}_{\text{out}} = \sum_{\lambda'} P(\lambda \to \lambda' \mid \mu) C_\lambda C_\mu F_{\lambda\mu}(1). \tag{9.2.24}$$

For the case of no energetic interactions, $P(\lambda \to \lambda' \mid \mu)$ reduces to $P(\lambda \to \lambda')$. The 'in'-term is analogous.

We turn now to the true bimolecular steps. For the triple (lmn) we make the *ansatz*

$$\rho^{(2z+1)}\left(\sigma_l\sigma_n\sigma_m; \{\sigma\}_l^{z-1}; \{\sigma\}_n^{z-1}\right)$$

$$\Rightarrow \rho^{(3)}(\sigma_l\sigma_n\sigma_m) \prod_{i=1}^{z-1} \frac{\rho^{(2)}(\sigma_l\sigma_i)}{\rho^{(1)}(\sigma_l)} \prod_{j=1}^{z-1} \frac{\rho^{(2)}(\sigma_n\sigma_j)}{\rho^{(1)}(\sigma_n)}. \tag{9.2.25}$$

Let l and n be the nearest neighbours in this triple on which the bimolecular step takes place. Now we must distinguish between two cases in which m is or is not the nearest neighbour of site l. Using this expression we get for the case that l and m are not the nearest neighbours

$$\left.\frac{d[C_\lambda C_\mu F_{\lambda\mu}(l-m)]}{dt}\right|_l^{\text{bi}}_{\text{out}}$$

$$= \sum_n \frac{\alpha_{l,n}}{z} \sum_{\sigma'_l \sigma'_n \sigma_n} K(\sigma_l\sigma_n \to \sigma'_l\sigma'_n)\rho^{(3)}(\sigma_l\sigma_n\sigma_m). \tag{9.2.26}$$

Otherwise we get

$$\left.\frac{d[C_\lambda C_\mu F_{\lambda\mu}(1)]}{dt}\right|_l^{\text{bi out}}$$

$$= \sum_n \frac{\alpha_{l,n}}{z} \sum_{\sigma'_l \sigma'_n \sigma_n} K(\sigma_l \sigma_n \to \sigma'_l \sigma'_n \mid \{\sigma_m\}_l) \rho^{(3)}(\sigma_l \sigma_n \sigma_m), \qquad (9.2.27)$$

where

$$K(\sigma_l \sigma_n \to \sigma'_l \sigma'_n \mid \{\sigma_m\}_l)$$

$$= \sum_{\{\sigma\}_l^{z-2}} \sum_{\{\sigma\}_n^{z-1}} \mathfrak{K}\left(\sigma_l \sigma_n \to \sigma'_l \sigma'_n \mid \{\sigma\}_l^{z-1}; \{\sigma\}_n^{z-1}\right) \times$$

$$\times \prod_{i=1}^{z-2} \frac{\rho^{(2)}(\sigma_l \sigma_i)}{\rho^{(1)}(\sigma_l)} \prod_{j=1}^{z-1} \frac{\rho^{(2)}(\sigma_n \sigma_j)}{\rho^{(1)}(\sigma_n)}, \qquad (9.2.28)$$

and $\{\sigma_m\}_l$ means that the state of one neighbour of site l is fixed in the state σ_m.

Therefore we can write for the 'out'-terms of the bimolecular steps of equation (9.2.20) (using the notation given in that equation)

$$\{\blacksquare\blacksquare\}_{\text{out}} = \frac{1}{2} \sum_{\{\sigma\}_l^{z-1}} \sum_{\{\sigma\}_m^{z-1}} \sum_{\sigma'_l \sigma'_m} \mathfrak{K}\left(\sigma_l \sigma_m \to \sigma'_l \sigma'_m \mid \{\sigma\}_l^{z-1}; \{\sigma\}_m^{z-1}\right) \times$$

$$\times \rho^{(2z)}\left(\sigma_l \sigma_m; \{\sigma\}_l^{z-1}; \{\sigma\}_m^{z-1}\right). \qquad (9.2.29)$$

Here we can use the approximation equation (9.2.14) to simplify the $2z$-point probabilities. The 'in'-terms can be obtained in an analogous manner to the 'out'-terms. Finally we get equation (9.1.16), for more details of the derivations readers are referred to in the previous Section 9.1.1.

The written above equations for the temporal evolution contain three-point probabilities, i.e., we obtain a hierarchy of equations. We must truncate the infinite set of master equations in order to obtain a finite system of non-linear equations. To this end we use the Kirkwood superposition approximation (see Section 9.1.1).

9.2.1.5 The definition of a standard model

Several simulation models have been introduced in the past for surface reactions, including energetic interactions [16, 35–37]. These models are very difficult to compare to each other because they use very different descriptions of the individual reaction steps. Furthermore, steps which are considered to be independent of energetic interactions in one paper, show a dependency in another paper. Therefore it is necessary and useful to introduce a kind of the *standard model* [38] in order to compare different models in the future.

First of all, we define the transition rates for our stochastic model using an *ansatz* of Kawasaki [39, 40]. In the following we use the abbreviation X for an initial state (σ_l for mono- and $\sigma_l \sigma_n$ for bimolecular steps), Y for a final state (σ'_l for mono- and $\sigma'_l \sigma'_n$ for bimolecular steps) and Z for the states of the neighbourhood ($\{\sigma\}_l^{z-1}$ for mono- and $\{\sigma\}_l^{z-1}; \{\sigma\}_n^{z-1}$ for bimolecular steps). If we study the system in which the neighbourhood is fixed we observe a relaxation process in a very small area. We introduce the normalized probability $W(X)$ and the corresponding rates $\Re(X \to Y \mid Z)$. For this (reversible) process we write down the following Markovian master equation

$$\frac{dW(X)}{dt} = \Re(Y \to X \mid Z) W(Y) - \Re(X \to Y \mid Z) W(X). \quad (9.2.30)$$

For $t \to \infty$ we expect to get an equilibrium state

$$W(X) = W_{\text{eq}}(X \mid Z) = \frac{\exp\left(-\frac{H(X\mid Z)}{k_B T}\right)}{\exp\left(-\frac{H(X\mid Z)}{k_B T}\right) + \exp\left(-\frac{H(Y\mid Z)}{k_B T}\right)}, \quad (9.2.31)$$

where $H(X \mid Z)$ is the corresponding energy of the system in the state X. (For an open system we must also take the chemical potential into account.) It is important to note that equation (9.2.30) describes a reversible process. In the case that one rate $\Re(X \to Y \mid Z) = 0$ (for example the adsorption or desorption process does not exist in the model) we are not able to say anything about the energetics of the system. Therefore we must describe irreversible processes with rates which are independent of Z. This is the first condition of the standard model.

For the reversible processes we can define the ratio of the transition rates using equation (9.2.30):

$$\frac{\Re(X \to Y \mid Z)}{\Re(Y \to X \mid Z)} = \exp\left(-\frac{\delta H}{k_B T}\right),$$
$$\delta H = H(Y \mid Z) - H(X \mid Z). \tag{9.2.32}$$

In order to define the rates we need an additional condition. We use the *symmetric* condition (which constitutes the second condition of the standard model)

$$\Re(X \to Y \mid Z) = Q(X \mid Y) W_{eq}(Y \mid Z),$$
$$Q(X \mid Y) = Q(Y \mid X), \tag{9.2.33}$$

where $Q(X \mid Y)$ is a factor which is independent of the neighbourhood Z. The neighbourhood is only taken into account by $W_{eq}(Y \mid Z)$. Normally one writes this in the form [40]

$$\Re(X \to Y \mid Z) = \frac{1}{2\tau}\left[1 - \tanh\left(\frac{\delta H}{2k_B T}\right)\right], \tag{9.2.34}$$

which is identical to our representation with $1/(2\tau) = Q(X \mid Y)$. The choice of the standard form equation (9.2.33) offers two advantages. First, the rates are bounded by $Q(X \mid Y)$, and second, we can use the normalizable probability $W_{eq}(Y \mid Z)$. In principle the non-symmetric *ansatz*

$$\Re(X \to Y \mid Z) = Q(X \mid Y) \exp\left(-\frac{\delta H}{k_B T}\right),$$
$$\Re(Y \to X \mid Z) = Q(X \mid Y) \tag{9.2.35}$$

can be used but we would have here the problem that the transition rate is not bounded. Furthermore steps like the diffusion should be symmetric which is not possible with the non-symmetric *ansatz*.

Now let us study the energetic aspect in more detail. For the monomolecular steps we obtain

$$H(X \mid Z) \equiv H(\sigma_l \mid \{\sigma\}_i^z)$$
$$= \left(\varepsilon(\sigma_l) - \chi(\sigma_l; T)\right) + \sum_{i=1}^{z} E_{\sigma_l \sigma_i}^j, \tag{9.2.36}$$

where $E_{\lambda\nu}$ is a matrix which contains the interaction energies, $\varepsilon(\lambda)$ is particle energy and $\chi(\lambda;T)$ is a chemical potential which depends on the temperature T. For vacant sites ($\lambda = 0$), $\varepsilon(0) = \chi(0;T) = E_{0\nu} \equiv 0$. In the standard model we obtain for the one-point transition rates (using equation (9.2.33))

$$\mathfrak{P}(\sigma_l \to \sigma'_l\{\sigma\}^z_l)$$
$$\equiv Q(\sigma_l \mid \sigma'_l) \frac{\exp\left[-\frac{H(\sigma'_l|\{\sigma\}^z_l)}{k_B T}\right]}{\exp\left[-\frac{H(\sigma_l|\{\sigma\}^z_l)}{k_B T}\right] + \exp\left[-\frac{H(\sigma'_l|\{\sigma\}^z_l)}{k_B T}\right]}. \quad (9.2.37)$$

On the other hand, for the bimolecular steps we obtain (analogous to equation (9.2.36))

$$H(X \mid Z) \equiv H\left(\sigma_l\sigma_n \mid \{\sigma\}^{z-1}_l;\{\sigma\}^{z-1}_n\right)$$
$$= E_{\sigma_l\sigma_n} + (\varepsilon(\sigma_l) - \chi(\sigma_l;T)) + (\varepsilon(\sigma_n) - \chi(\sigma_n;T)) +$$
$$+ \sum_{i=1}^{z-1} E_{\sigma_l\sigma_i} + \sum_{j=1}^{z-1} E_{\sigma_n\sigma_j}. \quad (9.2.38)$$

One gets using equation (9.2.33) for the two-point transition rates

$$\mathfrak{K}(\sigma_l\sigma_n \to \sigma'_l\sigma'_n \mid \{\sigma\}^{z-1}_l;\{\sigma\}^{z-1}_n)$$
$$= Q(\sigma_l\sigma_n \mid \sigma'_l\sigma'_n) W_{eq}(\sigma'_l\sigma'_n \mid \{\sigma\}^{z-1}_l;\{\sigma\}^{z-1}_n). \quad (9.2.39)$$

9.2.1.6 Examples of the standard model

We would like to demonstrate now the flexibility of the standard model. We introduce a model which includes mono- and bimolecular steps as discussed in Section 9.1.1. The previous model assumed no energetic interactions. By introducing the model described in this Section for an empty lattice we must obtain the previous model, which will be shown in this Section.

Let us begin with the monomolecular steps like the creation of a particle A with the rate $P(0 \to A) = p_A = p_A(T)$ and an annihilation process $P(A \to 0) = k_A = k_A(T)$ which is normally presented in the form $k_A = \kappa_A \exp(-E_A/(k_B T))$ with the activation energy E_A and the frequency factor κ_A. In the following we use the abbreviation $\{0\}^z_l$ for the empty neighbourhood of site l. For this case we obtain

$$H(A \mid \{0\}^z_l) = \varepsilon(A) - \chi(A;T), \quad H(0 \mid \{0\}^z_l) = 0. \quad (9.2.40)$$

From equation (9.2.37) it follows

$$\mathfrak{P}(0 \to A \mid \{0\}_{\tilde{l}}^{z}) \equiv Q(A \mid 0)\frac{\omega}{1+\omega} = p_A,$$

$$\mathfrak{P}(A \to 0 \mid \{0\}_{\tilde{l}}^{z}) \equiv Q(A \mid 0)\frac{1}{1+\omega} = k_A \qquad (9.2.41)$$

with

$$\omega = \exp\left(-\frac{[\varepsilon(A) - \chi(A;T)]}{k_B T}\right), \qquad (9.2.42)$$

i.e., $\omega = p_A/k_A$ and $Q(A \mid 0) = (p_A + k_A)$ hold. We get for the difference $[\varepsilon(A) - \chi(A;T)]$ the physical interpretation

$$\varepsilon(A) - \chi(A;T) = k_B T \ln\left(\frac{k_A}{p_A}\right) = k_B T \ln\left(\frac{\kappa_A}{p_A}\right) - E_A. \qquad (9.2.43)$$

The last equality holds for $p_A = \text{const}$. Therefore our model can take into account the adsorption energy and also the frequency factor.

Next we introduce the creation and annihilation process for a bimolecular step. In the previous model without energetic interactions discussed in Section 9.1.1 we have had the two-point transition rates $K(\sigma_l\sigma_n \to \sigma'_l\sigma'_n)$. Let us study now the processes $K(BB \to 00) = k_B$ (desorption) and $K(00 \to BB) = p_B$ (adsorption). In the new model we get for an empty surface

$$H(BB \mid \{0\}_l^{z-1}; \{0\}_n^{z-1}) = 2(\varepsilon(B) - \chi(B;T)) + E_{BB},$$

$$H(00 \mid \{0\}_l^{z-1}; \{0\}_n^{z-1}) = 0. \qquad (9.2.44)$$

The transition rates are given by

$$\mathfrak{K}(00 \to BB \mid \{0\}_l^{z-1}; \{0\}_n^{z-1}) = Q(00 \mid BB)\frac{\omega}{1+\omega},$$

$$\mathfrak{K}(BB \to 00 \mid \{0\}_l^{z-1}; \{0\}_n^{z-1}) = Q(00 \mid BB)\frac{1}{1+\omega} \qquad (9.2.45)$$

with

$$\omega = \exp\left(-\frac{2[\varepsilon(B) - \chi(B;T)]}{k_B T} - \frac{E_{BB}}{k_B T}\right). \quad (9.2.46)$$

We obtain that $\omega = p_B/k_B$, $Q(00 \mid BB) = (p_B + k_B)$ and

$$\varepsilon(B) - \chi(B;T) = \frac{1}{2} k_B T \ln\left(\frac{k_B}{p_B}\right) - \frac{1}{2} E_{BB}$$

$$= \frac{1}{2}\left\{k_B T \ln\left(\frac{\kappa_B}{p_B}\right) - E_B - E_{BB}\right\}, \quad (9.2.47)$$

for the case that $k_B = \kappa_B \exp(-E_B/k_B T)$ and p_B=const.
For a diffusion process we obtain for an empty surface

$$\mathfrak{K}(A0 \to 0A \mid \{0\}_l^{z-1}; \{0\}_n^{z-1})$$
$$\equiv \frac{1}{2} Q(A0 \mid 0A) = D = D_0 \exp\left(-\frac{E_{\text{diff}}}{k_B T}\right) \quad (9.2.48)$$

and $Q(A0 \mid 0A) = 2D$, where D is a transition rate for this process for the case without energetic interactions.

9.2.2 The $A + \frac{1}{2} B_2 \to 0$ reaction with energetic interactions

9.2.2.1 Introduction

Let us study now a stochastic model for the particular $A + \frac{1}{2} B_2 \to 0$ reaction with energetic interactions between the particles. The system includes adsorption, desorption, reaction and diffusion steps which depend on energetic interactions. The temporal evolution of the system is described by master equations using the Markovian behaviour of the system. We study the system behaviour at different values for the energetic parameters and at varying diffusion and desorption rates. The location and the character of the phase transition points will be discussed in detail.

In order to understand the details of surface reactions computer simulations are a powerful tool. For $A + \frac{1}{2} B_2 \to 0$ reaction Ziff, Gulari and Barshad have done the Monte Carlo simulations (the ZGB-model) [2]. The model takes into account the following steps, described by equations (9.1.39) to (9.1.41).

In order to develop a more realistic model, additional aspects have been taken into account. The CO desorption can be modeled by equation (9.1.43). The additional aspect of CO desorption leads to the disappearance of the CO-poisoned state [15] because at every value of Y_{CO} adsorbed CO molecules are able to leave the surface. Many other investigations under various conditions have been performed, like energetic interactions [16], the aspect of physisorption and reaction via the Eley–Rideal mechanism [17].

In the model of Section 9.1.2 we use a stochastic *ansatz* which describes the system by using master equations (under the assumption that the system is Markovian). This *ansatz* takes explicitly into account correlations in the particle distribution on the lattice. The results turn out to be in overall good agreement with the ZGB-model.

In order to get a more realistic description of surface reactions energetic interactions must be taken into account. We introduced in Section 9.2.1 a general model which is able to handle systems which include mono- and bimolecular steps like adsorption, desorption, diffusion and reaction [38]. Here we apply this model to an extended version of the ZGB-model which incorporates particle diffusion and desorption [41].

9.2.2.2 The model

We use a square lattice with coordination number $z = 4$. In principle the model can be described by equations

$$\frac{dC_A}{dt} = p_A C_0 - k_A C_A - K C_A C_B, \qquad (9.2.49)$$

$$\frac{dC_B}{dt} = p_B C_0^2 - k_B C_B - K C_A C_B, \qquad (9.2.50)$$

where C_λ is a density (concentration) of particles of type λ on the surface ($\lambda = 0, A$ and B); the sum rule $C_0 + C_A + C_B = 1$ holds. p_A and p_B are adsorption and k_A and k_B are desorption rates for the A and B particles, respectively. K is a reaction constant. In both the equations written above the first terms describe the creation of particles, the second ones represent the desorption and the last terms describe the reaction between an A and a B particle. In a mean-field model k, p and K are assumed to be constants which do not depend on the particle distribution. In the case that diffusion and desorption steps are taken into account (but neglecting energetic effects), some of the parameters become dependent on the particle distributions:

$$p_A = p_A^0, \quad k_A = k_A^0,$$
$$p_B = p_B^0 F_{00}(1), \quad k_B = k_B^0 F_{BB}(1) \qquad (9.2.51)$$

and $K = R F_{AB}(1)$, where the variables with superscript 0 do not depend on the particle distribution. $F_{\lambda\mu}(1)$ with $\lambda, \mu \in \{A, B, 0\}$ is a correlation function between the nearest neighbours. In the extended ZGB-model $p_A^0 = Y_{CO}$, $p_B^0 = 2(1 - Y_{CO})$ and $R \to \infty$. For the case of energetic interactions all parameters appearing in equation (9.2.51), p_A, p_B, k_A, k_B, depend on the particle distributions and therefore on the correlation functions. In order to solve these equations for the temporal evolution of the system we use the *standard model* discussed in Section 9.2.1 and an improvement of the Kirkwood approximation discussed in Section 9.1.2 to handle the problem which is necessitated by the fact that R goes to infinity.

9.2.2.3 Results for repulsive interactions

We start with the system without B-desorption ($k_B = 0$). This case is realistic for the oxidation of CO because the O atoms (B particles) are strongly bound to the metal surface and desorption does not occur at the temperatures used for this reaction. In this case the parameter E_{BB} does not play a role because the transition probabilities for the A and B particles depend on k_B^0/p_B^0 and on k_A^0/p_A^0, respectively. Therefore we have reduced the number of energetic parameter to two (F_{AA} and E_{BB}). In the following we present all values of $E_{\lambda\mu}$ in $k_B T$ units.

In Fig. 9.16 the coverages of A and B and the production rate R_{CO_2} are shown as a function of Y_{CO}. We assume a large desorption rate $k_A^0 = 0.05$ and no diffusion ($D = 0$). For curve 1 we have $E_{AA} = E_{AB} = 0$, which corresponds to the case of the ZGB-model incorporating A-desorption. The value of y_1 is not shifted by the desorption because at this point too few A particles are present. Complete coverage of the lattice by A does not occur because at every time step A particles have the chance to desorb from the surface. Both facts are in agreement with the corresponding Monte Carlo simulations [15].

In curve 2 we switch on the A–A interaction ($E_{AA} = 1$ and $E_{AB} = 0$), which corresponds to a repulsion between the A particles. Due to the repulsion the phase transition at y_2 nearly disappears and we obtain a very

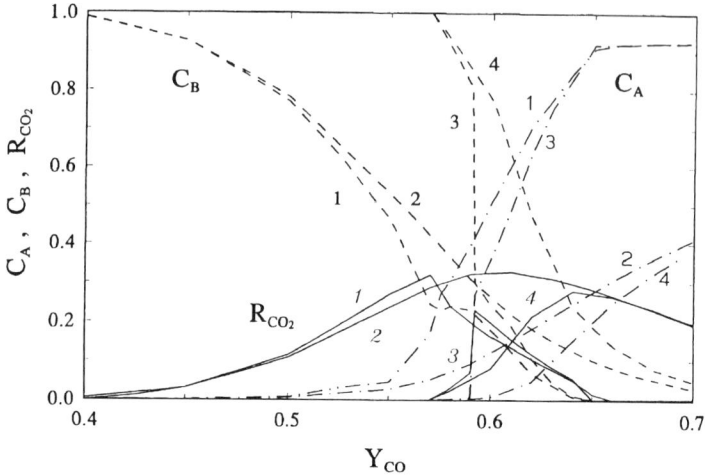

Fig. 9.16. The densities C_A, C_B and the production rate R_{CO_2} are shown as a function of Y_{CO} for $k_A^0 = 0.05$ and $D = 0$. The energetic parameters are: $E_{AA} = E_{AB} = 0$ (curve 1), $E_{AA} = 1$, $E_{AB} = 0$ (curve 2), $E_{AA} = 0$, $E_{AB} = 1$ (curve 3) and $E_{AA} = E_{AB} = 1$ (curve 4).

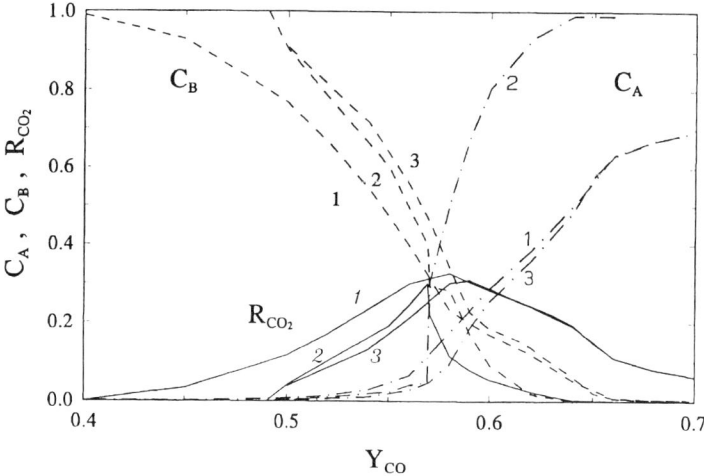

Fig. 9.17. The densities C_A, C_B and the production rate R_{CO_2} are shown as a function of Y_{CO} for $k_A^0 = 0.01$ and $D = 0$. The energetic parameters are: $E_{AA} = 1$, $E_{AB} = 0$ (curve 1), $E_{AA} = 0$, $E_{AB} = 1$ (curve 2) and $E_{AA} = E_{AB} = 1$ (curve 3).

smooth transition. The coverages are not influenced by E_{AA} for small values of Y_{CO} which arises from the small number of A particles in this region.

Curve 3 shows the diagram for $E_A = 0$ and $E_{AB} = 1$. The parameter E_{AB} shifts the critical point y_1 dramatically to larger values of Y_{CO} which means that the complete occupation of the lattice by B easily takes place. This is understandable from the fact that the A-adsorption probability is decreased by the repulsion with the B particles, which are the dominant species on the lattice. Therefore the tendency of the B particles to form large clusters is enhanced. This can be seen from R_{CO_2} which is for $Y_{CO} < y_2$ nearly zero and rises sharply at $Y_{CO} \approx y_2$.

In curve 4 we have $E_{AA} = E_{AB} = 1$. This results in a smooth transition at y_2 for the A-density (caused by E_{AA}) and a phase transition at y_1 at large values of Y_{CO} (caused by E_{AB}). The parameter E_{AA} is important for large values of C_A (i.e., for large values of Y_{CO}) and E_{AB} dominates the system behaviour at large values of C_B (i.e., for small values of Y_{CO}).

In Fig. 9.17 we study the case of a small desorption rate of A particles ($k_A^0 = 0.01$) neglecting diffusion. The energetic parameters are $E_{AA} = 1$ and $E_{AB} = 0$ (curve 1), $E_{AA} = 0$ and $E_{AB} = 1$ (curve 2) and $E_{AA} = E_{AB} = 1$ (curve 3). Here the system behaviour is strongly influenced by the desorption. Due to the smaller desorption rate (compared to the previous figure) the phase transition point y_2 is shifted to lower values of Y_{CO} and the transition turns out to be sharper. The phase transition point y_1 is also shifted to lower values of Y_{CO} for curves 2 and 3 which means that B poisoning is more difficult to achieve due to the reaction with A particles sitting on the surface. This effect is especially important in the case of curve 3 (cf. with the curve 4 from the previous figure, where the critical point y_1 is located at larger values of Y_{CO}). Here more A particles are present and the desorption plays a significant role.

How does the system behaviour change due to diffusion? In Fig. 9.18 the results are shown for $k_A^0 = 0.05$, $E_{AB} = 0$ and for $E_{AA} = 0, D = 0$ (curve 1), $E_{AA} = 0, D = 10$ (curve 2), $E_{AA} = 1, D = 0$ (curve 3) and $E_{AA} = 1, D = 10$ (curve 4). The diagram is plotted for large values of Y_{CO} only because for small values the diffusion plays no role. Under the influence of diffusion, the phase transition point y_2 is shifted to larger values of Y_{CO} because the fast moving A particles remove small B-clusters via reaction from the lattice. On the lattice cleaned in this way the B_2 particles have a larger probability to find two adjacent free sites to adsorb and build large clusters (the density of B particles increases). Therefore the complete A coverage takes place at larger values of Y_{CO} compared to the case without

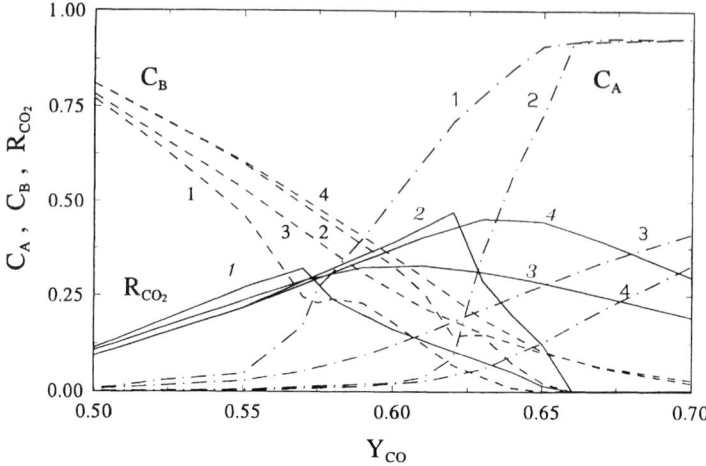

Fig. 9.18. The densities C_A, C_B and the production rate R_{CO_2} are shown as a function of Y_{CO} for $k_A^0 = 0.05$ and $E_{AB} = 0$. The parameters are: $E_{AA} = 0$, $D = 0$ (curve 1), $E_{AA} = 0$, $D = 10$ (curve 2), $E_{AA} = 1$, $D = 0$ (curve 3) and $E_{AA} = 1$, $D = 10$ (curve 4).

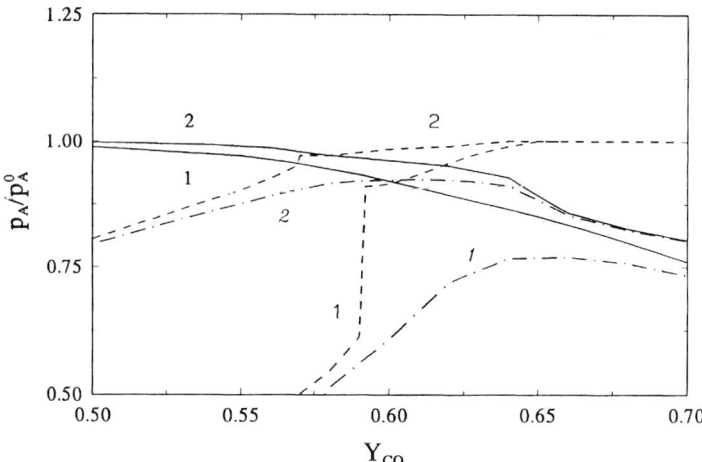

Fig. 9.19. The effective sticking coefficient p_A/p_A^0 as a function of Y_{CO} for $k_A^0 = 0.05$ (curve 1) and for $k_A^0 = 0.01$ (curve 2). The energetic parameters are $E_{AA} = 1$, $E_{AB} = 0$ (solid); $E_{AA} = 0$, $E_{AB} = 1$ (dashed) and $E_{AA} = E_{AB} = 1$ (dot-dashed).

diffusion. This result is in agreement with the Monte Carlo simulations [3]. The maximum value of the production rate is also shifted to larger values of Y_{CO} in accord with the shift of the A density to larger values of Y_{CO}. In the limiting case of a well-stirred system we would obtain that the maximum rate is located at the stoichiometric gas phase composition ($Y_{CO} = 2/3$) [3]. The parameter E_{AA} acts in the same way as discussed for the previous figure.

Next we turn to the details of the reaction and study the effective sticking coefficient, the effective desorption rate and the reaction constant. We denote by solid curves the energetic parameters $E_{AA} = 1$, $E_{AB} = 0$; by dashed curves $E_{AA} = 0$, $E_{AB} = 1$ and lastly, by dot-dashed curves $E_{AA} = E_{AB} = 1$.

In Fig. 9.19 the ratio of p_A/p_A^0 (which represents an effective sticking coefficient for the A particles) is shown as a function of Y_{CO}. In curve 1 the desorption rate is $k_A^0 = 0.05$ whereas in curve 2 – $k_A^0 = 0.01$. For $Y_{CO} < y_2$ only a small number of A particles is observed on the lattice (see above). From the solid curves we therefore conclude that the effect of blocking and the repulsion with other A particles plays no role and the sticking coefficient is nearly unity. Increasing Y_{CO} the sticking coefficient decreases due to larger number of A particles sitting on the lattice. The effect of a larger desorption rate enhances this effect because A particles located in unfavourable configurations desorb immediately. The repulsion between A and B particles (dashed curves) is important only in the region where many B particles are present (at lower values of Y_{CO}) resulting in a dramatic decrease of the sticking coefficient at lower values of Y_{CO}.

For the case that the A–A and the A–B interactions are both repulsive we obtain a composite system behaviour resulting from the behaviour discussed above.

In Fig. 9.20 the effective desorption rate k_A/k_A^0 is shown as a function of Y_{CO}. The parameters are: $k_A^0 = 0.05$ (curve 1) and $k_A^0 = 0.01$ (curve 2). The A–A repulsion is important for large values of Y_{CO} resulting in an increase of the desorption rate. The smaller k_A^0 the greater is the effective desorption rate affected by the A–A repulsion. The A–B interaction plays no role for the A-desorption because in our model AB pairs cannot exist on the lattice due to the infinite reaction rate. The dot-dashed curve shows the intermediate situation of the two cases discussed above.

The reaction constant K is shown in Fig. 9.21. Curve 1 shows the behaviour for $k_A^0 = 0.05$, $D = 0$, curve 2 for $k_A^0 = 0.01$, $D = 0$ and lastly curve 3 for $k_A^0 = 0.05$, $D = 10$. The reaction rate is small in the parameter region where large clusters are formed because reaction can occur only

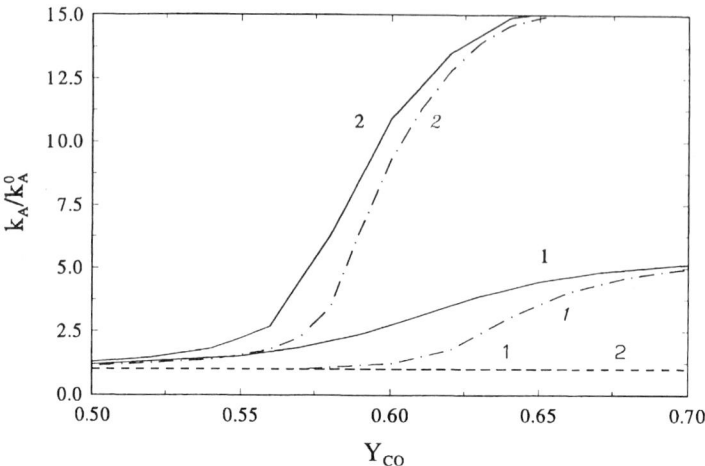

Fig. 9.20. The effective desorption rate k_A/k_A^0 as a function of Y_{CO} for $k_A^0 = 0.05$ (curve 1) and $k_A^0 = 0.01$ (curve 2). The energetic parameters are $E_{AA} = 1$, $E_{AB} = 0$ (solid); $E_{AA} = 0$, $E_{AB} = 1$ (dashed) and $E_{AA} = E_{AB} = 1$ (dot-dashed).

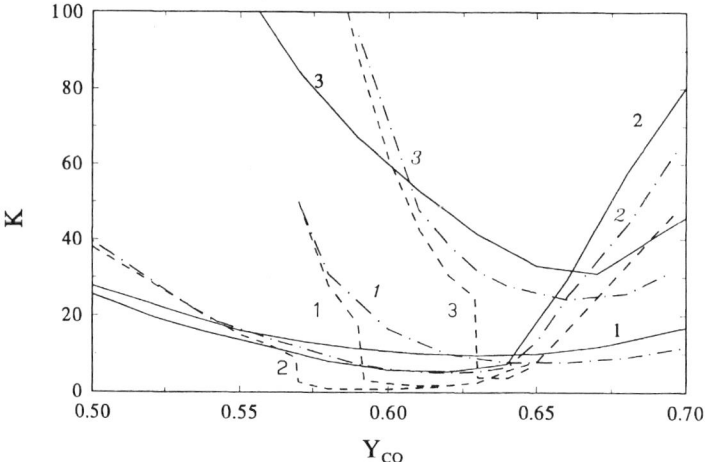

Fig. 9.21. The reaction constant K as a function of Y_{CO} for $k_A^0 = 0.05$, $D = 0$ (curve 1); $k_A^0 = 0.01$, $D = 0$ (curve 2) and for $k_A^0 = 0.05$, $D = 10$ (curve 3). The energetic parameters are $E_{AA} = 1$, $E_{AB} = 0$ (solid); $E_{AA} = 0$, $E_{AB} = 1$ (dashed) and $E_{AA} = E_{AB} = 1$ (dot-dashed).

at the border of the particle islands. This is true below $Y_{CO} < 0.64$. The diffusion increases the reaction rate due to the fact that the A particles are much more reactive. There is an interesting effect of the desorption rate in the absence of diffusion. Below $Y_{CO} = 0.64$, where large clusters arise, a smaller desorption rate reduces the reaction constant K probably because a larger desorption rate enhances the adsorption which mixes non-reactive A particles with reactive ones. Above $Y_{CO} = 0.64$ a small desorption rate enhances K compared to a larger desorption rate because the stoichiometric particle composition on the lattice is reached at lower values of Y_{CO} for the smaller desorption rate.

It should be stressed that K is not constant and changes over a factor of one hundred in the present case.

9.2.2.4 Results for attractive interaction

So far we have treated repulsive interactions. Now we turn to the investigation of systems with *attractive* interactions ($E_{\lambda\mu} < 0$). In Fig. 9.22 the particle densities and the production rate are shown as a function of Y_{CO} for $D = 0$ and $k_A^0 = 0.05$. The energetic parameters are: $E_{AA} = E_{AB} = 0$

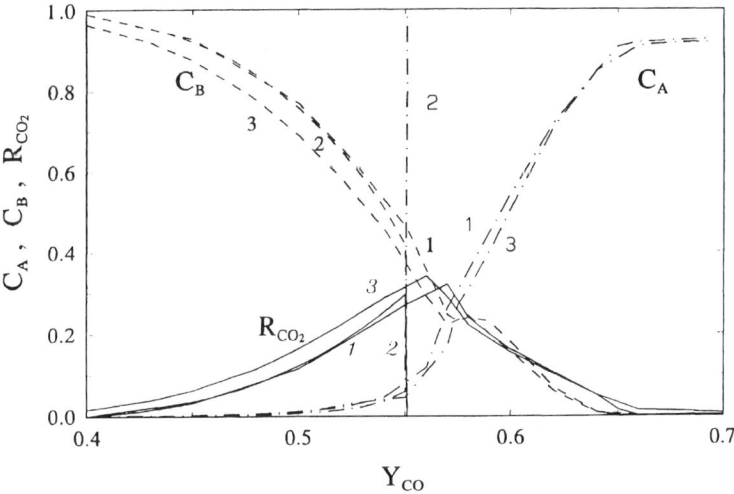

Fig. 9.22. The particle densities and the production rate are shown as a function of Y_{CO} for $D = 0$ and $k_A^0 = 0.05$. The energetic parameters are: $E_{AA} = E_{AB} = 0$ (curve 1); $E_{AA} = -1$, $E_{AB} = 0$ (curve 2) and $E_{AA} = 0$, $E_{AB} = -1$ (curve 3).

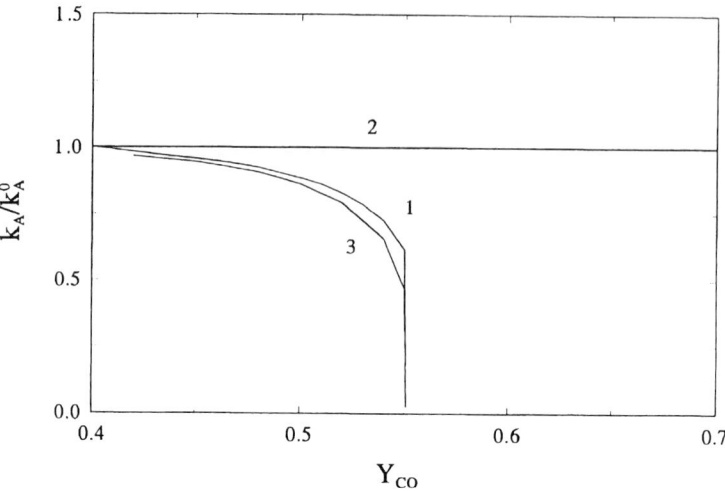

Fig. 9.23. The effective desorption rate is shown as a function of Y_{CO} for $D = 0$ and $k_A^0 = 0.05$. The energetic parameters are: $E_{AA} = -1$, $E_{AB} = 0$ (curve 1); $E_{AA} = 0$, $E_{AB} = -1$ (curve 2) and $E_{AA} = E_{AB} = -1$ (curve 3).

(curve 1), $E_{AA} = -1$, $E_{AB} = 0$ (curve 2) and $E_{AA} = 0$, $E_{AB} = -1$ (curve 3). Due to the attractive interaction between A particles (curve 2) the phase transition at y_2 changes its character for the first order and the value of y_2 is significantly lowered. At lower values of Y_{CO} this effect plays no role due to the small number of A particles. An attractive interaction between A and B particles (curve 3) smoothes the transition again and the resulting system's behaviour is very similar to the case without energetic interactions for larger values of Y_{CO}. The difference is a small decrease in C_A (for larger values of Y_{CO}) and a small decrease in C_B (for smaller values of Y_{CO}). This results from the attraction of A and B particles which enlarges the probability of reactive events.

In Fig. 9.23 the effective desorption rate is shown as a function of Y_{CO} for $D = 0$ and $k_A^0 = 0.05$. The energetic parameters are: $E_{AA} = -1$, $E_{AB} = 0$ (curve 1), $E_{AA} = 0$, $E_{AB} = -1$ (curve 2) and $E_{AA} = E_{AB} = -1$ (curve 3). The desorption rate is not affected by E_{AB} (curve 2) whereas attractive interactions between the A particles decrease the desorption rate. For smaller values of Y_{CO} no A particles exist on the surface.

9.2.2.5 The effect of B-desorption

We now take into account the effect of B_2-desorption. In the following we use the parameters: $D = 0$, $E_{BB} = 1$ and $k_A^0 = k_B^0 = 0.01$. The remaining energetic parameters are: $E_{AA} = E_{AB} = 0$ (curve 1), $E_{AA} = 1$, $E_{AB} = 0$ (curve 2), $E_{AA} = 0$, $E_{AB} = 1$ (curve 3) and $E_{AA} = E_{AB} = 1$ (curve 4).

In Fig. 9.24 the particle densities and the production rate are shown as functions of Y_{CO}. The phase transition at y_1 disappears due to the effect of B desorption (which is equivalent to the effect of A-desorption at y_2). The repulsive interaction of A particles (curve 2) smoothes the phase transition at y_2. $E_{AB} = 1$ (curve 3) shifts y_2 to lower values of Y_{CO}. This comes from the fact that the A particles can easily adsorb on the free sites created by B-desorption. The increased number of A particles on the lattice leads to an increased B-desorption due to the repulsive A–B interaction (see Fig. 10 and 11). In curve 4 the repulsion between the A particles leads to to shift of y_2 to larger values of Y_{CO}. Smoothing of the phase transition (compare curve 2) can be observed only at the beginning of the phase transition. At larger values of Y_{CO}, C_A increases abruptly. This can be explained by the

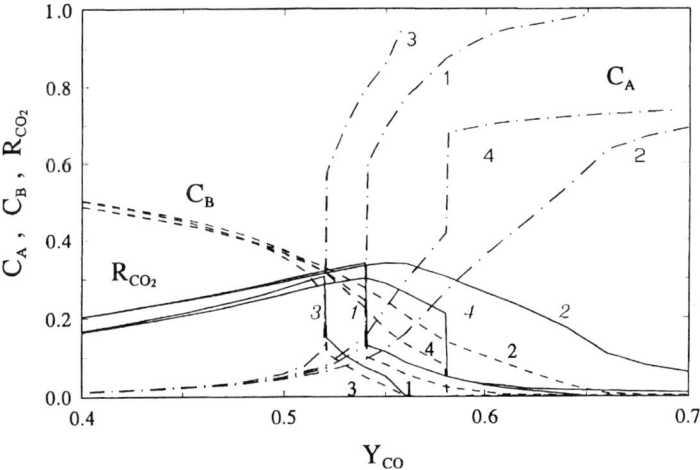

Fig. 9.24. The particle densities and the production rate are shown as a function of Y_{CO} for $D = 0$, $E_{BB} = 1$ and $k_A^0 = k_B^0 = 0.01$. The remaining energetic parameters are: $E_{AA} = E_{AB} = 0$ (curve 1); $E_{AA} = 1$, $E_{AB} = 0$ (curve 2); $E_{AA} = 0$, $E_{AB} = 1$ (curve 3) and $E_{AA} = E_{AB} = 1$ (curve 4).

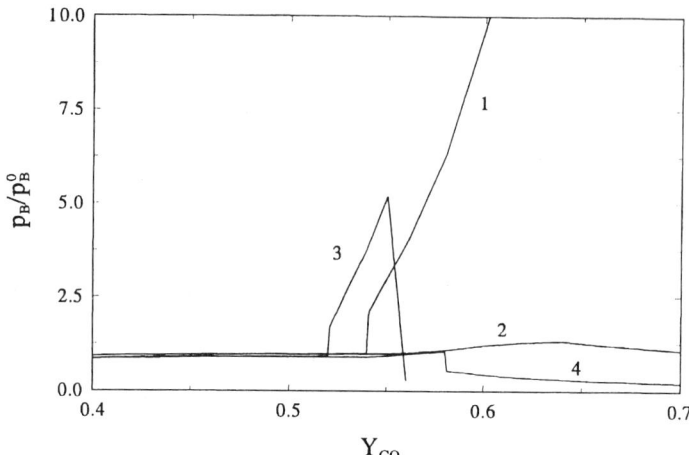

Fig. 9.25. p_B/p_B^0 is shown as a function of Y_{CO} for $D = 0$, $E_{BB} = 1$ and $k_A^0 = k_B^0 = 0.01$. The remaining energetic parameters are: $E_{AA} = E_{AB} = 0$ (curve 1); $E_{AA} = 1$, $E_{AB} = 0$ (curve 2); $E_{AA} = 0$, $E_{AB} = 1$ (curve 3) and $E_{AA} = E_{AB} = 1$ (curve 4).

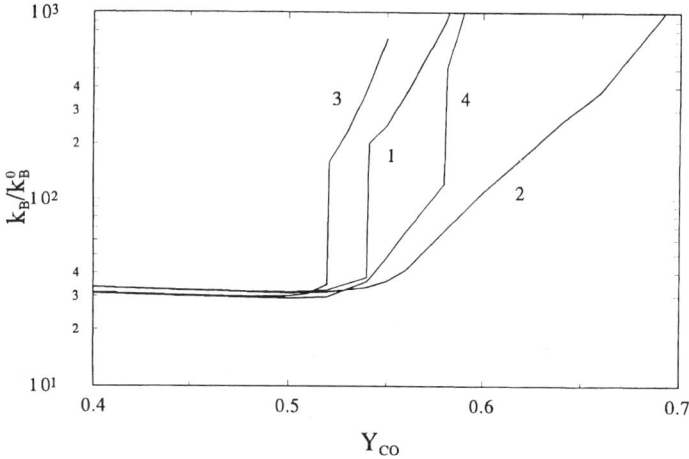

Fig. 9.26. k_B/k_B^0 is shown as a function of Y_{CO} for $D = 0$, $E_{BB} = 1$ and $k_A^0 = k_B^0 = 0.01$. The remaining energetic parameters are: $E_{AA} = E_{AB} = 0$ (curve 1); $E_{AA} = 1$, $E_{AB} = 0$ (curve 2); $E_{AA} = 0$, $E_{AB} = 1$ (curve 3) and $E_{AA} = E_{AB} = 1$ (curve 4).

fact that at this point nearly all B particles left the lattice which leads to a decrease of the repulsion on the A particles.

With the variation of the energetic parameters the value of y_2 can be shifted over a very large range of Y_{CO}; the system's behaviour at small values of Y_{CO} is much more stable against changes of the energetic parameters.

The effective adsorption and desorption rates for B particles are much more difficult to interpret than for the A particles because they are strongly influenced by structural correlations. Even without energetic interactions they are of the form $p_B = p_B^0 F_{00}(1)$ and $k_B = k_B^0 F_{BB}(1)$, respectively.

Therefore one observes only an effect of a product of the adsorption (or desorption) rate and the correlation functions where the correlation may be very large (see Section 9.1.2). In Fig. 9.25 p_B/p_B^0 is shown as a function of Y_{CO}. $F_{00}(r)$ is large in the parameter region where large segregation occurs. Energetic interactions decrease the rate. In Fig. 9.26 k_B/k_B^0 is shown as a function of Y_{CO}. For $Y_{CO} > y_2$ the few B particles arrange mainly in islands. Therefore $F_{BB}(1)$ is very large leading to large values of k_B/k_B^0.

9.2.2.6 Conclusion

We have introduced a stochastic model which is useful to describe surface reactions which include mono- and bimolecular steps such as particle adsorption, desorption and diffusion in the presence of energetic interactions. We have presented here the equation of motion in a form of master equations for the one- and two-point probabilities. The appearance of higher-order correlation functions which arise in the equations of motion is avoided by making use of a generalized superposition approximation and by suitable averaging procedures which are guided by a correspondence principle. According to this principle the equations should be reduced to the previously developed model (Sections 9.1.1, 9.1.2) when energetic interactions are neglected. Because the model takes into account explicitly spatial correlations of the particles on the lattice, the results are in good agreement with computer simulations in the absence of energetic interactions. The same is expected to be the case with energetic interactions.

At first glance the appearing equations seem to be very complex. But the numerical solution of the equations is a process which can be done with a computer program. The analytical model offers several advantages compared to simulations. Since such a theoretical *ansatz* needs only a small amount of computing time, more complex systems can be studied. Moreover our models are not restricted to small lattices which are inavoidably used in computer simulations. This is of particular interest at phase transition points of the first and second order or in the case of strong energetic interactions where large-scale particle correlations appear. All these advantages hold also for the model described above.

To demonstrate this, in Section 9.2.2 we have studied a stochastic model for an extended ZGB-model including diffusion, desorption and energetic interactions as additional steps. We have used different values of the diffusion and the desorption rates and different values for the energetic parameters. In the case of repulsive interactions the system's behaviour is strongly influenced by E_{AA} for large values of Y_{CO} and by E_{AB} for small values of Y_{CO}. The former parameter leads to a smooth phase transition at y_2 and the latter to a sharp transition at y_1. The sharpness and the location of the phase transitions depend also on the diffusion and desorption rate of the A particles. The A-diffusion leads to an increase of the value of y_2 due to the higher reactivity of the A particles. At lower values of Y_{CO} the system behaviour is nearly not influenced by the diffusion. The A-desorption increases the values of the critical points and smoothes the phase transition at y_2. This effect becomes very important if C_A is large.

Attractive interactions between A particles change significantly the system's behaviour at y_2, resulting in a decrease of the value of y_2 and change of the character of the phase transition to the first order (second order for the case of no energetic interactions and with desorption). We face here a case similar to a condensation. This is mainly the effect of the reduced desorption rate of the A particles which leads to the same character of the phase transition at y_2 as without the effect of desorption. The parameter E_{AB} enlarges the effective diffusion rate but the system's behaviour is not significantly influenced by this change.

The effect of B-desorption leads to the disappearance of the phase transition at y_1 which is equivalent to the effect of the A-desorption at y_2. The effect of the energetic parameters is very complex. Interpretation of the results is difficult due to the complicated adsorption and desorption rules for the B_2 particles. The value of the phase transition y_2 can be shifted over a large range of Y_{CO} by different model parameters.

The model we have presented here is well-suited for the description of surface reactions with energetic interactions. The use of realistic parameters for these interactions (which are not well-known today) should lead to a more realistic description of surface reactions in the nearest years.

References

[1] M. Ehsasi, M. Matloch, O. Frank, J.H. Block, K. Christmann, F.S. Rys and W. Hirschwald, J. Chem. Phys. 91 (1989) 4949.
[2] M. Ziff, E. Gulari and Y. Barshad, Phys. Rev. Lett. 56 (1986) 2553.
[3] J. Mai, W. von Niessen and A. Blumen, J. Chem. Phys. 93 (1990) 3685.
[4] J. Mai and W. von Niessen, Chem. Phys. 165 (1992) 65.
[5] Z. Ra'cz, Phys. Rev. Lett. 55 (1985) 1707.
[6] D. Ben-Avraham and S. Redner, Phys. Rev. A: 34 (1986) 501.
[7] Z.Y. Shi and R. Kopelman, Chem. Phys. 167 (1992) 149.
[8] I.M. Sokolov, H. Schnörer and A. Blumen, Phys. Rev. A: 44 (1991) 2388.
[9] K. Lindenberg and B.J. West, Phys. Rev. A: 42 (1990) 890.
[10] J. Mai, V.N. Kuzovkov and W. von Niessen, Phys. Rev. E: 48 (1993) 1700.
[11] J.G. Kirkwood, J. Chem. Phys. 76 (1935) 479.
[12] E.A. Kotomin and V.N. Kuzovkov, Rep. Prog. Phys. 55 (1992) 2079.
[13] J. Mai, V.N. Kuzovkov and W. von Niessen, J. Chem. Phys. 98 (1993) 10017.
[14] M.B. Cutlip, AIChE J. 25 (1979) 502.
[15] J. Mai, W. von Niessen and A. Blumen, in: Dynamical Processes in Condensed Molecular Systems (World Scientific, Singapore, 1990) p. 159.
[16] H.-P. Kaukonen and R.M. Nieminen, J. Chem. Phys. 91 (1989) 4380.
[17] J. Mai and W. von Niessen, Chem. Phys. 156 (1991) 63.
[18] J. Mai and W. von Niessen, Phys. Rev. A: 44 (1991) 6165.
[19] J. Mai, V.N. Kuzovkov and W. von Niessen, Physica A, 203 (1994) 298.
[20] V.N. Kuzovkov and E.A. Kotomin, Phys. Scr. 47 (1993) 585.
[21] H. Mamada and F. Takano, J. Phys. Soc. Jpn 25 (1968) 675.
[22] J. Mai, V.N. Kuzovkov and W. von Niessen, J. Chem. Phys. 100 (1994) 6073.
[23] S. Havlin, in: The Fractal Approach to Heterogenous Chemistry, ed. D. Avnir (Wiley, New York, 1989) p. 251.
[24] P. Pfeifer, Chimia, 39 (1985) 120.
[25] A. Casties, J. Mai and W. von Niessen, J. Chem. Phys. 99 (1993) 3082.
[26] J. Mai, V.N. Kuzovkov and W. von Niessen, J. Chem. Phys. 100 (1994) 8525.
[27] R. Kopelman, in: The Fractal Approach to Heterogenous Chemistry, ed. D. Avnir (Wiley, New York, 1989).
[28] J. Mai, A. Casties, V.N. Kuzovkov and W. von Niessen, J. Chem. Phys. 102 (1995) 5037.
[29] G. Ertl, Critical Reviews in Solid State and Material Science (CRC Press, Boca Raton, 1982).
[30] P. Stoltze and J.K. Norskov, J. Catal. 110 (1988) 1.
[31] P. Stoltze and J.K. Norskov, Phys. Rev. Lett. 55 (1985) 2502.
[32] G. Haase, M. Asscher, R. Kosloff, J. Chem. Phys. 90 (1989) 3346.
[33] G. Ertl, Nachr. Chem. Tech. Lab. 31 (1983) 178.
[34] G. Ertl, Angew. Chem. 102 (1990) 1258.
[35] J.J. Luque, F. Jimenez-Moralez and M.C. Lemos, J. Chem. Phys. 92 (1992) 8535.
[36] J.W. Evans, H.H. Madden and R.J. Imbihl, J. Chem. Phys. 96 (1992) 4805.

[37] J. Satulovsky and E. Albano, J. Chem. Phys. 97 (1992) 9440.
[38] J. Mai, V.N. Kuzovkov and W. von Niessen, J. Phys. A: (1996) (submitted).
[39] K. Kawasaki, Phys. Rev. 1450 (1966) 224; 148 (1966) 375; 150 (1966) 285.
[40] K. Binder and D.W. Heermann, Monte Carlo Simulation in Statistical Physics (Springer, Berlin, 1988).
[41] J. Mai, V.N. Kuzovkov and W. von Niessen, J. Phys. A: (1996) (submitted).

Chapter 10

General Conclusion

> Half-truth is the most dangerous lie.
> Hebrew proverb

This book is the first attempt to summarize, probably from our subjective point of view, the state of the art in a very rapidly developing theory of many-particle effects in bimolecular reactions in condensed matter, which up to now was a subject of several review papers only [1–10]. We have focused mainly on several basic bimolecular reactions trying not to cover all possible cases (e.g., more complicated reactions, cooperative processes in alloys under irradiation [11] or initial *macroscopic* separation of reactants, etc.) but to compare critically results and advantages/limitations of numerous approaches developed in the last years. We focused on processes induced by point particles (defects) only; the effects of *dislocation* self-organization are discussed in [12–16] whereas diffusion-limited particle aggregation with a special attention to fractal cluster formation has extensive literature [17–21].

In our opinion, this book demonstrates clearly that the formalism of *many-point particle densities* based on the Kirkwood superposition approximation for decoupling the three-particle correlation functions is able to treat adequately all possible cases and reaction regimes studied in the book (including immobile/mobile reactants, correlated/random initial particle distributions, concentration decay/accumulation under permanent source, etc.). Results of most of analytical theories are checked by extensive computer simulations. (It should be reminded that many-particle effects under study were observed for the first time namely in computer simulations [22, 23].) Only few *experimental* evidences exist now for many-particle effects in bimolecular reactions, the two reliable examples are accumulation kinetics of immobile radiation defects at low temperatures in ionic solids (see [24] for experiments and [25] for their theoretical interpretation) and pseudo-first order reversible diffusion-controlled recombination of protons with excited dye molecules [26]. This is one of main reasons why we did not consider in detail some of very refined theories for the kinetics asymptotics as well as peculiarities of reactions on fractal structures ([27–29] and references therein).

In conclusion, we tried to demonstrate that theory of many-particle effects presented here could find many new very interesting applications, like reactant self-organization in a course of *surface* catalytic reactions discussed in the last Chapter 9, which are important from both fundamental and applied points of view.

ACKNOWLEDGEMENTS

Authors are greatly indebted to N. Agmon, P. Argyrakis, A. Ben-Shaul, A. Blumen, W. Frank, W. von Niessen, and A. Burstein for many stimulating discussions.

References

[1] Ya.B. Zeldovich and A.S. Mikhailov, Sov. Phys. Usp. 30 (1987) 977.
[2] V.N. Kuzovkov and E.A. Kotomin, Rep. Prog. Phys. 51 (1988) 1479.
[3] A.S. Mikhailov, Phys. Rep. 184 (1989) 302.
[4] A.A. Ovchinnikov, S.F. Timashev and A.A. Belyi, Kinetics of Diffusion-Controlled Chemical Processes (Nova Science, New York, 1989).
[5] A. Szabo, J. Phys. Chem. 93 (1989) 6929.
[6] V.L. Vinetsky, Yu.H. Kalnin, E.A. Kotomin and A.A. Ovchinnikov, Sov. Phys. Usp. 30 (1990) 1.
[7] Proceedings of the Conference on Models of Non-Classical Reaction Rates, Special issue of J. Stat. Phys. 65(5/6) (1991).
[8] E.A. Kotomin and V.N. Kuzovkov, Rep. Prog. Phys. 55 (1992) 2079.
[9] A. Okninski, Catastrophe Theory, Comprehensive Chemical Kinetics, Vol. 33 (Polish Scientific Publishers, Warszawa, 1992).
[10] N. Agmon and R.D. Levine (eds), Dissipative Dynamics, Special issue of Chem. Phys. 180(2/3) (1994).
[11] P. Bellon and G. Martin, Solid State Phenom. 30/31 (1993) 107.
[12] L.P. Kubin and G. Martin (eds), Non-Linear Phenomena in Materials Science, Trans. Tech. Publications (Acdermannsdorf, Switzerland, 1988).
[13] E.S. Aifantis, Solid State Phenom. 3/4 (1988) 397.
[14] A. Seeger and W. Frank, Solid State Phenom. 3/4 (1988) 125.
[15] V.A. Davydov, V.S. Zykov and A.S. Mikhailov, Sov. Phys. Usp. 34 (1991) 665.
[16] W. Frank (ed.), Synergetic Ordering in Solids, Special issue of Appl. Phys. A 57(2) (1993); 58(1) (1994).
[17] L. Pietronero and E. Tosatti (eds), Fractals in Physics (North-Holland, Amsterdam, 1986).
[18] H.E. Stanley and N. Ostrowsky, On Growth and Form (Nijhoff, Boston, 1986).
[19] R. Jullien and R. Botet, Aggregation and Fractal Aggregates (World Scientific, Singapore, 1987).

References

[20] J. Feder, Fractals (Plenum, New York, 1988).
[21] A. Bunde and S. Havlin (eds), Fractals and Disorded Systems (Springer, Berlin, 1991).
[22] G. Lück and R. Sizmann, Phys. Status Solidi 6 (1964) 263; 14 (1966) K61.
[23] I.A. Tale, D.K. Millers and E.A. Kotomin, J. Phys. C: 8 (1975) 2366.
[24] B.F. Faraday and W.D. Compton, Phys. Rev. A: 138 (1965) 893.
[25] V.N. Kuzovkov and E.A. Kotomin, J. Phys. C: 17 (1984) 2283.
[26] D. Huppert, S.Y. Goldberg, A. Masad and N. Agmon, Phys. Rev. Lett. 68 (1992) 3932.
[27] A. Blumen, J. Klafter and G. Zumofen, in: Optical Spectroscopy of Glasses, ed. I. Zschokke (Reidel, Dordrecht, 1986) p. 199.
[28] S. Havlin and D. Ben-Avraham, Adv. Phys. 36 (1987) 695.
[29] G. Zumofen, J. Klafter and A. Blumen, J. Stat. Phys. 65 (1991) 1015.

Author Index

Abell, G.C., 230
Aboltin, D.E., 233
Agmon, N., 170, 298, 594
Agmon, N., *see* Edelstein, A., 298
Agmon, N., *see* Huppert, D., 298, 595
Agmon, N., *see* Pines, E., 170
Agmon, N., *see* Szabo, A., 297
Agranovich, V.M., 230, 297
Aguilar, M., 464
Agullo-Lopez, F., 464
Agullo-Lopez, F., *see* Aguilar, M., 464
Aifantis, E.S., 594
Albano, E., *see* Satulovsky, J., 592
Aluker, E.D., 168
Alvarez Rivas, J.L., *see* Hodgson, E.R., 464
Anacker, L.W., 464
Andrews, A.S., 383
Andronov, A., 135
Anlauf, J.K., 297
Antonov-Romanovskii, V.V., XVI, 169, 230, 462
Arbuzov, V.I., 232, 233
Argyrakis, P., 298, 299, 383
Argyrakis, P., *see* Sokolov, I.M., 463
Asscher, M., *see* Haase, G., 591

Babloyantz, A., *see* Nicolis, G., 136
Bachmann, K., 168, 232, 384, 464
Balagurov, B., XV, 52, 136, 297
Balding, D.J., 298
Balescu, R., 52, 137, 296, 385
Baranov, P.G., 233
Baranovskii, S.D., 465
Barshad, Y., *see* Ziff, M., 591
Bassagnani, G., *see* Sonder, E., 464
Baxter, R., 296
Bazikin, A.D., 136, 513
Becker, D.E., 170

Becker, O.M., 51
Bellon, P., 594
Belyi, A.A., 136
Belyi, A.A., *see* Ovchinnikov, A.A., XV, 136, 170, 296, 384, 594
Ben-Avraham, D., 297, 298, 464, 591
Ben-Avraham, D., *see* Burschka, M.A., 299
Ben-Avraham, D., *see* Doering, C.R., 299, 464
Ben-Avraham, D., *see* Havlin, S., 383, 595
Ben-Avraham, D., *see* Lin, J.-Ch., 299
Ben-Avraham, D., *see* Schoonover, R., 298
Ben-Shaul, A., *see* Becker, O.M., 51
Ben-Shaul, A., *see* Silverberg, M., 52
Benderskii, V.A., 135
Benson, S.W., XIV, 52, 134
Berezhkovskii, A.M., 136, 297, 298
Berlin, Yu.A., 231
Binder, K., 592
Birnbaum, H.K., *see* Murch, G.E., XIV
Bishop, J.E.L., 383
Bishop, J.E.L., *see* Andrews, A.S., 383
Bishop, J.E.L., *see* Searle, T.M., 383
Bishop, R.F., 384
Bixon, M., XVI, 138
Blanche, A., *see* Hanusse, P., 383
Block, J.H., *see* Ehsasi, M., 591
Bloor, D., *see* Hunt, I.G., 297
Blumen, A., XV, 297, 383, 463, 595
Blumen, A., *see* Agmon, N., 298
Blumen, A., *see* Klafter, I., 297
Blumen, A., *see* Luding, S., 296, 383
Blumen, A., *see* Mai, J., 591
Blumen, A., *see* Schnörer, H., 52, 136, 296, 383, 462
Blumen, A., *see* Sokolov, I.M., 383, 463, 591
Blumen, A., *see* Zumofen, G., 298, 383, 595
Bochkanov, P.V., 384

Bodunov, E.N., see Ermolaev, V.L., 135
Bogachev, L.V., see Berezhkovskii, A.M., 298
Bogans, Ya.R., 465
Boning, K., see Nakagawa, M., 463
Borckmans, P., see Dewel, G., XVI, 136
Borckmans, P., see Walgraef, D., 136
Borg, R.J., XIV
Botet, R., see Jullien, R., 594
Bramson, M., 297, 298, 384
Braunstein, L., 299
Brebec, G., see Laskar, A.S., XIV, 169
Brikenshtein, V.Kh., see Benderskii, V.A., 135
Brocklehurst, B., 233
Brout, R., 296
Brush, S.G., see Kikuchi, R., 137, 296
Bunde, A., 595
Burlatsky, S.F., XVI, 136, 138, 298, 383, 463
Burlatsky, S.F., see Oshanin, G.S., 136, 138, 298
Burlatsky, S.F., see Ovchinnikov, A.A., 136, 464
Burschka, M.A., 299
Burstein, A.I., 232
Burstein, A.I., see Doktorov, A.B., 230, 231, 232, 385
Burton, J.J., see Nowick, A.S., XIV
Butler, W.H., see Yoo, M.H., 232, 384
Butlers, P., see Kotomin, E.A., 169, 231
Byakov, V.M., 232
Bykov, V., see Yablonskii, G., XV

Calef, D.F., XV, 138
Camagni, P., see Sonder, E., 464
Careri, G., XVI, 51
Carragher, B.O., see Comins, J.D., 464
Casties, A., see Mai, J., 591
Chaturvedi, S.J., see Gardiner, C.W., XVI
Chen, R., 169
Chernoustan, A.I., see Burlatsky, S.F., 383
Chernov, A.S., see Kotomin, E.A., 384
Chernov, S.A., see Aluker, E.D., 168
Christmann, K., see Ehsasi, M., 591
Clayton, I.H., see Andrews, A.S., 383

Clement, E., 464
Clifford, P., see Mozumder, A., 135, 169, 232
Collins, F.C., 298
Comins, J.D., 464
Compton, W.D., see Faraday, B.F., 169, 462, 595
Corbett, J.W., see Peak, D., 169, 231, 298
Cost, J., see Murch, G.E., XIV
Cutlip, M.B., 591

Davis, N.A., see Mott, N.F., 169
Davydov, V.A., 594
Dean, P., 168, 464
Debye, P., 137, 230, 385
Dederichs, P.H., 168
Deich, R.G., see Aluker, E.D., 168
Deigen, M.F., see Grachev, V.F., 168, 231
Delbecq, C.J., 232, 233
Delgado, A., see Hodgson, E.R., 464
Delyon, F., 297
Dettmann, K., 137, 462, 463
Dettmann, K., see Schröder, K., 170, 232, 384
Deutch, J., see Calef, D.F., XV, 138
Deutch, J., see Felderhof, B.F., XVI
Deutch, J., see Nitzan, A., XVI
Deutz, J., see Dederichs, P.H., 168
Dewel, G., XVI, 136
Dewel, G., see Walgraef, D., 136
Dexter, D.L., see Delbecq, C.J., 233
Dienes, G.J., see Borg, R.J., XIV
Dishon, M., see Havlin, S., 297
Doering, C.R., 299, 464
Doering, C.R., see Ben-Avraham, D., 464
Doering, C.R., see Burschka, M.A., 299
Doering, C.R., see Lin, J.-C., 299, 464
Doi, M., XV, 138
Doktorov, A.B., 168, 230, 231, 232, 233, 385, 463
Doktorov, A.B., see Burstein, A.I., 232
Doktorov, A.B., see Kotomin, E.A., XV, 169, 231, 298, 384, 464
Donsker, M.D., 297
Duesing, G., 463
Dvorschak, F., see Becker, D.E., 170
Dzhumanov, S., 168

Author index

Ebeling, W., 134, 296, 512
Ebeling, W., *see* Feistel, R., 512
Eberlein, E., *see* Schröder, K., 170
Ebner, C., *see* Jiang, Z., 298, 384
Edelson, D., 512
Edelstein, A., 298
Edelstein, A., *see* Agmon, N., 298
Efros, A., *see* Shklovsky, B., 169
Eggert, J.R., 383
Eglitis, R., *see* Kotomin, E.A., 170, 231
Eglitis, R., *see* Popov, A.I., 170
Ehsasi, M., 591
Ekmanis, Yu.A., *see* Pirogov, F.V., 465
Ekmanis, Yu.A., *see* Schwartz, K.K., 384, 463
Elango, M.A., 168, 463, 465
Elango, M.A., *see* Lushchik, Ch.B., 463
Elderfield, D., XVI
Elokhin, V., *see* Yablonskii, G., XV
Emtsev, V.V., 168, 462
Ermolaev, V.L., 135
Ertl, G., 591
Eshelby, I.D., 168, 384
Evans, J.W., 591
Eyring, H., XIV, 135, 230, 384

Fabrikant, I.I., 231, 232
Fabrikant, I.I., *see* Kotomin, E.A., 168, 170, 231, 384, 465
Faraday, B.F., 169, 462, 595
Feder, J., 595
Fedorenko, S.G., *see* Burstein, A.I., 232
Feistel, R., 512
Felderhof, B.F., XVI
Field, R.J., 512, 513
Field, R.J., *see* Edelson, D., 512
Flynn, C.P., XIV
Franck-Kamenetskii, D.A., 134
Frank, O., *see* Ehsasi, M., 591
Frank, W., 594
Frank, W., *see* Kotomin, E.A., 383, 463
Frank, W., *see* Scheu, W., 168, 232, 384
Frank, W., *see* Seeger, A., 594
Fricke, T., *see* Wendt, D., 384
Fuks, N.A., 232

Gaididei, Yu.B., XVI, 137, 138
Gailitis, A.A., 231, 232
Gailitis, A.A., *see* Tale, I.A., 231, 233
Gailitis, A.A., *see* Vitol, I.K., 232
Galanin, M.D., *see* Agranovich, V.M., 230, 297
Gardiner, C.W., XIV, 52, 135, 169
Gavrilov, V.V., *see* Aluker, E.D., 168
Gektin, A.V., 464
Ghez, R., XIV
Gillespie, T.D., 383
Goldberg, S.Y., *see* Huppert, D., 298, 595
Goldhirsch, I., XVI
Gorban', A., *see* Yablonskii, G., XV
Gordon, I., *see* Andronov, A., 135
Gösele, U., 135, 169, 230, 232, 297
Grabovskis, V.Ya., 232, 233
Grabovskis, V.Ya., *see* Vitol, I.K., 231, 232, 233
Grachev, V.F., 168, 231
Grassberger, P., 136, 297
Green, N.J.B., *see* Balding, D.J., 298
Green, N.J.B., *see* Mozumder, A., 135, 169, 232
Green, N.J.B., *see* Pimblott, S.M., 232
Grinfeld, A.U., *see* Aboltin, D.E., 233
Grishkin, V.L., *see* Byakov, V.M., 232
Gromov, V.V., 463
Groote, J.C., *see* Seinen, J., 463
Grossmann, S., 137
Grynberg, M.D., *see* Braunstein, L., 299
Guillot, G., *see* Mercier, E., 463
Gulari, E., *see* Ziff, M., 591
Gutin, A.M., XVI, 138, 463

Haan, S.W., 297
Haase, G., 591
Haken, H., XVI, 51, 134, 135, 169, 512
Hantley, F.A., *see* Gösele, U., 230
Hanusse, P., 135, 383, 513
Hartog, H.W., *see* Seinen, J., 463
Hasegawa, A., XVI
Havlin, S., 297, 383, 591, 595
Havlin, S., *see* Bunde, A., 595
Havlin, S., *see* Schoonover, R., 298

Havlin, S., *see* Weiss, G.H., 298
Heermann, D.W., *see* Binder, K., 592
Hermandes, J., *see* Rubio, J., 464
Hermann, F., 465
Hirai, M., *see* Kotomin, E.A., 170
Hirayama, F., *see* Inokuti, M., 297
Hirschwald, W., *see* Ehsasi, M., 591
Hobbs, L.W., 169, 462
Hodgson, E.R., 464
Hong, K.M., 231
Horsthemke, W., XVI
Hubbard, J.B., *see* Kayser, R.F., 297
Hubbard, J.B., *see* Rasaiah, J.C., 299
Hughes, A.E., 462, 464
Hughes, A.E., *see* Hobbs, L.W., 169, 462
Hunt, I.G., 297
Huppert, D., 298, 595
Huppert, D., *see* Pines, E., 170

Imanaka, K., 232
Imbihl, R.J., *see* Evans, J.W., 591
Inokuti, M., 297
Ito, A., *see* Tomita, K., 513
Itoh, N., 167, 168, 170, 384, 462, 464
Itoh, N., *see* Saidoh, M., 384
Itoh, N., *see* Tanimura, K., 230
Itoh, N., *see* Tashiro, T., 233
Ivachenko, E.L., *see* Baranovskii, S.D., 465

Jain, U., 463
Jaque, F., *see* Aguilar, M., 464
Jaque, F., *see* Agullo-Lopez, F., 464
Jasnow, D., XVI
Jiang, Z., 298, 384
Jimenez-Moralez, F., *see* Luque, J.J., 591
Jonson, R.A., XIV
Jullien, R., 594

Kabler, M.N., 169, 231, 384, 462
Kalnin, Yu.H., XV, 137, 230, 297, 462, 463, 465
Kalnin, Yu.H., *see* Vinetsky, V.L., XV, 169, 296, 384, 462, 594
Kanders, U.K., 233

Kang, K., XVI, 136, 298, 383, 464
Kang, K., *see* Redner, S., 297
Kangro, A.R., *see* Arbuzov, V.I., 232, 233
Kangro, A.R., *see* Vitol, I.K., 233
Kantorovich, L., XV, 169
Kantorovich, L., *see* Kotomin, E.A., 169, 232
Kanzaki, H., *see* Ueta, M., 167, 462
Kapral, R., 297
Karplus, M., *see* Lee, S., 298
Karpushin, A.A., *see* Kipriyanov, A.A., 232
Kaukonen, H.-P., 591
Kawasaki, K., 592
Kayal, A.H., *see* Imanaka, K., 232
Kayser, R.F., 297
Kelman, I.V., *see* Vinetsky, V.L., 463
Khabibulaev, P.K., *see* Dzhumanov, S., 168
Khairutdinov, R.F., XV, 231
Khairutdinov, R.F., *see* Doktorov, A.B., 233
Khairutdinov, R.F., *see* Zamaraev, K.I., 168, 231
Kiefer, J.E., *see* Havlin, S., 297
Kikuchi, R., 137, 296
Kimball, G.E., *see* Collins, F.C., 298
Kipriyanov, A.A., 232
Kipriyanov, A.A., *see* Burstein, A.I., 232
Kipriyanov, A.A., *see* Doktorov, A.B., 230, 231, 385
Kirkwood, J.G., 137, 591
Kirsanov, V., 168, 462
Kirshon, O.M., *see* Kalnin, Yu.H., 465
Kirshon, O.M., *see* Vinetsky, V.L., 462
Kitshara, K., *see* Nicolis, G., 136
Klafter, J., 297
Klafter, J., *see* Blumen, A., XV, 297, 383, 595
Klafter, J., *see* Zumofen, G., 298, 383, 595
Klinger, M.I., 168, 464
Kobayashi, K., *see* Ueta, M., 167, 462
Kondepudi, D., *see* Horsthemke, W., XVI
Kondrachuk, A.B., 465
Kondrachuk, A.B., *see* Vinetsky, V.L., 462
Kopelman, R., XV, 298, 591
Kopelman, R., *see* Anacker, L.W., 464
Kopelman, R., *see* Argyrakis, P., 298, 299, 383

Kopelman, R., *see* Clement, E., 464
Kopelman, R., *see* Lindenberg, K., XV, 136, 298, 463
Kopelman, R., *see* Schoonover, R., 298
Kopelman, R., *see* Shi, Z.Y., 591
Kopelman, R., *see* Weiss, G.H., 298
Korepanov, V.I., *see* Bochkanov, P.V., 384
Kórös, E., *see* Field, R.J., 513
Koschnick, F.K., *see* Spaeth, J.M., 168
Kosevich, A.M., 168
Koshkin, V.M., 168, 464
Koshkin, V.M., *see* Zabrodskii, Yu.P., 168, 464
Kosloff, R., *see* Haase, G., 591
Kotomin, E.A., XV, 52, 137, 138, 168, 169, 170, 230, 231, 232, 296, 298, 383, 384, 463, 464, 465, 513, 591, 594
Kotomin, E.A., *see* Doktorov, A.B., 168, 230, 463
Kotomin, E.A., *see* Fabrikant, I.I., 231, 232
Kotomin, E.A., *see* Kalnin, Yu.H., XV, 463
Kotomin, E.A., *see* Kantorovich, L., XV, 169
Kotomin, E.A., *see* Kuzovkov, V.N., XV, 51, 52, 137, 169, 230, 231, 296, 383, 384, 385, 462, 463, 464, 513, 591, 594, 595
Kotomin, E.A., *see* Popov, A.I., 170
Kotomin, E.A., *see* Rogulis, U.T., 233
Kotomin, E.A., *see* Tale, I.A., XV, 296, 464, 595
Kotomin, E.A., *see* Vinetsky, V.L., XV, 169, 296, 384, 462, 594
Krikis, Yu.Yu., *see* Kalnin, Yu.H., 465
Krivads, E.A., *see* Aboltin, D.E., 233
Kronghaus, V., *see* Tale, I., 231
Kronmüller, H., *see* Schaefer, H.E., 384
Kronmüller, H., *see* Scheu, W., 168, 232, 384
Kuba, J., 231
Kubin, L.P., 594
Kulis, P., 231
Kulis, P., *see* Kotomin, E.A., 169, 231
Kulis, P., *see* Tale, I., 231
Kuzovkov, V.N., XV, 51, 52, 137, 169, 230, 231, 296, 297, 383, 384, 385, 462, 463, 464, 513, 591, 594, 595

Kuzovkov, V.N., *see* Kantorovich, L., XV, 169
Kuzovkov, V.N., *see* Kotomin, E.A., XV, 52, 137, 138, 230, 298, 383, 463, 465, 513, 591, 594
Kuzovkov, V.N, *see* Luding, S., 296, 383
Kuzovkov, V.N., *see* Mai, J., 591, 592
Kuzovkov, V.N., *see* Schnörer, H., 52, 296, 383, 462

Laidler, K.J., XIV
Lamm, G., *see* Szabo, A., 298
Laskar, A.S., XIV, 169
Le Gall, J.-F., 297
Lebowitz, J.L., *see* Bramson, M., 297, 298, 384
Lee, S., 298
Lee, S.H., *see* Rasaiah, J.C., 299
Lefever, R., *see* Prigogine, I., 135
Leibfried, G., XV, 52, 137, 169, 230
Leibfried, G., *see* Dettmann, K., 137, 463
Lemarchard, H., *see* Vidal, C., 169, 512
Lemos, M.C., *see* Luque, J.J., 591
Leontovich, E., *see* Andronov, A., 135
Levine, R.D., *see* Agmon, N., 594
Levy, P.W., 464
Leyvraz, F., 383
Lidiard, A.B., 463, 465
Lidiard, A.B., *see* Jain, U., 463
Lifshitz, E.M., 52, 137
Lifshitz, I.M., 297
Light, J.C., *see* Tyson, J.J., 135, 513
Lin, J.-C., 299, 464
Lin, S.M., *see* Eyring, H., XIV, 135, 230, 384
Lindenberg, K., XV, 136, 298, 463, 591
Lisitsyn, V.M., 168
Lisitsyn, V.M., *see* Bochkanov, P.V., 384
Lopez, F.J., *see* Agullo-Lopez, F., 464
Lorenz, E.N., 135
Lotka, A.J., 135, 513
Lück, G., 169, 296, 462, 595
Luding, S., 296, 383
Luding, S., *see* Blumen, A., 463
Luding, S., *see* Schnörer, H., 383
Luque, J.J., 591

Lushchik, A., see Lushchik, Ch., 135, 168, 231, 462
Lushchik, Ch., 135, 168, 231, 462, 463
Lushchik, Ch., see Klinger, M.I., 168, 464
Lushnikov, A.A., 138, 298

Ma, S., 136
Machovets, T.V., see Emtsev, V.V., 168, 462
Madden, H.H., see Evans, J.W., 591
Magee, I.L., 230
Mai, J., 591, 592
Mai, J., see Casties, A., 591
Makhnovskii, Yu.A., see Berezhkovskii, A.M., 136, 297, 298
Malek-Mansour, M., see Nicolis, G., XVI, 136
Mamada, H., 591
Manamura, E., see Ueta, M., 167, 462
Mandelbrot, B.B., 383
Manning, J.R., 169
Mansel, W., see Nakagawa, M., 463
Martin, G., 463
Martin, G., see Bellon, P., 594
Martin, G., see Kubin, L.P., 594
Martin, H.O., see Braunstein, L., 299
Masad, A., see Huppert, D., 298, 595
Mashovets., T.V., see Klinger, M.I., 168, 464
Matloch, M., see Ehsasi, M., 591
Mattuck, R.D., 52
Mayer, A., see Andronov, A., 135
Mazgar, A.C., see Imanaka, K., 232
Mazo, R., 136
McConnell, H.M., see Torney, D.C., 298
McNeil, K., see Mori, H., XVI
Meakin, P., 298, 384
Mercier, E., 463
Mikhailov, A.I., 465
Mikhailov, A.S., XV, XVI, 136, 138, 169, 594
Mikhailov, A.S., see Davydov, V.A., 594
Mikhailov, A.S., see Gutin, A.M., XVI, 138, 463
Mikhailov, A.S., see Polak, L.S., 135
Mikhailov, A.S., see Zeldovich, Ya.B., XV, 136, 169, 462, 594

Mikhnovich, V.V., see Emtsev, V.V., 168, 462
Millers, D.K., see Tale, I.A., XV, 233, 296, 464, 595
Mitchell, J.S., see Andrews, A.S., 383
Molchanov, S.A., see Berezhkovskii, A.M., 298
Molchanov, S.A., see Zeldovich, Ya.B., 136
Montroll, E.W., 230
Monty, C., see Laskar, A.S., XIV, 169
Moreno, M., 168
Mori, H., XVI
Morozov, V.A., see Burstein, A.I., 232
Mott, N.F., 169
Movaghar, B., see Hunt, I.G., 297
Mozumder, A., 135, 169, 230, 232
Mozumder, A., see Abell, G.C., 230
Mozumder, A., see Pimblott, S.M., 232
Mrowec, S., XIV, 169
Murch, G.E., XIV
Murrieta, H., see Rubio, J., 464
Muthukumar, M., XV, 138

Nakagawa, M., 463
Nicolis, G., XVI, 51, 135, 136, 137, 296, 512
Nieminen, R.M., see Kaukonen, H.-P., 591
Nitzan, A., XVI
Noks, R., 135, 230
Noolandi, J., see Hong, K.M., 231
Norskov, J.K., see Stoltze, P., 591
Nouilhat, A., see Mercier, E., 463
Nowick, A.S., XIV
Nowick, A.S., see Murch, G.E., XIV
Noyes, R.M., XV
Noyes, R.M., see Edelson, D., 512
Noyes, R.M., see Field, R.J., 513

O'Shaughnessy, B., 464
Ogrinsh, M., see Kalnin, Yu.H., 465
Ohta, T., see Tomita, K., 513
Ohtsuki, T., 298
Okninski, A., 594
Onipko, A.I., 137, 230, 297, 384
Onipko, A.I., see Gaididei, Yu.B., XVI, 137, 138
Onsager, L., 170, 230

Orlov, A.N., *see* Jonson, R.A., XIV
Ornstein, L.S., 52, 137
Ortoleva, P., *see* Nitzan, A., XVI
Ortoleva, P., *see* Schmidt, S., 513
Oshanin, G.S., 136, 138, 298
Oshanin, G.S., *see* Burlatsky, S.F., 298
Ostrowsky, N., *see* Stanley, H.E., 594
Ovchinnikov, A.A., XV, 52, 136, 170, 296, 298, 384, 385, 464, 594
Ovchinnikov, A.A., *see* Belyi, A.A., 136
Ovchinnikov, A.A., *see* Burlatsky, S.F., XVI, 136, 138, 298, 383, 463
Ovchinnikov, A.A., *see* Byakov, V.M., 232
Ovchinnikov, A.A., *see* Oshanin, G.S., 136, 138
Ovchinnikov, A.A., *see* Vinetsky, V.L., XV, 169, 296, 384, 462, 594
Ovchinnikov, A.A., *see* Zeldovich, Ya.B., XV, 135, 136, 298

Palagashvili, E.I., *see* Pirogov, F.V., 463, 465
Parus, S.J., *see* Kopelman, R., 298
Patashinskii, A., 136
Peak, D., 169, 231, 298
Peisl, H., *see* Bachmann, K., 168, 232, 384, 464
Perevalova, O., *see* Kalnin, Yu.H., 465
Pfeifer, P., 591
Pietronero, L., XV, 594
Pilippov, P.G., *see* Benderskii, V.A., 135
Pilling, M.J., 231
Pilling, M.J., *see* Rice, S.A., XV, 168, 231
Pimblott, S.M., 232
Pimbott, S.M., *see* Mozumder, A., 135, 169, 232
Pinard, P., *see* Hermann, F., 465
Pines, E., 170
Pirogov, F.V., 463, 465
Pirogov, F.V., *see* Kalnin, Yu.H., 230, 465
Pitaevskii, L.P., *see* Lifshits, E.M., 52, 137
Plekhanov, V.G., *see* Aboltin, D.E., 233
Pokrovskii, V., *see* Patashinskii, A., 136
Polak, L.S., 135
Poland, D., *see* Song, S., 384
Pólya, G., 52

Pooley, D., *see* Hobbs, L.W., 169, 462
Pooley, D., *see* Hughes, A.E., 462
Popov, A.I., 170
Popov, A.I., *see* Kotomin, E.A., 170, 231
Popova, L.B., *see* Arbuzov, V.I., 233
Prasad, J., *see* Kopelman, R., 298
Prigogine, I., 52, 135
Prigogine, I., *see* Nicolis, G., XVI, 51, 135, 136, 296, 512
Procaccia, I., *see* Goldhirsch, I., XVI
Procaccia, I., *see* Grassberger, P., 136, 297
Procaccia, I., *see* O'Shaughnessy, B., 464
Pronin, K.A., *see* Burlatsky, S.F., XVI, 138
Pyttel, B.L., *see* Vitukhnovsky, A.G., 136, 383

Ra'cz, Z., 464, 591
Ramos, S., *see* Rubio, J., 464
Rasaiah, J.C., 299
Reddy, V.T.N., 136
Redner, S., 297
Redner, S., *see* Ben-Avraham, D., 591
Redner, S., *see* Kang, K., XVI, 136, 298, 383, 464
Redner, S., *see* Leyvraz, F., 383
Reshetnyak, Yu.B., *see* Zabrodskii, Yu.P., 168
Rice, S.A., XV, 168, 231
Rice, S.A., *see* Pilling, M.J., 231
Riste, T., XVI
Rogulis, U.T., 233
Rogulis, U.T., *see* Arbuzov, V.I., 232
Rogulis, U.T., *see* Vitol, I.K., 232
Roman, H.E., *see* Braunstein, L., 299
Romanov, N.G., *see* Baranov, P.G., 233
Romanovsky, Yu.M., *see* Vasilev, V.A., 512
Rosenberg, L., *see* Spruch, L., 232
Ross, J., *see* Nitzan, A., XVI
Rossel, J., *see* Imanaka, K., 232
Rubin, R.J., *see* Rasaiah, J.C., 299
Rubio, J., 464
Rudavets, M.G., 384
Ruzmaikin, A.A., *see* Zeldovich, Ya.B., 136
Rys, F.S., *see* Ehsasi, M., 591

Sagdeev, R.Z., 135
Saidoh, M., 384
Saidoh, M., *see* Itoh, N., 170, 384
Saidoh, M., *see* Tashiro, T., 233
Samarsky, A.A., 513
Sander, L.M., *see* Clement, E., 464
Sassin., W., *see* Duesing, G., 463
Satulovsky, J., 592
Schaefer, H.E., 384
Scheu, W., 168, 232, 384
Schilling, W., 167, 462
Schmalzried, H., XIV
Schmidt, S., 513
Schnörer, H., 52, 136, 296, 383, 462
Schnörer, H., *see* Agmon, N., 298
Schnörer, H., *see* Luding, S., 296, 383
Schnörer, H., *see* Sokolov, I.M., 383, 591
Schoemaker, D., 233
Schoonover, R., 298
Schröder, K., 167, 170, 231, 232, 384, 462
Schröder, K., *see* Dettmann, K., 137
Schuster, P., XVI
Schwartz, K.K., 384, 463
Searle, T.M., 383
Searle, T.M., *see* Andrews, A.S., 383
Searle, T.M., *see* Bishop, J.E.L., 383
Seeger, A., 169, 594
Seeger, A., *see* Gösele, U., 135
Seeger, A., *see* Kotomin, E.A., 383, 463
Seinen, J., 463
Shahverdov, T.A., *see* Ermolaev, V.L., 135
Sheu, W.S., *see* Lindenberg, K., 298
Shi, Z.Y., 591
Shiran, N.V., *see* Gektin, A.V., 464
Shklovskii, B., 169
Shklovskii, B., *see* Baranovskii, S.D., 465
Shluger, A., 168
Shluger, A., *see* Kantorovich, L., XV, 169
Shröder, K., *see* Dettmann, K., 463
Sibley, W.A., *see* Sonder, E., 167, 461
Silverberg, M., 52
Silverberg, M., *see* Becker, O.M., 51
Sipp, B., *see* Kuba, J., 231
Sizmann, R., *see* Lück, G., 169, 296, 462, 595

Smoes, M.L., 512
Smoluchowski, M., XV, 51, 137, 169
Sokolov, D.D., *see* Zeldovich, Ya.B., 136
Sokolov, I.M., 136, 383, 463, 464, 591
Sokolov, I.M., *see* Blumen, A., 463
Sokolov, I.M., *see* Schnörer, H., 136
Sokolov, I.M., *see* Vitukhnovsky, A.G., 136, 383
Sonder, E., 167, 461, 464
Song, K., 384
Song, K.S., *see* Williams, R.T., 168
Sonnenberg, K., *see* Schilling, W., 167, 462
Soppe, W., 463
Soppe, W., *see* Kotomin, E.A., 463
Souillard, B., *see* Delyon, F., 297
Spaeth, J.M., 168
Spitzer, F., 297
Spruch, L., 232
Stanley, H.E., 51, 296, 594
Stanley, H.E., *see* Meakin, P., 298, 384
Stefanovich, E., *see* Shluger, A., 168
Stoltze, P., 591
Stoneham, A.M., 137
Stratonovich, R., 136
Suna, A., XV, 137
Suris, R.A., *see* Berezhkovskii, A.M., 136, 297, 298
Suvorov, A., *see* Kirsanov, V., 168, 462
Sveshnikova, E.B., *see* Ermolaev, V.L., 135
Szabo, A., 297, 298, 594
Szabo, A., *see* Agmon, N., 298

Takano, F., *see* Mamada, H., 591
Takeuchi, S., *see* Tashiro, T., 233
Tale, I.A., XV, 231, 233, 296, 464, 595
Tale, I., *see* Kantorovich, L., XV, 169
Tale, I.A., *see* Kotomin, E.A., 169, 170, 231, 232, 465
Tale, V.G., *see* Kotomin, E.A., 169, 231, 232
Tanimura, K., 230
Tanimura, K., *see* Itoh, N., 168, 462
Tashiro, T., 233
Tayler, A.B., *see* Magee, I.L., 230
Timashev, S.F., *see* Belyi, A.A., 136

Timashev, S.F., *see* Ovchinnikov, A.A., XV, 136, 170, 296, 384, 594
Tolstoy, M.N., *see* Arbuzov, V.I., 232, 233
Tomita, K., XVI, 513
Torney, D.C., 298
Torrey, H.S., 232
Tosatti, E., *see* Pietronero, L., XV, 594
Toussaint, D., XV, 52, 136, 296, 383
Toyozawa, Y., 168
Toyozawa, Y., *see* Delbecq, C.J., 232
Toyozawa, Y., *see* Ueta, M., 167, 462
Troy, W.C., 513
Trushin, Yu., *see* Kirsanov, V., 168, 462
Tyson, J.J., 135, 512, 513

Ueta, M., 167, 462
Uporov, I.V., *see* Mikhailov, A.S., XVI, 136
Usikov, D.A., *see* Sagdeev, R.Z., 135

Vaks, V.G., *see* Balagurov, B.Ya., 297
Valbis, Ya.A., *see* Bogans, Ya.R., 465
Van den Bosch, A., 465
van Kempen, N.C., 137
van Nypelseer, A., *see* Nicolis, G., 136
Varadhan, S.R.S., *see* Donsker, M.D., 297
Vasilev, V.A., 512
Vax, V., *see* Balagurov, B., XV, 52, 136
Veschunov, Yu.P., *see* Baranov, P.G., 233
Vidal, C., 135, 169, 512
Vinetsky, V.L., XV, 169, 296, 384, 462, 463, 594
Vinetsky, V.L., *see* Kalnin, Yu.H., 465
Vink, A.T., 168
Vitkovskii, N.A., *see* Emtsev, V.V., 168, 462
Vitol, I.K., 231, 232, 233
Vitol, I.K., *see* Arbuzov, V.I., 232, 233
Vitol, I.K., *see* Gailitis, A.A., 231
Vitol, I.K., *see* Grabovskis, V.Ya., 232, 233
Vitol, I.K., *see* Lushchik, Ch., 463
Vitukhnovsky, A.G., 136, 383
Voevodskii, V.V., 465
Volterra, V., 135, 513
von Niessen, W., *see* Casties, A., 591
von Niessen, W., *see* Kuzovkov, V.N., 385
von Niessen, W., *see* Mai, J., 591, 592
Vvedensky, D.D., *see* Elderfield, D., XVI

Waite, T.R., XV, 52, 137, 169, 230, 298, 384
Walgraef, D., 136
Walgraef, D., *see* Dewel, G., XVI, 136
Weerkamp, J.R.W., *see* Seinen, J., 463
Weiss, G.H., 298
Weiss, G.H., *see* Ben-Avraham, D., 298
Weiss, G.H., *see* Havlin, S., 297
Weiss, G.H., *see* Schoonover, R., 298
Weiss, G.H., *see* Szabo, A., 298
Wendt, D., 384
West, B.J., *see* Lindenberg, K., XV, 136, 463, 591
Wilczek, F., *see* Toussaint, D., XV, 52, 136, 296, 383
Williams, R.T., 168
Wollenberger, H., *see* Becker, D.E., 170

Yablonskii, G., XV
Yakhno, V.G., *see* Vasilev, V.A., 512
Yakobson, B.I., *see* Doktorov, A.B., 232
Yashin, V.V., *see* Gutin, A.M., XVI, 138, 463
Yashin, V.V., *see* Mikhailov, A.S., XV, 138
Yaskovets, I.I., *see* Vinetsky, V.L., 463
Yatsyuk, S.S., *see* Kondrachuk, A.V., 465
Yi-Cheng, Z., 136
Yoo, M.H., 232, 384
Yuster, P.H., *see* Delbecq, C.J., 232, 233

Zabrodskii, Yu.P., 168, 464
Zabrodskii, Yu.P., *see* Koshkin, V.M., 168, 464
Zaiser, M., *see* Kotomin, E.A., 463
Zakis, Yu., *see* Kantorovich, L., XV, 169
Zamaraev, K.I., 168, 231
Zamaraev, K.I., *see* Doktorov, A.B., 233
Zamaraev, K.I., *see* Khairutdinov, R.F., XV
Zaslavsky, G.M., *see* Sagdeev, R.Z., 135
Zeldovich, Ya.B., XV, 135, 136, 169, 298, 462, 594
Zeldovich, Ya.B., *see* Ovchinnikov, A.A., XV, 52, 136, 298, 385
Zernike, F., 52
Zernike, F., *see* Ornstein, L., 52, 137
Zhabotinsky, A.M., XV, 52, 135, 512
Zhang Yi-Cheng, 463

Zhdanov, V.P., 296
Zhdanov, V.P., *see* Khairutdinov, R.F., XV
Zhdanov, V.P., *see* Zamaraev, K.I., 168, 231
Zhu, J., *see* Rasaiah, J.C., 299
Zhuo, J., 299
Ziff, M., 591
Ziman, J.M., 51, 137, 296
Zozulenko, I.V., *see* Gaididei, Yu.B., 137

Zumofen, G., 298, 383, 595
Zumofen, G., *see* Blumen, A., XV, 297, 383, 595
Zumofen, G., *see* Klafter, I., 297
Zwanzig, R., *see* Bixon, M., XVI, 138
Zwanzig, R., *see* Haan, S.W., 297
Zwanzig, R., *see* Szabo, A., 297
Zykov, V.S., *see* Davydov, V.A., 594

Subject Index

Accumulation
- curve 457
- kinetics 456, 593

activated sites 562
activation energy
- for a hop 194
- for diffusion 388

active sites 549
aggregate
- average number of particles 419
- size 419

aggregation criterion 334
α-Al_2O_3 196, 552
annihilation radius 55
aperiodic motion (chaos) 487
approach
- macroscopic X
- mesoscopic X
- microscopic X

Arrhenius law 149
asymmetric
- particle mobility 347
- recombination region 452

asymptotic laws 12
asymptotical solution 57
auto-oscillating reactions 59
autowave processes 1

$Ba_3(PO_4)_2$ 196
back-coupling of dynamics 238
balance equations 67, 147
Becquerel's law 218, 222
Belousov–Zhabotinsky reactions 468
bifurcation point 86
bimolecular reactions XIV, 94, 526, 593
birth–death operators 133
black reaction sphere 153

black sphere
- approximation 371
- model 140, 244, 274, 479

Boltzmann ordering principle 5
boundary condition 24
box model 439
Brownian motion 15
Brusselator 66, 469

Catalytic
- CO oxidation 51
- NH_3 formation 51
- process 46
- reactions 22

cell formalism 116
cellular automata 550
colour centre 62, 99
- Ag^0 142, 221
- Ag^{2+} 220, 221
- E_1^- 224
- F 387
- F' 164
- F_2 388, 408, 429
- H 143, 356, 387
- H_A 221
- I 164
- Na^0 221
- Tb^{3+} 224
- Tl^0 142, 218
- V_k 143, 218

Chapmen–Kolmogorov master equation 93
chemical
- clock 48, 63
- oscillatory reactions 469
- processes 1

clustered particle distribution 292
CO oxidation 515

colloid formation 357
complexity of the system 12
computer
– experiments 42
– simulation 446
concentration
– dynamics 1, 42, 175, 478
– oscillations 57
conservative system 62
continuous
– diffusion 216
– Markov process 89
control parameters 482
correlated
– annealing 162
– distributions of traps 287
– particle creation 408
correlation
– dynamics 1, 42, 175, 236, 269, 472, 479
– form 124
– function 30, 260, 301, 363, 480, 536, 537, 561
– length X, 32, 182, 240, 262, 275, 303, 336, 340, 406, 539
– radius 32, 112, 301
Coulomb
– attraction 142
– catastrophe 381
coupled cluster approach 353
critical
– dose rate 420
– exponent X, 186, 270, 290, 303, 328, 336, 340
– phenomena 6, 31
– point 5, 186, 581
crossover 79
Cu 460
cumulant expansion 279

Debye radius 77
Debye–Hückel theory 251, 371
decay law 14, 77
defect
– accumulation 387
– – efficiency 145

– aggregation 147
– anisotropy 225
– annihilation 141
– hopping length 194
– interaction 142
– migration 143
– recombination mechanisms 141
– rotation 219
defects in solids 3, 139
particle density
– expansion 277
– fluctuations 26
desorption process 535
diagrammatic formalism 134
diffusion
– coefficient 15
– equation 69
– length 70, 78, 184, 330, 334, 359
– tensor 206
diffusion-controlled reactions IX
dipole–dipole interaction 182
distribution function 94
domain structure 71, 334
doped alkali halide crystals 218
drift in the potential field 17, 156
dynamical
– aggregates 75
– interaction 3, 38
– phase transition 295

Effect of excluded volume 27
effect of statistical screening 490
effective
– reaction radius 158, 191
– recombination radius 208, 209, 359
efficiency of a contact reaction 191
elastic
– attraction 415
– interaction 142, 143, 202, 356
electron traps 145
Eley–Rideal mechanism 528, 578
energy transfer process 56
exciton annihilation 56
excluded volume effect 27
extended ZGB-model 539

Fick law (first) 67
field theory 434
field-theoretical formalism 129, 414
first-order phase transition 5
flight motion 145
fluctuation
 – dispersion 110
 – kinetics 4
 – spectrum 41
fluctuation-controlled
 – kinetics 238
 – reactions X
fluctuation-dissipative theorem 85
focusing collisions of atoms 140
Fokker–Planck equation 84, 85
Frenkel defects 14, 22, 55, 139, 460
frequency factor 145

Geminate pairs 150, 288
 – of defects 91
generalized turbulence 103, 512
Gillespie algorithm 347
glass: $Na_2O \cdot 3SiO_2$-Tb^{3+} glass 224
grey boundary condition 295
grey sphere 190
 – model 335

Hanusse theorem 66
Hanusse, Tyson and Light theorem XIII
hidden parameters X
hole self-trapping 198
hopping
 – activation energy 143
 – kinetics 210, 218
 – length 143, 218
 – recombination 207
hyperchain equations 41

Initial conditions 24, 171
intermediate
 – asymptotics 93
 – exponent 303, 328
 – order 25
 – products XIII

interpolating schemes 179
irreversible bimolecular reactions 10
Ising model 125, 253

Joint concentration correlators 90
joint correlation function 30, 110, 381, 389, 454, 510

KBr 143, 204, 223, 356
KCl 204, 218, 334, 408
kinematics of defect encounter 207
kinetic
 – equations XIV
 – phase transition 45, 48, 527
Kirkwood superposition approximation 50, 124, 397, 593
 – complete 174, 235, 238, 295
 – modified 527
 – shortened 127, 128, 237, 295

Langevin equation 85
Laplace operator 93
Laplace transform 155
lateral interaction X
lattice
 – diffusion equation 164
 – model 352
 – walks 16
law of mass action 3, 54, 371
limit cicle 58
local
 – concentration 23
 – density fluctuations 32
 – equilibrium concept 108
long range
 – order 27
 – recombination 34, 301
long-wavelength approximation 434
Lorenz model 66
Lotka
 – model XIII, 46, 59, 473
 – reaction 51

Lotka–Volterra
- model XIII, 61, 99, 473
- reaction 51
lower-bound estimate 203, 260, 269, 328

Maltus law 490
Mamada–Takano approximation 534
many-particle
- densities X, 126
- effects 593
many-point
- densities 50, 128, 391
- density formalism XIII
marginal complexity XIII, 467
Markov stochastic processes 94, 116
master equation 93, 515, 518, 550
Mayer's group expansion 13
mean-field
- model 536, 558
- theory 38
mechanisms of defect recombination 141
microscopic
- inhomogeneity 31
- self-organisation 472
molecular
- (mean) field theory 7
- dynamics 42
- field approximation 273
moment method 79
Monomolecular reactions 55, 94, 145
Monte Carlo
- simulations 348, 437, 528, 577
- method 42
multicomponent chemical systems 10
multipole interaction 182, 265, 315

NaCl 416, 420
$Na_2O \cdot 3SiO_2-Tb^{3+}$ glass 224
Na-salt of DNA 196
nearest-available-neighbour approximation 320
NH_3 synthesis 515
non-equilibrium
- charge screening 382
- extended systems 467
- processes XIII
- screening 251, 371
non-Poisson
- density fluctuations 334
- fluctuation spectrum 303, 338
- fluctuations 26, 106

Occupation numbers 257
Onsager radius 159, 187, 201, 252, 371
order parameter 4
Oregonator model 469
Ornstein–Zernike theory 44
oscillation period 484

Pair
- potential 152
- problem 50
partial lightsum method 218
particle
- accumulation on fractals 430
- birth–death processes 53
- creation 120
- motion 121
- recombination 118
- segregation 542
- survival in cavities 81
particle interaction 53, 187, 577
- Coulomb 53
- – screened 371, 415
- – unscreened 187
- elastic 53, 205
- – anisotropic 205
- repulsive 579, 585
- van der Waals 53
pattern formation X, 34
percolation theory 48
Percus–Yevick
- approximation 124
- equation 41
phase
- diagram 557
- portrait 484, 485
- space 61
- transition 587

Subject Index

phenomenological approach 7
Poisson
– distribution 25, 248, 303
– equation 251, 441
probabilistic models 438
probability density 16, 455
production rate 587
pseudo-first order reaction 56
Pt 515

Radial distribution function 45
radiation boundary condition 190
random
– process 93
– walks 14
reactant
– concentrations 1
– density fluctuations 3
– dynamical interaction 249
reaction
– $A + B \to 0$ 22, 34, 173, 235, 267, 310, 352, 387, 480, 515
– $A + \frac{1}{2}B_2 \to 0$ 526
– $A + A \to B$ 10, 269, 352
– $A + B \to B$ irreversible 81
– $A \rightleftarrows B + B$ reversible 70
– $A + B \to C$ irreversible 10, 70, 73
– $A + B \rightleftarrows C$ reversible 73
– depth 255, 267, 328, 480
– immobile particles 243
– rate IX, 340, 350, 506
reaction-induced segregation 34
reactions
– on critical percolation clusters 354
– on fractals 309
recombination
– centres 145
– profile 228
– rate 16
reversible diffusion-controlled reactions 288, 593
rough system 106

Saddle-point method 203
saturation concentration 389, 438

scaling analysis 430
secondary quantization formalism 132
self-organisation XIII
– phenomena 468
– microscopic 2
several-stage reactions 51
short-order parameter 27
Sierpinski gasket 310, 430
silica glasses 218
site percolation 46
– threshold 545
small parameter concept 240
Smoluchowski
– boundary condition 153, 247
– equation 16
spatially
– extended systems 67, 107
– nonuniform systems 45
spatio-temporal structures XI, 53, 114
stable
– focus regime 486
– limit cycle 63
– node 61
– stationary state 10
standard
– chemical kinetics IX, 174
– model 573, 579
standing waves 501
static
– reactions 188
– recombination 153
stationary
– point 57
– solution 98
statistical screening effect 389, 490
steady state 185
steepest descent method 339
stochastic
– differential equation 84
– model 554
– particle generation 90
– Winer process 87
structure factor 111
surface
– coverage 557

surface disorder effects 544
– reactions 515
survival probability 16, 187, 269, 277
synergetics XI, 63, 461
systems with particle source 50

Target problem 281
thermal ionization 147
– energy 145
third-order kinetics 294
three-particle
– correlation function 41
– density 127, 172
transient kinetics 185, 219, 226
transition probability 104
trap-controlled kinetics 145
trapping (reaction) radius 24
trimolecular reactions 54, 472
triplet–triplet energy transfer 182
tunnelling recombination X, 141, 182, 259, 302, 316, 453

two-particle
– densities 126
– problem 149
two-point density 114

Upper estimate 202

Variational
– principle 199, 318
– procedure 206
virial expansion 13
virtual particle configurations 391

Waite's equations 179
Waite–Leibfried theory 43
waiting time 143
white noise 92
Wiener trajectories 283

Ziff, Gulari and Barshad (ZGB) model 527
– extended 539